MANUFACTURING

Charles E. Merrill Publishing Company
A Bell & Howell Company
Columbus, Ohio

PROCESSES

Vernon L. Stokes

Tarrant County Junior College

Published by
Charles E. Merrill Publishing Company
A Bell & Howell Company
Columbus, Ohio 43216

*This book was set in Times Roman.
The Production Editor was Marilyn Neyman Schneider.
The cover was designed by Will Chenoweth.*

International Standard Book Number: 0–675–08758–9

Library of Congress Catalog Card Number: 74–14368

1 2 3 4 5 6 7 8 9 10 — 84 83 82 81 80 79 78 77 76 75

Printed in the United States of America

To those technicians, engineers, and scientists who are diligently searching for utilization of new forms of energy while maintaining harmless levels of pollution.

Preface

Manufacturing processes are undergoing change, the change having been very pronounced during the last few years. Because of these changes and the increasing need for better qualified manufacturing technicians, this text has been prepared. Even though conventional manufacturing methods are basic to a society's needs, new ways of producing things have been developed. Much of this development is related to the discovery of new methods and techniques in cutting, in fabrication, and in corrosion prevention, along with the pressing requirements for increased production of all kinds of products for worldwide distribution. *Manufacturing Processes* is presented in a practical manner sufficient to help prepare the new technician in his urgently needed industrial occupation. Technicians must be technically capable of converting raw materials into usable consumer products, therefore, they must understand the numerous details in processing industrial materials. At no time can materials be separated from the processes, consequently, the changing mechanical properties of the specific materials are constantly related to the process throughout this text.

Because materials and processes go hand in hand and because they must be thoroughly understood by students entering the technical and engineering fields, a comprehensive and effective coverage of each subject cannot be accomplished in a single text for a semester's work. The author has discussed *Industrial Materials* as a separate text for one semester's work which could be considered prerequisite or very helpful in understanding *Manufacturing Processes.* In this text, however, both subjects have been handled sufficiently to provide the student with a working knowledge and understanding of both materials and processes. It is the engineer and technician who are primarily concerned with the series of processes from the raw materials to the finished products. However, it is the whole manufacturing team which is charged with profitable production within the limits of pollution control and the conservation of materials along with the search for effective utilization of new forms of energy. It is then with these objectives in mind that the story of *Manufacturing Processes* is told.

A secondary reason for preparing this text has been to provide a practical approach to manufacturing processes so that engineers and technicians can effectively place their ideas into profitable production. Basically, then, this text has been completed for use by students in the first two years of colleges and in technical institutes. It is equally applicable to all students entering the engineering fields where processes are used. Also, this material has been prepared to enable the beginner to effectively grasp the concepts in manufacturing even though he has not had any experience. On the other hand, the student is taken from the beginning to the end in all major areas of industrial manufacturing with the objective of understanding the constant relationship between material and process.

It is my firm belief that technical and engineering students must thoroughly know the subject matter involved with manufacturing processes, therefore, I have attempted to present the field of processes as you will see it. Many thanks go to the numerous individuals, professional organizations, and industrial enterprises who have furnished me with data for this text. Recognition of those participating enterprises is pointed out in proper places throughout the text. Also, a special thanks go to Professor Joseph Cipriano and Professor Joe Waldinsperger for their helpful comments resulting from their review of the manuscript. For typing the manuscript and helping me produce this text in final form, I am indeed grateful to my wife Dorothy.

Vernon L. Stokes

Contents

Manufacturing Processes

1

Introduction to Manufacturing Processes

Manufacturing processes are performed for the purpose of producing some part, assembly, or product with a view of making a profit. The processes which are subsequently described are mainly concerned with the metals and those nonmetals, such as the plastics and ceramics, used in engineering applications. These materials come from the elements and compounds of the earth, the manufacturing processes being responsible for producing them in usable raw material form or as finished parts or assemblies.

Types of Industries

Industrial manufacturing consists of the many primary, secondary, and tertiary industries. Earth materials such as ores, woods, oils, water, and gases are treated, processed, and refined by the *primary industries* into the thousands of different raw materials such as metal castings and wrought bars, the numerous plastic materials, and the many powders of metals and nonmetals. *Secondary industries*, in turn, take the basic raw materials and process them into usable items such as hardware, tools, fixtures, parts, machines, and other assemblies. The *tertiary industries* provide the vast amount of services and repairs to those products having been previously manufactured by the secondary industries. Much of the discussion in this text is concerned with the products of the secondary industries. The production of basic raw materials such as steel and aluminum are discussed, however, to present the whole story of industrial manufacturing. Functions of the tertiary industries are also pointed out and include such activities as the bending and shaping of aluminum and repairs of steel castings by welding.

Engineers and technicians must be cognizant of the whole industrial process in order to more effectively function in their assigned capacities. Without understanding the history of a steel casting, such as the chemistry and mechanical properties, technical personnel are limited and are actually less effective in assigning a use for this metal. Therefore, those individuals who function in the several engineering areas of industry must be aware of the origin of materials and the capabilities of materials before they can study, calculate, design, and operate a manufacturing process. A knowledge of materials provides the engineer and technician with data concerning the material to be processed into a product, and it also enables the engineer and technician to arrange the production line in a profitable way.

The Need for Built-In Skills in the Machine

As the process of manufacturing becomes more technical and sophisticated, larger production outputs occur and fewer personnel are needed in the vicinity of the machines. Built-in skills within the machines allow the use of some less skilled personnel with regard to what they do with their hands. But the overall knowledge and skill of the technicians who supervise these machines are steadily increasing. As an example, the regular lathe operator turns a metal to its desired diameter by hand operation. As the lathe becomes more oriented toward production, the need for and skill in use of micrometers are reduced because automatic and semiautomatic mechanisms allow the cutting process to continue until the preset stop is contacted, the diameter being exactly as predetermined. Raw materials are fed into the path of the cutting tool, cutting occurs until dimensions are established, the part is ejected automatically, and the process repeats while an operator or technician observes. In other words, several machines are observed by the operator. Periodic sampling verifies proper dimensions of the parts and satisfactory functioning of the machines. Figure 1–1 is an example where an operator's skills are built into a numerically-controlled cutting machine.

THE NEED FOR PERSONNEL SKILLS

In order for mechanization to exist, highly knowledgeable and properly skilled personnel are needed to design and construct the machines and equipment. Knowledge of the basic principles of engineering and technology provides the technicians with capabilities to foresee the end product of production and to be able to back up all along the processing steps to the beginning so that the several machine controlled steps will function properly in sequence and produce the desired operations (fig. 1–2). Machine and tool designers, along with manufacturing engineers and technicians, pool their efforts into a manufacturing device that appears to think and act accordingly until a satisfactory job is done (fig. 1–3). The marvels of technology have allowed men to transfer their skills to the machine and, in turn, a less skilled operator is needed to attend a partially mechanized process, such as periodic or intermittent removal of chips, replacement of cutting tools, or some other kind of assistance. Machines do not think, however, and occasionally they need repair or adjustment. On the other hand, machines perform their functions rather faithfully and generally only require oiling from time to time. The expertise of the technician is important as he relies on his capabilities to analyze any problems with the machine and to act effectively.

FIGURE 1-1 A numerically controlled cutting machine. Electrical impulses from the control mechanism are converted to mechanical power at the machine enabling the machine to repeat a sequence of operations. (Courtesy of Clausing Corp.)

FIGURE 1-2 An automated boring machine which has data and mechanisms built into it in order to relieve the operator of time-consuming operations. (Courtesy of Heald Machine Co.)

FIGURE 1-3 A welding manipulator with dual welding heads and integrated controls for automatic welding of road tramping rollers. (Courtesy of Ransome Co.)

The Technician

Manufacturing technicians assist the engineer in plant operation and perform technical tasks to keep the processes running. Their knowledge, skills, and capabilities result from intensive education and training in the needed elements of manufacturing. A study of the principles of mechanical engineering and technology, together with specialized studies in materials science, especially metallurgy, mechanical design, mechanization, and automation, equip the technician with effective plant support capabilities. He is cognizant of materials' properties and is able to detect inferior materials. Further, he is able to recognize improper operating techniques in a production line and to take appropriate action to correct the deficiency. Because manufacturing processes are so varied and complicated, a thorough understanding of the complete production cycle is essential. Generalities are inadequate in solving problems. Technicians, therefore, are capable of working with fine details in solving production problems and maintaining the flow of finished products.

The Engineer

The manufacturing engineer, or tool engineer, is responsible for the overall operation of the production line. He is responsible for the layout and equipping of the process so that profitable production will result. His knowledge of the subjects pertinent to manufacturing enables him to not only understand the process to be performed, but also to be able to turn the theory of manufacturing into realistic operations. His advice is sought in relation to the product's design, cost, and probable profit. He analyzes the contemplated process and works with the technicians to establish processes and procedures.

The Manufacturing Team

The Manufacturing team is aware of the ever presence of material and labor costs, time at work functions, design requirements, and consumer demands. The team includes the engineers, technicians, operators, supervisors, and inspectors who are primarily responsible for the production process. Manufacturing is not simply the making of things, but it is the making of things in conjunction with the functions and requirements of production so that a reasonable profit will be made. Production is the primary goal, and for it to continue, an integration of people, materials, and machines must exist. Profit must feed back into the stabilization of the production facility to assure a long-time process and satisfied employees. Constant contact with the purchasing and selling markets is essential so that production facilities can be modified. In time, older machines will require replacement. Therefore, long-range plans and assets must be provided for effective transitioning. Failure by the administration and production team to look ahead may be disastrous. The team knows how to plan and implement the manufacturing operation and when effectively supported with logistical assistance and competent research, profits continue and morale is high.

Manufacturing Costs

Costs of the total production facility include land and building costs, establishment of the production line, procurement of raw materials and support materials, allowances for depreciation of the plant and production line facility, personnel salaries and benefits, utilities, and market and production research. Other costs are pooled into a miscellaneous fund and include provisions for training personnel, visits to conventions, interest, taxes, insurance premiums, consultants, and one-time expenditures.

One system of accounting for costs arranges the expenses and initial outlay of funds into direct and indirect costs and fixed costs. *Direct costs* are involved with the operation of the production line and include such items as materials in the product and salaries of personnel who are directly engaged in the manufacturing process. A technical breakdown of the direct costs would include power and lighting along the process line and such smaller costs as lubrication. *Indirect costs* are involved with the maintenance of the production line and include replacement parts, lubricants, and coolants; plant lighting other than that required in the actual production; heating and cooling; all other salaries such as supervisors, inspectors, administrative, and sales;

and all those costs not classified as direct or fixed. *Fixed costs* then would be the remaining ones, such as the equipment and initial facility costs together with the costs of interests, taxes, insurance, and related items. Because a profit must be made, the selling price is then established after an analysis of all these factors is accomplished.

Obviously, miscalculations can endanger the security of an operation. Therefore, plans account for such costly events, but these unforseen miscalculations must not become habitual or too large. Consequently, administrative and production personnel join with the overall team and engage in continuous surveillance of an operation to make sure that costs are controlled.

Facilities

Production line facilities must be previewed before buying to confirm that they will produce the product in the most economical way. Questions must be asked, such as whether the proposed machine is capable of future modifications and if it is properly constructed to sustain the heavy loads incurred during processing the product. For example, a small milling machine equipped with numerical control along the three axes is a poor initial investment when hundreds of parts are scheduled for production each day. A heavy duty industrial type machine is most desirable, even though the cost is twice or three times that of the smaller machine. Analyses of the several processes dictate the kinds and magnitudes of loads and forces which will be placed onto the machines. For continuous operation over long periods of time, heavily constructed machines are needed. As the size of the machine is increased, its power consumption also increases, but its life does not decrease, as in the case of the smaller machine.

TYPES OF PROCESSES

An analysis of the production line will indicate whether advanced mechanization or automation is needed. But first, the manufacturing process must be divided into all of its component parts. At the very beginning, the decision is made as to whether the plant is to completely produce the raw materials such as castings, ingots, or rolled sheets or purchase them in one of these initial shapes and then perform the remaining operations. Most of the major processes included are the refinement and production of castings or ingots; rolling, drawing, or extruding; forging or pressing; forming, such as bending, trimming, stamping, or coining; cutting; powder pressing and sintering; cleaning and plating; or welding or heat treating. As an example, an automated welding machine, shown in figure 1-4, is capable of automatically welding a very large pressure vessel along its circumferential seam in an exact straight line.

Due to the shape, use, and cost of the proposed part, one kind of manufacturing process is more suitable than another. A base for a large generator can be easily cast, but not machined. The piston rod in a diesel engine is forged because if cast, it may break. When precision is required in a shaft, machining is performed because no other process can produce the part and comply with the dimensional tolerances. But when thousands of identical spur gears are needed for lightly stressed mechanisms, metal powders are pressed into the exact shape and then sintered for strength. On the other hand, when lightly stressed and quiet-running gears are needed, nylon plastic is used. And finally, tungsten carbide and aluminum oxide cutting tools are used to machine metals faster than any other production cutters.

FIGURE 1–4 A manipulator with integrated welding system and vertical head support. The machine is electronically controlled to maintain constant arc height on inside of hemispherical heads. (Courtesy of Ransome Co.)

Once the types of processes are determined, the schematic of the production line can be established to show the main stations and what is to be done at those stations. Then, connecting conveyor systems are sketched in to provide complete mechanization, whereby the raw material moves from the starting point to each processing machine along a conveyor system. If tapes or computers control the completely mechanized line, a type of automation is provided. Such a situation requires a minimum of personnel along the line because the skills are built into the machines and handling facilities. *Automation* is what it implies, a mechanical process of pro-

FIGURE 1-5 A manufacturing plant arranged to produce large quantities of parts with a minimum of personnel at the machines. (Courtesy of Cincinnati Milacron)

duction. In this respect, many plants have specialized machines to produce precision cutting operations automatically (fig. 1-5).

Product Materials

Products are produced from many different types of materials ranging from the iron-bearing, or ferrous, metals to the nonferrous and nonmetallic. Because all materials are different, their differences must be recognized and evaluated prior to instigating production. The characteristics of materials, or engineering properties, involve the capabilities of the materials' combined chemical, physical, and mechanical properties. Engineering properties include a set of fixed and variable factors pertinent to the specific material. The fixed properties are chemical and physical, while the variable factors are mechanical. In other words, materials include certain quantities of specific elements and compounds, these elements and compounds establishing fixed physical properties in the material, such as color and electrical resistance. In turn, the mechanical or strength properties are influenced and potentials set with reference to the materials' abilities to do their jobs, such as holding a given load under certain circumstances. In conjunction with the engineering properties of a material, its cost must be reckoned with. Basically, the most economical material that will satisfactorily do the job should be chosen. The actual processing capabilities of the material must be investigated to determine if the material is capable of being shaped into the product.

MECHANICAL PROPERTIES

Mechanical properties of the material are usually the most important to consider.

Once a material is found that has the strength potentials for the product, then the physical properties are investigated to find out if the material has the desired physical properties such as corrosion resistance. As an example, when a part is to be used in *tension,* pulling forces are exerted on it. When pushing loads are placed onto a material, *compression* loading occurs. When cutting loads act on the material, the part is in *shear.* (Shear strengths are often about one-half of the tensile strengths.) The most important other mechanical properties are the modulus of elasticity and impact resistance values. The *modulus of elasticity* pertains to the rigidity factor in a metal, for example, steel is three times stiffer than aluminum. *Impact resistance* pertains to the ability of the material to resist suddenly applied loads which tend to fracture the material quickly.

Examples of mechanical properties include hardness, such as the hardness of a file measuring approximately Rockwell C (RC) 63, or the softness of a sheet of steel which is to be cupped into a hemisphere and measures RB 79. A cutting tool measures RC 65, while the metal being cut often measures in the vicinity of RB 80, while a good steel spring will measure RC 45. Each of these hardness values is convertible into tensile strengths according to table 1–1.

RC 63 = more than 300,000 psi
RB 79 = approximately 70,000 psi
RC 65 = more than 300,000 psi
RB 80 = 72,000 psi
RC 45 = 214,000 psi

A study of the above values shows that RC 63 and RC 65 materials will break without bending because these materials are very hard, but brittle in impact, yet will hold extremely high loads in tensile or compression stresses. The Rockwell hardness values are based on the penetration of a cone-shaped diamond or hardened steel ball into a material's surface. The softer the material, the deeper the penetration. Values measured in the "B" scale of the machine are soft, running from B 0 to B 100. Then, harder metals begin at C 24 which is approximately equivalent to B 100 and increase to maximum hardness at RC 70. With respect to the RB 79 and RB 80 hardness values, metals with these values will bend without breaking because these materials are soft, ductile, and weak in tensile strengths. In impact resistance, RB values bend without fracturing and have little toughness. Materials being formed or cut should normally have RB hardness scale values or less, while the materials or tools doing the forming or cutting are much harder, as pointed out. The RC 45 value represents good elasticity and a high degree of toughness when measured in foot pounds of absorbed energy. A steel bolt, for example, may measure RC 24 with a tensile strength of 117,000 psi, while a part of a steering mechanism may measure RC 38 with a tensile strength of 171,000 psi. The RC 24 material is softer, weaker, more ductile, and less tough than the RC 38. Basically, parts which are to be fabricated by forming and cutting must be fairly soft. The greater the deformation to be made, the more ductile and plastic the material must be. (Properties of industrial materials were discussed and illustrated in a previous text.)

TABLE 1-1 Hardness-Tensile Strength Conversion

C	A	15-N	30-N	Knoop	Br'l	Tensile Strength
150 kg Brale	60 kg Brale	15 kg N Brale	30 kg N Brale	500 gr & over	3000 kg 10 mm Ball	
Rockwell	*Rockwell*	*Rockwell Superficial*	*Rockwell Superficial*	*Knoop*	*Brinell (Hultgren Ball)*	*Thousand lb. psi*
70	86.5	94.0	86.0	972	—	Inexact and only for steel
69	86.0	93.5	85.0	946	—	
68	85.5	—	84.5	920	—	
67	85.0	93.0	83.5	895	—	
66	84.5	92.5	83.0	870	—	
65	84.0	92.0	82.0	846	—	
64	83.5	—	81.0	822	—	
63	83.0	91.5	80.0	799	—	
62	82.5	91.0	79.0	776	—	
61	81.5	90.5	78.5	754	—	—
60	81.0	90.0	77.5	732	614	—
59	80.5	89.5	76.5	710	600	—
58	80.0	—	75.5	690	587	—
57	79.5	89.0	75.0	670	573	—
56	79.0	88.5	74.0	650	560	—
55	78.5	88.0	73.0	630	547	301
54	78.0	87.5	72.0	612	534	291
53	77.5	87.0	71.0	594	522	282
52	77.0	86.5	70.5	576	509	273
51	76.5	86.0	69.5	558	496	264
50	76.0	85.5	68.5	542	484	255
49	75.5	85.0	67.5	526	472	246
48	74.5	84.5	66.5	510	460	237
47	74.0	84.0	66.0	495	448	229
46	73.5	83.5	65.0	480	437	221
45	73.0	83.0	64.0	466	426	214
44	72.5	82.5	63.0	452	415	207
42	71.5	81.5	61.5	426	393	194
40	70.5	80.5	59.5	402	372	182
38	69.5	79.5	57.5	380	352	171
36	68.5	78.5	56.0	360	332	162
34	67.5	77.0	54.0	342	313	153
32	66.5	76.0	52.0	326	297	144
30	65.5	75.0	50.5	311	283	136
28	64.5	74.0	48.5	297	270	129
26	63.5	72.5	47.0	284	260	123
24	62.5	71.5	45.0	272	250	117
22	61.5	70.5	43.0	261	240	112
20	60.5	69.5	41.5	251	230	108

TABLE 1-1 (continued)

B	F	30-T	E	Knoop	Br'l	
100 kg 1/16" Ball	60 kg 1/16" Ball	30 kg 1/16" Ball	100 kg 1/8" Ball	500 gr & over	3000 kg / D.P.H. 10 kg	Tensile Strength
Rockwell	*Rockwell*	*Rockwell Superficial*	*Rockwell*	*Knoop*	*Brinell*	*Thousand lb. psi*
100	—	82.0	—	251	240	116
99	—	81.5	—	246	234	112
98	—	81.0	—	241	228	109
97	—	80.5	—	236	222	106
96	—	80.0	—	231	216	103
95	—	79.0	—	226	210	101
94	—	78.5	—	221	205	98
93	—	78.0	—	216	200	96
92	—	77.5	—	211	195	93
91	—	77.0	—	206	190	91
90	—	76.0	—	201	185	89
89	—	75.5	—	196	180	87
88	—	75.0	—	192	176	85
87	—	74.5	—	188	172	83
86	—	74.0	—	184	169	81
85	—	73.5	—	180	165	80
84	—	73.0	—	176	162	78
83	—	72.0	—	173	159	77
82	—	71.5	—	170	156	75
81	—	71.0	—	167	153	74
80	—	70.0	—	164	150	72
79	—	69.5	—	161	147	Even for steel, tensile strength relation to hardness is inexact, unless determined for specific material.
78	—	69.0	—	158	144	
77	—	68.0	—	155	141	
76	—	67.5	—	152	139	
75	99.5	67.0	—	150	137	
74	99.0	66.0	—	147	135	
72	98.0	65.0	—	143	130	
70	97.0	63.5	99.5	139	125	
68	95.5	62.0	98.0	135	121	
66	94.5	60.5	97.0	131	117	
64	93.5	59.5	95.5	127	114	
62	92.0	58.0	94.5	124	110	
60	91.0	56.5	93.0	120	107	
58	90.0	55.0	92.0	117	104	
56	89.0	54.0	90.5	114	101	
54	87.5	52.5	89.5	111	*87	
52	86.5	51.0	88.0	109	*85	
50	85.5	49.5	87.0	107	*83	

TABLE 1-1 (continued)

B	F	30-T	E	Knoop	Br'l	Tensile Strength
100 kg 1/16" Ball	60 kg 1/16" Ball	30 kg 1/16" Ball	100 kg 1/8" Ball	500 gr & over	3000 kg / D.P.H. 10 kg	
Rockwell	*Rockwell*	*Rockwell Superficial*	*Rockwell Superficial*	*Knoop*	*Brinell*	*Thousand lb. psi*
48	84.5	48.5	85.5	105	*81	Even for steel, tensile strength relation to hardness is inexact, unless determined for specific material.
46	83.0	47.0	84.5	103	*79	
44	82.0	45.5	83.5	101	*78	
42	81.0	44.0	82.0	99	*76	
40	79.5	43.0	81.0	97	*74	
38	78.5	41.5	79.5	95	*73	
36	77.5	40.0	78.5	93	*71	
34	76.5	38.5	77.0	91	*70	
32	75.0	37.5	76.0	89	*68	
30	74.0	36.0	75.0	87	*67	
28	73.0	34.5	73.5	85	*66	
24	70.5	32.0	71.0	82	*64	
20	68.5	29.0	68.5	79	*62	
16	66.0	26.0	66.5	76	*60	
12	64.0	23.5	64.0	73	*58	
8	61.5	20.5	61.5	71	*56	
4	59.5	18.0	59.0	69	*55	
0	57.0	15.0	57.0	67	*53	

SOURCE: Courtesy Wilson Instrument Division, ACCO
*Below Brinell 101 tests were made with only 500 kg load and 10 mm ball.

Questions

1. Describe the major differences among the three main types of metals industries.
2. In reference to a machine, what is a built-in skill?
3. As machines become more automatic, why can less skilled operators be used?
4. Discuss the three basic costs in operating a manufacturing plant.
5. Discuss the needed capabilities of a manufacturing team.
6. List and describe the major metals' processes.
7. What is automation?
8. Describe several mechanical properties and give some examples.
9. Describe the major difference between production and mass production with reference to the facility.
10. What is your concept of a manufacturing technician?

11. Differentiate between the need for knowledge of details of the engineer and the technician.
12. Explain why the initial cost of a heavy duty machine may be more economical in the long run than a lower priced machine.
13. If automation is to be used in the production of a product, how can it be justified?
14. Give an example of an impact load.
15. List some advantages of using the hardness-tensile conversion chart (table 1–1).
16. According to the hardness-tensile conversion chart, what is the relationship between hardness and tensile strength?
17. If the hardness of a steel is Rockwell C 50, what is its approximate tensile strength?
18. If the Rockwell hardness value is B 85, what are the Knoop and Brinell values?
19. Name the three important properties of any material.
20. Why is the technician more concerned with specific situations rather than general?

Metal Production

The production of metals is one of the most important efforts accomplished by man, because it is the metals which support the industries and serve man in many of his activities. Metals are used in some way in nearly all aspects of advanced societies. The transportation systems use metals for railroad rails and train wheels, automobile bodies, and most of an aircraft's structure. Machines and tools and hundreds of different household items are made of metals, because metal is the strongest and stiffest of the fabricating materials. Most of the elements from which all things are made are metals and, in turn, each metal has its particular characteristics. It is these distinct characteristics that enable one metal to be able to perform in a task while another instantly fails. Once the metals are released from their associated materials in the ores, they are available as raw materials for the dozens of industries.

Production of Metals

Most of the earthly elements are metals, but their occurrence is nearly always accompanied by other materials in the form of ores. Seldom is a fragment of metal found isolated from the soils and minerals of the earth. (Gold, however, is an exception; pieces of gold often exist in river beds where gold ore is located.) Iron, for instance, is accompanied with compounds of other elements, especially oxygen. Iron ore is polluted with the soils from which it is dug. In fact, approximately five percent of the rocky crust of the earth is iron. Aluminum constitutes an even greater percentage of the earth's crust, but this metal is much more difficult to refine and produce than iron. Silicon, a nonmetal, is very abundant, constituting more than one-fourth of the earth's rocky surface, and is mixed in with the metals. Oxygen, a gas, is found throughout the earth's crust, and it is oxygen that accounts for numerous compounds in the crust and in the ores, especially iron ore. Silicon and oxygen are therefore closely associated with the refining and manufacture of steels. The metal copper is also infiltrated with other metals and minerals in such a way that several separate processes are required to isolate the copper. Much of the magnesium comes from sea water and requires chemical treatments and electrolysis to refine it, while titanium ore demands a very complex processing. Eventually, a metal must be extracted from its ore in order to use it. Once extracted, most metals are again mixed, this time by man under controlled conditions, to provide the numerous pure metals and alloys.

MINING AND PROCESSING THE ORES

Metals are mined in many areas of the world either from open pits or from shafts dug deep into the earth. When an ore concentration is economically feasible to mine, facilities are provided to dig, crush, wash, size, analyze, and grade the ore for shipment to processing mills. Because pieces of mined ores vary greatly in size, huge crushers with screens produce a more uniform size of ore to facilitate further processing. When some ores become dust, they are agglomerated, because most processes cannot efficiently use a raw material which is too fine. As an example, the blast furnace will be choked by lack of air should raw materials be too fine and block the flow of air.

Some ore concentration methods include washing away unwanted gangue and concentrating the remaining materials by means of gravity, magnetism, or frothing. Many ores are separated by means of pyrometallurgical processes such as roasting ovens, smelters, blast furnaces, cupolas, and electric and open hearth furnaces. More complex separation procedures use electricity in the electrolytic processes. Often, during the metal's extraction process, several additional metals such as gold or silver will be extracted, an example being nickel refinement. The end product of most ore processing is a raw material to be used in future manufacturing processes. For example, iron ore that has been beneficiated or improved in size and concentration for handling provides more efficiency in the blast furnace as it mixes with coke and limestone. Attainment of the ore is then the first major step in the processing procedures which culminate in the production of the pure metal or alloy.

Processes involved in metal production vary from the simple procedure of flotation used in gold extraction to the complex chemical and electrical procedures used in obtaining aluminum from bauxite ore. In one way or another, the produced metal goes through a melting stage prior to attaining its final shape. During the melting and solidification phases, certain inherent characteristics remain in the metal. It is these characteristics that are important to the engineer and technician.

FERROUS METAL PRODUCTION

The two basic forms of metals include the iron base, or *ferrous*, and the noniron base, or *nonferrous*. Steel is a metal which is basically an alloy of iron and carbon. However, four other elements—silicon, manganese, sulfur, and phosphorus—are also included in its composition. In order to produce steel, a sequence of procedures must be completed, beginning with changing the ore into pig iron for subsequent refinement into steel. *Pig iron*, the product from iron ore, is then the first product to be obtained in the steel production process and is the primary raw material for refinement into steel. Other ferrous metals such as wrought iron and cast iron are also produced from pig iron.

Production of Pig Iron

Iron ore must be processed through a heat and chemical procedure in order to remove the iron from the ore. The best known method for this type of metal extraction is the *blast furnace* (fig. 2-1). This type of furnace is constructed in such a manner that raw

FIGURE 2-1 A modern blast furnace. Raw materials are carried to the top and dumped into a specially designed receiving system. (Courtesy of Lone Star Steel)

materials enter the top and pig iron is drained off at the bottom in a continuous process, extending into months of operation and even into years without shutdown. Because the blast furnace uses coke and limestone in addition to iron ore, some blast furnaces are located within a reasonable distance from the basic raw materials. On the other hand, many thousands of tons of materials are imported, especially for the steel-making furnaces.

THE BLAST FURNACE

The blast furnace is a cylindrically shaped furnace, often 200 feet high, which is lined with refractory bricks. The stack portion is covered with a steel shell. A schematic diagram of a typical blast furnace is illustrated in figure 2-2. Due to the nature of ore reduction, the top part of the furnace contains a system of movable hoppers, or *bells,*

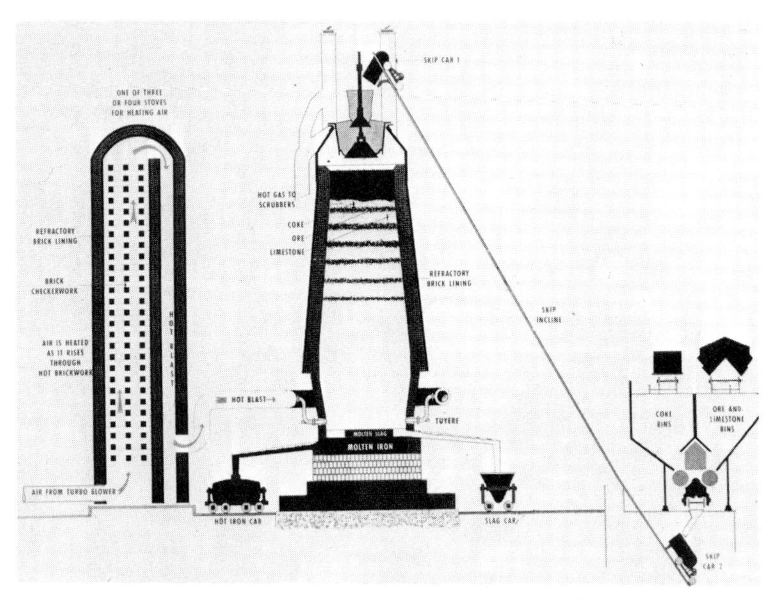

FIGURE 2–2 Schematic diagram of a blast furnace. (Courtesy of American Iron and Steel Institute)

which allow entry of raw materials but prevent the escape of important gases which rise to the top. When one bell is open the other is closed, therefore, after an accumulation of a raw material around the top bell, the bell lifts and allows the material to drop to the inner bell. When the inner bell discharges its contents into the furnace, the top bell is closed and gases are prevented from escaping the system.

At the bottom of the furnace is located the *hearth*, often 25 feet in diameter. This portion of the furnace is where the molten iron accumulates. The hearth must be strong enough to sustain the total load in the furnace and also be able to resist the extremely high temperatures generated as pig iron is formed. The hearth is lined with refractory bricks, either an acid or basic type, depending on the kind of pig iron being produced. Notches in the hearth provide means for slag and iron removal. Because temperatures of approximately 3500° F. are often required to remove the iron from the impurities, assistance is provided by high pressure air nozzles, or *tuyeres*, which are located just above the hearth's periphery. Near this zone is the *bosh*, and it is in the bosh that ore reduction occurs. This zone receives the hot blast from the tuyeres and reduces the ore. The bosh is slightly enlarged so as to enable the raw materials, as they move downward in the stack, to be more effectively reduced to slag and iron. Adjacent to the bosh is the *stack*, and it is in the stack that temperatures around 500° F. begin to heat the cold materials.

At the top of the stack is the skip car facility for delivering coke, ore, and limestone to the furnace. The production of one ton of pig iron requires approximately one and one-half tons of iron ore, nearly a ton of coke, several hundred pounds of limestone, and often a smaller proportion of scrap steel and pig iron. During the reduction process, more than four tons of air are used. Even though more than 50% of the materials entering the furnace is air, more than 75% of the materials leaving the furnace is gas due to the conversion of solid oxides and carbon to gases.

The blast furnace is supported with additional facilities such as dust catchers and also stoves for heating the air to be discharged at the tuyeres. Gases forming at the top of the furnace are utilized in several plant heating operations, one being the heating of the stoves. The stoves are constructed of special brick and in such a pattern that hot air circulates in and out of the many passageways as it increases in temperature on the way to the tuyeres.

ACID AND BASIC PIG IRON

Most of the pig iron produced in the United States is basic and, in turn, the basic pig iron is processed into basic steel. Basic and acid irons constitute the two types of pig iron. The acid bessemer process cannot remove phosphorus, therefore, low phosphorus content raw materials must be used. In the basic process, higher phosphorus contents in the raw materials can be used, and this factor justifies the predominant usage of basic processes. The main differences between the acid and basic processes are the type of *flux* used during reduction of the ore and the type of brick which lines the hearth. The acid process requires silica in the reduction of manganese, silicon, carbon, some of the sulfur, and little or none of the phosphorus. The acid reactions require that the lining of the hearth be silica brick. In the basic process, the flux is lime, and this type of flux also reduces the manganese, silicon, and carbon, along with a great reduction in the sulfur and phosphorus contents. Furnace linings supporting the basic process are magnesite or dolomite bricks.

The Reduction Process

Different iron ores produce different chemical reactions during the reduction process. The basic furnace's burden, or *load,* is provided with alternate layers of coke, iron ore, and limestone from near the top of the furnace. Hot air enters the hearth from the tuyeres and coke ignition begins, causing higher temperatures to occur. Hot gases move upwards and are subsequently removed from the furnace for further utilization in the heating process. As the gases move upwards in the furnace, moisture is removed from the materials, and iron oxides and other earthy materials, called *gangue,* begin to fall toward the hearth and are captured in the fluxing slag which floats on top of the molten iron. Carbon monoxide generated in the process is greatly responsible for this action. During reduction, iron oxide in the ore plus carbon monoxide produce iron and carbon dioxide, while iron oxide plus carbon produce iron and carbon monoxide. As oxides are reduced to slag, iron from the ore becomes spongy, absorbs carbon from the coke, and eventually liquifies on the hearth. As molten pig iron is formed, the burden in the furnace descends.

Because manganese, silicon, phosphorus, and sulfur are present, each of these elements remains in the iron in larger than desired quantities. However, reactions occurring in the bosh and hearth help reduce the quantities of these elements. Most of the oxides of phosphorus are reduced, the larger percentage of manganese oxide is reduced, silicon is controlled in quantity, and nearly all the sulfur enters the slag. Even though chemical reactions occur as a result of materials and temperatures, the furnace operator maintains a reasonable control of events and final analysis of the iron. Several times a day, slag and iron are drained into ladles. The slag is cast from the ladle in a manner appropriate with the slag's future use, for example, granulated

FIGURE 2–3 Pouring pig iron into pigs. Railroad car facilities are required to handle the heavy equipment and molten metal. (Courtesy of Lone Star Steel)

particles for use in masonry blocks. The iron is cast into small molds which results in numerous small castings called *pigs* (fig. 2–3). In the pig iron condition, the metal is nearly useless, except for its main purpose as the material to be refined into useful objects of wrought iron, cast iron, and steel.

All ferrous metals reflect their initial manufacture from the blast furnace by containing smaller quantities of silicon, manganese, carbon, sulfur, and phosphorus arranged in some manner throughout the base material of ferrite. *Cast iron* contains more than twice the carbon content in steels, whereas *wrought iron* contains a very small amount of carbon mixed with a sizeable amount of slag. The cast pigs are stored along with other raw materials for future use in the steel refinement processes. However, it is often profitable to pour the molten pig iron into ladles of larger steel-making furnaces for conversion into steel when the steel refinery is nearby.

Production of Steel

Steel is one of the end products of the refinement of pig iron. The production of steel is not too unlike the production of pig iron because both processes use high temperatures and some similar raw materials. Because the acid and basic processes are used

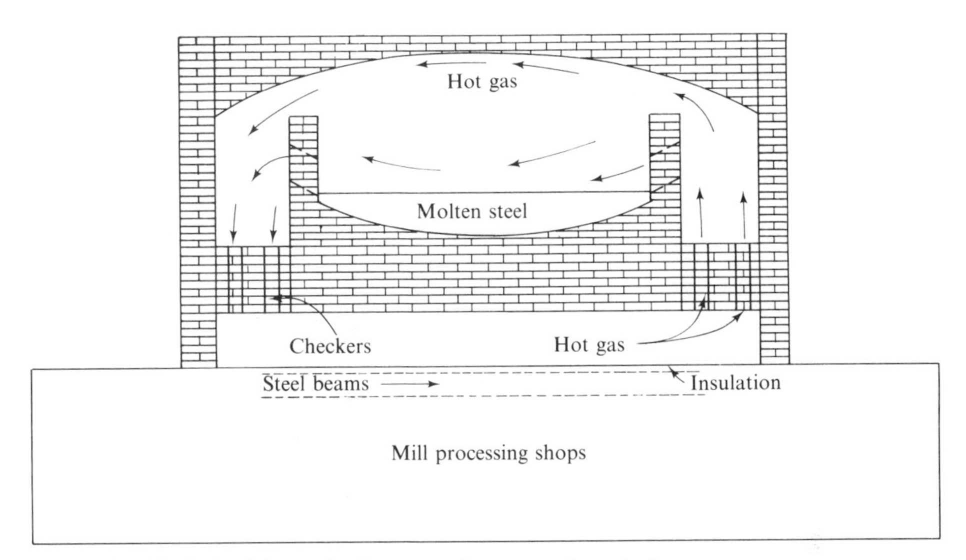

FIGURE 2–4 Schematic diagram of an open hearth furnace.

in steel production, furnaces are constructed to suit the requirements for each method. As has been pointed out, the flux and furnace lining determine the type of steel to be produced. The actual process of producing steel is accomplished by several methods such as the open hearth, basic oxygen, acid bessemer, and electric. By far the greatest tonnage is produced in the basic open hearth, but recent uses of oxygen are increasing production in new kinds of oxygen furnaces. The basic open hearth process is used for steels having carbon contents up to about one percent.

THE OPEN HEARTH FURNACE

Because large quantities of steel must be produced in order for the process to be profitable, the open hearth furnace is constructed to resemble a large covered shallow bowl. These furnaces have output capacities of several hundred tons. Hearth linings reflect the type of steel being produced, while the roof is arched to cause the heated air to take a controlled path over the hearth. Modern hearths are suspended over the plant floor (fig. 2–4), in order to more effectively expedite the manufacturing process. Because furnace gases must be utilized for heating the charge, the shape of the furnace roof causes heat reflection onto the hearth in a reverberatory manner and, at the same time, heated gases pass in and out of a special brick checker system whereby bricks are heated to help produce higher furnace temperatures. The regenerative action occurs in the *checkers*. Periodically, the gas flow is reversed in order to cause the incoming cold air to be heated to very high temperatures. Commercial gas and air together with the regenerative process provide the high temperatures needed in this steel-making process. Provisions are included in the hearth for entry of raw materials and for collection of the slag and pouring the steel.

Charging the Furnace

Some basic furnace procedures vary with regard to the charging of the raw materials. In figure 2–5, an open hearth furnace is being charged. In other words, a *charge* may consist of solid pig iron and steel scrap, solid steel scrap, or molten pig iron and solid

FIGURE 2-5 Floor side of an open hearth furnace being charged. Some of the raw materials are on the floor. (Courtesy of Lone Star Steel)

steel scrap. Sometimes, iron ore is used with the scrap. Either way, a layer of fluxing limestone is laid on the hearth and ore and scrap are dumped onto the limestone. However, some furnace operators reverse this procedure. Also, cold pig iron is added along with various kinds of mill scrap, from rolling mills and ingot production areas, for example. The charge then is a heterogeneous mixture of ferrous products and limestone. The application of high temperatures from the checker systems starts chemical reactions in the charge. Approximately one-twentieth of the total charge is flux. A large furnace may be charged with more than one-half million pounds of materials.

Melting and Reduction

Because molten pig iron will be quickly solidified if dumped onto solid scrap, the usual procedure is to liquify the scrap and then add the molten iron. This mass of material rests on the furnace's hearth. If molten iron is not used, the entire cold charge is heated from the solid. Unlike the blast furnace and many open hearths, some open hearths are equipped with roof lances to eject oxygen into the charge for the purpose of hastening the oxidation of solid materials. Hearths without roof lances often inject excess oxygen into the flames of the blasting gases. When melting occurs, chemical reactions begin as the result of high temperatures. Oxidation of the liquid metal which is contained within the hearth starts with silicon removal followed by manganese and iron oxide removal from the iron to the slag. Carbon commences its oxidation to carbon monoxide when the manganese and silicon contents are sufficiently removed. Because of the rapid evolution of carbon monoxide, the molten metal is agitated sufficiently to cause the ore to boil, and this boiling helps the reduction of phosphorus and sulfur to the slag. Calcination of the limestone follows the ore boil. Temperatures increase as the carbon is decreased and the bath becomes severely

FIGURE 2–6 Pit side of an open hearth furnace being tapped. Metal is flowing from the drainage side of the hearth, similar to the illustration in figure 2–4. (Courtesy of Lone Star Steel)

agitated through the bubbling actions of carbon dioxide. A more stable slag forms and more impurities leave the iron and enter the slag. Upon completion of the lime boil, the analysis control period begins, and it is by this process that the metal's analysis of chemicals is nearly completed in the hearth according to specification requirements.

An analysis of the production period indicates that approximately two to three hours are normally required to produce a uniform melt, while the ore boil period consumes another two or three hours. The lime boil is much shorter in duration, but when coupled with the subsequent analysis control, or testing period, the total time elapsed is another two or three hours. Consequently, from six to ten hours are consumed during the processing period. An additional hour or two can be saved when increased oxygen contents are directed at the bath from roof lances or through increased quantities in the air blasts.

Figure 2–6 shows the pit side of an open hearth furnace during tapping. As the steel enters the ladle, a temperature of approximately 3000° F. exists. Before slag appears at the pouring spout of the ladle, deoxidizing materials, recarburizers, and alloys are added, and because of the turbulence in the ladle, a homogeneous mass of molten

FIGURE 2–7 Ladles are large insulated containers for holding the molten metal until the time is appropriate for pouring into an ingot or casting. (Courtesy of The Timken Co.)

steel exists. Slag subsequently covers the metal in the ladle and acts as a protection from the atmosphere. Small samples of the liquid are quickly analyzed and adjustments are made. Because of the floating slag, basic pouring ladles are usually of the bottom pour type. The ladle is basically a large vessel to temporarily hold the metal until pouring time, figure 2–7. Major additions and analyses pertain to the furnace.

Pouring the Ingot

An *ingot* is a large casting weighing from a few hundred pounds to more than seventeen tons. This mass of metal results from solidification of the ladle's liquid metal. Molten metal is poured into the mold which has the shape best desired for the steel's future fabrication process (fig. 2–8). These shapes vary from corrugated and fluted to round and many sided. Also, the mold may be of the big-end-up or big-end-down type. Mold shape is a factor in the history of the metal because a metal's history begins at solidification, even though the solid is a partial reflection of the liquid from a given chemical analysis and homogeneous condition of constituents.

Due to slag at the top of the ladle, most pouring is done from the ladle's bottom by means of a stopper and nozzle. A continuous and smooth stream of metal is discharged until the mold is filled. Care is exercised to assure the entry of no foreign

FIGURE 2-8 Pouring steel into ingot molds. Molten metal leaves the bottom of the ladle to avoid slag at the top of the ladle. (Courtesy of Lone Star Steel)

matter into the molten metal. Splashing can also cause surface defects in the resulting ingot, therefore, experience dictates a good flow and proper rise rate in the mold. When metal solidification occurs, the mold is lifted from the ingot.

THE BASIC OXYGEN PROCESS

A pear-shaped basic brick-lined container similar to the acid bessemer is used to refine pig iron and scrap into steel. The charge includes selected steel scrap and a higher percentage of molten pig iron. Contents of several hundred tons are processed in a much faster time than that required in the open hearth process. In fact, this process is capable of refining large quantities of steel in approximately one hour per tap. A typical furnace charge weighs approximately 100 tons, while some weigh up to 150 tons.

Oxygen Injection

As soon as the molten iron is poured into the converter, a unique lance descends into the vessel and injects a high pressure stream of oxygen into the charge. Oxidation begins in the molten charge, and impurities are reduced to the fluxing slag. The color of the molten charge changes as refining proceeds. Flux and other necessary materials such as *scale* are added at this time and the oxygen blow is continued to completion of the process and is then terminated. Completion is signalled by a flame change in color. Recarburizers are then added to bring the melt to a desired analysis. High carbon melts are produced as carbon is refined, while continued refining of carbon is indicated by flame color and sounds of reactions. The temperature of the melt is approximately 3000° F. and an analysis is made. After correct analyses of constituents are established, the melt is positioned for pouring. Molds are filled without entry of slag in a manner similar to open hearth mold pouring practices.

THE ACID BESSEMER PROCESS

Only a small quantity of acid Bessemer steel is produced in the United States, possibly not exceeding one percent of the total steel output. The *converter* is also a pear-shaped vessel, being silica brick lined and holding up to approximately 24 tons of materials. Because liquid iron enters and exits the mouth of the converter, a tilting system is provided. Compressed air or oxygen is blown through the bottom of the vessel, causing oxidation and reduction of the molten pig iron into steel.

The Air Blow

After molten iron is poured into the converter, air is exhausted through the molten mass under pressures sufficient to cause oxidation. Excess quantities of silica and manganese commence oxidation and reaction to the acid slag, along with some iron oxide. Air pressure causes large showers of sparks to be ejected from the converter, accompanied with flames and very strong fumes, an indication of the refining of silicon and manganese. As silicon is refined, the flame's color changes to yellow, and as carbon is refined, the white flame appears. Violent boiling occurs until element refining has been completed, the air blast being reduced as conversion proceeds. Addition of manganese and other additives completes the conversion. The molten steel is poured into ingots.

ELECTRIC FURNACE STEEL PRODUCTION

The use of electricity to melt ferrous metals is rapidly increasing because of several factors, bearing particularly on the speed of iron refinement, the character of furnace charge, and the quantity of steel production. One type of electric furnace uses an induction process, whereby the melt is part of the electrical transformer system. Furnaces of this type are limited in capacity, but have ability to produce high quality steel. The induction furnace is used primarily to remelt steels. The other type of electric furnace is the electrode type which uses the arc to melt the charge. This type of furnace presently has capacities up to 200 tons and can produce a pour several times a day. Again, these furnaces may be acid or basic, however, the acid furnace is being used less frequently.

The Basic Electric Furnace

Shaped like a large bowl with a removable top and mounted on swivels, the basic electric arc furnace uses electricity as the heating medium. The furnace uses three electrodes of carbon or graphite ranging in sizes from a few inches to approximately two feet in diameter. Three-phase voltage drives a tremendous quantity of current

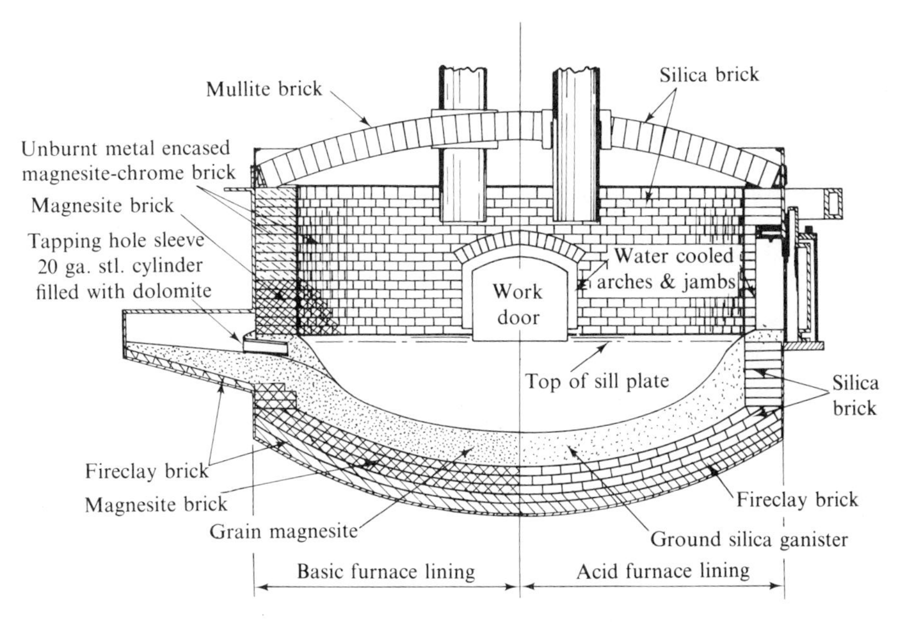

FIGURE 2–9 Schematic diagram of an electric arc furnace showing basic or acid furnace linings. The electrodes are lowered onto the charge for melting. (Courtesy of American Iron and Steel Institute)

through the electrodes, some furnaces using enough current to provide service to hundreds of houses. The charge acts as a resistance to the current and, therefore, melts. When the arcs are struck, blinding flashes of light occur along with thunder-like blasts from the arcs. The walls and hearth of the basic furnace are lined with magnesite brick (Figure 2–9 illustrates the main parts of an electric furnace.) The charge is provided through the removable top. Swivels allow the melt to be exhausted at the spout into the awaiting ladle. An electric arc furnace is shown in figure 2–10.

The Melting Process

The charge in the furnace consists mainly of selected scrap and a small quantity of flux. In all steel-making processes, scrap selection is essential so that undesirable elements will not be introduced into the melt. Once some elements are dissolved in iron, they are difficult to remove, but can be diluted. In addition to scrap, certain ferroalloys are added, along with an occasional addition of iron ore, for the main purpose of reducing the carbon content.

When the charge is completed, the roof swings into position and the electrodes are lowered within arcing distance of the charge. Arcing at the electrodes increases the metal's temperature rapidly until melting occurs. Oxidation begins and undesirable quantities of impurities and elements are reduced to the slag. During the melting and oxidation periods, electricity consumption is very high as silicon, manganese, phosphorus, and carbon are reduced to the basic slag. Desulfurization occurs at a later stage of refinement. As carbon monoxide is generated as a reaction between iron oxide and carbon, the melt boils. The use of additional oxygen in modified furnaces at this point serves to promote cleaner steel.

FIGURE 2–10 A typical electric arc furnace. Current to the three electrodes is provided from cables at the right. (Courtesy of The Timken Co.)

Basic open hearth and electric arc basic processes are similar. Samples of the melt are analyzed and additions made to bring the melt in line with the specifications. Some procedures use a one-slag process, whereas other processes use a two-slag method, that is, the original slag is drained off and a second slag is formed. Higher grades of steels use the two-slag method. When the melt is ready according to requirements, the furnace is tilted and ingot molds are filled.

CONTINUOUS CASTING

A new process, continuous casting, is being readied in an effort to provide a faster and more economical ingot production process. Figure 2–11 schematically illustrates the sequence of continuous casting from the molten metal to parted solid sections of hot metal ready for rolling into usable shapes. Basically, molten metal is discharged from the furnace into a large ladle. Liquid metal flows by gravity to a reservoir and subsequently into a shaped duct where cooling occurs and the mushy stage forms. Primary rolls pull the plastic metal into their turning direction which results in a greatly cooled solid mass of plastic metal being forced into reducing and shaping rolls. The solid ingot is bent to an approximate 90° angle, but with a very large radius, in order to further shape and straighten the continuously moving mass of red-hot metal. After straightening, the shaped section is parted by a flying saw.

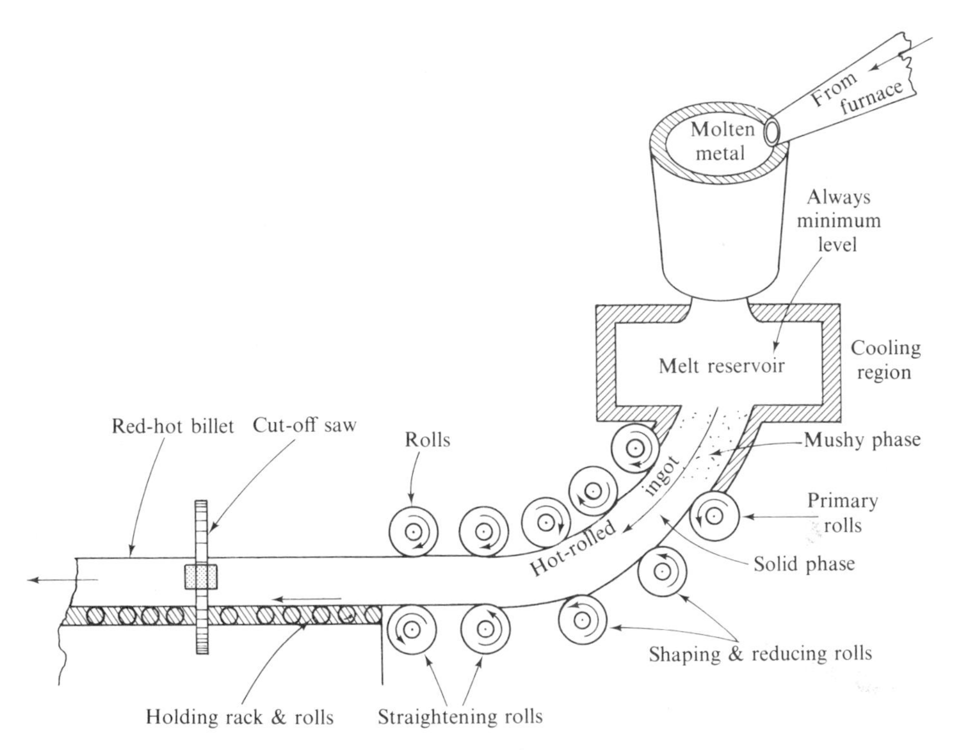

FIGURE 2-11 A continuous casting process that produces a continuous ingot which is parted at any desired length by a flying saw moving in sequence with the hot ingot.

Deoxidation of Steel

During the manufacture of steel, oxides are dissolved in the iron and when pouring occurs, evolution of gases occurs, sometimes leaving surface and subsurface irregularities. As the gas leaves the partially solid metal, a cavity may be formed. Should the cavity be at the metal's surface, oxide again forms and sets the stage for trouble ahead. Should the cavity form below the surface, harm may not result if no oxide has formed. Solid metal and sound metal are the metallurgist's goals. Subsequent hot working of the metal does not always weld or fuse these internal cavities, therefore, steps known as deoxidation processes are frequently used to help reduce unsound metal.

High quality steel is deoxidized to the point whereby it is known as *killed steel*, (fig. 2–12*a*). Deoxidation is accomplished with ferroalloys known as *ferrosilicon* and *ferromanganese*. These additives are introduced into the melt just prior to pouring or are added in the ladle of molten metal. Furnace additives are most effective in producing clean and solid steel, while semikilled steels are produced by additives in the ladle. Small quantities of aluminum also help the deoxidation process.

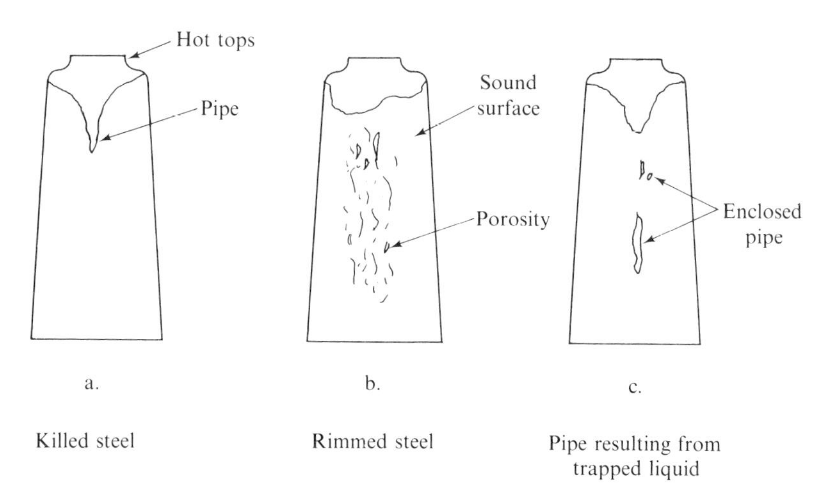

FIGURE 2–12 Cast Steel ingots.

When profits become close, a more economical deoxidation process results in *rimming steel,* or partially deoxidized steel (fig. 2–12*b*). Again, manganese causes a cleaning action on the metal's surface, resulting in a sound skin, but with a slightly porous interior. Because blowholes within the ingot are not in contact with the air, internal cavities will usually weld together when the solid metal is hot worked. A killed steel is superior to the partially killed because internal flaws are less existent. Holes near the top or in the center of an ingot are known as *pipe* and are illustrated in figure 2–12*c*.

The Ingot

Liquid steel, as other liquid metal, must be poured into a container as it solidifies. The shape of the container or mold is determined by the future use of the casting. If the casting is to be used in the cast shape, then no further treatment may be necessary. When used in this condition, the casting is called *as cast*. Frequently, however, heat treatment or machining follows the casting process. On the other hand, when the casting is to be reshaped by some type of pressure-forming process, the casting then becomes the raw material for subsequent fabrication processes. In this respect, the casting is called an ingot.

Depending on the particular shaping process, ingots are produced in many shapes and sizes. In all castings, when the mold is removed, the resultant metal is shaped according to the mold's cavity. Consequently, the cavity of the mold must be accurate in all respects. Unfortunately, castings do not always solidify according to standards or specifications. From a given chemistry of the melt, solidification of the metal does not guarantee a homogeneous chemical arrangement of the constituents, nor is there assurance that the casting will be sound. Because solid metal is smaller in volume than the melt, shrinkage occurs. Combined with the hazards of gas evolution and shrinkage, the resultant casting begins its history in the solid state.

Iron and carbon form the long series of ferrous metals from the nearly pure irons, to wrought iron to steels, and to cast irons. Because the basic ferrous metal carbon steel has limited strength uses, the alloys of steel are produced in great quantities. While refined carbon steel is being processed from the furnace, alloys such as nickel are added to the molten metal. Sometimes, these alloys are added in the ladle just before pouring. The resulting metal, plain carbon or alloy, is then ready to be processed from its ingot shape into numerous shapes needed in industry. The end product of refinement is the ingot, the steel casting, or in the case of cast iron, it is also the casting.

Steel Alloys

Numerous alloys of steel are available. Each added element provides its particular effect in the metal. Nickel induces toughness to the iron, or ferrite, chromium promotes hardness and depth hardness in the alloy, manganese promotes hardness and toughness, vanadium helps assure fine-grained metal, tungsten supports red hardness as is frequently displayed in the high speed steels, and molybdenum resists fatigue while it promotes hardness and strength to the alloy. Ingots, ferrous and nonferrous, may be rolled into shape immediately following casting and subsequent reheating, or they may be cooled and placed in storage areas for use at a later time.

Degassing of Steel

Considerable research has shown that gases dissolved in steels may cause problems, for example, hydrogen embrittlement. Any method that is capable of removing dissolved gases such as hydrogen is essentially a good method. When liquid steel fills a mold which is in the presence of a vacuum, hydrogen, oxygen, nitrogen, and other gases are removed because gas solubility decreases as pressure decreases. Cleaner steels result, and a more homogeneous and ductile metal is produced. The ability of a process to melt steel in the presence of a vacuum enables the production of a very high quality steel. Usually, the induction furnace is used with a basic lining. Melting occurs in the presence of a very low surface pressure, whereby only desirable elements and compounds remain in the finished steel. This type of steel production process presents increased usage potentials due to increasing demands for cleaner and sounder metals.

USES OF STEEL

Steel remains the number one metal in industrial applications because it has the strongest strength properties of all the commercially used metals. In the cast condition, the tensile strengths of the many steels vary from 57,000 psi to more than 180,000 psi, depending on the type of steel and its heat treatment. Steel castings are used for all kinds of machinery and machine parts such as bases and supports. All basic metal industries use steel in their fabrication facilities. When steel ingots are subsequently rolled, extruded, or forged, tensile strengths increase, because in most instances, wrought products are superior in strengths to the cast products. This statement is also true for the nonferrous metals.

When properly alloyed and heat treated, all kinds of steels are used for tools, fixtures, fittings, hardware, household appliances, toys, automobile bodies and parts, the railroad and shipbuilding industries, landing gears and engine mounts of aircraft, agricultural implements, armament parts, furniture, bicycles, pipe, columns and beams for buildings and bridges, and numerous housings and parts for power transmission parts and assemblies. Tensile strengths in excess of 300,000 psi are being used in numerous applications. In the aircraft industry, structural strengths from 125,000 psi to 280,000 psi are common.

Production of Wrought Iron

The alloy of iron and slag, known as *wrought iron*, may be decreasing in importance as it is being replaced with other metals such as the low carbon steels. Wrought iron has the advantage of a slag stringer fiber that has been elongated in the solid by pressure forming. In the production of wrought iron, a slag composed of siliceous materials and iron oxide is prepared and drained into a ladle. Subsequently, molten iron is poured into the molten slag which is at a lower temperature. Consequently, turbulence occurs, the slag and iron physically combining and mixing into numerous small masses. As solidification occurs, a mass of the metal is squeezed into shapes. Slag stringers elongate in the direction of pressure. Wrought iron is similar to low carbon steel, except it contains from two to three percent slag and a carbon content of approximately 0.02 percent.

Wrought iron is used for decorations such as balcony and fence architecture because it can be bent with ease. The electrical industry uses it in some transformer parts, while the pipe industry takes advantage of its high corrosion resistant properties.

Production of Cast Iron

The *cupola* (fig. 2–13) is the furnace most used in ferrous melting processes, and it is primarily used for the production of the several cast irons. *Cast iron* is also a product of pig iron or of melted scraps of all kinds of ferrous metals. Included in cast iron are the same elements found in steel, only in different quantities. Carbon, the most critical element, is always higher in content than in steel. The iron obtains its name because it must be cast from the liquid to attain its shape due to the higher carbon content. The metal is brittle in the gray and white forms, but ductile in other forms. Cast iron can be thought of as clean pig iron.

As mentioned, cast iron is produced from a furnace known as a cupola. Basically, some of the same reactions occur in the cupola as occur in the blast furnace. However, the purpose of the cupola is different from that of the blast furnace. Raw materials used in the production of cast iron include coke, pig iron, scrap iron, steel, and limestone. These materials are placed in the cupola through a door approximately halfway to the top of the furnace. Heat required to melt the raw materials is provided from the burning coke. Tuyeres located around the cupola's periphery provide blasts of compressed air needed for the high temperatures required for melting the constituents.

CUPOLA OPERATION

Unlike the blast furnace, the cupola is small, often only one-third the width of a blast furnace, but it is efficient with respect to its purpose. Cupolas resemble tall steel

FIGURE 2–13 A typical cupola used in the production of cast iron. There is some similarity between the cupola and the blast furnace. (Courtesy of Lone Star Steel)

cylinders standing on their ends. Refractory bricks form the inner linings, and products of combustion are exhausted through the top. The hearth is compacted sand on top of a steel bottom. In the bottom of the cupola is a door which us used to enable the cupola to be cleaned periodically. Coke is laid on the sand bottom and maintained at a certain height commensurate with good foundry practice. The coke bed is in the combustion chamber where the tuyeres force air through the coke, and it is here that much of the chemical reactions occur. Some cupolas are equipped with air preheaters which results in greater efficiency. Iron, scrap steel, and limestone are placed on the coke bed in alternate layers. As the charge is reduced to iron and slag, additional layers of coke, iron, and limestone are added to maintain a minimum coke bed height needed for good combustion. It requires approximately one pound of coke to produce eight to ten pounds of melted iron, while approximately three percent of the charge is fluxing material such as limestone. Some cupola slags are acid and some are basic. In these respects, ductile iron is produced in basic cupolas.

Combustion uses the oxygen from the air blast. Carbon dioxide along with nitrogen is forced upwards through the mass of materials. Temperatures around 3300° F. exist in the melt zone. The coke region above the combustion zone is a reducing zone, and metal in this zone is prevented from oxidizing. As carbon dioxide moves upwards, some is converted to carbon monoxide. Nitrogen, carbon monoxide, and carbon dioxide exit through the top of the stack. As the reactions continue, analyses of stack

FIGURE 2–14 Pouring cast iron pipe in a centrifugal casting machine where the mold spins at several hundred revolutions per minute. (Courtesy of Lone Star Steel)

gases imply cupola chemical reactions and chemistry of the iron. Percents of silicon, manganese, carbon, sulfur, and phosphorus are established according to specifications. In regard to construction, a fore-hearth type of cupola enables easier additions to be made to the melt.

Slag and iron may be drained separately or by special construction at the tap hole. The two products accumulate separately due to slag floating on the liquid metal. Separate tap holes are provided in the ladle for tapping slag and metal independently. As iron is needed, it is drained into ladles from the reservoir while the slag is retained. Ladles are carried to the molds and the metal is cast. Different types of cast irons are subsequently produced by means of heat treating the castings. A method of producing cast iron pipe is shown in figure 2–14.

USES OF CAST IRON

Because of its ability to flow freely into the intricate regions of molds, cast iron is used in large tonnage. This metal is frequently used in the gray cast condition, but often it is found in the harder areas of machine tools, such as the surfaces of ways on the lathe. Gray iron is soft and brittle, while white iron is hard and brittle. Tensile strengths of cast irons range from around 22,000 psi to over 160,000 psi, depending on the kind of iron and its heat treatment. Malleable iron is arranged in its microstructure so that malleability results due to carbon being arranged in spheres within the ferrite.

Cast iron is used for low strength and high strength parts which are not subject to impact loads. In this respect, the ductile iron withstands shock, however, its tensile strength is below the higher strength steels. Gratings, drainage fittings, grilles, machine parts such as housings, automotive parts such as blocks and piston rings, pipe, and hundreds of simple and complex castings are produced from the several kinds of cast iron.

NONFERROUS METAL PRODUCTION

Some nonferrous metals include iron in their analyses, but iron is the alloy and not the base, or *matrix*, metal. A great similarity between the two metals is found in their original forms, that is, in their ores. Most metals are produced from earthy materials, and it is in these ores that the metals are locked and trapped among other materials. Extraction processes used in producing nonferrous and ferrous metals have similarities as well as unlike procedures. However, both types of ores are normally processed so as to concentrate, or beneficiate, the metallic materials as much as is feasible in order to help expedite the total process. Grinding, washing, screening, and sorting of the ores are accomplished in some manner, followed by heating and chemical treatments. Some of the chemical processes include the introduction of aluminum ore into a strong solution in order to help release aluminum from some of its impurities. On the other hand, chemical reactions are caused by means of heat, such as the reduction of iron from iron oxide in the blast furnace. Because oxygen has a great affinity for many materials, numerous chemical compounds have been formed in the earth's crust. Iron and oxygen, for example, have produced iron oxide, while aluminum and oxygen have produced aluminum oxide. These compounds created by nature are broken down into their elements during chemical reactions which occur during processing. Basically, the nonferrous processes of ore reduction are much more complex than the ferrous production processes.

Production Equipment

Because numerous types of mechanical and complicated chemical processes are needed in nonferrous ore reduction, only a general description can be made. However, some types of processing equipment used in nonferrous procedures are similar, and often identical, to those used in ferrous processing. Both types of ores must be extracted from the earth, transported to assembly areas, grouped, crushed, washed, sorted, and heated. Following initial preparation of the ores, specific procedures are then necessary to relieve the ore of its metals. As an example, the blast furnace provides heat and chemical processes in the reduction of iron and some nonferrous metals. Roasting and sintering ovens, converters, and furnaces are commonly utilized in several preliminary processes during nonferrous metal production, such as in the production of copper by smelting which helps remove iron from the copper ore.

On the other hand, aluminum oxide reduction requires more complicated and costly procedures to separate the oxygen and aluminum. In addition to initial aluminum ore processing, bauxite requires soaking pits containing chemicals, pressure chambers and tanks, and electrolytic processes to remove the metal. Electrolysis uses large quantities of electricity in conjunction with chemicals. Other metals require vacuum conditions during processing. Much of the magnesium used today comes from the sea; consequently, initial processing is unlike the processing of ores.

Basically, nonferrous metals' production incorporates more complicated processes than are used in the production of ferrous metals.

Production of Aluminum

Aluminum is the third most abundant element in the crust of the earth, but it is so contaminated with iron, silicon, oxygen, titanium, and water that a long and complicated process is required to produce the element. Aluminum comes from *bauxite* ores which vary in quantities of different minerals mixed with the earth's soils. Ores are often a dirty red in color, but many are found as yellow and brown. Some are found in sands while other types are found in clays. These ores are mined in similar ways to the mining of iron ore, only bauxite exists at the earth's surface while iron ore may also be found deep in the earth. Processing plants prepare the ore for subsequent treatments.

CHEMICAL PROCESSING

After crushing and drying, the bauxite ore is introduced into large tanks containing a caustic soda solution. Chemical reactions along with heat and pressure dissolve the materials and cause precipitation of compounds and other materials, resulting in a sodium aluminate. During subsequent processing, hydrated alumina is added. The new crystals are heated in kilns at approximately 2000° F., resulting in a highly refined alumina. The newly produced white powder is a compound of aluminum and oxygen.

ELECTROLYSIS

In order to remove the oxygen from the alumina, electrolysis is necessary as the bond between oxygen and aluminum is extremely strong. In this process, a special type of carbon-lined furnace is used, the carbon acting as the cathode. Alumina is placed in the tanklike furnace along with a solvent for the aluminum oxide and aluminum flouride. Direct current passes from the carbon anode and penetrates the mixture to the carbon cathode. Melting temperatures are generated due to the passage of current through the dissolved alumina. Reduction occurs, which causes molten aluminum to form along the cathode. The electrolytic cell is actually the furnace where a continuous process of aluminum production exists. The slag at the molten surface supplies protection for the melt below. Pure aluminum, better than 99% pure, is drained from the processing furnaces and poured into ladles or retaining furnaces for subsequent pouring into ingots, or castings. Raw materials are added to the retaining furnace as aluminum is extracted. Should purer aluminum be required, special refining furnaces are subsequently used.

What begins as a dirty appearing ore ends up as a bright metal. In the ingot form, the metal is weak in its various strengths, but is very resistant to corrosion. In order to be commercially useful as structural material, the ingot is melted along with certain other elements to result in a high strength engineering alloy. Aluminum becomes the base metal for a large series of aluminum alloys, each having its own peculiar mechanical properties.

ALUMINUM ALLOYS

Alloys are produced as a continuous process immediately following the production of the pure metal. Special elements are added to the molten aluminum, along with fluxing materials. These elements primarily are copper, silicon, manganese, zinc, and magnesium. Each alloying element provides a special property to the base metal. In addition, different gases, such as chlorine and nitrogen, are bubbled through the melt to float oxides and other impurities to the surface for removal. Some additional refinement and alloy content control follows, and then the new metal is cast into ingots. The size and shape of the ingot varies, depending on future use of the metal.

USES OF ALUMINUM AND ITS ALLOYS

Pure aluminum is weak in tensile strength and is soft. The low range of its strength may be 11,000 psi in tensile, but when alloyed with approximately four percent copper, its tensile strength increases to 71,000 psi after heat treatment. The pure metal is used for corrosion resistant purposes and for nonstressed parts such as household appliances and low strength fittings. As an alloy, strengths are increased to those of the carbon type of structural steel. In this high strength category, beams and columns are made for bridges and buildings, in addition to the numerous castings, forgings, extrusions, and rolled products.

One of the outstanding users of aluminum alloys is the aircraft and space vehicle industry. Its high strength–weight ratio makes it ideal for high strength construction. In the construction industries, aluminum and its alloys are used for buildings and facilities. The railroad industry uses aluminum alloys in some of its tank cars, and the automotive and shipbuilding industries find these metals very satisfactory for parts of automobiles and hulls of ships. Other uses include furniture, window and door frames, ladders, barbecue grills, window curtains, fabrics for clothing, and toys.

Production of Titanium

Titanium is classified as a light metal, being nearly half the weight of steel. This metal has gained extensive use only during the latter part of this century, possibly because of its difficult production procedures. The metal is found in several parts of the world, but is always locked in tightly with various elements in compound form.

The ore *rutile* is a titanium dioxide compound. Chlorination of the ore is one of the beginning processes in titanium production. When mixed with some form of carbon in the presence of chlorine at an elevated temperature, a colorless liquid, titanium tetrachloride, forms. In the presence of liquid magnesium within a heated and pressure-tight vessel and a controlled atmosphere, titanium tetrachloride produces sponge titanium in pellet form. The inert atmosphere prevents oxidation. Sponge titanium is then leached with an acid after grinding. The *leaching* process is a cleaning process for the purpose of eliminating traces of foreign materials such as magnesium. The metal is subsequently formed into electrodes and melted under vacuum to produce ingots. Once melted, the metal is produced as the element and also as an alloy. Additions of alloying elements provide the numerous available alloys in ingot form.

USES OF TITANIUM

Titanium is one of the newer metals to enter the industrial market. Because of its high strength and light weight, it has found numerous applications in many industries. As an example, titanium has been used in space vehicles due to a combination of excellent properties which include high strength, low weight, and ability to perform in cryogenic temperatures and in heat up to 1250° F. An alloy of titanium, aluminum, and molybdenum, for example, is heat treatable to tensile strengths of some structural steels, but has subzero use capabilities beyond that of most steels. This unusual metal finds numerous applications in the construction of stressed parts such as missile components, pressure vessels, rocket nozzles, structural shapes, aircraft parts which include jet engine compressor blades and hardware, and numerous welded structures. Titanium is formed into the many shapes being found among the steels, magnesiums, and aluminums. Selected alloys have high resistance to impact loading at subzero temperatures and are dimensionally stable under stresses of centrifugal forces at elevated temperatures.

Production of Copper

The amount of copper in the earth's crust is much smaller than that of aluminum, iron, magnesium, and titanium. However, it is less difficult to produce than is aluminum, magnesium, and titanium because these three elements require intricate and complex mechanical, chemical, and electrical operations before the element is produced. The refinement of copper by means of the blast furnace process is similar to iron. However, several methods are used in the production of copper, depending on the kind of ore available.

A common method used in the smelting of sulfide-bearing copper ores is the concentration of the ore, similar to other ore concentration methods. The concentration is smelted and results in a copper *matte*. This matte is mostly unwanted material, but approximately one-third is copper sulfide. Useful elements such as iron are extracted from the residue of waste materials during the several operations. In order to remove the sulfur and other constituents, the matte is placed in a converter vessel, similar to the bessemer converter, and low pressure air is blown through the liquid. Oxygen and sulfur combine into sulfur dioxide, iron is oxidized, and a blister copper results. This is a high purity copper, approximately 98% pure. Again, air is used to oxidize the liquid blister copper in an oxidizing furnace. The remaining sulfur and oxygen are removed, and the resulting high purity copper is cast into anode bars. Because further purity may be desirable for some industrial uses, copper anodes are transformed electrolytically to copper. Further melting and refinement produce a 99.9% pure copper which is cast into ingots.

Several other methods are used in the production of copper, just as other methods are available in the production of other elements; ore contents determine the extraction method. Should a high purity copper ore be needed, a procedure similar to blast furnace iron reduction is used. Copper oxides are reduced to copper in a general manner similar to iron oxide reduction to iron. Another method requires a high concentration of copper in its ores in order to economically refine it. Sulfuric acid leaches low copper-bearing ores to a suitable raw material for refinement in a procedure patterned after the smelting and electrolytic processes. The end product of copper ore reduction is the copper ingot.

Because copper has a limited usage, alloys are produced. Copper alloys consist of the numerous brasses and bronzes. Also, copper is an alloy in other base metals such as aluminum; one of the strongest aluminum alloys contains a small amount of copper. When approximately 30% zinc is added to molten copper, a yellow alloy is produced called *brass*. When a small quantity of tin and often smaller percentages of elements such as silicon, manganese, phosphorus, or aluminum are added to the molten copper, an alloy called *bronze* is produced. Copper, brass, and bronze castings are subsequently processed into their final shapes.

USES OF COPPER AND ITS ALLOYS

As an element, copper finds extensive use in the electrical transmission industries. It is soft and ductile, but low in tensile strength. In structural applications where higher strengths are required, copper is alloyed with chromium, beryllium, phosphorus, and other elements. Tensile strengths of some copper alloys exceed 200,000 psi. Brass and bronze parts are used extensively in the water pipe industries because of high corrosion resistant properties. Machine bearings are produced from the several bronzes, while numerous marine fittings use brass, bronze, and copper. Several items of household hardware are made from brass, mainly because of good resistance to corrosion and a pleasing appearance, however, brass hinges and door locks have sufficient strengths to carry the necessary loads. Because copper alloys are adaptable to shaping in any commercial form, the metals are therefore produced in many cast and wrought forms. Standard items of hardware are produced, as well as the many common fittings in the electrical industry.

Production of Nickel

Most of the nonferrous metals are difficult to remove from their ores. In the ore, nickel is found as a type of sulfide in which copper, iron, and other elements are present. As in the iron ores, the rocky materials are crushed and treated magnetically to separate the high nickel-bearing ores from the remainder. Both concentrations are treated and smelted separately to produce a basic matte. Part of the minerals are run through the crushing and flotation process to obtain the matte, some having high nickel sulfide contents. In order to obtain the pure metal, the matte is subjected to further refinement in special converters whereby anodes are produced for electrolytic refinement. As in other electrolysis processes, the converted material becomes part of the electrolyte and the current moves to the cathode where a high purity nickel is formed. Subsequent melting and casting provide different shapes for industrial uses.

USES OF NICKEL

Nickel is a shiny metal which is extremely resistant to corrosion. Its melting temperature is more than twice that of aluminum. These two properties, corrosion resistance and a high melting temperature, make nickel useful in the electrical resistor industries. Nickel, along with chromium, makes fine wire heating elements in home appliances and industrial furnaces. When used as a plate such as in eletroplate, a highly corrosion resistant surface results. One of the largest uses of nickel is in the production and fabrication mills and foundry industries. Nickel is used in castings, often alloyed with other elements for special strength purposes. The metal can be rolled, drawn, and shaped as required from the ingot. High strength steels such as a common structural steel having a tensile strength of 125,000 psi are alloyed with nickel. Nickel is alloyed with other elements such as molybdenum and chromium in the production of a large tonnage of high strength steels. Because of its heat resistant qualities, nickel is used in many parts which function in elevated temperature ranges. In engineering designs, nickel is used to help produce the strongest of steels, some being more than 300,000 psi in tensile strength.

Production of Magnesium

Magnesium is the lightest in weight of the commonly used engineering metals, which gives it an excellent strength-weight ratio. The source of the metal is sea water and the minerals *dolomite* and *magnesite.* Because of its abundance in the oceans of the world, magnesium ranks within the world's top ten elements in regard to abundance. Production of magnesium from sea water by electrolysis is being used extensively, even though the process is complicated. Several chemical reactions occur between the moment the sea water is treated with slacked lime to when it is scooped from the electrolytic cell. After the lime treatment, a milk of magnesia is formed which is then converted to a magnesium chloride. Further processing changes the concentration of magnesium chloride to a condition which is suitable for electrolysis.

The electrolysis process is similar to other electrolytic systems whereby the anode and the cathode are emerged in a solution to complete an electric circuit. The flow of current through the material moves particles from the solution, causing liquid magnesium to be deposited at the cathode where it is dipped and poured into molds. Being a product of electrolysis, it is pure. When alloys are needed, the pure magnesium is melted under a blanket of flux and specified elements added, such as aluminum. The molten alloy is cast into molds for use as permanent castings or as ingots for subsequent pressure-forming operations.

USES OF MAGNESIUM

Because magnesium is easily cast, rolled, and extruded, numerous shapes are available. The metal's weight is approximately two-thirds that of aluminum and is formed into structural parts, consumer products in the appliance industries, automotive and aircraft wheels, optical fittings, containers, machine covers, hardware fittings, castings, anodes, welding electrodes, and flares.

Being hexagonally closely packed in its atomic structure, the metal requires heating to a few hundred degrees when bending is desired, that is, cold magnesium will crack if attempts are made to bend it to a small radius. When die cast, magnesium alloys have superior mechanical properties with respect to their weight; tensile strengths of 31,000 psi are typical. Another unusual property of magnesium is its ability to catch fire and burn at a white temperature. Very small chips and powder are highly susceptible to ignition when oxygen is present. Sulfur dioxide, sand, or most granular materials will extinguish the fire. If water is used, the fire may increase. Water is never used to extinguish magnesium fires unless the fire is small.

Production of Zinc

Zinc is one of the most important of industrial metals, due to the ease with which it can be shaped and to its ability to prevent corrosion. Zinc is fourth to steel, copper, and aluminum in regard to consumption. Having a bluish white color as the element, it changes toward yellow as the brownish colored copper is added to produce brass. Zinc tends to be brittle after cooling from the liquid, but when reheated to approximately 300° F., it regains its softness and ductility, and it can be rolled into sheets. Zinc is used as alloys of zinc and some other element. It lends itself favorably to die castings such as automotive and consumer parts, is coated onto steel for production of galvanized iron, and is used to prevent corrosion on the hulls of ships by being anodic to iron and steel.

Unlike many deposits of iron ore, zinc is mined from deep veins beneath the earth's surface and is combined with other materials such as lead, and often silver and gold. The first major step in zinc processing is the rolling and crushing of the ores so that a separation process can be used to drop out the small metallic particles by gravity or float the metallic flakes in a froth for subsequent removal and drying. The dried zinc concentrate, being mainly zinc sulfide, is subsequently roasted to remove most of the sulfur. In chemical reactions, white-hot sulfur dioxide is produced along with zinc oxide. Further refinement of the oxide by carbon reduction is one method of producing zinc. Refinement includes the high temperature heating of zinc oxide and carbon in special furnaces, or *retorts*, whereby boiling occurs, followed by condensation of high purity zinc. Another process, the electrolytic, produces nearly an equal amount of a very high quality zinc. Roasted ores which have been treated with sulfuric acid are subjected to electrolysis, whereby lead anodes and aluminum cathodes cause the zinc to adhere to the several cathodes. Subsequently, the high purity zinc is removed from the cathodes at intervals, melted, and cast into ingots or special shapes for further processing.

USES OF ZINC

The element zinc weighs nearly as much as steel and melts at 787° F. Besides its superior corrosion resistant properties, it is die cast as an alloy into hundreds of different parts. The zinc alloys are numerous, and some include small quantities of iron, magnesium, aluminum, copper, and lead. As is typical with alloys, the melting point is slightly less than the basic element, the exact temperature depending on the specific chemical analysis. Alloying increases mechanical properties and in some die cast alloys, tensile strengths range up to 47,000 psi.

Because zinc is so easily formed, a nearly endless number of cast and wrought parts are produced annually. Much of the grille and trim parts of automobiles are cast from zinc, in addition to the door handles and engine accessories and housings. Besides the automotive industry, numerous consumer parts and appliances include castings. Farm products and commercial hardware are frequently zinc coated, while thousands of toys, sporting goods, and machine parts are formed from zinc. Dry cell batteries also use zinc in their construction.

Questions

1. What is meant by beneficiation of ore?
2. Describe the basic differences among the three main ferrous metals.
3. Name a metal that is difficult to separate from its compound in nature.
4. What is pig iron, and how is it different from cast iron?
5. What is a slag? Describe its purpose.
6. Differentiate between the cupola and blast furnace.
7. Describe the difference between the open hearth and basic oxygen steel-making processes.
8. Compare the mechanical properties of steel and titanium alloy.
9. Name several unusual properties of magnesium.
10. Differentiate between chemical and electrolytic processing of metals.
11. What is the first major step in metal processing?
12. What is meant by characteristics of a metal?
13. What is the function of the blast furnace?
14. What are the main requirements of the furnace linings in making basic and acid steels?
15. Name the six basic elements in all ferrous metals.
16. Why are many steel-pouring operations equipped with a bottom-pouring capability?
17. What is an alloy steel?
18. Why is environmental temperature important during the use of a metal part?
19. Explain a main difference between titanium, magnesium, and steel.
20. What is meant by the statement, "copper and its alloys?"

Casting Processes

Casting a metal is the art of converting liquid metal to a solid. Once the liquid has solidified, the methods and techniques used in liquid conversion to the solid are inherent in the solid. Therefore, great care and preparation of the metal are essential while it is liquid and during pouring in order that the resultant solid will be acceptable. Scientific procedures control the processing of ore to liquid, the control of the liquid's chemistry, and some of the pouring techniques; however, much of the liquid-to-solid conversion process is controlled by man's observations and manipulative skills. As a science and an art, the casting of metals provides the solid products for numerous industrial applications.

Characteristics of Cast Metal

Cast metal has a fixed three-dimensional shape which occupies a given volume at a certain temperature, while a liquid's shape is dependent on the shape of the container. Once solidification of the liquid occurs, the three dimensions of the solid become partially fixed, but dimensional change continues in the solid from beginning of solidification to room temperature. Hot metal envelops more space than cold metal; consequently, these dimensional factors are considered in the planning and pouring of metals.

DENSITY AND TEMPERATURE

The contraction effect is shown in figure 3–1. As the liquid cools, density is reduced uniformly until solidification begins. At the solidification temperature of the metal, solid nuclei form in front of the receding liquid. This phase of liquid to solid depends on mold effects, time, and alloy analysis. It is the most critical period in the entire contraction period because of the short time duration and the unknown capability of

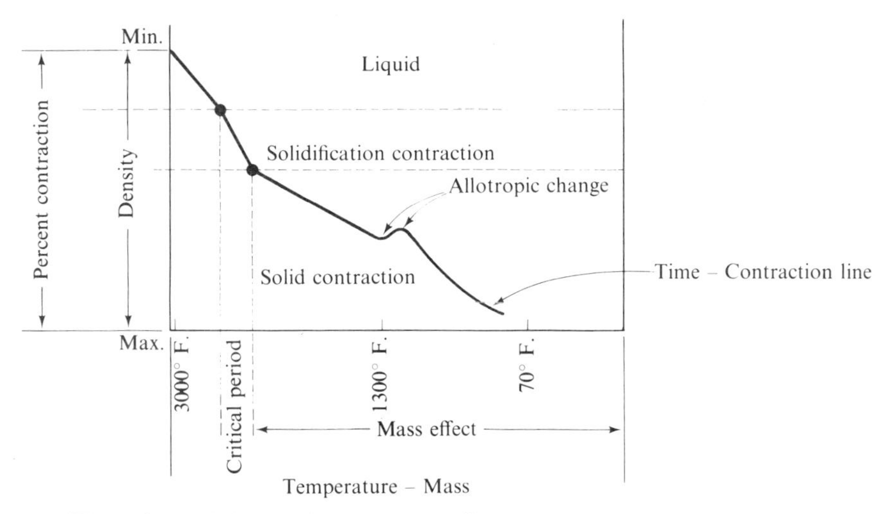

FIGURE 3-1 Shrinkage in steel on cooling.

the liquid to adequately feed the rapidly growing solid nuclei. When solidification of the liquid is complete, mass contraction occurs, this being mainly a function of mass of material. The inside regions are cooled from the outside areas. Contraction occurs uniformly to room temperature except in the case of some metals which have allotropic changes. In all body-centered structure steel, the allotropic change occurs below 1300° F., as shown by the rise and fall of the contraction line in the solid phase.

BIRTH OF DISCONTINUITIES

In regard to the resulting solid, it cannot be assumed that the volume represents the product of its dimensions. Often, several volumes of space, such as gas cavities, cracks, or a nonmetallic material, are contained within the solid. These discontinuities are unwanted. With respect to gas, its ability to escape as it leaves the liquid is not always possible. As illustrated in figure 3–2*a*, gas was trapped just as solidification occurred. The liquid released the gas at the moment solid nuclei formed; consequently, a small cavity occurred in the surrounding solid. Notice the time interval between the void and the wave of solid metal. Also, two additional discontinuities resulted, as illustrated in figure 3–2*b*, due to the liquid failing to properly feed the advancing solid wave. The solidification line shown in figure 3–2*c* illustrates a smoother joining of solid waves because the metal is pure, that is, it has one freezing temperature as compared to variable freezing temperatures of some alloys. Separations within a solid represent imperfect material, and it is this imperfection factor that concerns engineers, designers, and technicians. Therefore, scientific knowledge and application of skill are needed to more effectively control the conversion of liquid metal to solid.

Pattern and Cavity Relationship

When liquid metal is poured into a cavity, a casting results, and this casting reflects many previous processing efforts. Because a casting has shape and volume, it is

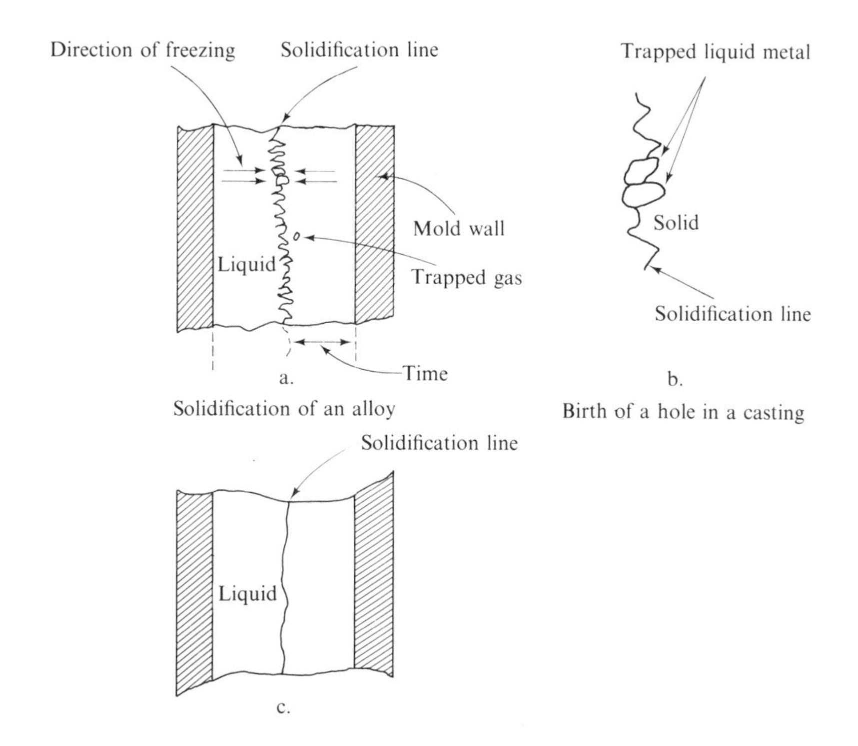

FIGURE 3–2 Solidification of metal.

evident that some exact means are necessary to provide the shape of the cavity into which the liquid metal is to be poured. Figure 3–3 illustrates the relationship between a pattern and a cavity when the pattern is placed into a box (*a*) which is filled with sand (*b*). When the pattern is removed, the cavity results. However, removal of the pattern is dependent upon the ability of the pattern to be removed without damaging the cavity. When the sides of the pattern are slightly tapered, usually a minimum of one degree, the pattern is removed with ease. Figure 3–4 illustrates this taper, or *draft*, in a pattern.

Shape and size refer to the resulting casting and not the shape and size of the cavity or the pattern which is used to produce the cavity. Even though a pattern is used to form the cavity, its dimensions are adjusted to account for the metal's shrinkage as temperature is reduced to room temperature. In other words, the pattern is somewhat larger than the dimensions of the resulting casting so that the oversize cavity will contain enough liquid metal in order for the contracting shape to result in the desired dimensions when the casting is cold. Accordingly, an object known as a pattern is used to establish a cavity within some kind of solid material. When the pattern is removed from the solid, a cavity results and it is this cavity that gives the cooling liquid its shape.

Patterns are made from wood, metal, and other materials which are compatible with the production process. Hardwoods last longer than softer woods, while a large volume of casting is accomplished with metal patterns. When the casting contains

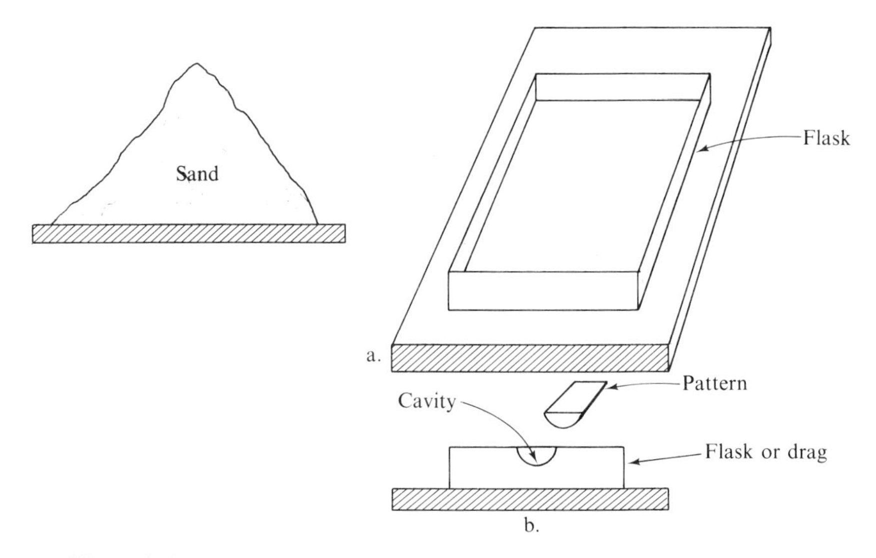

FIGURE 3–3 The pattern and the flask.

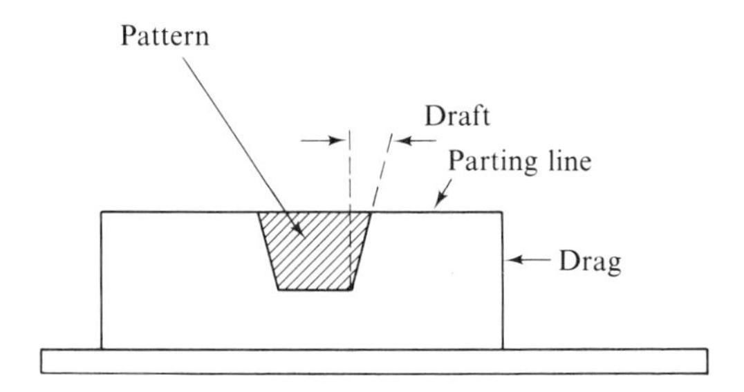

FIGURE 3–4 The draft angle at the parting line of a mold.

holes, an increased draft angle is used for the cores. Besides shrinkage and draft, the pattern maker must also consider a *finishing allowance* for subsequent machining, a *distortion allowance* to prevent overstressing of nonuniformly shaped parts, and a *rap allowance* which enables the pattern to be ejected from the mold. The most critical of these allowances is shrinkage. Some typical allowances for shrinkage are shown in table 3–1.

TABLE 3–1 Shrinkage Allowances

Metal	Shrinkage in./ft.
Aluminum alloys	1/8 to 3/16
Brass	3/16
Bronze	1/8 to 3/16
Cast iron	1/10 to 1/8
Magnesium	9/64 to 3/16
Malleable iron	1/8
Steel	3/16 to 1/4

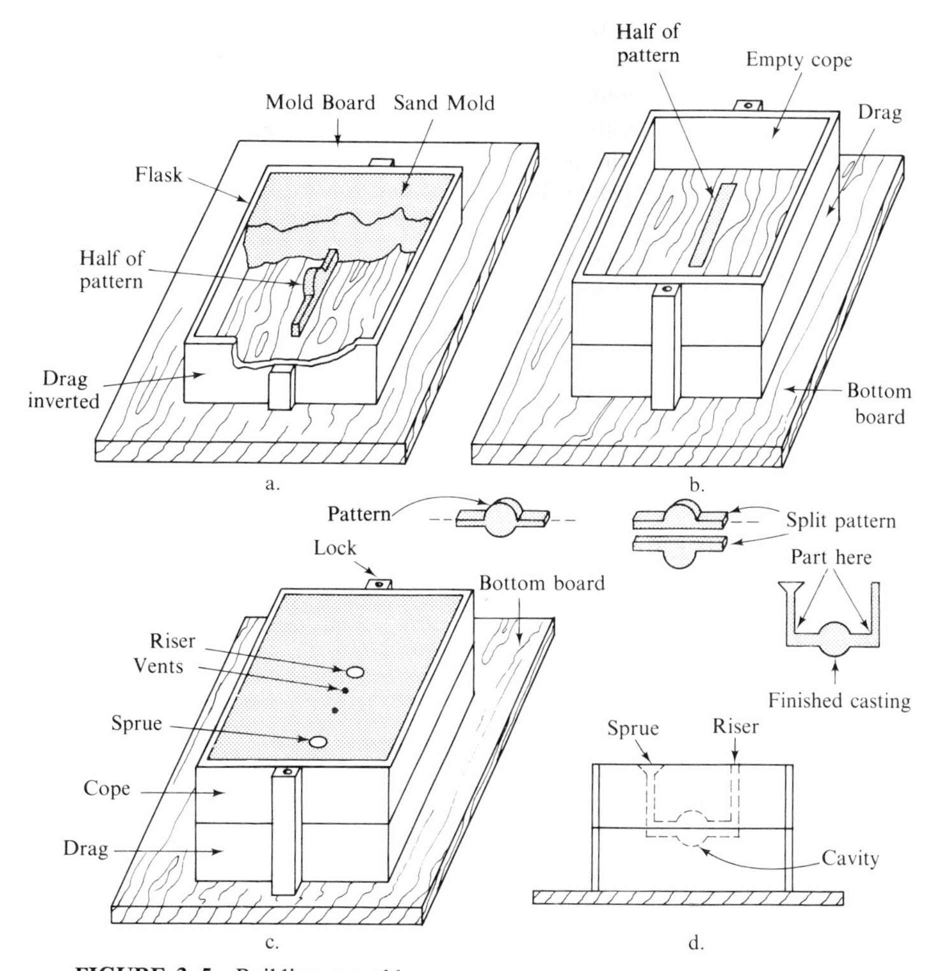

FIGURE 3-5 Building a mold.

Requirements for a Molded Casting

In order to produce a casting, several factors must be considered. The shape of the cavity is most important, as it is this cavity that shapes the casting. Therefore, the pattern must be properly shaped. Next, there must be some strong heat resistant material surrounding the cavity to assure rigidity when it is filled with heavy molten metal. There also must be a means of pouring the metal through the mold into the cavity. At the same time, provisions must exist to allow air and other gases to escape from the mold as the liquid consumes the cavity's space. Retention of gas in a casting is harmful. Air which is trapped at the surface changes the surface profile of the casting. As illustrated in figure 3-5*d*, the cavity is built within a sturdy wall of sand. In order to produce the cavity, steps *a*, *b*, and *c* are performed. These steps are discussed later in this chapter.

DEGASSIFIERS AND REMOVAL OF AIR

Hydrogen pickup can be disastrous to solid metal, especially in thin sections, as it fosters brittleness. Other gases, for example, oxygen, are also unwanted. Consequently, degassifiers are commonly used in the melt to exclude these harmful gases. Bubbling of nitrogen through the melt, for example, is one method used to eliminate hydrogen and oxygen when the solidification time factor and techniques allow gas escape. Also, air must be expelled from the cavity as metal enters. As the metal is being poured and air is being evacuated, a means must also be present to signal the presence of a full cavity of liquid metal. A reservoir is needed to accomplish this. In other words, once pouring is started, it must continue until the entire cavity is filled with molten metal and it begins to appear in a riser type of reservoir (fig. 3–5). The liquid drives the gases from the cavity as pouring continues. Once the cavity is filled, pouring ceases.

FLUXING

Some liquid metals require additional protection such as fluxing as long as the metal is liquid, in order to protect it from the atmosphere or for other puroses. Magnesium, for example, must be protected from oxygen through the use of a flux containing chlorides. Should fire commence, sulfur dioxide retains the fire until it is removed from the burning support of oxygen. Copper and its alloys often are fluxed with charcoal preparations. Aluminum and its alloys, on the other hand, are self-protected with an oxide slag much like the slag on ferrous metals. Other metals are protected from oxide contamination according to the chemistry of their slags.

LIQUID TO SOLID CONVERSION

Conversion from liquid to solid occurs in the cavity, the solid metal gradually contracting in conformance with predetermined coefficients of contractions relevant to volume. Contractions occur due to less space being required for the metal's vibrating atoms as cooling continues. As temperatures are further reduced, atoms get closer together because their maximum vibration tolerances are less. As atomic vibrations

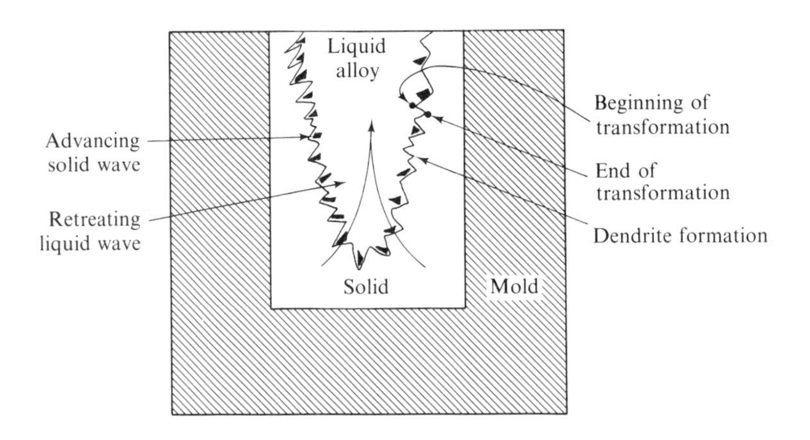

FIGURE 3–6 The cooling of an alloy.

decrease in relation to temperature, and as the solid metal begins to appear, rows of atoms form in organized lattice networks which promote rigidity, and solid metal results. The liquid-to-solid phase is schematically illustrated in figure 3–6. As the molten alloy cools, waves of solid metal reach into the liquid and form nuclei of the highest melting solution. Then the lower melting solutions become attached to the growing and cooling dendrite. Because time is a function between beginning of metal solidification and complete solidification, some regions of solid metal are richer in a constituent than other regions. The longer the time between beginning and ending of dendrite formation, the greater the segregation and the more undesirable the casting becomes.

METALS ARE CRYSTALLINE

Crystallization, or grain structure, is established upon solidification as all metals are crystalline. Dendrite grain, or crystal formation, is illustrated in figure 3–6. When the

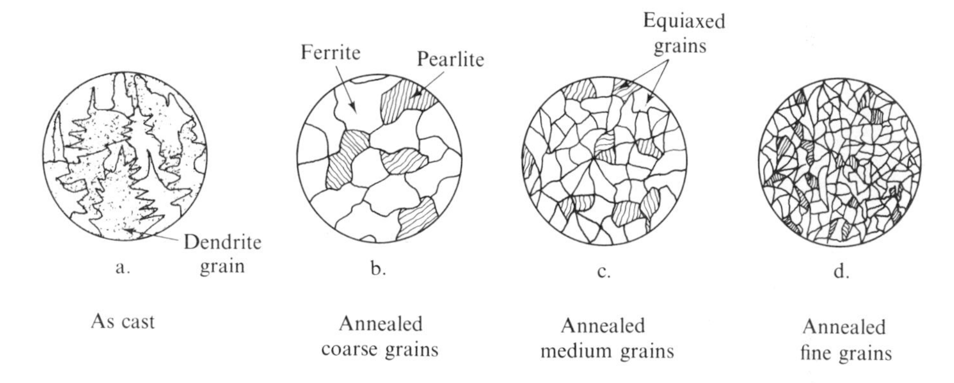

FIGURE 3–7 Cast structural steel followed by annealing at different temperatures.

casting has solidified, and after certain treatments, the dendritic structure shown in figure 3–7*a* is eliminated and new grains are produced, as illustrated in *b, c,* and *d.* An explanation of these treatments is given in subsequent chapters.

CONTRACTION CAUSES STRESS

Continued decreases in temperatures force the atomic network to contract further, resulting in dimensional contraction. Contraction occurs according to the effects illustrated in figure 3–1. Equilibrium is eventually attained between atomic movements and dimensions at a given temperature; consequently, stabilization occurs on cooling to room temperature. Prior to reaching room temperature, however, the hot solid is often strained due to contraction stresses. During this phase of contraction, the cavity's shape and mold material must allow shrinkage to occur in order to prevent tearing or cracking of the solid metal and to reduce the type and quantity of residual stresses.

Shrinkage from the liquid occurs in relation to the atomic structure of the metal. As an example, molten brass shrinks about 3/16 inch per foot of metal. Even though shrinkage occurs inward from three-dimensional surfaces, the linear factor is used in expressing contraction. The entire volume is considered when allowances are calculated. However, an exception to some shrinkage rules applies to a type of gray cast iron in which the precipitated graphite expands as the ferrite contracts. Skrinkage factors then mean that the pattern must be made somewhat larger than the needed dimensions of the casting. As a given metal cools and shrinks, the contour change occurs in accordance with temperature, cavity shape, and metal analysis.

Materials Used in Molding

Another factor that is considered in the casting process is the mold's material. Nonferrous molding sands are often finer than ferrous sands, however, all molds are not made from sand. Some molds are made from heat resistant metals and even nonmetals. Sand-to-metal may often provide for slower cooling than metal-to-metal contact and produce a different structure in the metal. Because the wall of the mold chills the molten metal, smaller grains are formed than those existing deeper in the casting (fig. 3–8). Because cooling is slower as the depth of solid metal increases, grains grow

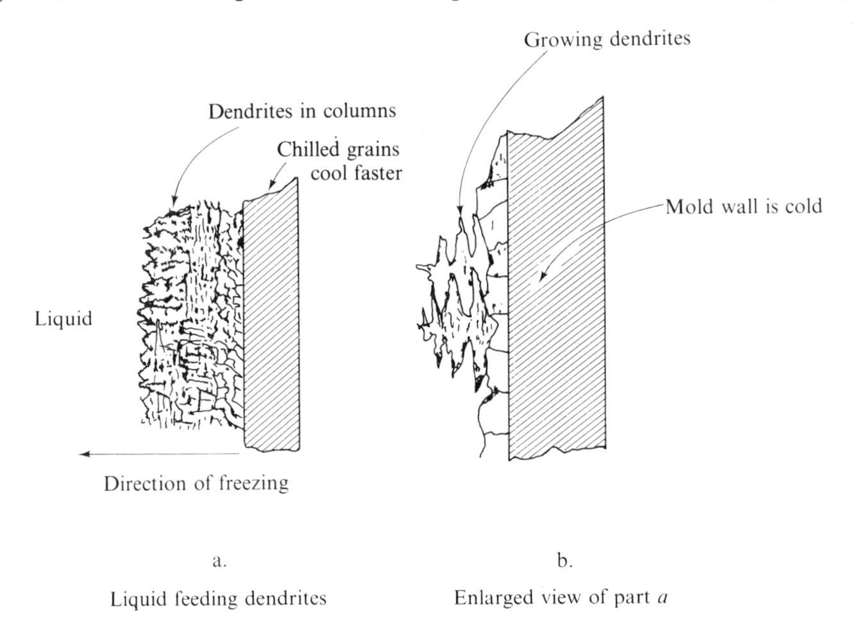

FIGURE 3–8 Schematic of growing dendrites in a cooling liquid.

in elongated shapes, maintaining a nearly vertical growth from the wall, as shown in figure 3–8*a*. The dendrite illustrated in *b* is an enlargement. Dendritic growth is normal. When a small steel ingot is fractured, the dendritic grain pattern is again evident (fig. 3–9).

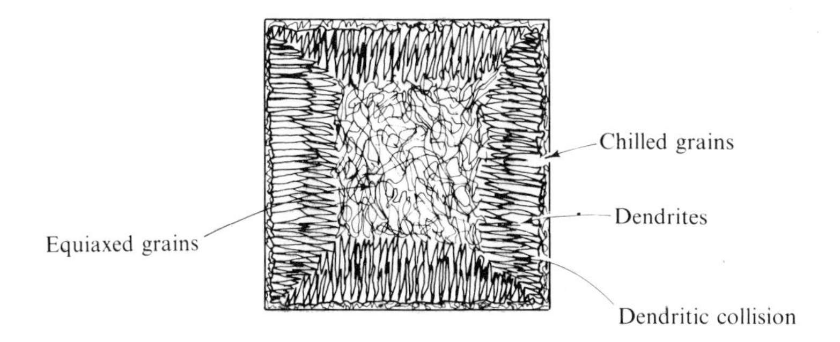

FIGURE 3–9 Cross section of a cast steel ingot. Such a structure must be destroyed and replaced with a more homogeneous structure.

The Finished Casting

When complete solidification occurs after pouring, there then must be a means of removing the casting from the mold. Also, a final cutting operation is necessary to remove excess metal from the casting (fig. 3–5). These excess areas include the metal that is part of the cavity's entrance and exit points and also fins and spines resulting in vented areas. In other words, the casting shows some kind of liquid overflow that occurred during pouring in order that the cavity would be completely filled. Some castings, however, are produced with little or no metal overflow.

The Molding Process

When liquid metal is poured into a mold, a *casting* results. Often the casting is poured from the ladle immediately following ore processing and refinement, whether ferrous or nonferrous. Sometimes, however, the refined metal is cast into some types of castings called ingots (fig. 3–10). The casting of metal is therefore accomplished for one of two purposes, that is, cast into an ingot for purposes of future rolling into various shapes or cast into a permanent casting (fig. 3–11).

Ingot molds are usually large and are uniformly shaped according to the rolling processes to be subsequently performed on the casting. Steel ingots, for example, frequently weigh several tons. When solidification occurs after filling the mold, the mold is lifted from the ingot. Giant tongs then lift the white-hot ingot from its base and place it in a soaking or holding pit at white temperatures until it is ready to be rolled into shape. The shaping of ingots is discussed in chapter 4.

Castings, on the other hand, are usually smaller and are of numerous shapes, depending on the use of the casting. However, some castings weigh as much as 300 tons. The casting processes used in ingot and casting production are related, in that molten metal is poured into a mold's cavity and solidification occurs. Killed ingot pouring and casting processes encourage procedures that help assure sound castings by providing an extra quantity of molten metal at the top of the mold. This reservoir of liquid metal feeds the needed volumes of the cavity as solidification and contraction

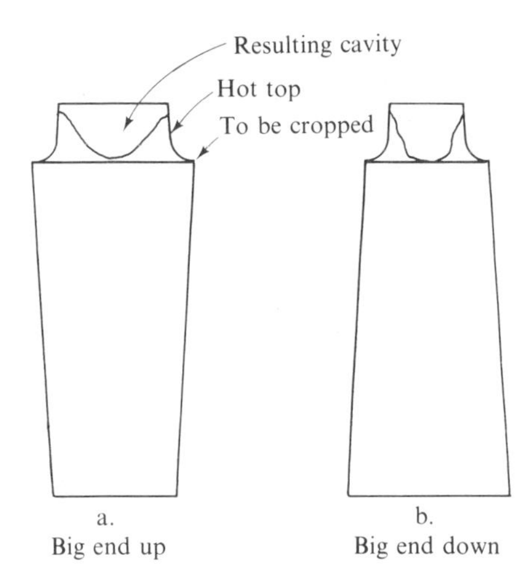

FIGURE 3-10 Cast steel ingots. Cropped tops will be remelted while the remainder will be rolled into shapes. The last metal to solidify is in the specially designed top region.

FIGURE 3-11 A very large casting. This cast iron casting was poured into a pit mold which contained cores and weighed 62,000 pounds at pouring. (Courtesy of H. C. Macaulay Foundry Co.)

occur. If the last metal to solidify is not part of the final casting, then the remaining *pipe,* or hole, due to volume collapse on solidification is not part of the casting. Figure

3–10 illustrates collapsed liquid at the top of the ingot casting. Mold shape and pouring techniques allow controlled solidification, whereby the last metal to solidify is in the top of the casting. The collapsed portions of the castings are removed and remelted. In this respect, the casting, as cast, is still part of the casting system, that is, all solid metal that is attached to it (fig. 3–5). All excess parts are removed from the casting.

From Liquid to Solid

From a metallurgical point of view, some control is used during the liquid solidification process. The metallurgist and technician are interested in *sound* metal being cast, that is, a casting that is free from voids, segregations, and excess stresses. Also, a grain pattern that reflects directional solidification is most desirable because this mold design technique encourages the final shrinkage void to occur near the cavity's entrances and outside the cavity. Without some control, liquid metal will freeze in its natural way, all regions cooling concurrently but unevenly, whereby chance controls the resulting soundness of the solid metal. A mold wall provides a very cold area to liquid metal and, consequently, rapid cooling and solidification occur first in these cold regions. Solid nuclei of metal grow at 90° to the cooling medium and continue to grow vertically and then laterally until collision occurs with other growing crystals. The crystalline pattern is dendritic, or treelike, that is, a central column forms from the mold's wall and crystalline branches radiate at near right angles to the column. Solid crystals grow in the direction of the temperature gradient until the rapidly freezing metal contacts other freezing crystals. Small liquid pools, caught between two or more freezing regions, cause volumetric density differentials and, consequently, voids appear in the solid. When the solid metal blocks the rise of escaping gas, a hole will also result in the solid. This often occurs concurrently with trapped pools of metal. Further, a freezing and expanding region of plastic and rippling crystals reduces the volume of metal, but volume reduction occurs in the presence of uneven temperatures. Consequently, chance often allows metallic separations in the solid, because enough liquid is not available to quickly fill the pending void.

NEED FOR GOOD MOLD DESIGN

It is the liquid-to-solid transformation phase that frequently causes trouble within the solid mass of metal. The planning of molds and the design of the pattern can often cause directional freezing of the liquid, whereby crystals form in the direction of the remaining liquid. Hopefully, final solidification occurs in the riser, or reservoir, as a result of controlled cooling. Hot and cold areas of a mold's wall can greatly accelerate the directional growth of solid grains of metal.

Improperly cooled castings are often inferior in quality. Metal which cools too slowly provides excessive segregation of some constituents, an enlarged grain pattern, and nondirectional properties. Normal segregation is expected in alloys because the dendritic pattern is a segregated pattern of constituents. The chemistry of the dendrite or segregation differs across its volume. Excessive chemical differences are undesirable and are often permanent, therefore, proper cooling is essential. Because thin sections cool faster than thicker sections, care is exercised in the mold's design to allow the liquid to feed the extremities in proportion to a uniformly decreasing temperature, moving toward the highest point of liquid. As solidification occurs, large temperature differentials sometimes cause excessive pulling within the partially solid metal and, in turn, residual stresses remain in the solid. Any mold design which enhances directional cooling toward the reservoir of liquid at the casting's high point is a well designed mold in this respect.

PURE METALS AND ALLOYS

Under usual foundry conditions, liquid metal solidifies and thickens from the mold's walls at an approximate rate relative to the square root of time. In other words, from a given metal and a given mold condition, solidification increases inward at a rate relative to the time lapse from the beginning of liquid cooling to crystal birth. Pure metals and alloys solidify differently. The beginning and ending of solidification in a pure metal is a function of time, as melting and freezing occur at the same temperature. The pure metal produces limited dendritic grains which merge quickly due to the lack of a partial mushy stage of metal as exists in most alloys. Solidification occurs more uniformly throughout the cooling liquid because no alloys alter the constant freezing temperature. The liquid moves to the solid uniformly as temperatures are decreased.

In alloys, however, temperature differentials exist throughout the cooling liquid because a three-phase condition exists as compared to a two-phase condition in pure metals. Different alloying elements added to the base, or pure, metal create nuclei with growth differentials. Consequently, a liquid, a mushy or combination, and a solid phase exist concurrently. The mushy stage provides the time differentials in freezing and, therefore, the dendritic pattern and volumetric differentials between liquids and solid are established. The alloy provides the temperature differences between the *liquidus* and *solidus,* the middle phase being the *variable.*

Sand Castings

The older and most common type of casting process is sand casting. In a way, it is the simplest because only a hole surrounded by sand needs to be filled with molten metal. The nature of sand enables it to be packed firmly around some pattern. When the-pattern is removed, a cavity results. Silica is the main constituent in a molding sand. It is hard, refractory, many sided, and easily mixes with a small amount of clay and water. This basic mixture is *green* and is often used in many casting processes. The green sand mold is a mold containing damp sand made primarily of silica and clay. This type of mold collapses quickly to prevent stressing the casting. The dry sand mold, on the other hand, requires additional drying. Due to its greater strength, intricate molds are not used with dry sand. Special additions to green sands include flour, cereals, powdered coal, and very fine sand. The finer the sand, the smoother the surface of the casting, but the permeability factor decreases. When properly mixed according to special uses, these sands produce quality molds. Many formulas exist for making various mixtures, but all good molding sand should be sufficiently cohesive to retain its shape when properly packed. Besides its ability to withstand high casting temperatures, it must be permeable enough to allow the escape of gases feeding into the mold's walls. Also, the bonded sand must have a compromise strength whereby it holds the heavy metal until solidification is complete and then is able to collapse after solidification as the contracting stresses in the metal demand dimensional change. These requirements originate from the overall demands of the sand mold. When properly bonded, the mold must be sufficiently hard and strong to absorb the shock of the hot metal. This strength factor depends on the sizes of the sand grains, moisture content of the mold, and how well the pattern of grains is bonded.

RECLAIMED SAND

When large quantities of sand are used, they can be reclaimed after casting by introducing the sand's original properties. A large mixer called a *muller* is used to grind, mix, and restore original properties to the sand crystals. Basically, each grain is processed through the clay, water, binder, and aeration procedure and made ready for reuse when it is considered economically feasible.

CORES

Sand is also used for making cores which are placed in the cavity of the mold. The purpose of the core is to form a hole in the casting. Because cores are exposed to situations often unlike the molding sand, they must meet certain requirements which are sometimes basically different from those of the mold. Cores are compacted into the shape of the resultant cavity in the casting, including a slightly increased draft. When many cores are needed, core machines are used. The cores are made from clean silica sand mixed with an oil such as linseed and formed into a paste with flour and water. The shape is baked dry. Depending on the kind of core materials, temperatures as high as 425° F. are used in baking. When unusual loads are to be exerted onto the core, metal wires are inserted in the soft shape before it is baked. Cores must also be able to collapse when the solid metal shrinks.

Molds

Molds for sand castings are numerous in shape, size, and purpose. Sizes of the castings vary from a few ounces to several hundred tons. Large castings are cast in or on the foundry floor and require large cranes to handle the cope and drag operations. Because most molds contain at least two parts, the *cope* and the *drag* (fig. 3–5), the

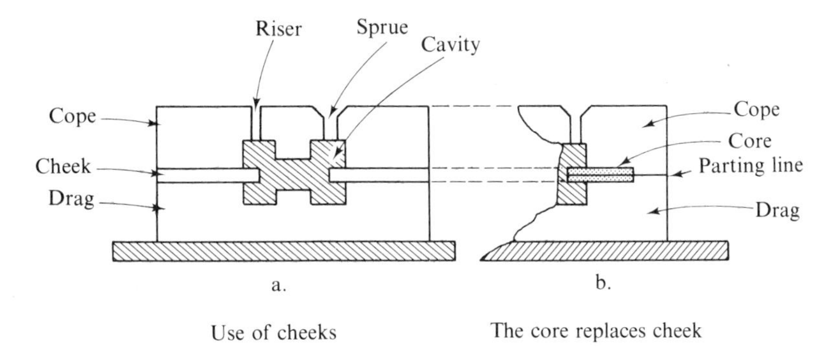

FIGURE 3–12 Parts of a sand mold. The mold may be arranged into numerous designs.

principle of mold preparation is basically the same for small handmade molds and large machine manipulated ones. However, some oddly shaped castings require an extra section in the flask between the cope and drag. These parts are known as *cheeks*. Observations show that the cheek is mated properly to the pattern and, also, a core can be used sometimes to replace the cheek (fig. 3–12).

RELATIONSHIP OF PATTERN AND MOLD

Patterns and molds are usually more complicated than two-piece assemblies. More sophisticated patterns consist of several parts and may be constructed of several materials. As has been pointed out, any solid such as wood, plastic, sand, metal, and even wax or frozen mercury may be used for the pattern. The use of a metal pattern increases pattern life so that it can be used thousands of times. Basically, the solid or one-piece pattern is the simplest. However, a split pattern facilitates the molding process by eliminating handwork in the mold. With the split pattern, half of the pattern is fitted into the cope while half is fitted into the drag, the top and bottom parts of the mold, respectively (fig. 3–5). Figure 3–13 illustrates the relationship between the pattern and the cope side of a special mold.

FIGURE 3–13 Sand casting pattern and cope side of snap mold. (Courtesy of Texas Bronze–Division of Anadite, Inc.)

When larger quantities of parts are needed, a metal pattern *match plate* system is frequently used, whereby match plates (fig. 3–14) are attached to the two halves of the mold so that more than one casting is produced in each mold. Patterns are also made with the use of *chills* along the sides of the mold. Chills cause faster cooling of the liquid in contact with the chill, and a harder- or finer-grained metal will result (fig. 3–15). When making symmetrical castings such as large wheels or bells, a contoured, or sweep, pattern with one fixed axis is rotated in the mold to form the cavity.

FIGURE 3-14 Match plate for an aluminum casting. (Courtesy of Texas Bronze-Division of Anadite, Inc.)

FIGURE 3-15 The use of chills in a mold for a large sand casting. (Courtesy of Meehanite Worldwide)

HANDMADE MOLDS

The procedure described below is frequently used to produce simple sand castings of various metals. First, a boxlike flask is procured. The flask consists of two parts, the cope being the top part and the drag being the bottom. When pinned together, they maintain their proper alignment. When the drag is inverted and placed on a mold board (fig. 3–5), an empty box with no top is formed. Half of the pattern, or any portion of the pattern, is then placed with its severence line adjacent to the board. Care is exercised to assure that enough space remains for placement of entrance and exit holes for the liquid metal to enter and leave the volume occupied by the pattern. Next, the drag is filled with specially prepared filtered sand to form the mold. A *riddle* is then used (fig. 3–16) to screen the sand as it fills the drag, while an air *rammer* (fig.

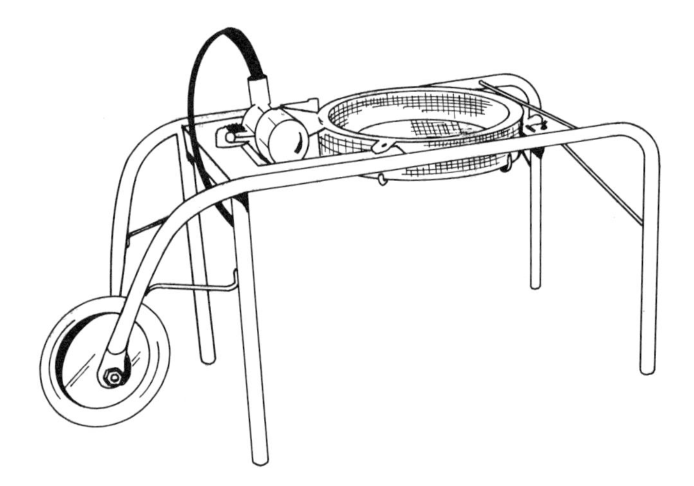

FIGURE 3–16 Mechanically operated riddle. The riddle screens sand for proper size. (Courtesy of Meehanite Worldwide)

3–17) packs the bonding sand tightly around the pattern and within the drag. The excess sand is then struck away, and a bottom board is placed on the drag. When the drag is inverted, the mold board is removed.

At this time, the cope is aligned on the drag. The remainder of the pattern is placed on its mating part, and a *sprue* pin for the pouring hole is secured in the sand. The solid sprue pin will result in a hole through which the liquid metal will enter the cavity from the top of the mold. At the same time, a *riser* pin is also placed so that the cavity metal will rise above the top portion of the cavity and assure that the cavity is filled. Next, a dusting compound is sprinkled on the drag so that sticking of cope and drag will be prevented. Then molding sand is riddled into the cope, being sure that the sand is firmly and tightly rammed around the pattern and sprue and riser pins. Continued filling of the mold is accomplished while ramming. Excess sand is struck from the mold. A vent wire is pushed through the sand in several places to the cavity surface. The cope is then removed carefully.

The area near the pattern is moistened with water and the pattern is withdrawn. The sprue and riser connections are completed and the cavity is inspected. Repairs

FIGURE 3–17 Rammer operator prepares a sand mold for a gray iron casting. (Courtesy of Trinity Valley Iron & Steel)

are made as are necessary. Slicks, spoons, and trowels, in addition to a bellows, help repair ruptured areas (fig. 3–18). If cores are necessary, as in castings having holes, they are inserted in place. The bellows is used to gently blow sand particles from the cavity. Next, the cope is placed on the drag, guiding pins are aligned, and the two halves of the mold are locked in place.

An examination should reflect a boxlike mold filled with sand with only one or two large holes in the top, the pouring hole, or sprue, and the reservoir, or riser. (Figure 3–5 illustrates each step in the preparation of the mold.) Also, several small vent holes should exit from the cavity. The mold is subsequently prepared for the pouring operation. To avoid steam, many molds are dried, especially those used in casting steel. Ovens or torches are used to dry the molds. If too much moisture is present, the mold will explode as molten metal enters.

When metal is poured into the sprue, it runs into the main cavity. When the cavity is filled, gravity in the sprue causes the melt to appear in the riser, thus indicating a filled cavity and a casting. When metal solidification occurs, castings are shaken from their molds, thus destroying the molds.

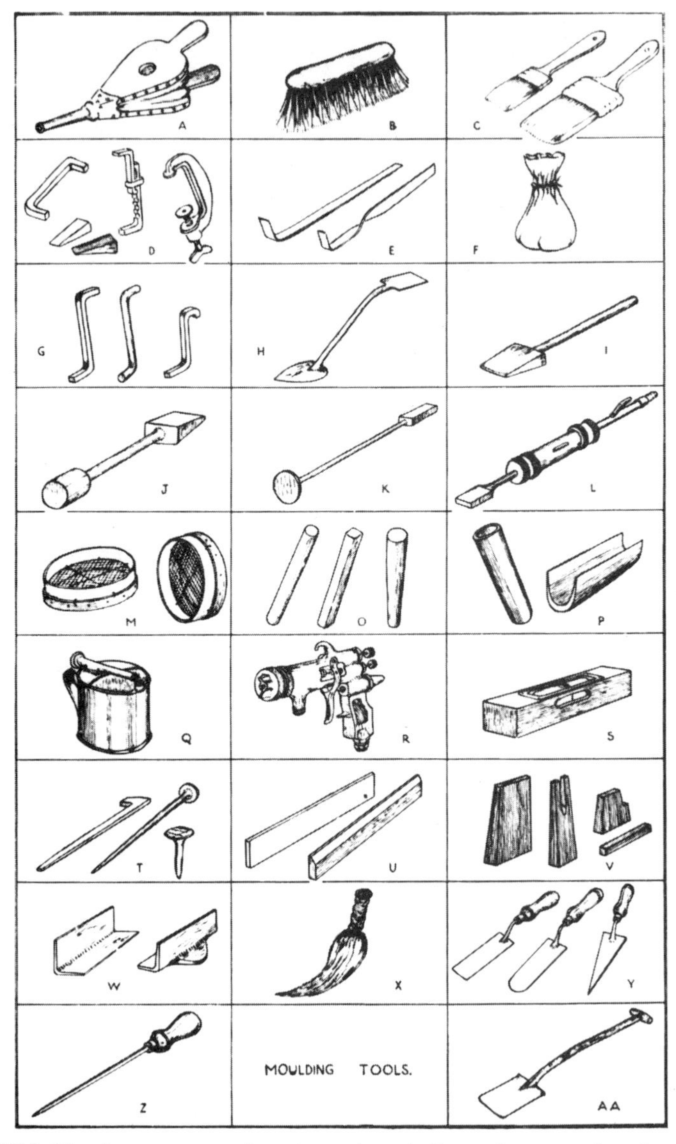

FIGURE 3–18 Some common foundry tools: *a*, bellows; *b*, pattern brush; *c*, swab brush; *d*, mold clamps; *e*, mold repair lifter; *f*, dusting bag; *g*, sand reinforcement rods; *h*, mold finishing tool; *i*, rammer; *j*, hand rammer; *k*, floor type rammer; *l*, pneumatic rammer; *m*, riddles; *o*, runner peg; *p*, sprue cutter; *q*, spray can; *r*, spray gun; *s*, spirit level; *t*, sprigs & nails; *u*, straightedge; *v*, wood supports; *w*, sleek; *x*, swab; *y*, trowels; *z*, vent wire; *aa*, shovel. (Courtesy of Meehanite Worldwide)

MACHINE MADE MOLDS

Because only a limited quantity of parts are produced by means of handmade processes, the machine processes are found in large foundries. These machines range from man assisted to automated and vary in size according to purpose. Molding apparatus are classified according to the type of ramming operation they perform. Machines ram sand into the drag and cope by either jolting action (fig. 3–19), squeezing (fig. 3–20), or slinging (fig. 3–21). Also, these rammers have combination capabilities along with their highly mechanized abilities. Because drags require in-

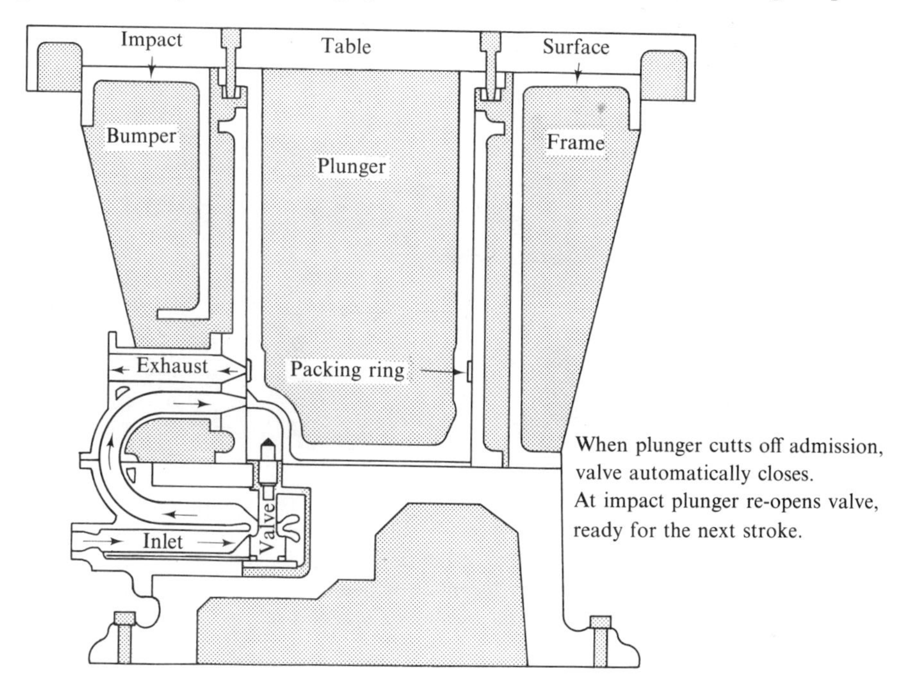

FIGURE 3–19 Jolting action sand rammer. (Courtesy of Meehanite Worldwide)

verting, special machines also roll the drag after squeezing its sand tightly in place. More capable equipment even removes the pattern, accomplishes the preparation of a complete mold, and then readies itself for the pouring operation. When large production quotas are required, these special machines are placed in the production line. The name of the molding machine indicates its capabilities.

SEMIAUTOMATED MACHINES

Many semiautomated molding machines respond to an operator who presses buttons which control specific operations. As an example, a crane places the large parts of the mold in place, the pattern is placed, a sand slinger fills the mold from a chute and sand box, a squeezer board is laid in place, and pressure is exerted on the board which tightly packs that portion of the mold. Air rammers are equally used instead of squeezers to pack the sand (fig. 3–17). When a button is pressed, the crane lifts the mold. In other words, when man cannot easily lift a mold, he uses machines. When the processes are repetitive, man makes the machines automated in varying degrees to expedite the process. A typical semiautomated foundry operation is shown in figure 3–22.

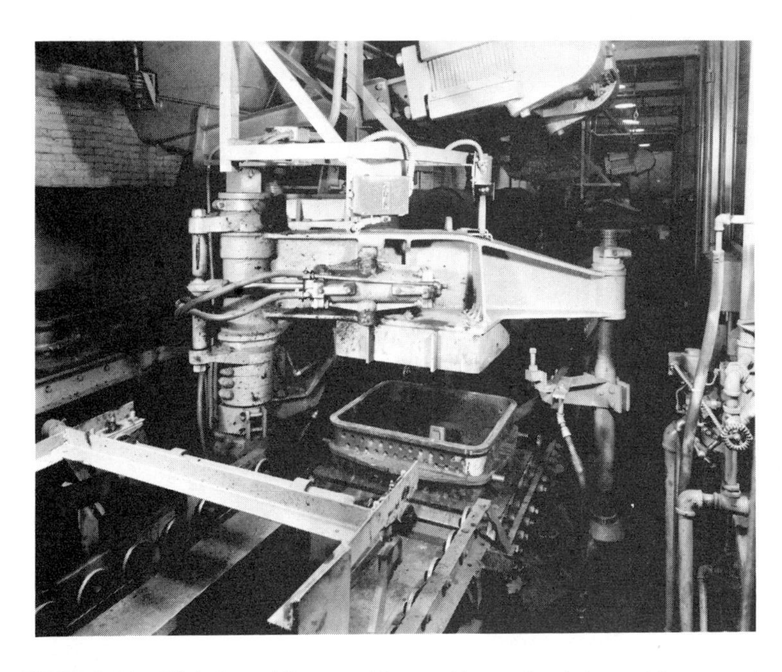

FIGURE 3–20 Nichols molding machine on hi-speed unit in position to make a drag mold. The flask has been pushed into place by the mechanism in the foreground, on the rollers, and lowered on the pattern in readiness for filling with sand. An air cylinder squeezes sand in the mold which was dropped into the mold. (Courtesy of Campbell, Wyant and Cannon Foundry Co.)

FIGURE 3–21 A sand slinger. (Courtesy of Meehanite Worldwide)

Inherent in sand casting processes are the dangers of harmful dust. Personnel engaged in these processes must protect themselves from silicosis and should check to see that they are subjected only to allowable concentrations of different dusts. Silica dust causes serious hazards to the body over a long time period. Also, eye protection from flying sand particles and liquids is essential. Proper ventilation will help to exhaust fumes and remove dust.

FIGURE 3–22 Gray iron being poured into a large pit mold. (Courtesy of H. C. Macaulay Foundry Co.)

EXAMPLES OF SAND MOLDED PARTS

Because sand will allow exceptionally high temperature casting, most metals that are castable can be sand cast. Gray iron is sand cast in large tonnage, as well as steel. Iron and steel castings are used in the machine industry for various parts of machines such as housings, footings, mounts, supports, and other structural parts normally used in compression loading. A large sand casting for a machine part is shown in figure 3–23. Also, moving parts such as wheels, gears, sliding ways, containers, automotive pistons and cylinder blocks and heads, gratings, and specialized fittings are cast from gray iron and steel. The pipe industry uses thousands of tons of gray iron. When mechanical properties are such that tensile strengths must be high, then cast steel is used in preference to cast aluminum, brass, or magnesium. Lower strengths are produced in the gray irons, but these are not used in tension due to brittleness. Aluminum and most of the nonferrous metals are also used in sand casting processes. Some of the largest castings ever made are sand castings, the size limit depending heavily on the quantity of molten metal available. Often, molds are made in the foundry floor, as shown in figure 3–24. Figure 3–25 shows some typical sand castings.

FIGURE 3–23 A sand cast lathe bed. (Courtesy of Meehanite Worldwide)

Shell Molding

The shell molding process is one of the newest casting processes and is adaptable to automated processes, because its procedures are sequential and repetitive within very close tolerances. Surface smoothness is superior to sand castings, and dimensions can be obtained to within 0.004 inch per inch in many cases. Basically, the casting of different metals utilizing this process involves no more than preparing a number of thin-walled shells of sand into which molten metal is poured (fig. 3–26). The sprue must be deep enough to allow for the collapsing solid at the top of the mold.

Shell molding consists of a sequence of procedures, beginning with the procurement of the two pattern halves which are usually made of metal. The pattern is heated to approximately 400° F. and placed on a container of sand and a binder. Thermosetting resin binders such as phenol-formaldehyde and silica, when heated, form a hard and refractory shell of about 1/8-inch thick. The container of treated sand is reversed to allow the sand to fall onto the heated pattern. A timed curing

FIGURE 3–24 Pit molding the bed for a large lathe. (Courtesy of Meehanite Worldwide)

period of short duration causes the thickening shell to form. Again, the container is reversed. Both pattern and shell are then baked in an oven at approximately 550° F. The shell is ejected from the pattern. Because of the unity of the process, the entire system of sprue, runner, gate, and other needed parts are included as part of the shell. The two halves of the shell are attached, gluing is used frequently, and are then placed in the flask as cope and drag, or as illustrated in figure 3–26. Foundry sand is used to fill the flask in order to give support to the shell. Standard pouring practices are used. After metal solidification, castings are shaken from the flask onto an oscillating conveyor, the original shell resin often being attacked by the molten metal.

Because this process has large volume production capabilities, improvements are being made, especially in the materials constituting the shell. Larger castings are being produced than those originally produced, however, for normal production purposes, castings vary in weight from a few ounces to several hundred pounds.

EXAMPLES OF SHELL-MOLDED PARTS

Basically, if a pattern can be produced, then a shell can be made to enclose it. Therefore, this process is extending to numerous types of parts and fittings in the structural

FIGURE 3–25 Some typical sand cast parts. (Courtesy of Meehanite Worldwide)

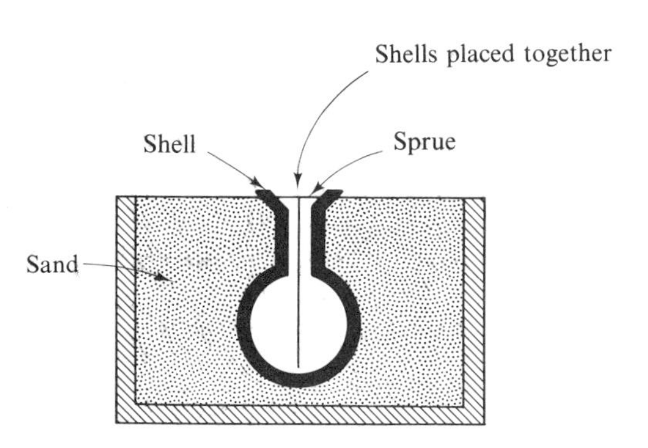

FIGURE 3–26 A shell mold. In some instances, only the shell becomes the complete mold.

and nonstructural categories. Small and large gears, cylinders, supports, heavy earth-moving parts, housings, plates, guides, and brackets are frequently shell molded. Any of these parts may range up to several hundred pounds in weight, however, a great proportion of castings are in the structural fitting category. Figure 3–27 shows some typical shell molds.

FIGURE 3–27 Shell molds being readied for firing. (Courtesy of Texas Bronze–Division of Anadite, Inc.)

Permanent Mold Casting

Because both sand and shell casting processes require fracturing of the molds each time a casting is produced, the saving of the mold is not possible. Therefore, a more costly mold is produced from ferrous alloys which has a permanent shape and can be used repeatedly several thousand times. Permanent mold processes are used where production quotas are very high and accuracy of dimensions are important. Tolerances from 0.005–0.010 inch are practical. In sand castings dimensions vary enough to require machining. Sometimes, however, variance in dimension is not significant in sand castings. In regard to permanent molds, only when the cost of the mold can be justified is permanent molding economically feasible. Even though many castings of close tolerances and smooth finishes are produced by this method, the nature of the process limits the size and weight of the casting. On the other hand, when a sequence in procedures can be established, then repetition is possible and the process makes itself available for automation. Many thousands of castings are made by this process, some individually, some by automation, and some by semiautomation.

Much like shell molding, the permanent mold includes the functional unit, that is, the sprue, runner system, gate, riser, and a means for gas exhaust. When a refractory wash is placed on the metal mold's surface, or a carbon wash for iron castings, the incoming melt flows smoothly without sticking to the mold. Should holes be needed in the casting, metal or sand cores are used. When sand cores are used, the mold is semipermanent as new cores are required for each new casting. Molten metal flows through the system and fills the cavity by means of gravity, similar to the sand casting procedure. Cooling time is often less than that in sand casting and, in turn, finer grained castings are produced. After metal solidification, the mold is swung or slid open, the two halves allowing ejection of the casting. Because the casting is attached to the cast system, the unwanted sprue, runners, gate, and riser are removed and remelted for further use.

Permanent mold casting has its set of unique procedures. As an example, when metal solidification occurs, the mold is quickly opened to allow shrinkage to occur. In this respect, cores must also be removed as cooling and shrinkage occur. Because gravity feeds the liquid system, intricate castings are not feasible due to the inability of the liquid to deeply penetrate extended regions. However, a semi-pressurized system will force the liquid into more complex spaces. When this procedure is justified in lieu of die casting, then it is used. Also, special venting is necessary, often at flash spots along the mold's closing areas, in order to produce void-free casting.

When the casting procedure has been established and thousands of duplicate castings are needed, a specialized production line is instigated. Molding machines open their molds, wash the cavities' surfaces, pour the metal, eject the castings, move the castings along the lines, and then repeat the process. A permanent molding machine is shown in figure 3–28.

FIGURE 3–28 A permanent molding machine in the open position for molding small parts. (Courtesy of Texas Bronze–Division of Anadite, Inc.)

A modified permanent mold process is known as *slush molding*. Should the cast product be hollow, then only a shell-type shape is needed. When liquid metal is poured into a permanent mold, allowed to solidify only at the surface, and then the mold is quickly emptied of the remaining liquid, a hollow casting results. By careful timing while in the mold, numerous slush castings are produced. The slush that is ejected leaves a rough interior which reflects dendrite growth. Only the outer surfaces need be smooth, therefore, much metal is saved. Slush casting lends itself favorably to the jewelry and ornamentation industries.

EXAMPLES OF PERMANENT-MOLDED PARTS

This process is used extensively when simple parts of small to medium sizes are needed. These parts are produced from aluminum, zinc, magnesium, copper, and some types of gray iron. Many lightly stressed fittings on aircraft and in the automobile industry are cast by this method. Household appliances such as the base of the electric iron and waffle irons are easily cast. Heads on four-cycle engines, machine tools, and numerous other similar parts are produced. The permanent mold process is practical when many hundreds of parts are to be made to compensate for the high cost of the mold.

Die Casting

The die casting process is a modified permanent mold process that uses pressure instead of gravity to fill the mold's cavity. Also, die casting is common to the nonferrous metals industries, whereas sand, shell, and permanent mold processes are applicable to ferrous and nonferrous metals. The mold is called the *die* because liquid metal is suddenly placed under pressure and takes the shape of the mold. The solid casting is quickly ejected. Speed of production and very close tolerances are the main advantages of this process.

In order to produce a die casting, a manual or automated sequence is established in a cycle, and the cycle repeats itself, producing thousands of identically shaped parts. A disadvantage of the process is the complex procedure required to insert and remove cores whenever the cast part contains holes. The first operation in the cycle requires a *ready* mold, that is, the two mold halves being closed with cores in place. Some external arrangement sweeps or slides the cores in place as the mold is closed. In this respect, clearances must be close enough to insure that liquid metal under pressure cannot be forced from the mold. Air and other gases escape between the die faces unless special vents are provided.

The complete die mold is attached to a mechanism that provides opening and closing of the parts, along with a metal injection apparatus. When step one is completed, liquid metal is forced into the die under a given and constant pressure for a timed period. These conditions allow the metal to solidify and fill the intricately shaped spaces with sound metal. Unless gases can be vented, porosity is formed. At the end of this operation, the dies are opened and the casting is ejected. Immediately, the cycle is repeated, continuing as long as necessary. Again, complexity exists in the cycle when cores are placed and removed. The placement and removal of cores are tied in with the closing and opening operations. When cores are of the piece type, additional complexity and a slowdown are experienced. Also, in some instances, harder metals such as steel bolts are often placed within the cavity prior to metal injection. Should the casting include bearings, these are frequently arranged in a manner similar to bolts or other internal and integral parts of the casting. Incoming metal solidifies around these preset parts. The dies in figure 3–29 illustrate the versatility of the casting process.

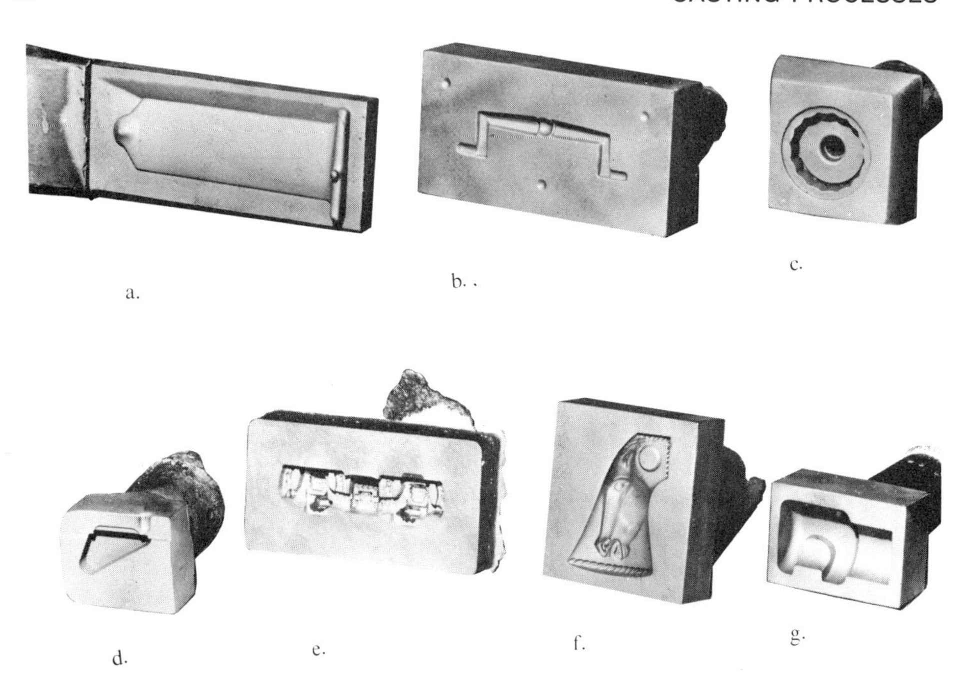

FIGURE 3–29 Examples of Shaw process, applications for the tool and die industry. *a*, blade forging die made in Carpenter 883 steel; *b*, diecasting die made for handle to be cast in zinc alloy: NuDie V; *c*, small die cavity for knob made in CSM #2 to be die cast in zinc alloy; *d*, plastic mold cavity made in NuDie V; *e*, high carbon, high chromium tool steel die cavity for railway train side truck; *f*, plastic mold cavity made of Super Samson; *g*, left-hand cavity for die cast cable cutter made in NuDie V. (Courtesy of Avnet Shaw)

Two main types of die casting machines are available, the cold chamber and the submerged piston type. Cold chamber machines (fig. 3–30) operate at lower temperatures than the submerged types, therefore, the liquid metal solidifies quickly as it is injected into the cold chamber. A carefully measured quantity of liquid metal such as copper alloy or aluminum alloy is ladled into the chamber, and the piston compresses the liquid into a solid which has the shape of the cavity. Immediately, the dies are opened, along with accessory operations, and the casting is ejected. Because some thin fins of metal are squirted from the dies at their closing areas, removal of this small quantity of metal is necessary.

The submerged piston type of machine is arranged so that the piston and chamber mechanisms are submerged beneath the pool of molten metal. Figure 3–31 shows a gooseneck type of die casting machine. Metals used for this type of operation are frequently the low melting types of zinc, lead, or tin. Reciprocation of the piston moves molten metal through a gooseneck arrangement into the die under pressures up to 2000 psi. Freezing of the liquid occurs differently in this type of machine, because parts of the machine are maintained at the temperature of the liquid, thus causing the sprue portion of the casting to remain with the casting until final ejection.

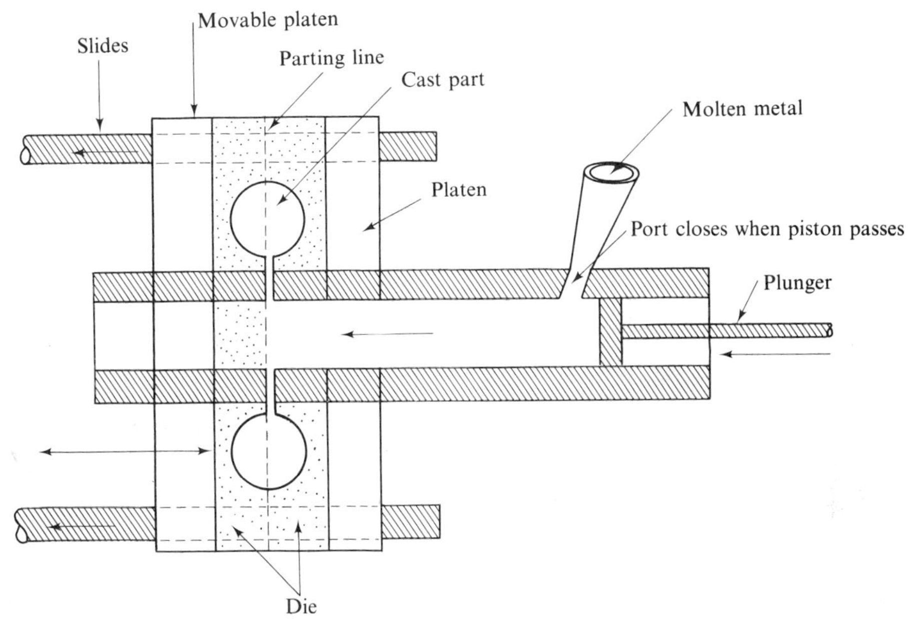

FIGURE 3–30 Cold chamber die casting machine schematically illustrated.

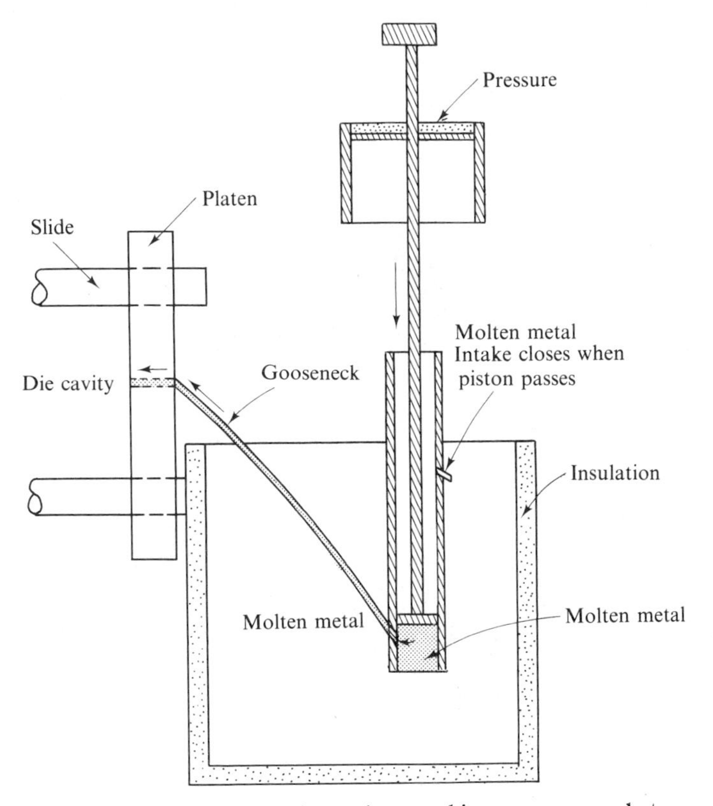

FIGURE 3–31 Hot chamber die casting machine, or gooseneck type.

EXAMPLES OF DIE CAST PARTS

If a cast metal can be removed from the die without destruction, then it can be die cast. Such a wide distribution of shapes makes this process especially adaptable to household hardwares; automotive and aircraft engine accessory housings, fixtures, and parts; lawn and garden tool components; and housings for small machines. The list of products is nearly endless, but most of these parts are not complicated in design. Figure 3–32 shows some typical die cast parts.

FIGURE 3–32a A die set showing a typical die cast part. (Courtesy of Paramount Die Casting)

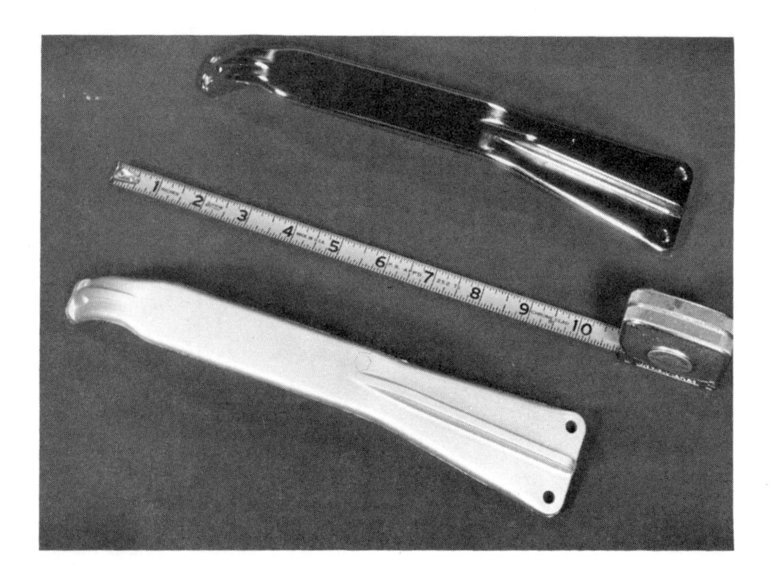

FIGURE 3–32b Two die cast parts. (Courtesy of Paramount Die Casting)

Centrifugal Casting

When liquid metal is allowed to freeze inside a spinning mold, the casting is centrifugally cast. The main advantages of this process include a tighter grained metal at the outer surface of the casting and the fact that no cores are needed to provide the hollow interior. A rapidly turning metal or sand-faced metal mold causes a liquid to move to the periphery under force of the rotation. A measured quantity of liquid, ferrous or nonferrous metal, is allowed to enter the rotating mold. Freezing occurs at the surface and in accordance with dendritic actions, and the wall is thickened proportionately to the amount of metal available. Outside surfaces take the smoothness and shape of the mold while inside surfaces are rough, sometimes necessitating machining. Centrifugal processes include true centrifugal casting, semicentrifugal casting, and centrifuging.

When horizontal casting is accomplished, various shapes such as round, out of round, or several-sided castings can be produced, depending on the mold's shape. Speed of rotation varies in proportion to shape, size, and density of material factors. As an example, pipe is produced in various lengths and diameters by rotations from a few hundred to several thousand revolutions per minute. The slower freezing regions of metal are pushed to the inside by dendritic action which opposes centrifugal action. When cross sectional strengths are critical, a thicker wall is cast, and the weaker inside regions of metal are machined away. Figure 3–33 shows a centrifugal casting machine.

FIGURE 3–33 Automatic centrifugal casting machine. (Courtesy of Meehanite Worldwide)

Vertical centrifugal casting is also a true centrifugal process, but the axis of rotation is 90° from the axis of rotation of the horizontal method. Speeds of rotation are directly proportional to the wall thickness and shape, because slower rotations produce a parabolic shape. Increased rotation forces the heavy metal upwards, changing the angle of wall thickness, the inner surface being the wall affected.

The semicentrifugal process is not a true method of centrifugal casting, but is a modification. As an example, some castings are shaped in such a way that the periphery requires as much or more metal than the inner regions. Large gears and wheels are produced as the result of molten metal being poured into a slowly rotating mold having characteristics of the cope and drag. A tighter grained structure is produced where most of the casting's work is to be performed. Modifications are also made in the central regions due to the need for a hub. Cores are used according to the need. When the cope and drag are arranged to allow slow rotation as molten metal flows through a system similar to the sand casting process, a centrifuge is established and provides a more dense grain pattern.

EXAMPLES OF CENTRIFUGAL CAST PARTS

Hollow parts requiring soundness of metal at the outside such as pipe or periphery of wheels, discs, and gears are commonly produced by this method. The function and geometry of the part dictate the type of casting process when special designs are needed. The mold may be a metal or nonmetal, but must allow the incoming metal the opportunity to quickly move toward the outer regions of the mold. An example is the tube which has extreme length but thin walls due to the hollow interior. Some hardware which have hollow centers such as nuts, bearings, bushings, and cylinders are often centrifugally cast.

Investment Casting

The investment casting process is one of the oldest casting processes, yet it has been extensively used only since World War II. Because the process lends itself very favorably to intricate and sophisticated designs, it is used in industrial applications where other processes fail or are not appropriate. *Investment* is a refractory powder that is used to fill a flask in which the pattern's cavity exists. When liquid metal is poured with conventional procedures, or modified ones such as centrifugal, into a heat resistant cavity, fine detail, close tolerances, and a smooth surface finish are possible. Investment casting is another term used for the *lost wax* process.

Several procedures of investment casting are available, however, a common procedure is to first secure a pattern from any source and then produce the master die. Sometimes the die cavity is carved within the die material, while other procedures involve the pouring of a metal around the pattern to form the die. Next, the die, which is a replica of the original pattern, is filled with molten wax or frozen mercury. Tiny crevices, undercuts, and extremely difficult regions to reach by conventional pouring processes are accessible with this casting process. Sprues and pouring systems are made at the same time as the wax patterns. When hardened, the wax pattern, sometimes in several pieces is assembled with its mating sprue system.

The entire system is then coated with a mixture of refractory material by repeated dippings in the finely ground slurry of refractory investment. Each dipping increases the wall thickness of the investment. When a sufficient thickness is obtained, usually 1/8 inch or thicker, the shell is placed in a flask in a manner similar to the sand casting process. After the flask is filled with investment and vibrated to assure a sound mold, time is allowed for drying. An advantage of the investment slurry is the ease with which it flows to all surfaces. When set, the wax is melted from the mold. Most of the wax is reclaimed, however, some is lost as it drips away, thereby creating the name of the process. When the cavity is free from the wax, observation shows the fineness of the cavity with all its details.

The mold is then heated to temperatures commensurate with the kind of metal to be poured. While the mold is hot, metal is poured into the sprue system, either gravity fed, pressure induced, or centrifugally cast. The method used depends on the intricacy of the cavity. As the liquid cools, so does the mold. In turn, contraction between the freezing metal and cooling mold are compatible. Fine and sound castings are produced by this method. When cooled, the investment is fractured and the casting removed. Dimensions within 0.005 inch per inch are obtainable, and no machining is required.

EXAMPLES OF INVESTMENT MOLDED CASTINGS

Many hard-to-machine parts are provided by this method, especially ones of highly alloyed metals such as chromium, cobalt, and molybdenum. Intricately shaped turbine blades for super-chargers, high pressure and high temperature nozzles, and parts reaching into all industrial applications are produced by this process. Recent research shows further use of this process in the area of the tough alloys, especially the heat resistant ones. Parts exposed to thermal stresses are precision made by investment casting. Also, the dental profession casts teeth and bridges with the use of investment casting. Many of these parts are molded in machines similar to the wax injection machine shown in figure 3–34*a*. Some typical investment molded parts are shown in figure 3–34*b*.

Plaster Mold Casting

A modification of the sand casting process uses a special mixture of temperature resistant plaster as the mold material which forms a syrup when mixed with water. The syrup may contain fibers because some plasters such as calcined plaster include fibers. When the thin syrup is poured onto the pattern, a nearly perfect cavity forms as the plaster hardens. Each part of the mold, if using sand casting procedures, is baked in ovens at temperatures up to 500° F. in order to set the plaster. Many of the nonferrous metals such as aluminum, zinc, and copper alloys are used in this process. If the mold material is made slightly permeable, a sounder casting results because gases are vented. When the mold is assembled and preheated, molten metal is poured in the sprue and a precision casting is produced. Usually, a vacuum is required during pouring, but sometimes pressure is used. The cope and drag process used in plaster molding is like that used in sand molding. Cores are made of the same material as the

FIGURE 3-34a A wax injection machine. (Courtesy of Texas Bronze–Division of Anadite, Inc.)

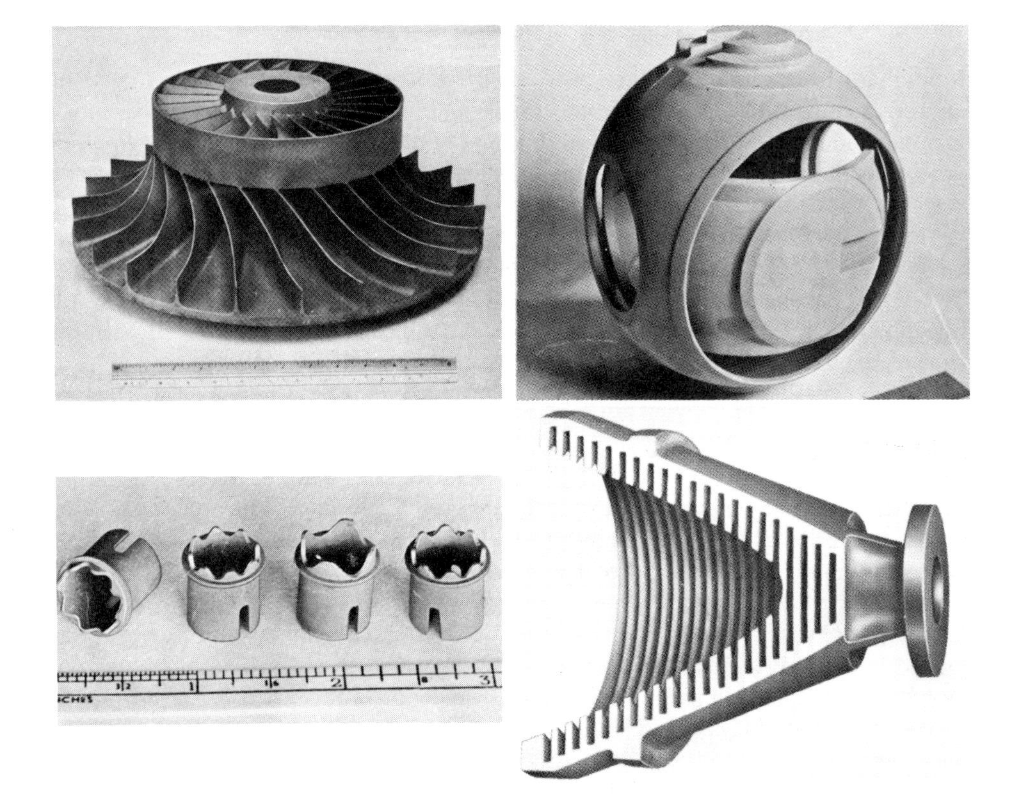

FIGURE 3-34b Typical investment molded parts. Liquid wax was injected into the mold cavity, the die was opened, and the wax replica was coated with investment. After the wax was melted, liquid metal was poured into the cavity. (Courtesy of Electronicast)

mold. Also, chills are used in the same manner that they are used in other casting processes to cause faster cooling of the metal in contact with the chill spot.

Ferrous metals are not cast into plaster molds because the calcium material deteriorates at temperatures beyond 2100° F. Ferrous pouring temperatures are above this value. Magnesium is not poured into some plaster molds due to dangers of explosion. Water moisture and other chemical combinations and liquid magnesium are incompatible.

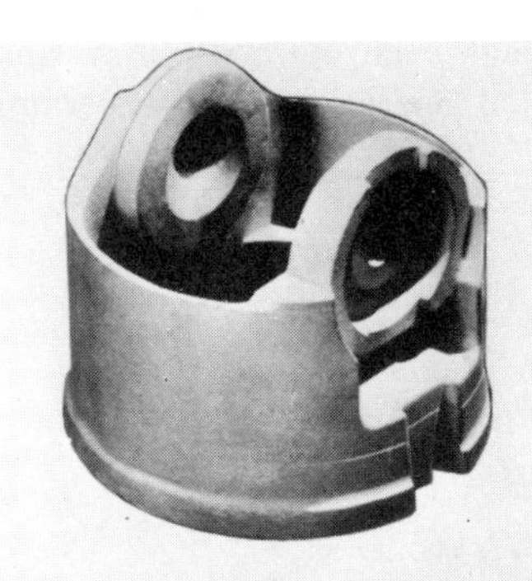

Complexly cored aluminum casting

Stainless steel complexly cored valve body

Tool steel injection molding die cavity and core

5% chromium hot work steel cast cavity and core for aluminum diecasting die

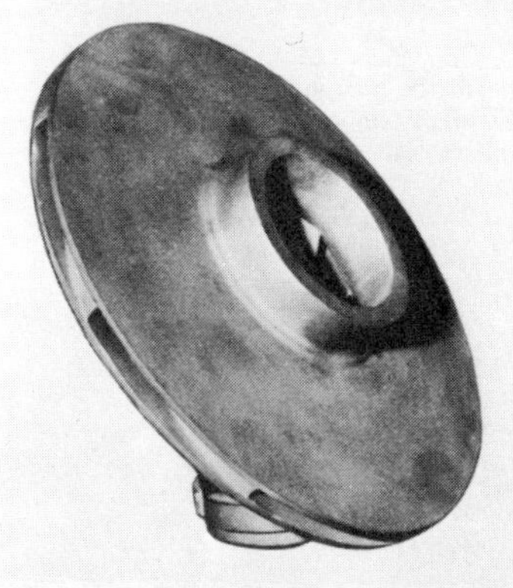

Stainless steel pump impellor with 3/8" web

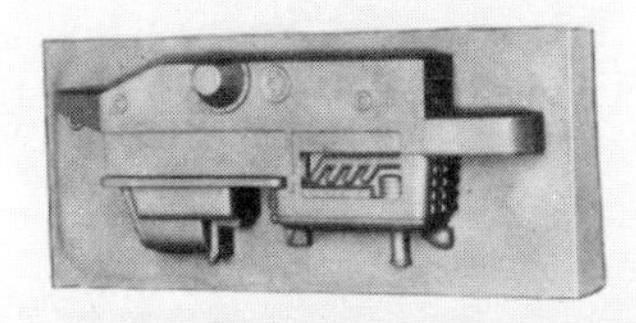

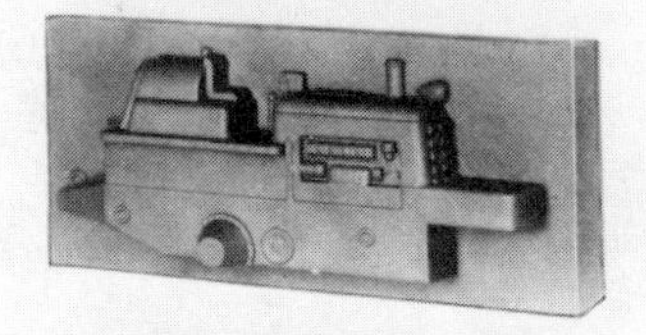

Left and right hand hobs in chromium vanadium Tungston hot work steel

FIGURE 3–35 Typical ceramic molded parts. (Courtesy of Avnet Shaw)

EXAMPLES OF PLASTER MOLDED CASTINGS

Any smooth-skinned object that can be arranged as a pattern and sand cast can be made by this process; the main difference is that liquid plaster is substituted for rammed sand. Obviously, plaster costs more than sand, so the cost factor is a governing point in its use. Plaster or ceramic molded parts are normally small and include fixtures, holders, housings, fittings, spiral blades, and some of the parts cast with the sand process. Typical ceramic molded parts are shown in figure 3–35, ceramic being the molds for these examples rather than plaster. As has been pointed out, various materials are available for this process. A sequence of operations in the ceramic casting process is illustrated in figure 3–36.

STEP 1

REFRACTORY – Is composed of a variety of specially blended groups of refractory powders.

STEP 2

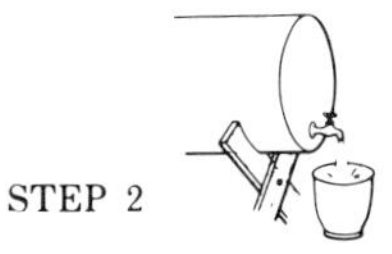

BINDER – The liquid medium is usually based on ethyl silicate and is specifically produced to proprietary formulations.

STEP 3

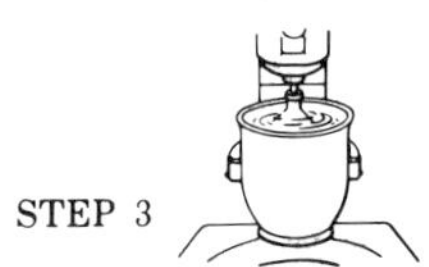

MIXING – A small percentage of gelling agent is added to the binder and mixed with the refractory powder to produce a creamy slurry.

STEP 4

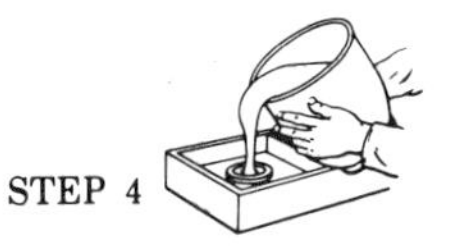

PATTERN – The slurry is poured over a pattern made of wood, metal, plaster, plastic, etc. It is then allowed to gel in about 2 to 3 minutes.

STEP 5

STRIPPING – The gelled refractory mass is stripped from the pattern by hand or a mechanical stripping mechanism.

STEP 6

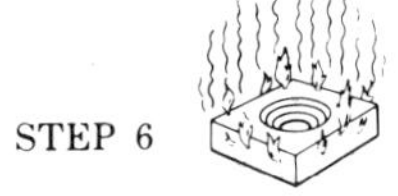

BURNOFF – The mold is ignited. It burns until all volatiles are consumed. This sets up the "microcrazed" structure.

STEP 7

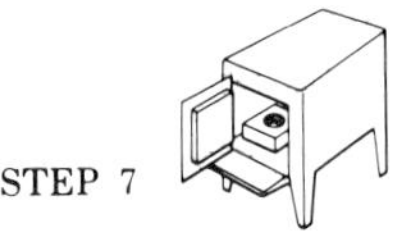

BAKING – The Shaw Mold, now immune to thermal shock, is placed in a high temperature oven until all vestiges of moisture are driven off.

STEP 8

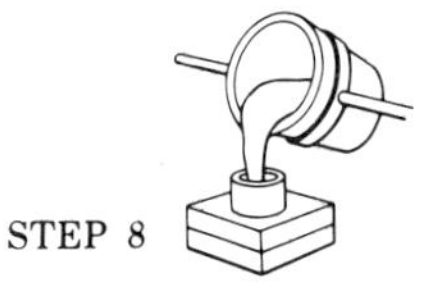

CASTING – Cope and drag mold pieces are assembled along with any necessary cores and poured.

FIGURE 3–36 Sequence of process operations used in ceramic molding. (Courtesy of Avnet Shaw)

Questions

1. Name and describe the main parts of a sand mold.
2. Explain why liquid metal consumes more space than solid metal.
3. Describe dendrite formation.
4. Explain the differences between the liquid-to-solid phase of a pure metal and of an alloy.
5. What is the purpose of a flux?
6. Which casting process is most productive and economical in producing automotive carburetors?
7. What circumstances justify the manufacture of a permanent mold?
8. Describe three kinds of discontinuities.
9. Why is hydrogen dangerous in steel?
10. What is the purpose of a riser in sand castings?
11. What is an inherent defect?
12. Why must shrinkage factors be considered prior to a casting process?
13. Why must patterns be very carefully made?
14. What is meant by the crystalline nature of metals?
15. Differentiate between an ingot and a casting.
16. What is the main difference between a pure metal and an alloy?
17. What is shell molding?
18. Compare the die casting process with permanent molding.
19. Explain a main advantage of centrifugal casting.
20. Explain an advantage of the lost wax process.

Wrought Processes

Many differently shaped parts are needed in the various metal working industries. Castings fulfill the need for only a part of the nearly endless quantity of sizes and shapes of metal parts. Most of the remainder of shapes are produced by *wrought* processes, that is, by various pressure forming methods. Some of these methods include rolling, forging, pressing, extruding, and drawing. By far, the rolling processes exceed all others in the production of pressure formed shapes.

Wrought processes are required when casting methods are not suitable or feasible. A long wire or a long wide flange shape, for example, cannot be economically produced by casting. Both of these items are easily and economically formed by pressure, however. When mass of material and contour vary along any axis, the casting process is often used. Also, when an exceptionally large part such as a 250 ton housing is needed, casting is used. But when a part having long straight lines along its axis is required, some form of a wrought procedure is necessary. Figure 4–1 illustrates the logic involved when choosing either the casting or wrought process. The can is produced by one quick stroke of the punch as shown in *b*, whereas the casting process in *a* would be much slower, even if more sophisticated casting processes were used.

Beginning of Wrought Processes

Inherently, wrought processes are preceded by the casting process because a mass of metal must first be available before it can be exposed to shaping. Therefore, a major function of casting is to produce the ingot for subsequent shaping. Ingots, both ferrous and nonferrous, are cast into many sizes and several shapes depending on their purposes. For example, slabs of metal are squeezed from larger rectangularly shaped ingots. Once solidification of the ingot has occurred, it is placed in oil- or gas-fired soaking pits so as to maintain its high temperature. Temperatures of hot ingots vary

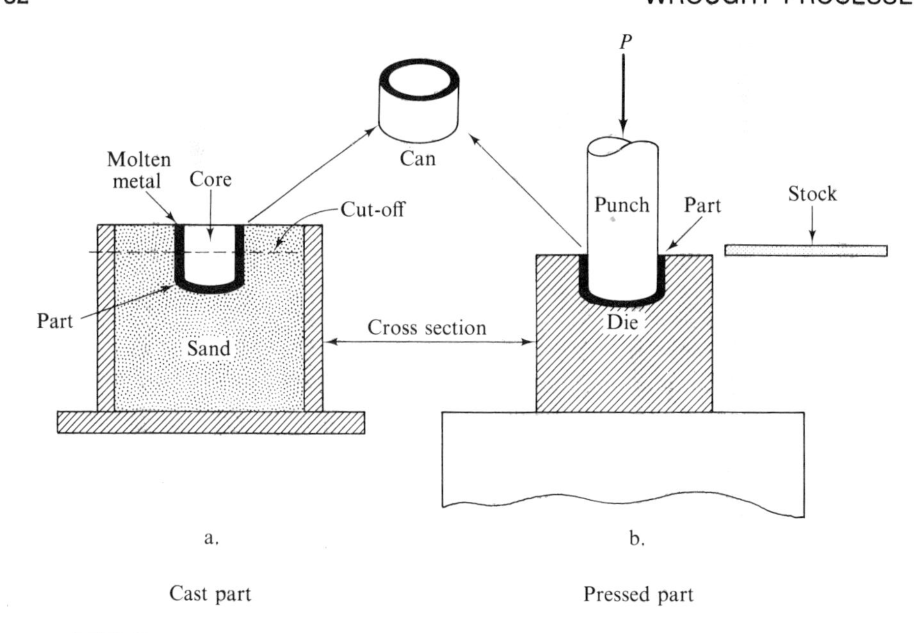

FIGURE 4–1 Cast or wrought process?

according to the metal, but all temperatures are maintained slightly below the metal's melting point or only a few hundred degrees below melting in the case of steel. Steel is soaked at 2200° F. until the cross section is uniform in temperature. Unless the centers of ingots are saturated with heat, tearing will result when they are rolled. Such large masses of metal require several hours of reheating when they are allowed to cool, and this is an extra expense. A 3½ ton ingot is presented in figure 4–2, the big end up. The design promotes sound metal in view of the top portion which is to be cropped prior to rolling.

Temperatures just below the melting point of the metal allow the metal to be plastically deformed without rupture. This is the purpose of maintaining high temperatures. Once the process of shaping is decided, the hot metal is quickly delivered to the shaping area and powerful machines work the metal, much like the kneading of bread dough. (Figure 4–3 shows an ingot in the rolling mill where work is continued until the new shape is completed.) Basically, the hot ingot or section of ingot is fed between two large rolls for rolling purposes, placed beneath the hammer in a forge machine, or pushed in front of a ram in an extrusion mill. The end product is quite unlike the original ingot. A roll of wire over a mile long, for example, is produced from a single mass of metal.

MOST METALS CAN BE FORMED WITH PRESSURE

Some metals are inherently soft while others are tough and hard. When alloys are produced by combining two or more elements, the resultant alloy is often harder than the materials from which it is made. Heating softens most, but not all, metals. As an example, alloys of chromium, molybdenum, tungsten, and cobalt are stiff even when white-hot. Extremely high forming pressures are required to shape some of these metals. When pressure forming cannot be accomplished, the part is cast or pressed

FIGURE 4–2 A white-hot ingot being removed from the soaking pit in preparation for rolling in the mills. (Courtesy of The Timken Co.)

into shape from powdered metals. Tungsten, for example, is *swaged* or hot drawn into shape.

Even though most metals can be formed hot or cold, some are formable only while very hot. The introduction of certain elements in the base metal frequently permits only hot forming. Body-centered cubic structures in metals, such as in most carbon and alloy steels, permit hot or cold deformation. Face-centered cubic structures such as austenitic steels, deform easily, but harden rapidly when cold worked. Annealing between forming operations is therefore frequently required. On the other hand, a hexagonal close-packed metal such as magnesium requires heating in order to deform. to prevent rupturing of the metal. Magnesium will deform at 400° F., for example, but it will split if bent cold.

Purpose of Wrought Processes

Obviously, the major purpose of pressure forming is to shape the metal according to its need. During shaping, secondary objectives are also obtained, including the-refinement of grain structure and the improvement of mechanical properties. In many instances, improvement of mechanical properties is the primary objective. Because shaping is accomplished either hot or cold and because many processes are performed

FIGURE 4–3 Hot steel ingot entering the two-high rolling mills. (Courtesy of Lone Star Steel)

by hot and cold shaping, some means must be provided to differentiate between the two processes. A basic rule, therefore, states that plastic deformation occurs while hot working with no increase in hardness, while plastic deformation occurs during cold working with an increase in hardness. Hot working precedes cold working, and work begins with the ingot. As shaping is subsequently performed, either the dendritic or the equiaxed grains are destroyed and new grains are created. During hot working, the hot and plastic metal flows in the direction of pressure, its grains tumbling over one another and deforming as the boundaries collapse and re-form. The tumbling or deforming of the grain structure occurs through multicompression loading during rolling, whereby layers of metal are squeezed together and slide along organized atomic planes.

Many forming operations, however, apply tensile forces and compression forces to the metal such as in cold bending (fig. 4–4). When a section of metal is bent cold, compression stresses are formed on the inside of the bend while tensile stresses form on the outside. Stress moves inward from the surface, therefore, a *neutral plane* exists where compression and tension stresses are zero. Because cold bending increases hardness, this factor must be considered when severe deformation is to be performed. Sliding actions of the deforming metal eventually destroy original grain boundaries, therefore, the grains are broken up and, in turn, finer grains result which help increase the various strengths of the metal.

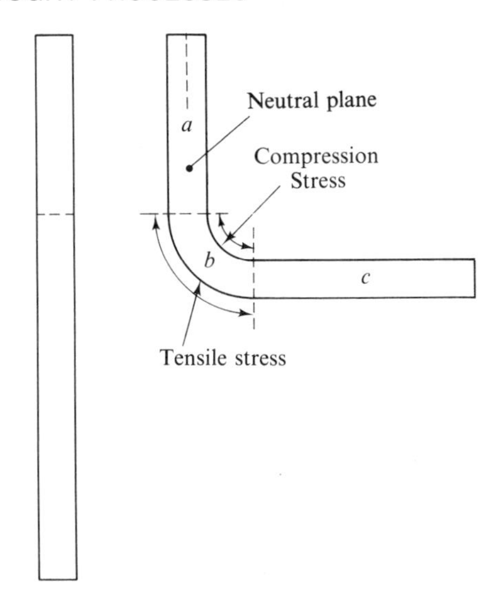

FIGURE 4–4 Bending of a metal causes tensile and compression stresses when bent cold and below the point of recrystallization.

As the grain structure is broken up by working processes, new patterns are established which subsequently provide direction in the grain pattern (fig. 4–5). Figure 4–5*a* illustrates the hot rolling of a steel ingot and figure 4–5*b* the cold rolling of a section of hot-rolled steel. Cast steel contains the original coarse grains of metal, either equiaxed or dendritic. When run between the rolls, the coarse grains are squeezed and broken up while, at the same time, recrystallization occurs because the

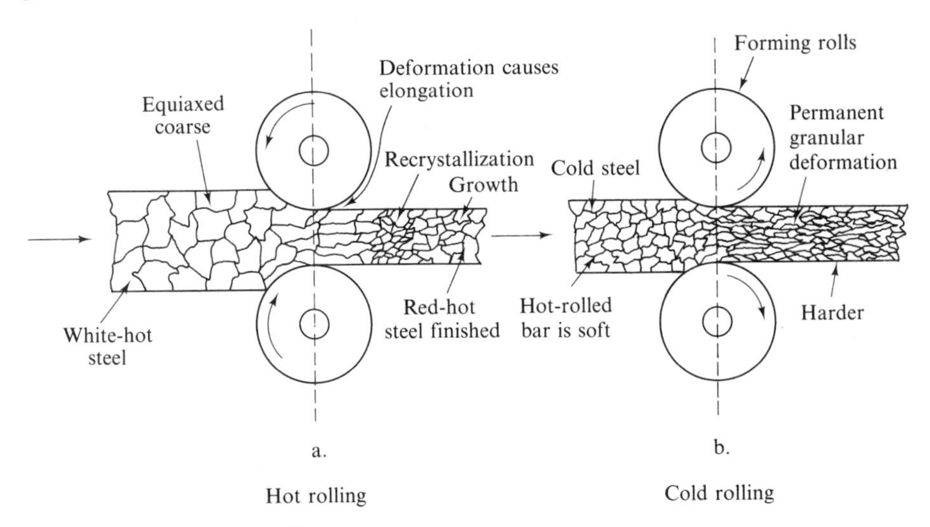

FIGURE 4–5 Working causes new grains to form.

temperature is above the metal's *critical*, or recrystallization, point. Also, due to the temperature being above the point of recrystallization, the newly formed grains grow in size until they conform to the size compatible with the rolling temperature. Hot rolling ceases before cold rolling commences, a main reason being to prevent forcing iron oxide scale intc the surface of the metal.

As pointed out in figure 4–5*a*, metallic grains are squeezed and elongated, recrystallization occurs, and then grain growth begins and eventually stabilizes at a certain size. Hot rolling refines the granular structure and eliminates the patterns imparted by casting. The main difference between hot and cold rolling is that the hot process is conducted above the metal's recrystallization temperature and cold rolling is performed at temperatures below the metal's recrystallization temperature.

Figure 4–5*b* illustrates the cold rolling of a hot-rolled plate. As the equiaxed grains of metal flow between the rolls, plastic deformation again occurs, but the change in dimension is permanent. Therefore, as elongation occurs, elongated grains remain and hardness increases. Cold rolling reduces the cross sectional area of material, as does hot rolling, but cold reduction induces strain. In turn, both hardness and increased strengths result from cold deformation. Mechanical properties of the metal are, therefore, directly improved.

Another purpose of pressure forming is to close blow holes in the ingot while hot working. If oxide has not formed on the surfaces of the hole, hot working often welds the void (fig. 4–6). However, if oxygen has been contacted by the hot metal at the hole, a dangerous life-long flaw is established. An oxide inclusion will not allow the metal to be welded. Rolling, for example, merely elongates the flaw and an inherent discontinuity is established. Holes in ingots occur during casting as metal solidifies at the moment the gas attempts escape. Because shear and compression stresses exist internally during rolling, virgin metal will weld itself at hot rolling temperatures, stresses then being immediately reduced.

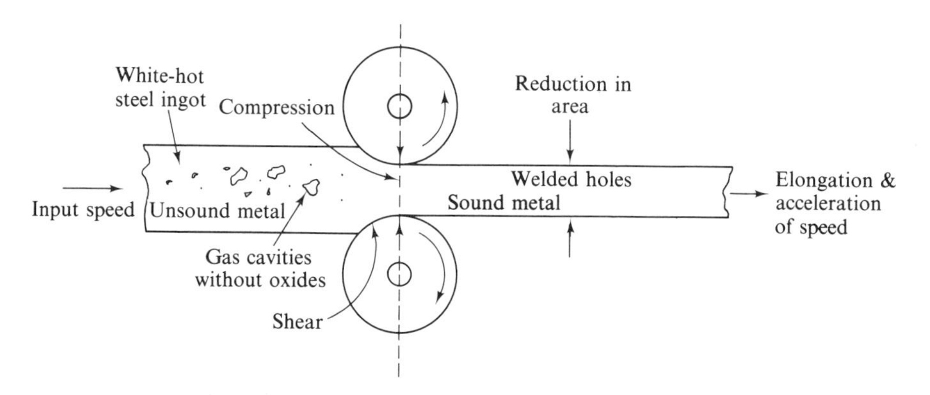

FIGURE 4–6 Holes in ingot closed by hot rolling.

Plasticity Allows Deformation

When metals are worked at temperatures above their recrystallization temperatures, stresses are removed because the deformed grains recrystallize immediately following deformation (fig. 4–7). As an example, as illustrated in figure 4–7, a 0.7% carbon steel is heated to its critical temperature, slightly above 1333° F. Between room temperature and the critical point, a given grain size remains constant, but at the critical temperature, grain boundaries collapse and new grains emerge, being at their finest

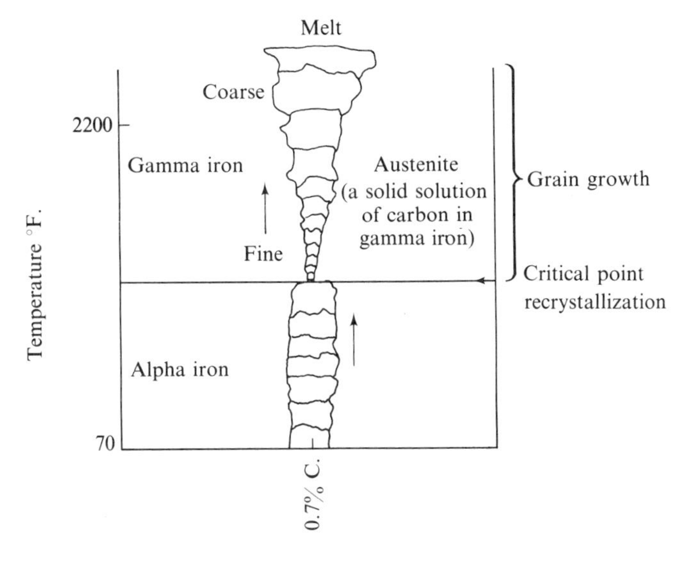

FIGURE 4–7 Recrystallization in steel upon heating.

size. As temperature increases above the critical, all grains grow in size by absorbing other grains. Growth parallels temperature in austenite, a grain size being associated with a temperature. This grain growth continues as temperature increases until the metal melts and all grain boundaries collapse.

An application of this grain growth process occurs during the hot rolling or other working processes in steel. As illustrated in figure 4–8, the hot and plastic metal is

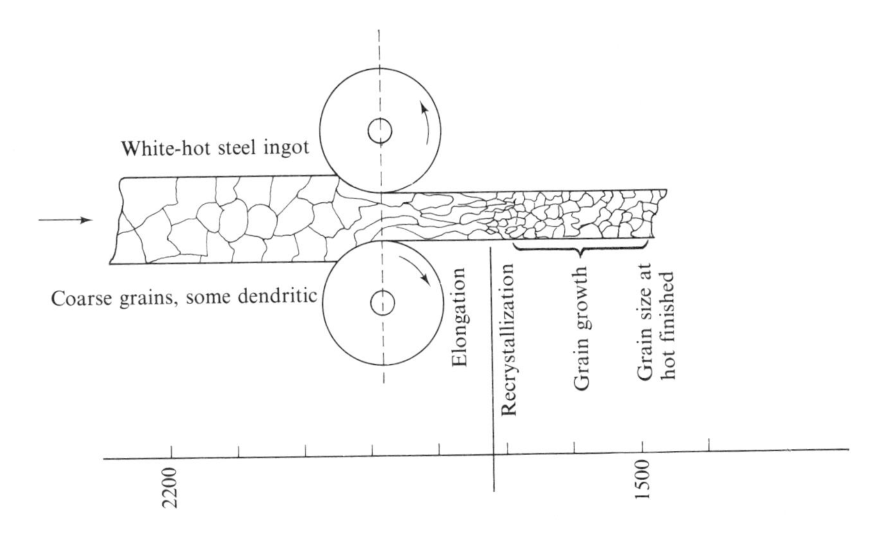

FIGURE 4–8 Grain refinement by hot rolling.

squeezed as the turning rolls force the metal inward and outward. Because a cross sectional reduction in area accompanies an elongation, grain boundaries are destroyed. Also, because metals are crystalline, new grains form from the elongated grains and, in turn, the newly formed grains grow into larger equiaxed grains, but remain smaller than in the original cast structure. The hot finished section of metal is a great improvement over the original casting.

DEFORMATION ABOVE THE RECRYSTALLIZATION TEMPERATURE

As hot working progresses, no energy buildup in the atomic pattern is established as no increased hardness or strength occurs. Consequently, the energy expended in causing plastic deformation during hot working is not an increasing effort, but only a continuous effort at a given energy output. Pressures on the metal are above the *yield*, the stress in the metal which causes permanent deformation, but must remain below the tensile so as not to rupture the metal. The *tensile strength* of a metal is its highest strength in tension before breaking. In this respect, the stress-strain relationship for metals being worked above recrystallization temperatures is a straight line from the yield. The yield strength of a hot metal is lower than that of the same metal at room temperature, consequently, less force is needed in hot working processes than would be needed at or near room temperature. This relationship is illustrated in figure 4–9. Work hardening or strain hardening does not occur in hot working processes.

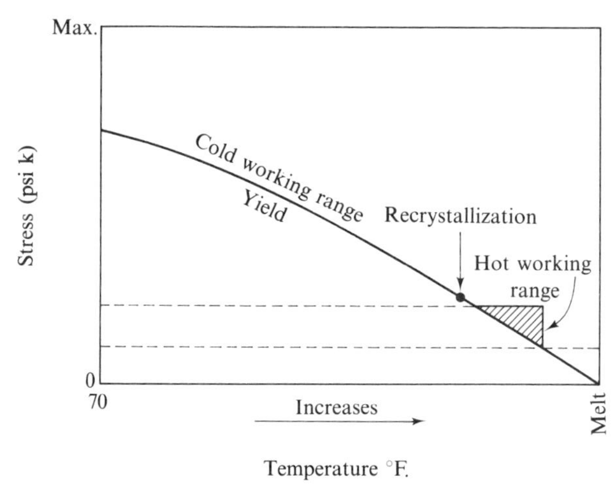

FIGURE 4–9 Yield strengths of a structural steel rod vs. temperature.

DEFORMATION BELOW THE RECRYSTALLIZATION TEMPERATURE

On the other hand, should deformation continue at temperatures below the metal's recrystallization temperature (fig. 4–10*a*), strain hardening occurs as the temperature drops below the recrystallization point. Figure 4–10*b* illustrates what happens when cold steel is loaded in tension to its fracture point. In this respect, compression and tension forces have similar reactions. The stress-strain curve moves upward as strain increases beyond the yield due to increased tensile force, and grains remain

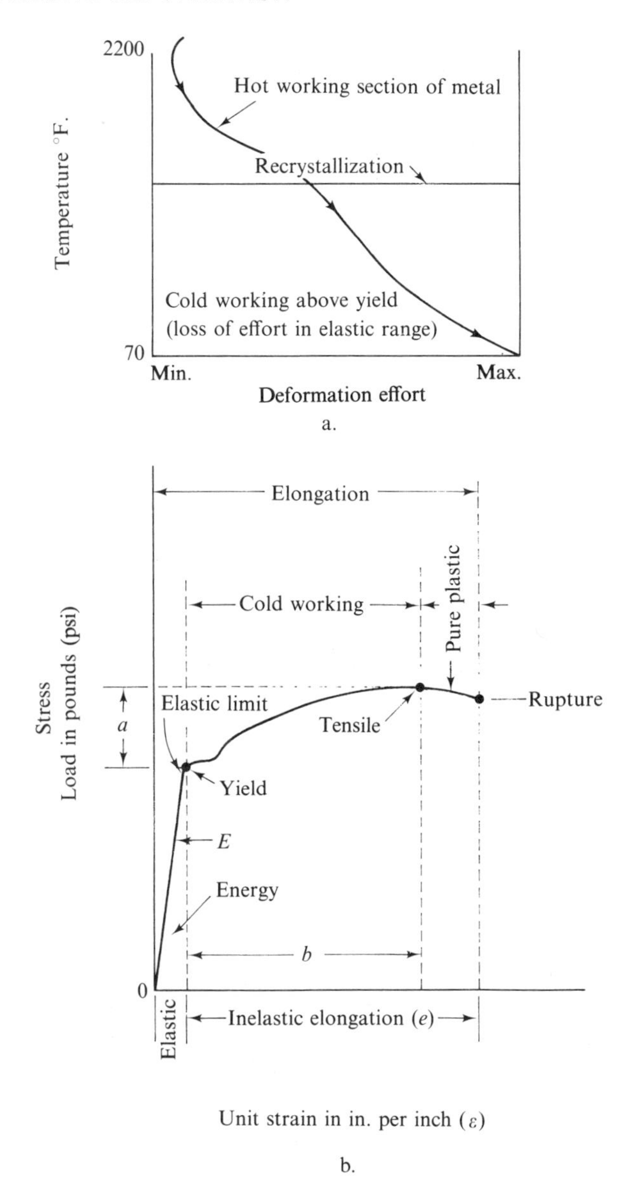

FIGURE 4–10 Deformation above yield increases when temperature is below recrystallization.

permanently deformed. Strain hardening is work hardening, the greater the deformation, the greater the hardness. An increased energy input is then required for further deformation because hardness is induced as the result of deformation. In other words, a given pressure on the metal beyond the yield induces a certain hardness and strength in the metal as deformation occurs. In order to further deform the metal, an increase in energy is required to overcome the metal's resistance to being further deformed. As energy of the machine is increased, deformation increases in the metal.

Resistance of the metal to deformation consequently increases again as each volume is deformed beyond its preceding deformation, thereby requiring an ever increasing energy input for increased deformation. Because plastic deformation is limited in metals, rupturing of the metal occurs when plasticity ceases. Therefore, metals can be work hardened by cold working up to a certain limit while hot working can be continued until the objective is obtained.

STRESS–STRAIN RELATIONSHIPS

A further observation of figure 4–10*b* illustrates the stress-strain relationship. As the load on the cold metal increases from zero, elongation increases proportionately to approximately the yield. But just below the yield is the *elastic limit*, therefore, elastic elongation occurs up to the elastic limit where increasing loads cause excessive elongations and permanent stretch occurs. In other words, at any load below the elastic limit, the metal will spring back to its original dimension when the load is removed. Beyond the elastic limit, or basically at the yield, permanent elongation occurs as the result of plastic deformation rather than elastic. Beyond the yield, a ductile metal stretches and is permanently strained until its tensile strength is reached. The increase in load required for continued elongation is due to work hardening because hardness and strength have relatively close relationships. Some very ductile metals, after reaching the *tensile*, the greatest load held during the pull before breaking, will rapidly elongate causing a reduction in load all the way to fracture. This is the pure plastic stage, as indicated by necking on the tensile specimen. Tensile specimens are especially machined so that the sample of material can be tested by pulling loads. As the yield is exceeded by a considerable amount of stress, some metals such as aluminum will suddenly increase in elongation, whereby the material decreases tremendously in cross-sectional area, giving a bottleneck effect as it breaks.

As pointed out in figure 4–10*b*, the equal ratio between stress (load) and strain (elongation) up to the yield is a straight line known as the *modulus of elasticity* ($E = \frac{S}{\varepsilon}$). However, beyond the yield, there is an unequal ratio between stress and strain, as pointed out by the load increase at *a* and the inelastic stretch at *b*. Further, the triangular area below the yield is an energy region in the metal, much like a powerful spring that has been stretched and awaits a sudden collapse with its release of energy.

HOT AND COLD WORKING

When metals are worked and plastically deformed at temperatures above their recrystallization temperatures, grain refinement occurs continually throughout the process of recrystallization. This method of plastically deforming metals is known as *hot working*; the recrystallization temperature varies as the metal varies, steel being in the region between approximately 1300° and 1500° F. When steels are hot worked, a heavy iron oxide scale forms on the surfaces, but it is loose and easily shatters. This scale, however, must not be rolled into the plastic metal. Aluminum and lower melting metals recrystallize at much lower temperatures and do not produce the heavy scale. On the other hand, some metals recrystallize at room temperature, therefore, they cannot be work hardened.

When metals are loaded beyond their yield strengths and worked at temperatures below their recrystallization temperatures, permanent deformation of grains occurs and hardness increases proportionally to the amount of deformation performed at that temperature (fig. 4–5*b*). This method of plastic deformation is known as *cold working*. Steel is sometimes cold worked at temperatures above room temperature, but below approximately 1200° F. Figure 4–11 illustrates the results of deforming a steel slug at room temperature and at a temperature in the hot working range. From part *a* to part *b*, the original hardness value of RB 69 is increased to RB 91 as the area changed from *b* to *b′* when cold compression occurred. This was accompanied by a decrease in length from *a* to *a′*. In part *c*, hot compressioin caused no increase in hardness and, consequently, no increase in stength. Actually, four points of hardness were removed from the original due to strain reduction.

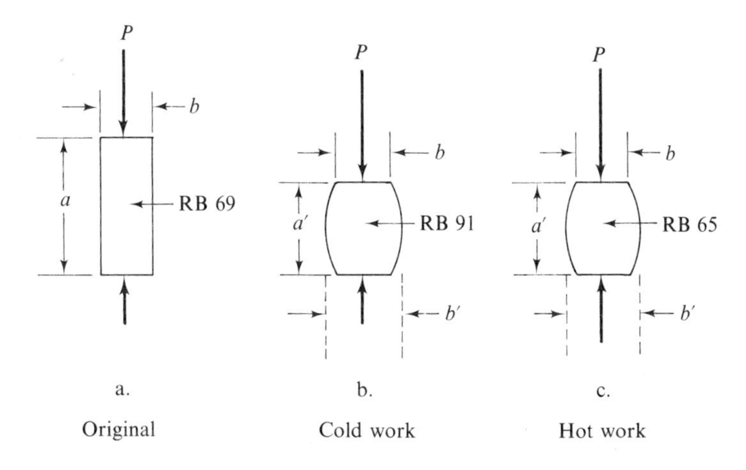

FIGURE 4–11 Hardness vs. cold and hot work.

The Three Main Engineering Forces

Structural materials, when fabricated into structures and subsequently loaded, are frequently exposed to several types of stresses, sometimes singularly and sometimes under multistress conditions. Stresses in the metal occur as the result of loads or forces (fig. 4–12.) Even though dynamic loads are sometimes part of the stressed structure, static forces play a more common role due to their constant presence in any loaded structure. Whether moving or stationary, a load causes reacting stresses in materials. When pulling loads occur (fig. 4–12*a*), the structure or part is in *tension,* and is accompanied with an internal load, or stress, reacting in an opposite direction to the load axis. When pushing loads exist (fig. 4–12*b*), the part is in *compression*, and when loads having cutting tendencies exist (fig. 4–12*c*), *shear* loading results. Tension, compression, and shear are, therefore, common stresses existing concurrently in many fabricated assemblies (fig. 4–12*d*).

ELASTICITY AND STRESS

Just as a bridge column is in compression, a steel ingot is also in compression when squeezed between two forging dies. The main difference in this example is that the cold steel column is never to be compressed beyond its yield strength, while the hot ingot must be in order for deformation and shaping to occur. Elastic deformation always occurs in the cold column of the bridge when loaded, while plastic deformation

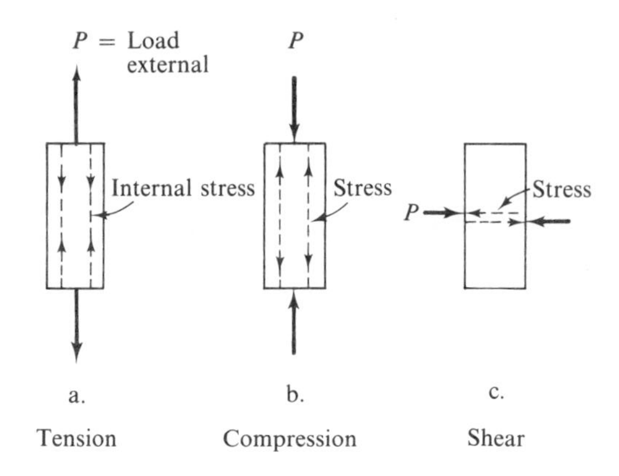

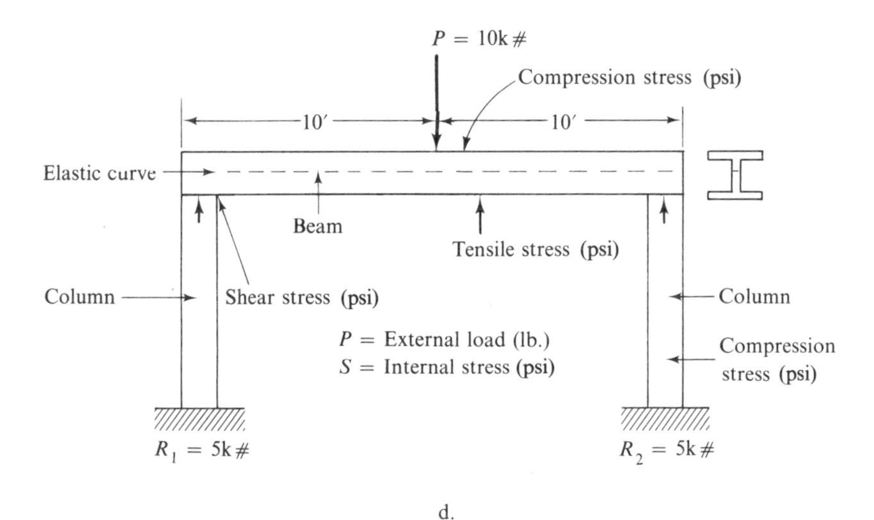

FIGURE 4–12 Loads cause stresses in a metal.

occurs in the hot ingot. In other words, as a cold metal is compressed or stretched, the load causes dimensional changes in the metal. The shape changes continually as the load increases (fig. 4–13). If the load is removed at a point below the elastic limit of the metal or, basically, below its yield strength, the metal will spring back to its original shape, as shown at dimensions *a* and *b*. If the load on the metal is continued beyond the metal's yield, permanent deformation occurs in dimensions *a* and *b*, and the metal will not return to its original shape. Part of this shape change is caused by *springback* and is illustrated in figure 4–14. The region *ab* is in the elastic range, but when the metal is stressed from *a* to *d*, both elastic and plastic deformation occur. When the stress is removed, the metal will spring back to *c* due to part of the-deformation being elastic. In cold working operations, springback must be considered, that is, deformation must be continued past the desired amount to allow for the reduced dimension. Should the metal be stressed to point *e*, fracture occurs. Consequently, calculations must assure that working is conducted from point *b* to a

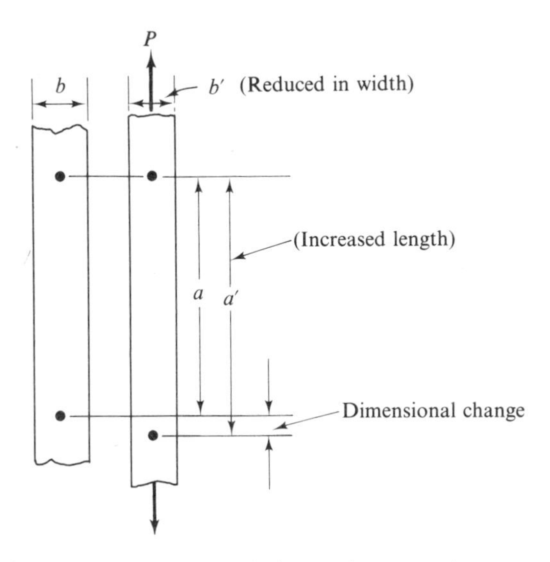

FIGURE 4–13 Load vs. dimensional change in a metal.

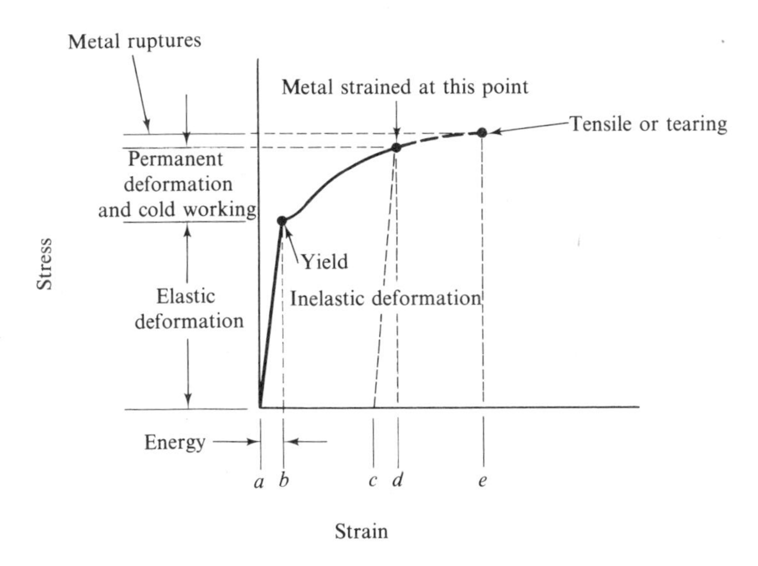

FIGURE 4–14 Springback in a metal.

point below *e*. A further explanation of these factors is illustrated in figure 4–15.

The *stress diagram* (fig. 4–15) is an illustrated comparison between stress and strain conditions inside a metal. An external force causes the internal conditions. When the annealed steel is placed in tension, strain increases proportionally to stress up to the yield. Below the yield, the metal functions in its modulus of elasticity capacity, about 30,000,000 psi. Beyond the yield, plastic instead of elastic deformation occurs, causing a nine point increase in hardness as the highest load, the tensile strength, was reached. Only because of a high ductility factor, did the RB 89

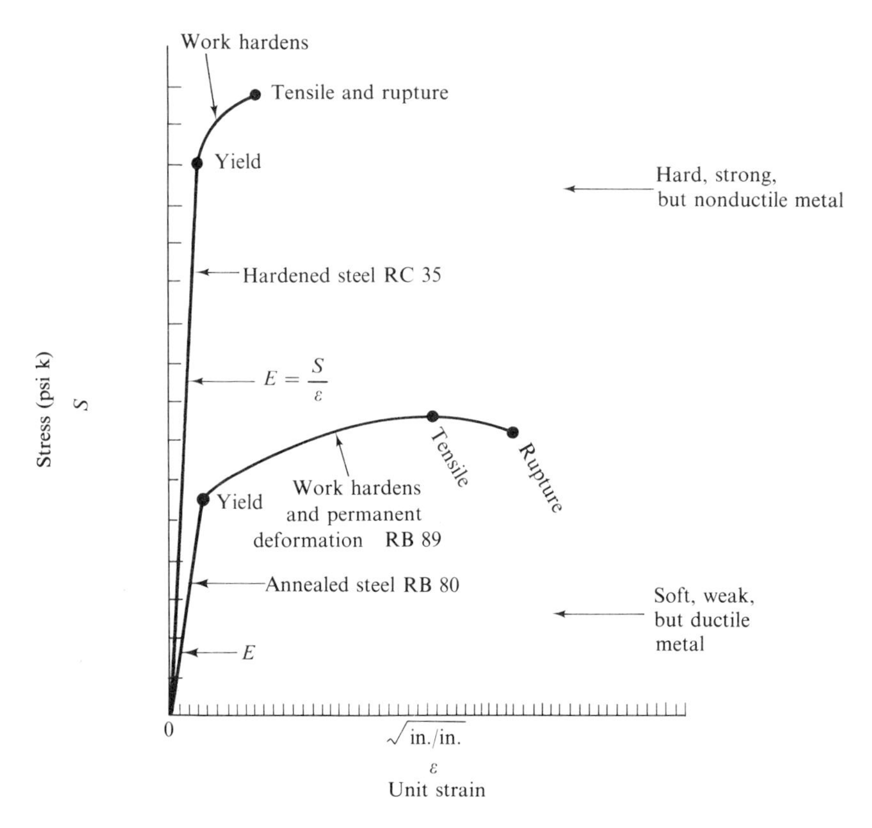

FIGURE 4-15 A stress-strain diagram for two steels.

specimen hold itself together during the long extension before rupture, but with a decreasing load factor beyond tensile, accompanied with severe necking. The other partially hardened specimen, RC 35, withstood more than twice the stress before it yielded. However, its percentages of elongation and reduction in area were severely curtailed. To gain strength, the designer and technician must expect an increase in hardness along with a decrease in ductility. Figure 4-15 was plotted while specimens were cold worked to destruction at room temperature. Ductility factors are determined by these formulas:

% reduction in area =
original area minus final area divided by original area multiplied by 100

% elongation = final length minus original length
divided by original length multiplied by 100

PLASTICITY AND TEMPERATURE

If the above tests had been conducted at 1800° F, the loads required to permanently deform the specimens would have been greatly reduced. Also, no work hardness or strains would have occurred. As a result, the hot working of metals is conducted beyond the metal's recrystallization temperature and the yield strength of the metal is lowered. In respect to varying temperatures, a metal's yield strength is lowered in value as temperature of the metal increases. As an example, if it requires a load of 70,000 psi of cross-sectional area to permanently deform a metal while cold (room temperature), then that same metal will plastically and permanently deform under a load of only a fraction of 70,000 psi at its recrystallization temperature. These principles are illustrated in figure 4–16. The load required to permanently deform a metal drops appreciably as temperatures approach the yield strength of the metal. Beyond the yield, the load drops drastically while the yield-elongation factors become a straight line.

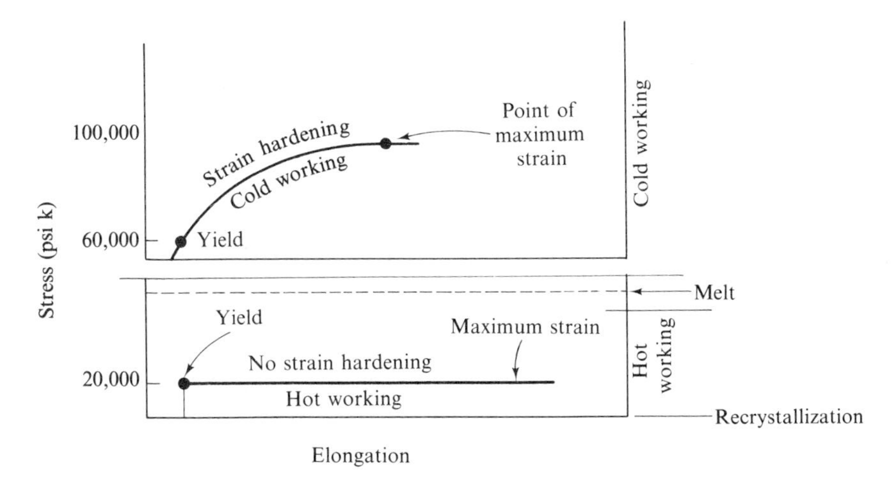

FIGURE 4–16 Working within a metal's recrystallization range induces no strain hardening even though the yield strength is reduced.

Rolling Operations

Rolling is one of the several wrought processes and is applicable to all metals which will plastically deform. Steel and aluminum roll easily, but cast gray iron will not roll hot or cold. Because most metals will deform either by hot or cold rolling forces, this process provides more shaped metals than any other.

Rolls are shaped according to the job to be done and the desired shape of the stock. Therefore, many differently shaped rolls are used in the rolling mills. Some of the standard shapes of rolls are presented in figure 4–17. When hot metal passes through round rolls, the metal deforms to the cross-sectional shape of the space inside the two mating rolls, figure 4–18. Flat stock such as blooms, slabs, billets, plates, and sheets are run through round rolls (fig. 4–19). Because metal elongates as its cross section is reduced, round rolls are also provided along the sides of the stock in order to maintain desired widths or to squeeze the stock to smaller dimensions. Side rolls are at 90° to the main rolls.

Rolls are made of extra strong metals so that spring is reduced to a minimum. When two rolls are in the process of reducing a large ingot, tremendous forces are used. Therefore, there is the tendency for the rolls to spring away from the material at the midpoint of the rolls (fig. 4–20). Also, the force on the metal being reduced reacts into the rolls and most of its supports. The end effect is elastic deformation within the entire mill, meaning an elastic movement away from the metal being worked. In order to overcome this situation, rolls and the entire mill are fabricated from huge masses of metal which are properly heat treated. As reactions caused from rolling occur, the springback within a roll's modulus of elasticity will be minimized.

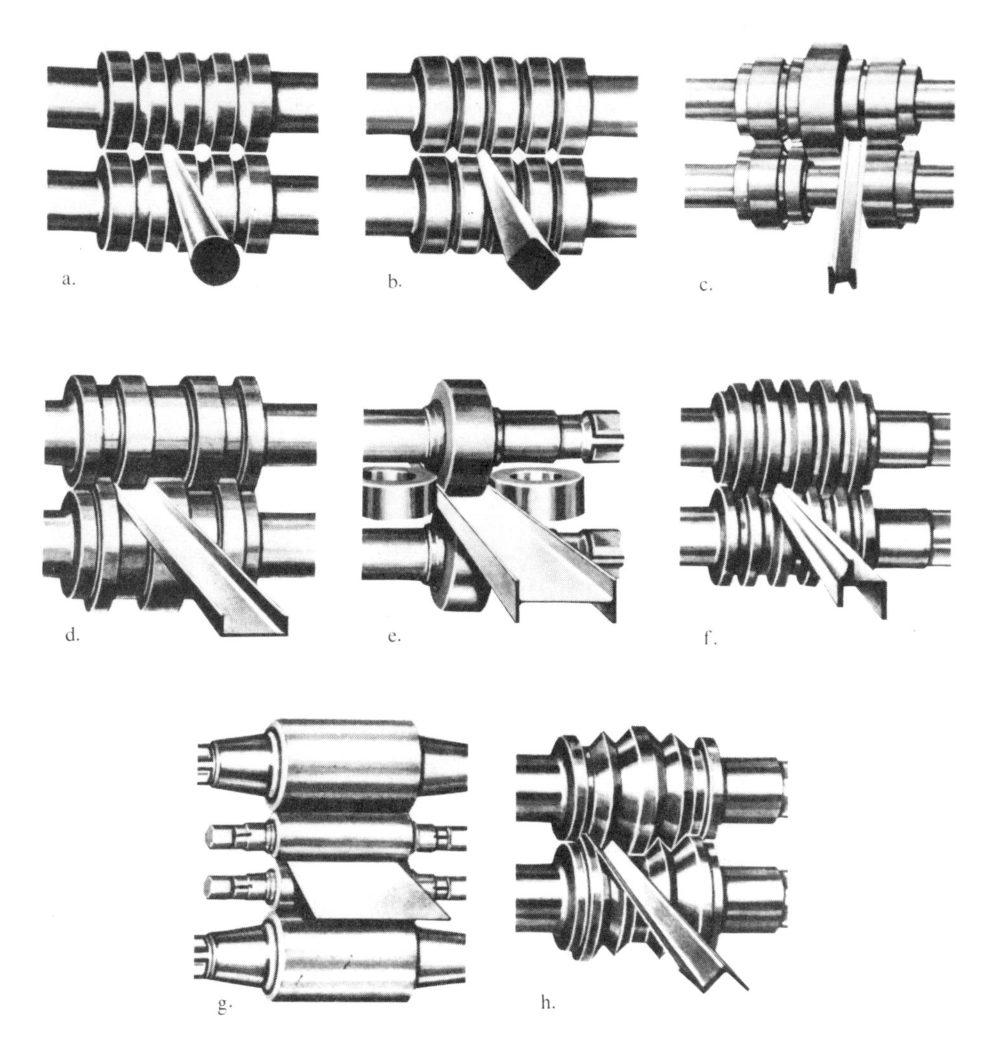

FIGURE 4–17 Standard shapes of rolls which produce *a*, rounds; *b*, squares; *c*, rails; *d*, channels; *e*, H-columns; *f*, special sections; *g*, sheets; *h*, zeebars. (Courtesy of American Iron & Steel Institute)

ROLLS

Rolls are driven by powerful electric motors. When more efficiency is needed from the rolling operation, a three-high mill is used (fig. 4–21). The center roll provides friction for each outer roll as metal is drawn in from one side of the lower rolls while a later pass causes it to be drawn in from the opposite side of the upper rolls. Such a procedure lends itself to automation when the moving ingot, bloom, billet, slab, or other shape is turned back into the receiving mill. Some mills are frequently arranged in a four-stand production system (fig. 4–22), while others are arranged in a manner to support the production operation. Larger diameter rolls allow greater pressures on

FIGURE 4–18 The rolling of a metal plate above the metal's recrystallization temperature produces hot worked metal. (Courtesy of Stellite Division–Cabot Corp.)

the ingot as spring is reduced. However, many mills use small diameter rolls to better knead the metal while larger backup rolls press inwards against the smaller rolls to give the inner roll proper stability. Special rolls such as the planetary system, or cluster rolls (fig. 4–23), increase rolling efficiency.

Form Rolls

Form rolls are used in producing finished products from the mill. These are arranged so that the cavity between the mating rolls is such that the resulting product will have the shape of the cavity (fig. 4–17). Several passes of the stock between the rolls are necessary while reducing the mass of metal to its finished dimensions. Tearing of the metal occurs if too high a load is applied in attempts to speed up deformation. The more complex the shape, the more passes are needed for shaping. Metal will only flow so far during one deformation. This distance depends on a combination of factors such as volume shortage and overage compromises during flow which occur when multistress conditions exist. An *H*-shaped column, for example, requires careful reduction practices by means of several passes so that rupturing of the metal will not occur. As proper lengths of the finished product pass from the mill, flying shears, or saws, part the length at the proper dimension and quickly move into position to repeat

FIGURE 4–19 Turning rolls having parallel surfaces with the metal being rolled produce flat sections of metal in a cold worked condition as forming is conducted below the metal's recrystallization temperature. (Courtesy of Stellite Division–Cabot Corp.)

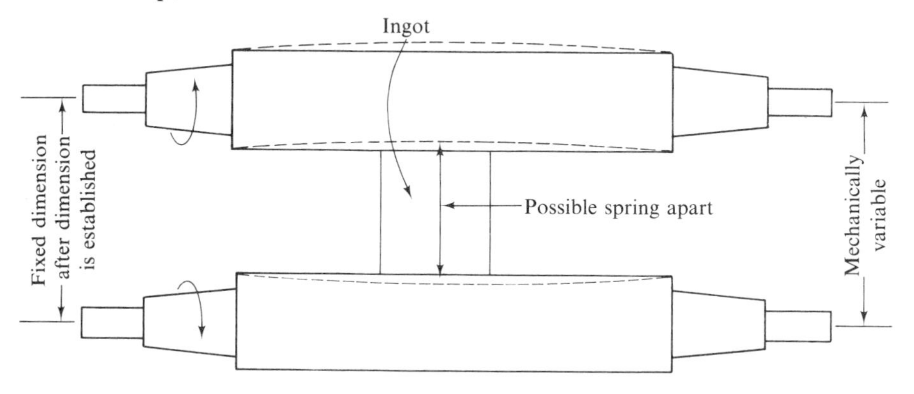

FIGURE 4–20 The springing of a roll during rolling.

the process. Standard lengths are 12, 16, and 20 feet. Cut lengths then move along a conveyor system for further processing and subsequent storage.

When plate and sheet stock are required, slabs are reduced to the desired thickness and width. As thickness is reduced and width is fixed, only the length changes. When the stock cools to lower temperatures (for example, 1400° F. for steel), it is reheated

FIGURE 4–21 A three-high rolling mill. (Courtesy of The Timken Co.)

for continued hot rolling processing. When bars or formed sections such as railroad rails are needed, blooms are used. Smaller sections are formed from billets. Figure 4–24 illustrates the reduction of a slab to plate whereby the thickness is reduced, width is maintained, and length is greatly increased.

Angular Rolls

Seamless tubing is produced by hot piercing and by extrusion. The extrusion process is discussed in a later part of this chapter. The piercing process involves the forcing of a hot metal rod over a pointed mandrel while angular type rolls maintain an even wall thickness. Several stages of rolling are subsequently required before the finished tube is produced. Figure 4–25 illustrates the piercing and rolling procedure for producing seamless tubing of varying wall thicknesses and outside diameters.

HOT ROLLING

Hot rolling is performed on metals while they are within their recrystallization temperatures. Because hot metal offers less resistance to deformation than cold metal,

FIGURE 4–22 This 22-inch mill is a four-stand cross-country mill producing rounds to 8 inches in diameter and equivalent size squares. (Courtesy of The Timken Co.)

hot rolling is normally commenced slightly below the melting point of the metal and continues to a point well above the recrystallization temperature. In carbon steel, ingots are released into the rolling mill at approximately 2200° F. Immediately, the controlling mechanisms guide the moving ingot into scale breakers, and powerful water jets wash away the loose scale. Figure 4–26 shows an ingot as it enters the rolls. The moving ingot is drawn into the turning rolls by frictional forces because the two rolls, turning in opposite directions, allow only an exact thickness of metal to pass through them. Some rolls are grooved, the grooves being 90° to the direction of rolling. Therefore, biaxial compression forces occur on the metal which reduces the ingot's cross-sectional area to the diameter between the contacting points of the rolls. The load on the metal depends on the kind of metal, its temperature and cross-sectional area, and rolling factors.

Because the rolls are turning at a given speed, metal leaving the mill moves at a faster speed than that entering since the given volume of metal elongates and moves away from the rolls as its cross-sectional area is decreased by the squeezing action of the rolls. Consequently, the incoming ingot's speed is slower than the portion of the ingot leaving the mill. There is only one point on the ingot's surface, therefore, that is moving at the same speed as the rolls. It is at this speed that no slip occurs on the

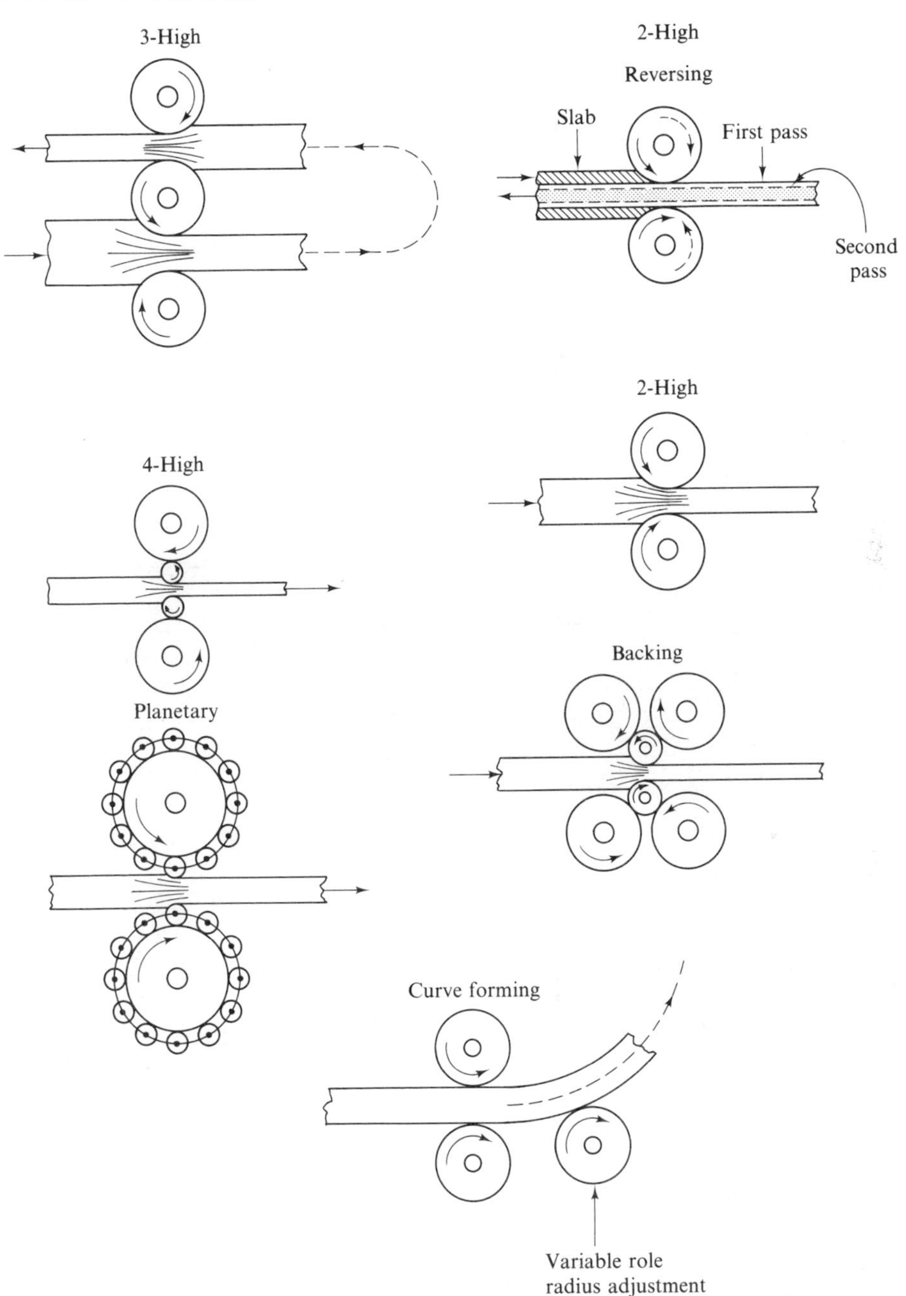

FIGURE 4–23 Typical forming rolls.

metal's surface, that is, the metal is moving at that point at exactly the same speed as the periphery of the rolls.

Metal entering the rolls is grabbed and pulled through the measured space by friction (fig. 4–27). Metal leaving the rolls offers resistance as it accelerates away from the

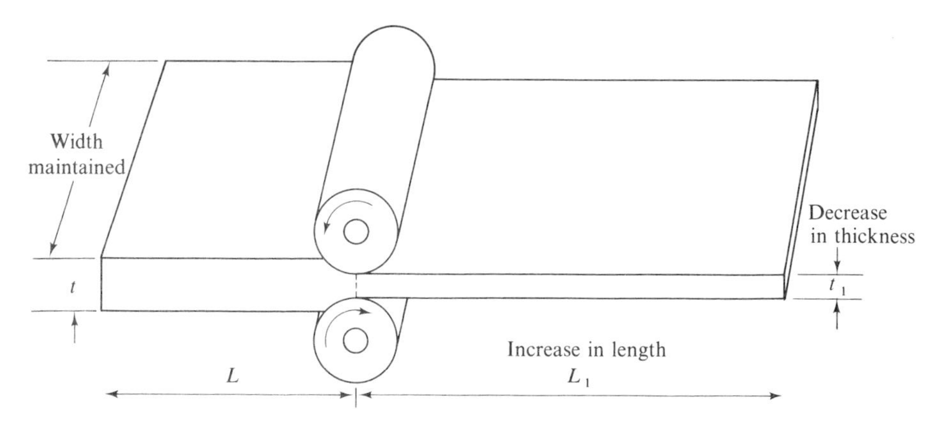

FIGURE 4–24 Slab reduction to plate.

FIGURE 4–25 Production of seamless tubing in a piercing mill in which the hot billet is forced over the plug. (Courtesy of The Timken Co.)

FIGURE 4–26 A steel ingot entering the rolls. Rollers move the scale coated and white-hot ingot in contact with the forming rolls whereby the scale is broken loose and the ingot is reduced in thickness, but increases in width and length. (Courtesy of Lone Star Steel)

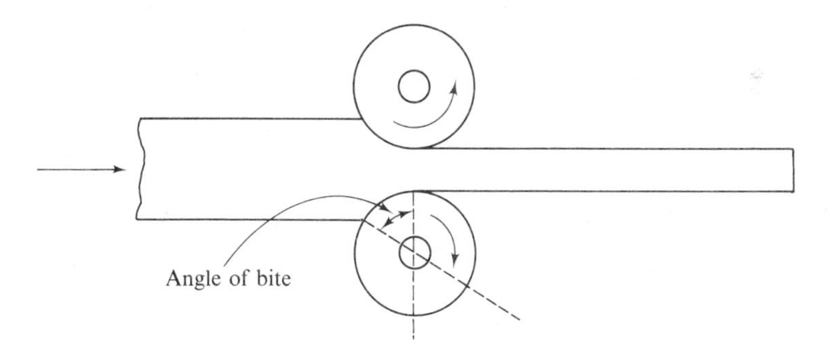

FIGURE 4–27 Angle of bite during the rolling of metal.

mill. It is the reduction in cross-sectional area of the metal that is controlled by the angle of bite. This angle must be great enough to effectively deform the metal; if the angle is too great, metal separation often occurs as it leaves the mill. Consequently, mill operators work the metal within reasonable tolerances.

Products of Hot Rolling

Hot rolling processes produce blooms, billets, and slabs. The *bloom*, being reduced in size from the ingot, has a much smaller cross-sectional area than the ingot, but is greater than six inches thick. Changes in direction of rotation of the rolls causes the ingot to move back and forth between the rolls as the space becomes smaller during each pass. This procedure is accomplished in a two-high reversing mill (fig. 4–23). Width and straightness of the hot mass of metal are governed by hydraulically operated manipulators which move inwards on command to maintain the desired shape (fig. 4–28). Conveyor rolls in the bed of many mills move the section of metal in either direction while the bed is moved upwards and downwards. Some mills are manually operated, but many are automated. In this respect, as the hot bloom cools, it shrinks and its dimensions are somewhat smaller in length and width than at a given set of dimensions while hot. Shrinkage, therefore, must be considered while rolling. When the bloom is further rolled into a *billet*, its cross section is obviously reduced. The billet may be square or rectangular, just as the bloom. The *slab*, however, is a flat reduction of the bloom or billet. Processing of the hot ingot then provides the raw materials for production of standard shapes (fig. 4–17). Rolls shown in figure 4–17*a* produce round rods. Further processing of long lengths of these round rods is accomplished by sawing operations, illustrated in figure 4–29.

FIGURE 4–28 An ingot being guided by manipulators as it moves into the rolls. (Courtesy of Lone Star Steel)

FIGURE 4–29 Round rods produced by the rolling mills are sawed to desired lengths. (Courtesy of Simonds Saw and Steel)

COLD ROLLING

Cold rolling is the method used to shape a metal when the metal's temperature is below its recrystallization temperature. If a metal can be plastically deformed below its recrystallization temperature without tearing, then it can be cold rolled. Incidentally, hot rolling always precedes cold rolling. As hot steel, for example, approaches the lower zone of its recrystallization temperature range, rolling is usually stopped in order that surface scale can be removed before cold rolling is performed. Steel stock is then pickled in sulfuric acid and neutralized with water and lime. Lime hardens on the metal's surface and serves as a rolling lubricant. The cleaned metal is then oven baked to remove hydrogen which was picked up during pickling. After baking, the metal is rolled into desired shapes. Greater forces are required in cold rolling than in hot rolling, the forces from the rolls being functions of metal analysis, cross-sectional area of the metal in contact with the rolls, and rolling temperature. Because cold working induces hardness, the metal may require annealing before specifications are established. Annealing induces softness and, in this respect, defeats some objectives of cold working. However, the final two passes through the rolls often provide the desired characteristics of cold rolling, including hardness.

Because hot rolling induces no hardness in the metal, many metals are finished by cold forming in order to improve mechanical properties. Reductions in cross-sectional areas are much smaller in the cold forming processes than in the hot. As cold deformation increases over previous deformation, hardness and strength in the metal

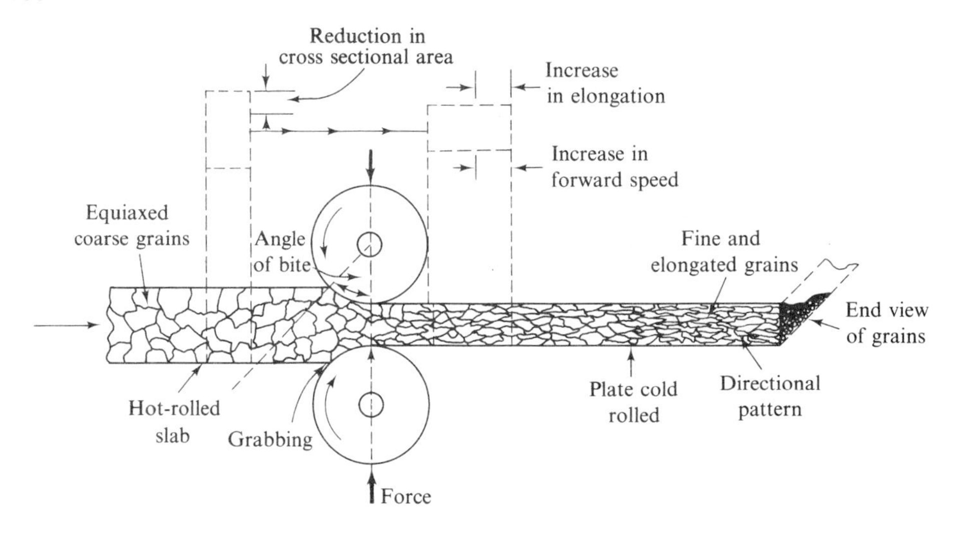

FIGURE 4–30 Grain structure produced in hot and cold rolling operations.

increase rapidly. Grain structure develops in a pattern and directional strengths are more pronounced than in hot working. Figure 4–30 illustrates the differences between grain structure in hot- and cold-rolled steel. As illustrated, the coarse and equiaxed grain structure of the previously hot rolled slab moves into the cold rolling mill and is grabbed by the rolls' angle of bite. The angle is measured from the point of desired diameter of metal along the roll's periphery to the point where the roll's periphery and slab's top surface meet. Too small an angle produces a waste of energy while too great an angle is damaging to equipment and stock. As the force between the rolls compresses the metal, its grains become elongated in the direction of rolling and remain elongated because the force is greater than the metal's yield, but below its tensile. Because a given volume of metal is reduced in cross-sectional area as it elongates, grains are compressed in several directions, bringing opposite and horizontal grain boundaries in close proximity. Much overlapping of grains occurs and, in turn, hardness and strength are increased, not by grain size or shape, but by a change inside the many grains. Rolling provides directional properties where grains are longer along the line of rolling, but are observed as smaller masses when microscopically viewed from the end of the stock.

Because reduction in area occurs between the rolls, stock leaving the rolls moves away at a speed relative to the elongation factor. Cold-rolled surfaces are smoother and cleaner than hot-rolled surfaces. Further, dimensional tolerances are closer in cold finished parts. Sheets and some rods are produced in the full-hard, half-hard, and quarter-hard condition, depending on the severity of rolling.

Products of Rolling Processes

The rolling of metals provides the specific raw materials such as ingots, blooms, billets, and slabs for differently shaped rolled parts. Because the finished part originates from a large mass of metal, previous rolling operations are required before any final processing can be accomplished. The railroad industry uses rolled rails, while the bridge and building industries use shaped beams and columns. Much pipe is rolled into a circular cross section by a combination of rollers in the proper sequence. Many shapes of bar stock are produced, including rounds, squares, rectangles, hexagons, and octagons. Also, thousands of tons of plates and sheet stock are made for numerous types of processing industries. The food industry uses a nearly endless quantity of tin-coated low carbon steel sheet in the production of cans. However, newer types of alloys are reducing the quantity of tin-coated steel needed for the food industry. Also, the automotive, railroad, aircraft, and shipbuilding industries use large quantities of plate, sheet, and strip stock. Plates are produced in thicknesses suitable for warship armor, while sheets are produced in increments of 0.001 inch, beginning at 0.001 inch in shim stock up to a certain size. Other than structural shapes, most rolled metal becomes the basic stock for subsequent forming and machining operations.

Forging

Forging is a squeezing process that involves triaxial compression forces on the metal causing material to flow laterally. Compression forces are applied by impact or by a slow squeezing procedure, depending on whether the forging hammer is dropped from a height or applied with slowly increasing pressure. The objective of forging is to shape the metal and, at the same time, induce desirable mechanical properties in the resultant part. Most forging processes are accomplished while the metal is hot and above recrystallization to better allow the plastic metal to flow into the different cavities of the forging dies. On the other hand, cold forging is also accomplished. Sometimes, special forging operations are required due to the geometry of the part and the chemistry of the material. A product of the specialized forge is shown in figure 4–31.

Some forging processes use a flat hammer and anvil which enable manipulation of the part by the operator, while other processes require dies in order to produce special shapes. Although forging is similar to rolling in that metal is plastically deformed into shape, each process is specifically different. Rolling produces stock in more uniform cross sections along the longitudinal axis of the metal, whereas in forging, intricate designs are produced as compressive forces squeeze the metal from opposite sides, causing lateral spreading in different directions. Forged metal results from the head-on collision of two opposing forces, as pointed out in figure 4–32. As in hot rolling, forging is more frequently performed while the metal is in its recrystallization temperature zone. A maximum of plasticity exists at these temperatures in addition to lower yield values in the metal. Consequently, less energy is required to form a mass of metal.

DROP FORGING

When a heavy hammer having a flat face or die attached to its end is dropped onto a heated metal, a reduction of the metal's cross-sectional area occurs in conjunction with the resulting lateral movement of the metal. The force required to cause permanent deformation is greater than the metal's resistance to permanent deformation. As an example, an average size small drop hammer weighing 2500 pounds will deform a hot section of steel whose dimensions are 5 inches by 10 inches in cross-sectional area. Several blows of the hammer are necessary to produce the desired thickness; the

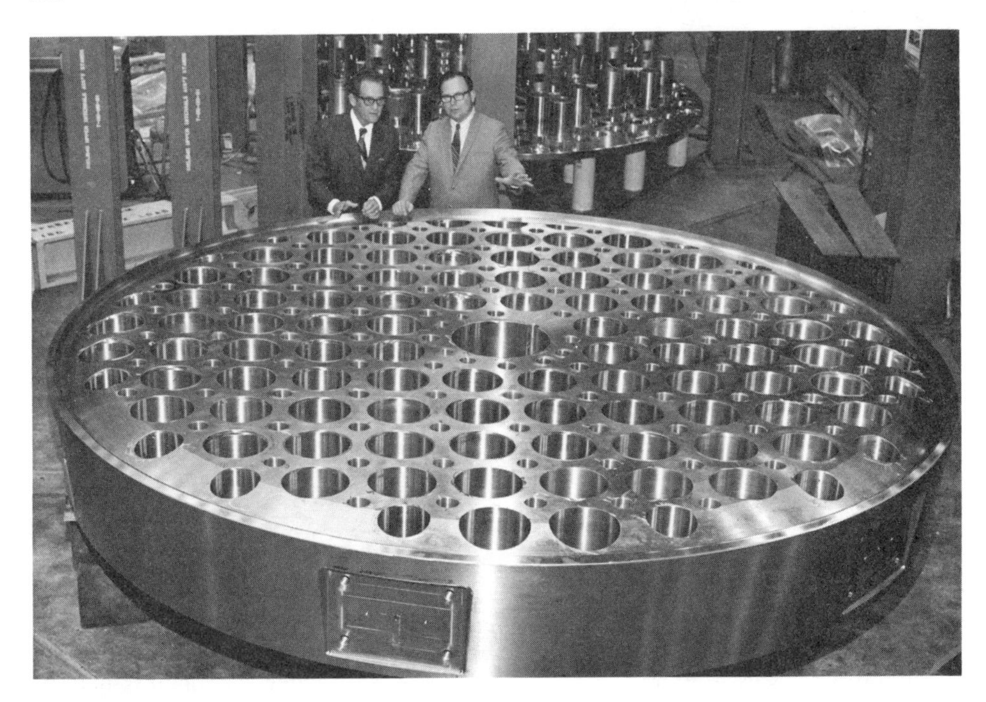

FIGURE 4–31 The largest forging of its type in the world, made of stainless steel and originally measured 20¼ inches thick and 152¼ inches in diameter before finishing, with a weight of 110,000 pounds. The forging forms the lower-core support of a nuclear reactor which positions and supports nuclear fuel in the reactor. (Courtesy of U. S. Steel)

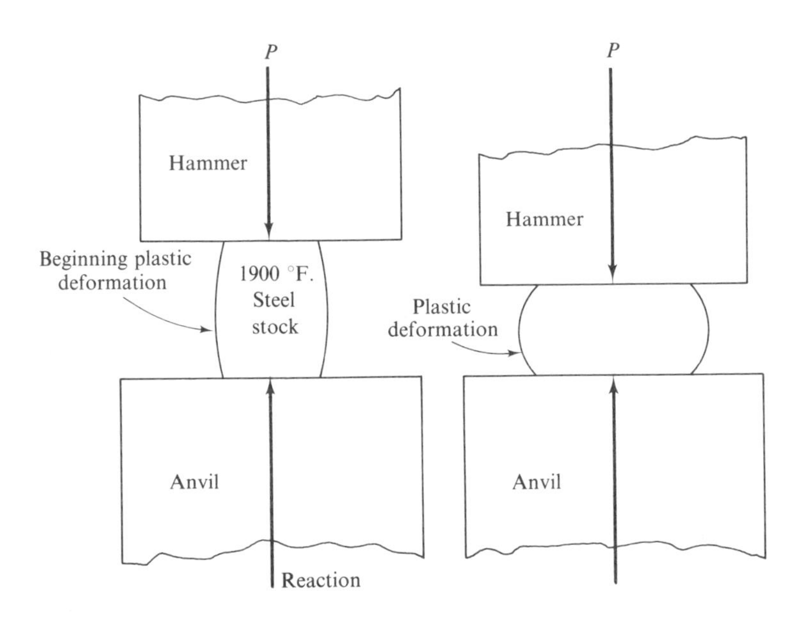

FIGURE 4–32 Compression forces are used in forging.

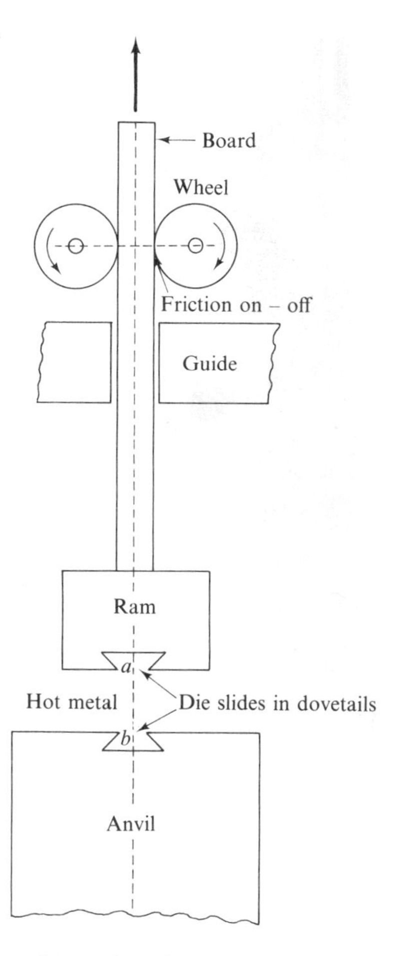

FIGURE 4–33 A board type drop hammer.

number of strikes depend greatly on the thickness and temperature of the metal. As the hammer and die strike the surface of the metal which rests on a die seated in the anvil, maximum work is applied and flow of metal occurs.

The board drop hammer is a common machine used for forging. Basically, two hardwood boards are attached to the hammer, as illustrated in figure 4–33. Friction type rollers are pressed into the boards near their other ends which causes the hammer to be lifted and then dropped. Gravity provides the impact load. Hammers weighing up to three tons are commonly used with a nonvariable impact load. Dimensional tolerances for thickness are good with the drop hammer, for example, a 50-pound forging can be produced within a 0.035-inch tolerance. When the steam hammer is used, variable loads are possible because there is operator control of the hammer. Intensity of the blow is increased when steam pressure is applied at the instant the hammer falls, or the load can be decreased by decreasing steam pressure. Some hammers are manipulated by a combination of short or long strokes or varying the striking force. At intervals, the hammer may even reciprocate without doing work. Such a variation in the hammer's capability allows the operator to change the

FIGURE 4–34 An air-gravity drop hammer. Weight of the hammer deforms the hot metal on impact. (Courtesy of Chambersburg Engineering Co.)

workpiece for maximum shaping effectiveness. Steam hammers weigh as much as 25 tons. Figure 4–34 shows an air lift hammer with stroke control.

A specialized forging machine is also produced that incorporates large flywheels to carry its energy to a ram which is equipped with upper and lower fixtures for the particular dies. This 12,000-ton mechanical forging press has two flywheels measuring 90 inches in diameter, whereas the ram weighs 118,000 pounds. The bed of the machine weighs 457,800 pounds. The air clutch provides eight million foot pounds of torque at an air pressure of 90 psi. Overall weight of the press is approximately 2,750,000 pounds (fig. 4–35).

When metal is placed on the anvil of any forging machine and force is applied, the metal is squeezed into the die cavity causing directional properties in the grain pattern. Directional flow of the granular structure is evident in the forged part illustrated in figure 4–36. Should the same part be machined from a forged or rolled blank, as illustrated in *a*, the resultant part will be much weaker in the areas marked *A* and *B*. Maximum strength is obtained when grain pattern has flow lines contoured with the surface. Several blows are often required in order to completely force the plastic metal into all the differently shaped spaces of the die cavity. Figure 4–37 presents the several stages of development in a forged part.

FIGURE 4–35 The world's largest mechanical forging press—12,000 tons. Its die seat measures 85½ inches x 101½ inches, the two motors are 373,000 W each, and its total weight is 2¾ million pounds. (Courtesy of Erie Foundry Company)

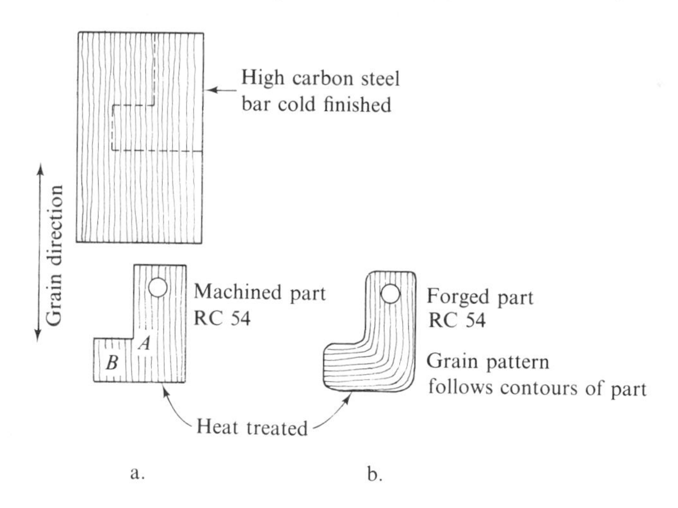

FIGURE 4–36 Grain pattern in a forging as compared to a machined part.

PRESS FORGING

Sometimes it is more practical to use hydraulically powered presses to shape a part by compression forces. The metal may be hot or cold. Presses apply an even but increas-

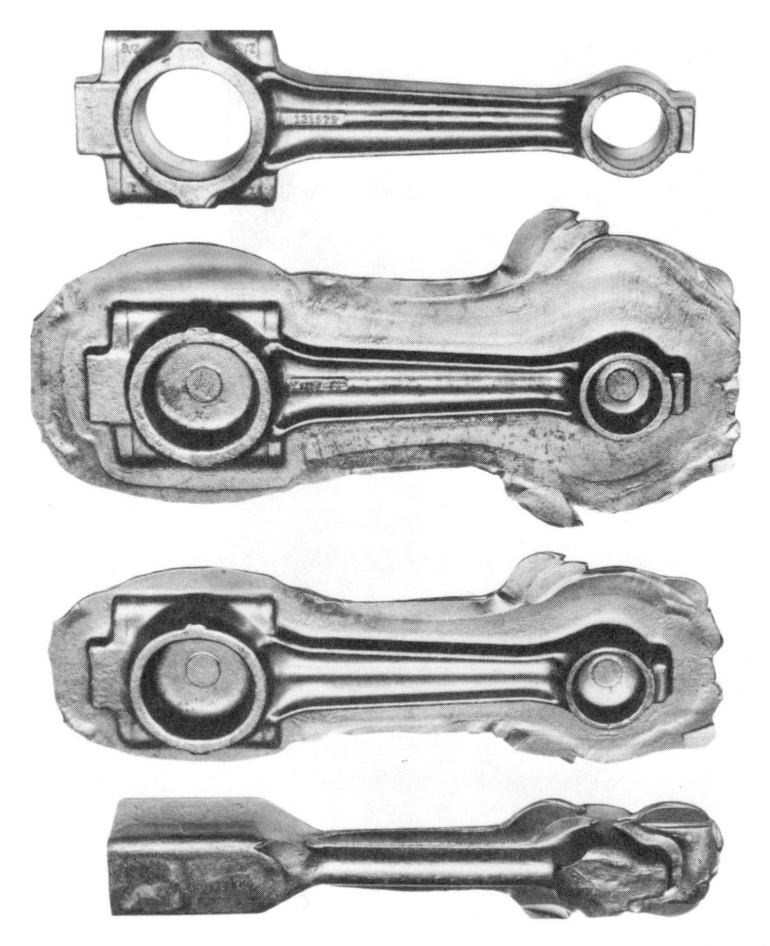

FIGURE 4–37 Successive stages in the development of a forged connecting rod. (Courtesy of The Ajax Manufacturing Co.)

ing compression load on the part, thereby giving tougher metals more time to flow into the die's cavity. The tougher the metal, the greater the need for hot forming. Press machines are available with force capacities rated into many thousands of tons. The press is often more effective than the drop forge due to less loss of energy at the anvil. Drop forge machines depend on the anvil's reaction to help shape the part, and the anvil reacts against the foundation. This energy is lost.

SIZING

Compressive forces are used to bring specific dimensions to forged parts or other ductile parts. Because forging and other pressure forming processes do not always produce close enough tolerances, specialized squeezing procedures force the metal within a smaller tolerance to fit the need. The sizing press slowly squeezes the oversize

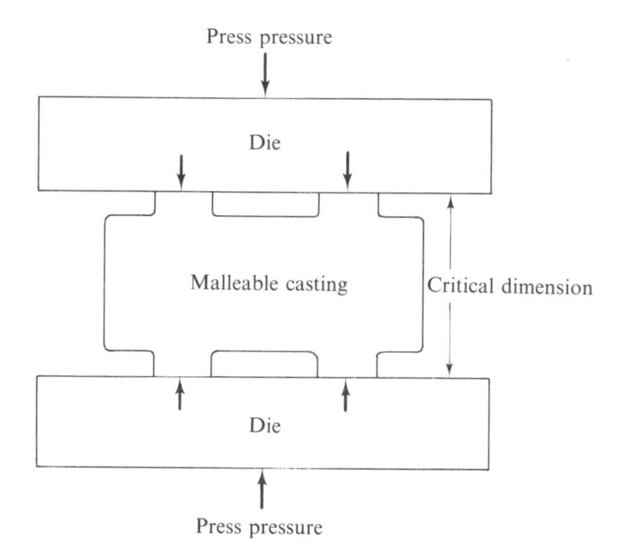

FIGURE 4–38 Sizing a malleable casting by deforming metal a few thousandths of an inch.

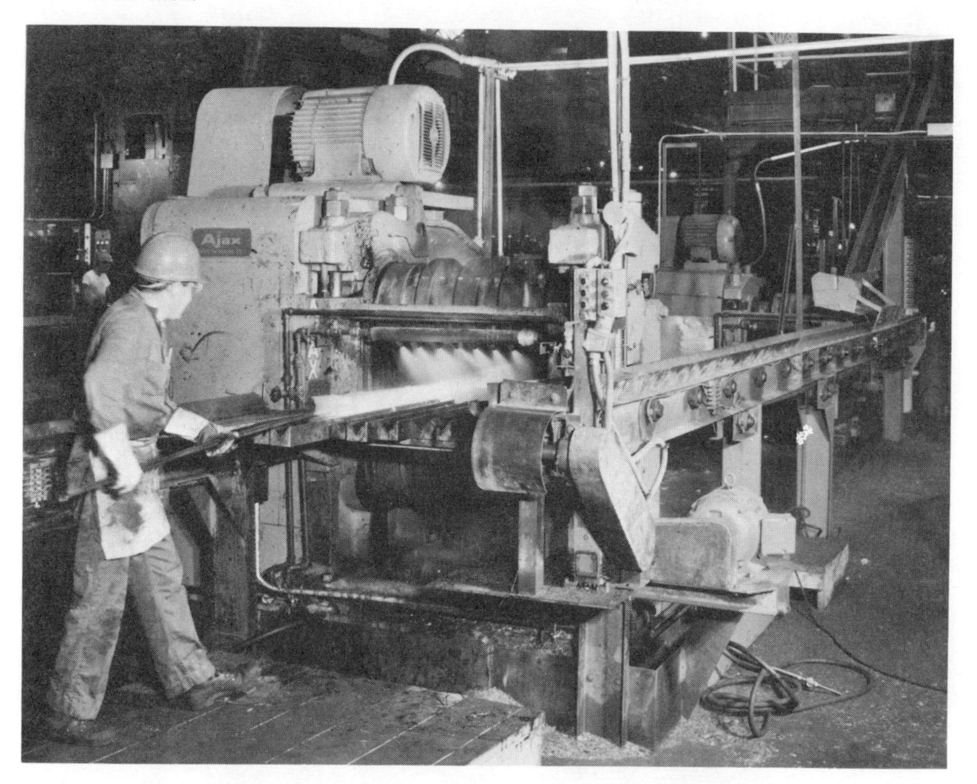

FIGURE 4–39 Preforming a blank with the roll forge for subsequent press forging. (Courtesy of Ajax Manufacturing Co.)

part to size while the metal is cold. In this respect, the forging may have a large tolerance because no mating parts are used along the dimension. However, some particular areas, such as a support base or *boss* on the part, often requires a much closer tolerance because of its mating part. The area of metal is then sized to the proper dimension (fig, 4–38).

ROLL FORGING

A modified forging machine that produces formed shapes along a longitudinal axis is called a roll forge (fig. 4–39). Straight or tapered shafts, for example, requiring formed cross sections are produced when metal passes through selected grooves in the machine. Greater efficiency of the machine results when hot rolling procedures are used. A similarity to the rolling mill exists, however, special shapes of smaller sizes are produced in the roll forge.

UPSET FORGING

A modification of impact forging is the use of compressive forces to upset the end of a part so as to enlarge the end. Hot and cold upsetting or hot and cold heading are used

FIGURE 4–40 Roll threading a steel bolt. (Courtesy of Elco Industries, Inc.)

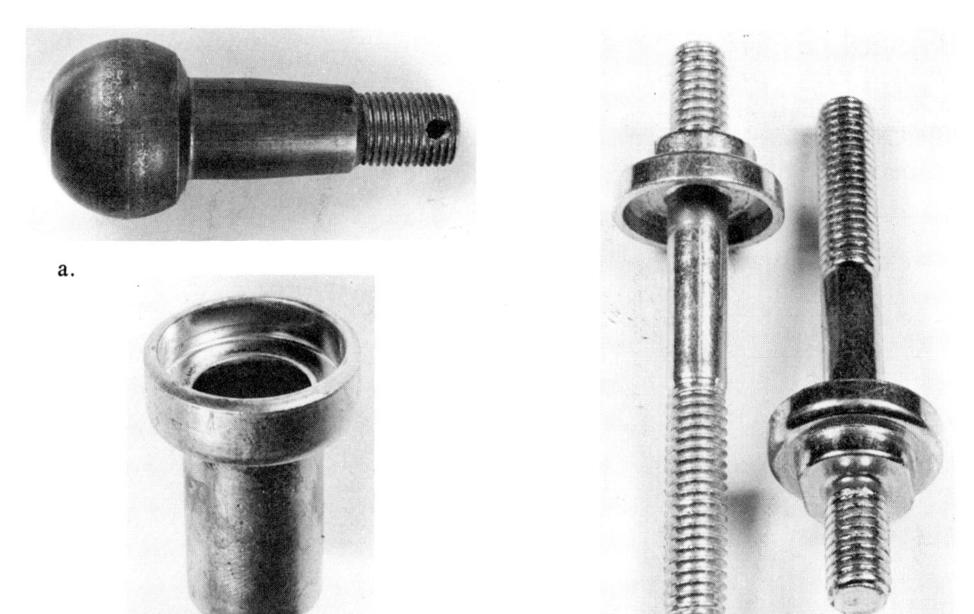

FIGURE 4–41 Cold-headed parts: *a*, an upset metal automobile suspension stud; *b*, auto radiator thermostat element cup cold formed of silicon bronze; *c*, cold-headed, round collar, lightning arrester terminal cap stud with rolled thread. (Courtesy of Elco Industries, Inc.)

to produce parts. Cold heading of a metal part results when its end is suddenly enlarged by an impact force while the temperature of the metal is below the point of recrystallization. As an example, bolt stock is upset or cold headed in order to produce the head. Subsequent operations roll threads on the cold shank of the bolt which is then sheared to length (fig. 4–40). The principle of upsetting requires that a bar, hot or cold, be inserted in a die and held tightly. When the enclosed end is rammed, metal flows to all parts of the cavity. Bars or other shapes may be upset anywhere along their length by causing deformation of the metal in a die at that place on the bar. Therefore, numerous types of supports and fixtures are produced by upsetting procedures. When a section of a metal part is suddenly pushed into a die cavity, the resulting deformation is changed into the new shape (fig. 4–41). Bars as large as nine to ten inches in diameter are commonly upset in these machines. This type of machine is fully automated whereby thousands of parts are produced in one day.

SWAGING

The swaging of metal utilizes an open-ended die having the desired shape of the finished part. The die may be fixed or adjustable to accommodate its purpose. Repeated blows or squeezes gradually bring the end of the part to the proper shape by means of rotary or reciprocating actions. Bolt heads can be quickly swaged to shape, or tapers can be formed at the ends of shafts. Swaging uses biaxial compressive forces to squeeze the metal into shape. A swaging machine is shown in figure 4–42. Fittings on steel cables are swaged so that the cable will not pull away from the fitting. Sometimes the ends of pressure cylinders such as entrance holes are swaged to shape. Hot or cold swaging is performed on several metals using the principles of plastic deformation. Figure 4–43 shows a typical swaged part.

FIGURE 4–42 A swaging machine connecting the attachment to the cable. (Courtesy of Esco Corporation)

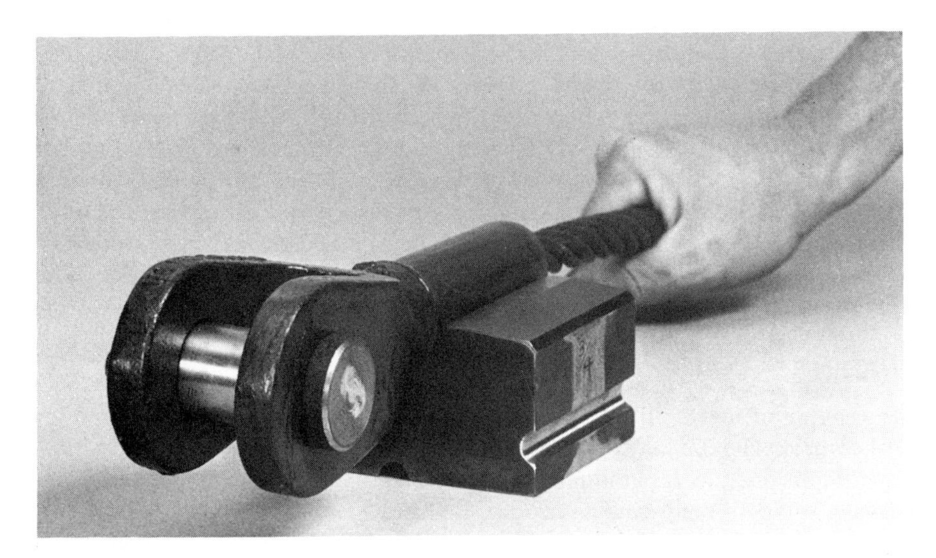

FIGURE 4–43 A typical swaged part. The cable is held securely to the connection so that the pin can hold a load. (Courtesy of Esco Corporation)

FORGING DIES

Forging dies are made of alloys that provide hardness and toughness in order that cavity contour will not be lost due to breakage or severe wear. Dies are very expensive, therefore, a compromise is provided in the metal's mechanical properties so that the die will wear before breaking. Contours and draft angles allow reworking after severe wear results. Alloys of the nickel-chromium-molybdenum steel series provide excellent dies when properly heat treated. Yield strengths of dies and rolls must be high enough to withstand the forces trying to deform them. Design of dies should provide adequate draft for quick separation at the parting line. This line should be located near the center of the die and be in one plane if possible. Because of impact, reinforcements must be placed at strategic points in the die. Also, surface contour changes should allow for adequate radii to help lessen fatigue failure.

A set of dies may consist of only two sections, each section fitting on the machine's parts to allow a cavity when the dies come together. In addition, some dies may be used to roughen the stock, while closer tolerance dies are used to finish the part. Because compressive forces are used in forging, thin fins of metal called *flash* are sometimes squeezed from the die's parting line. Flash also occurs at the parting line in castings. This excess metal is subsequently removed from the part.

Dies are attached to the hammer and anvil by standard fixtures, one being the sliding *dovetail*. This allows quick placement and removal while adequate backing metal is provided to absorb the forging shock on impact. In operation, dies are frequently heated to prevent stock temperature reduction on contact. The preheating temperature depends on the metal to be forged and the metal from which the dies are made. As an example, when AISI 4340 steel is used for the dies, care is exercised not to heat the die higher than its tempering temperature in order not to reduce the die's hardness. AISI 4340 is an extremely tough steel, even in the higher hardness values. It has exceptionally good impact resistance and resists fatigue failure. Other excellent die steels include the W1, O1, A2, D2, D4, D5, M2, and M3 series of tool steels.

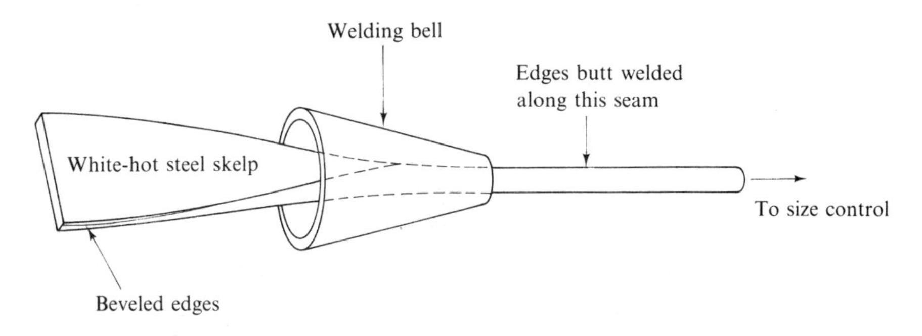

FIGURE 4–44 The welding bell process of butt welding pipe.

PIPE WELDING

Pipe welding is a fusion process that provides large quantities of pipe for the rapidly growing pipe industry. Pipes up to three inches in diameter are produced by butt welding processes using a conical shaped welding bell to cause fusion of the metal. This is a type of forge welding, as illustrated in figure 4–44. Steel pipes formed by this process are products of long thin sheets of metal called *skelp*. A production line is established when the long and narrow furnace is placed in line with the welding bell

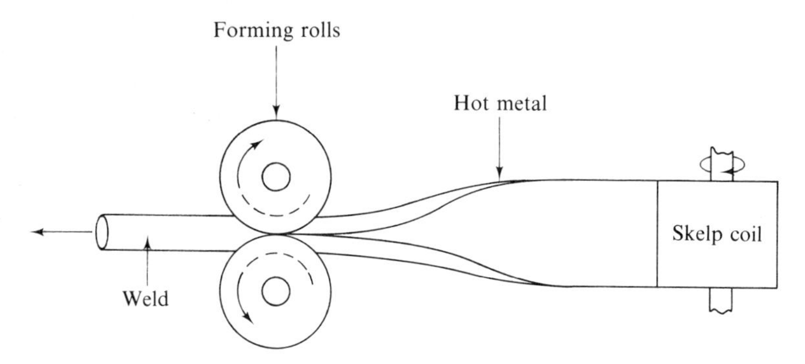

FIGURE 4-45 Roll forming welded pipe.

and pulling tongs. With a given wall thickness and diameter, the long strip is pulled by tongs from the furnace and then through the opening of the bell. The furnace quickly brings the metal up to welding temperature. As the very plastic metal is forced through the progressively round and closing opening of the bell, its two sides are gradually folded inwards until they are in compression contact. Being at the fusion

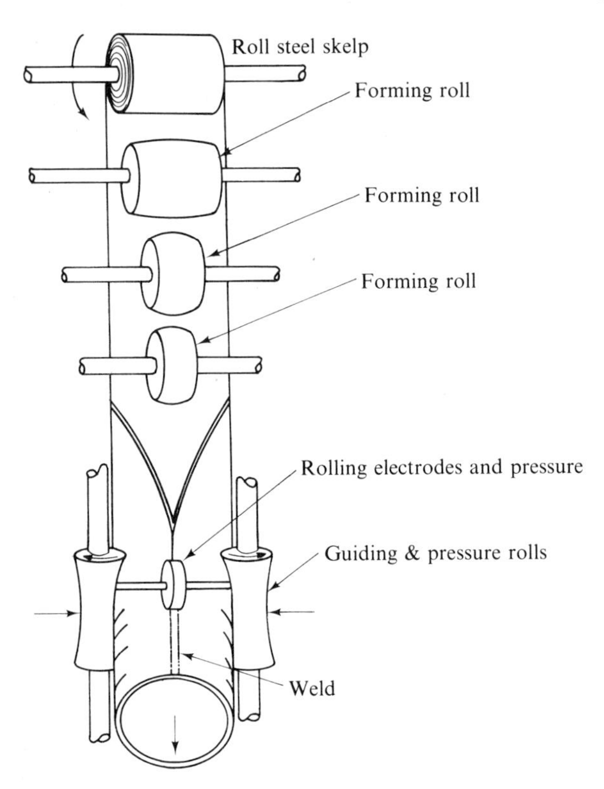

FIGURE 4-46 Pipe being welded by rolling electrodes.

temperature, compression force fuses the sides in a continuous operation.

Another method ot fusion welding also uses the furnace system to heat the skelp. Upon leaving the furnace, the metal is rolled through precision rolls which force the two sides into compression (fig. 4–45). A welded seam results, much like that made from the welding bell. The process is continuous, therefore, flying shears or saws part the pipe in standard lengths while the conveyor system carries the pipe to completion rooms and subsequent storage.

More sophisticated welding processes butt weld the edges of skelp into pipe from minimum sizes up to 16 inches in diameter. Large coils of sheet stock are unwound and placed in an automated production line. In sequence, the proper size beveled edge sheet or plate moves through tapered rolls to the resistance welding unit. Two rolling electrodes on each side of the joint fuse the joint without a filler rod as compression force is applied (fig. 4–46). Because this process is also continuous, flying shears or saws part the pipe in standard lengths. Along the production line, rolls maintain straightness in the pipe while it moves to further processing rooms. Part of pipe processing includes heat treatment. High quality welded steel pipe is subsequently air cooled from austenitic temperatures, that is, from 1600° F., to homogenize the metal's structure, especially at the weld, and also refine the grain pattern through recrystallization.

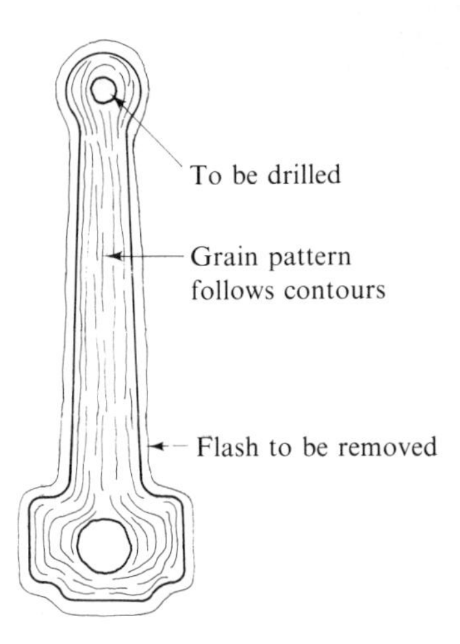

FIGURE 4–47 Grain flow in a forged part.

PRODUCTS OF FORGING PROCESSES

Because forging metals is very simple and yet very complicated, depending on the part, numerous products are produced. The small tool industry produces thousands of small tools such as wrenches, hammers, pliers, instrument parts, and vises. Commercial hardware such as connectors, fittings, bolts, and nuts are manufactured by the millions. Household appliance parts such as housings and frames are forged. In addition, motor supports and bases are formed by some of the forging processes. Engine parts such as piston rods have the proper grain flow when produced by forging. Also, forgings provide excellent mechanical properties (fig. 4–47). If a mass

FIGURE 4-48 A 393,000 lb. forged shaft for a 1,280,000 kVA generator to be operated at 1800 r/min (Courtesy of U. S. Steel)

FIGURE 4-49 A draw bench for drawing cold finished steel bars. (Courtesy of LaSalle Steel Co.)

of metal can be squeezed into shape, then it becomes a potential product of the forging process. In most instances, forged parts are superior parts when properly heat treated, due mainly to the stress resistant internal pattern of the granular network. A forged generator shaft is shown in figure 4–48.

Drawing

The drawing process is either a cold or hot forming process that reduces a metal's cross section as it is elongated. In the cold drawing process, hot-rolled bars or other

FIGURE 4–50 A steel wire mill showing drawing units. (Courtesy of U. S. Steel)

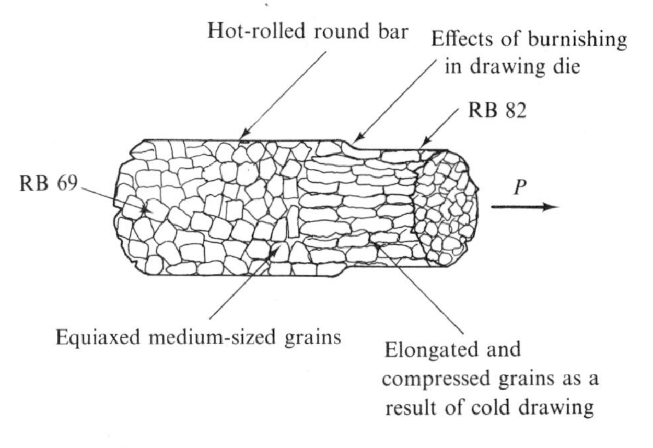

FIGURE 4–51 Grain flow in a cold-drawn bar.

shapes are pulled through a tungsten carbide die of a given size and shape. Lubrication of the die allows a sliding action of the metal through the die. Tension forces on the forward end of the bar cause biaxial compression and tensile forces on the bar as it is reduced in diameter. When this process is conducted on cold metal, permanent deformation occurs, hardness and strength increase, and a smoother surface with close tolerances results. Cold drawing is accomplished in a draw bench (fig. 4–49). Mill scale from the hot rolling process is removed prior to drawing. The cleaned bar is then elongated as its diameter is established by cold reduction, the amount of reduction being determined by several factors such as chemistry of the metal and the desired size. Before drawing can begin, steel stock must be pickled and limed; other metals are treated differently. Should increased softness be needed, the cleaning treatment is performed after annealing.

Figure 4–50 shows a steel wire mill with drawing units. Cold-drawn wire has similarities to cold-rolled bars in that the metal's structure shows elongated grains in the direction of the longitudinal axis of the bar (fig. 4–51). Rolling pushes grains of metal

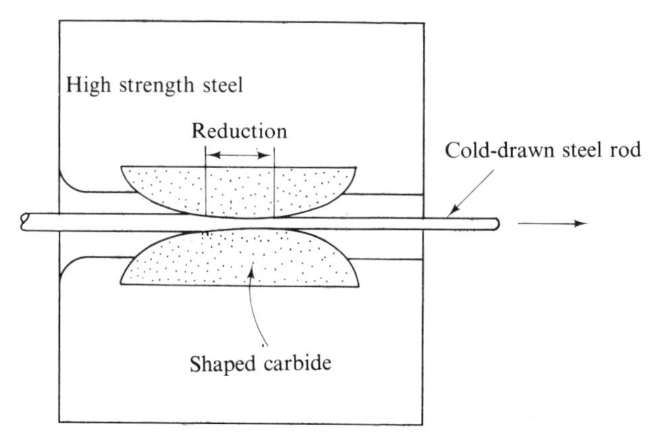

FIGURE 4–52 Cross section of a drawing die.

into and over other grains, whereas drawing pulls the granular structure by tension and compression forces. Elongation of grains results in both processes, an occurrence which greatly improves several mechanical properties such as hardness and strength. However, ductility is reduced, which means the metal is harder to bend. Pure metals are drawn easily, but some alloy steels can be reduced only a small percent. Figure 4–52 is a sketch of the cross section of a drawing die.

WIRE DRAWING

Other shapes are drawn with equal efficiency. The drawing of wire through the tungsten carbide die is a continuation of rod drawing. After proper cleaning of a large coil of 1/4-inch diameter rod, for example, the end is pointed, pushed through the first die and then threaded through succeedingly smaller orifice dies until the correct diameter is established (fig. 4–53). Powerful tongs pull the wire through the dies. Once the process is started, it is continuous and automated.

Large wires are often pulled through a single die and wound on reels. Small diameter wire is drawn through several dies, being wound and unwound on successive reels. As the wire diameter is reduced, the wire becomes longer and the reels turn faster. Dies are lubricated to allow smooth flow of the metal. The tongs apply a tensile load on the wire which is related to several drawing factors such as initial wire hardness and cross-sectional area. Strength properties are improved as a result of drawing.

FIGURE 4–53 Operator measures wire product on a wire drawing unit. (Courtesy of U. S. Steel)

Some wire is stocked in the *as rolled* condition after an application of corrosion preventative. Music wire, for example, is often paper-wrapped around a powder which covers the coil of wire. This wire is drawn hard, is polished, and is available in increments of a few thousandths of an inch. Music wire is used for making cold-formed or heat treatable springs, as well as for its other purposes in the music industry (fig. 4–54). Low carbon steel wire is used as produced, copper coated, or zinc coated. Brass

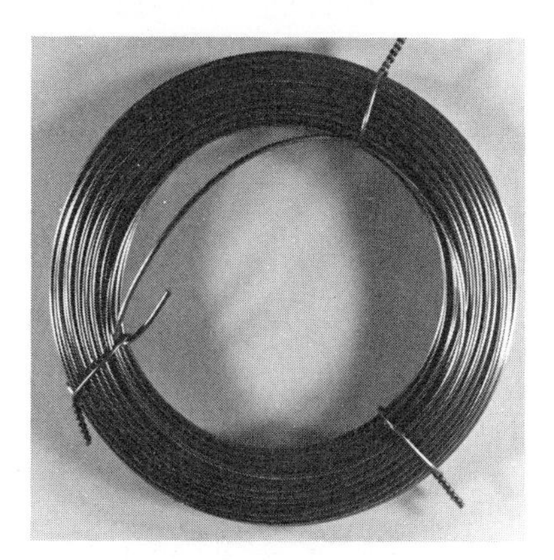

FIGURE 4–54 A coil of cold-drawn music wire.

wire is available, as is copper and aluminum, in the soft- and hard-drawn conditions. High strength steel wires are used in the manufacture of cables for suspension purposes and in crane mechanisms. In fact, cables are made from many different kinds of metal wire. The electrical transmission industry uses aluminum cables for power transmission due to the lightness in weight of the metal and the excellent electrical conduction properties.

HOT DRAWING

A deeply drawn cylinder having a closed end is hot drawn over a mandrel or rammed with a punch to a stage where further annealing operations are essential in order to finish the deeply cupped part. When a heavy wall thickness is essential for compressed gases entering tanks and boilers or for munitions and related seamless parts, the hot draw is useful. Being hot, the total load to cause deformation is less because of the plastic nature of the metal. Both tension and compression forces are involved in causing severe plastic flow of the metal. Hot drawing lends itself very favorably to the tougher alloys of steel where subsequent heat treating produces tremendous strengths.

Press Forming

Most press forming is performed on cold metal such as sheet, plate, strip, and an assortment of preflanged shapes and bar stock. In order to press the metal into the dies, or other types of deformation, the elastic limit of the metal must be exceeded, keeping in mind that the force required to do so is much greater for cold metal than for hot. The purpose of a press is to form or cut the metal into the desired shape. Some presses accomplish their purposes by smooth movements while others perform their jobs through impact. The press provides the controlled energy needed to pierce a metal, fold it, shear it, forge it, bend it, cup it, and many other related processes. These machines range from those with energy capacities satisfied by hand or foot power, as needed for folding a thin sheet of metal, to those exerting many thousands of pounds of pressure. Besides the hand or foot operated presses, there are mechanical, hydraulic, and pneumatic presses. Mechanical presses release their energy from huge flywheels or mechanical advantage systems. Electromechanical presses are more sophisticated and specialized and many are automated, being in lines of production. Hydraulic presses are very powerful and large, while pneumatic presses are more specialized. Figure 4–55 shows a forging press.

Due to the nature of the press, it must contain its own stresses, whereas the drop forge loses some of its effects through the anvil. The metal to be shaped or sheared rests between an action ram and a reaction base, or bed, whereby, in many work situations, the compression loaded part causes multistress forces in the frame of the press. This situation is much like that in rolling and forging equipment, but the nature of the press creates design problems not necessarily existing in other processing equipment. As an example, the upper die, when closing on the lower die, builds up compressive loads in the metal caught between the dies. As metal deformation begins, like forces build up in the machine's framework that holds the two dies in place. In other words, a squeezing action on the metal reacts into the overarm, frame, and base of the machine in a manner that springs the frame wider and tends to throw the dies out of alignment. This misalignment applies equally to the brake, punch and die, or any other tooling arrangement with similar uses. Misalignment can wreck dies and

FIGURE 4–55 A 3,000-ton forging press forges ingots into blooms and billets for later rolling. The manipulator handles the section of metal while the hammer descends to shape it. The operator controls movements of the manipulator and hammer. (Courtesy of The Timken Co.)

work areas and may cause permanent damage to the machine. An increase in the section modulus, or dimensional properties of the cross section of the metal part at the more highly stressed sections of the frame, will eliminate many design problems in these types of machines. This springing apart reaction is an elastic stretch, being within the metal's modulus of elasticity. Therefore, to overcome most of this stretch, heavier frames, cross arms, and bases are required. In this respect, machines are designed to operate within their assigned capacities and must not be used beyond their capacities. Exceeding this capacity can bring about misalignment problems.

TYPES OF PRESSES

Presses vary according to their purposes. Single-action presses extend a single ram, but several tools may be attached to its end. Double-action rams perform a holding and working effort in that the ram or other part of the machine's working apron moves to secure the stock while an inside ram performs the work, either shearing or deforming the metal. Multiple-action presses perform additional work, either simultaneously or sequentially. Reciprocation of the ram is rated in strokes per minute, whether pressing shapes or shearing out blanks or shearing along a length. The bed, or working area, varies according to the size of stock to be worked and the kind of work to be performed. A massive forging press is shown in figure 4–56.

FIGURE 4–56 This 50,000-ton hydraulic forging press is capable of providing pressure on metals sufficient to handle the biggest jobs. (Courtesy of Mesta Machine Co.)

A further classification of presses separates them by construction, that is, their structural design. The open-back press (fig. 4–57) allows working area in the vicinity of dies and punches. Gap-frame presses allow large pieces of stock to be placed in the machines (fig. 4–58). The turret press provides for quick acting tools in the action area, while a horn press (fig. 4–59) has the capability to expedite sheet metal work such as seaming and folding edges.

Many single-action ram presses perform cupping and other deep drawing operations as illustrated in figure 4–60. Very ductile sheet or plate stock is placed between the punch and die, and the punch is lowered. Clearance between the sides of the punch and die are such that drawing of the metal occurs instead of shearing. In this respect, when drawing clearances are too narrow, shearing of the metal occurs instead of deformation. Examples of deep-drawn parts include household appliance parts such as one-piece washing machine tubs, half-cylinders of air conditioning compressor tanks, turret covers, automobile body parts, large shell casings, and a large variety of other cupped parts, as illustrated in figure 4–61. In *a*, a can has been produced, while in *b*, the punch is ready to repeat the operation.

Deforming presses place biaxial tension and compression forces on a metal, and in several situations such as deep drawing, multistress forces occur concurrently in

FIGURE 4–57 A 150-ton double crank open-back press provides space behind the dies for larger sections of metals. (Courtesy of Rousselle Corp.)

different parts of the stock as it is being shaped. Figure 4–62 illustrates how the metal must flow in order to adequately fill the space between the punch and die. When deformation is too severe, rupture of the metal occurs, therefore, several annealings may be necessary between drawings. In order to deeply cup a sheet or plate of metal, the metal must be physically capable of being drastically deformed. As pointed out in figure 4–62, the 90° circumferential bend demands metal flow in several directions in order to fill the required wall thickness dimension. Too much metal accumulates in the compression areas while too little metal is immediately available in the tension areas where stretching occurs. A highly ductile metal will plastically flow a reasonable amount, but will not flow when rigidity reduces ductility an appreciable amount. According to the factors shown in figure 4–62, the punch must not extend more than approximately 60% of the can's diameter into the blank in one pass. Then the metal must be softened by annealing before further deformation occurs. This procedure is variable, due to the kind of metal, the shape of container, and the condition of the original blank. Too much plastic flow during one pass often reduces the wall thickness to zero near the bottom of the can. Yellow brass, for example, becomes very brittle when excessively deformed without annealing and frequently flies into small pieces as it shatters.

FIGURE 4–58 A 40-ton adjustable bed gap press provides pass-through capability in forming parts rapidly (Courtesy of Rousselle Corp.)

Shearing Presses

The shear press is foot or hand operated and is either manually powered or power operated. Rating of the shear is determined by the maximum thickness of a specific metal it will shear and the width of the shear. Small shears are produced in three and four foot widths, while power shears are available with capacities from three to ten feet wide and much larger. Small shop shears have shearing capacities from 1/16-inch to 1/8-inch thicknesses with 36-inch to 48-inch widths. Large shears sever 1/2-inch thick plate in large widths at a specified number of strokes per minute.

The shear operates in a two-step stroke, that is, the clamping bar precedes the blade in order to securely hold the stock as the blade moves along a linear path parting the metal. Because there is little clearance between the shear blades as they close, a shearing action occurs (fig. 4–63*a*). Shearing forces are sliding forces acting in opposite directions, whereas compressive forces are head-on forces causing squeezing of the part. In other words, shearing forces are noncollinear parallel forces applied to the metal's cross section. Compression forces are collinear axial forces being applied

FIGURE 4–59 A 25-ton adjustable bed horn press provides variable shaping capability with respect to irregular shapes and universal jobs such as punching, riveting, and seaming. (Courtesy of Rousselle Corp.)

at 90° to the cross section. Shearing forces are cutting forces, while compressive forces are collapsing forces. Some procedures call for shaving the sheared edges to tolerances of the reamer, such as within 0.001 inch. The reamer is a cutting tool designed for removing up to 0.030 inch of material from a round hole. Different types of tools perform this shaving operation.

Figure 4–63*a* points out the two fractures in the metal which occur at the end of the metal's plastic flow as stress is applied. When the fractures meet, shear is accomplished as the metal is parted. The load, or shearing force, on the blade must be greater than the metal's shearing strength in order for shear to occur. Shearing stresses vary as the metals vary and also vary as the metals' hardness varies. Clearance

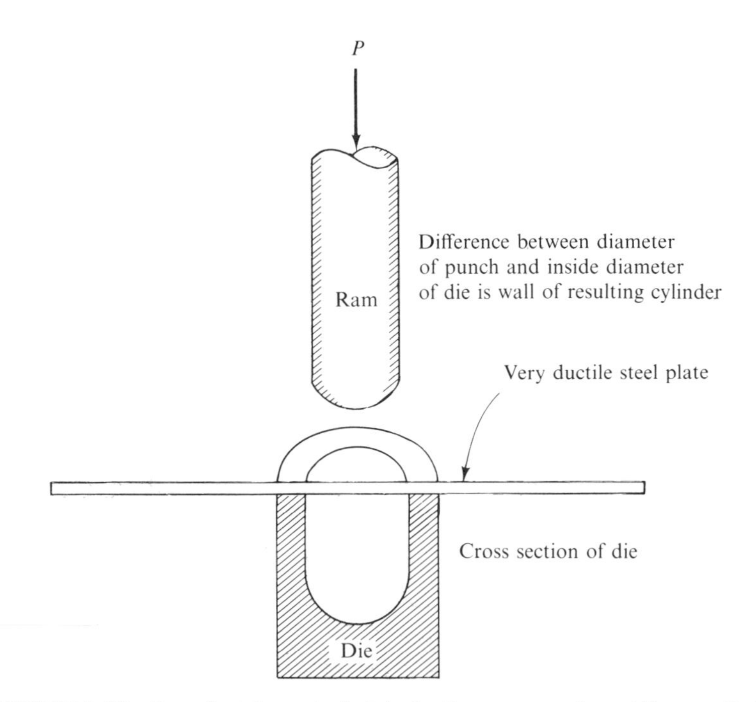

FIGURE 4–60 Deep drawing a steel plate for the purpose of providing a cylinder. As the ram moves into the die it carries the very ductile plate with it, the plate wrapping around the ram.

in shear is vital, too little damages the blade and tears the metal. Too great a clearance cups and tears the metal. In *b*, the punch is equipped with a shear clearance at the cutting edge, while in *c*, the die is equipped with shear. Piercing operations require shear on the punch, while shear on the die is provided in blanking procedures. When no shear is provided on the punch and die, higher shearing forces are required as the entire cutting edge is applied at the same time. In *d*, additional clearance is available for quick extraction of the punch. A metal shear is shown in figure 4–64.

Rotary punches provide cutting actions on sheet metal through punch and shear action, while ring and circle shears begin a cut internally on the blank and the off-cut remains flat (fig. 4–65). With respect to the punch and shearing stress, the load on the punch can be calculated by multiplying the shearing stress by the area of the metal to be sheared. In this respect, the sheared metal has the shape of a cylinder (fig. 4–66).

A good cutting ratio between hardness of punch and stock is shown in figure 4–66*a*, this ratio being approximately three-to-one in favor of the punch. In *b*, the punch has been lowered and has sunk into the soft metal, whereby four fractures have started, following collapse of plastic flow. Following plastic flow, a burnishing of the deformed surfaces occurs. A sequence of steps is recapitulated as follows: 1, elastic deformation begins as the punch lowers, the metal building up a higher stress reaction to the force; 2, plastic flow commences at the yield, causing an increase in hardness at the deformation zones; 3, fractures start as the ultimate stress is reached; 4, fractures

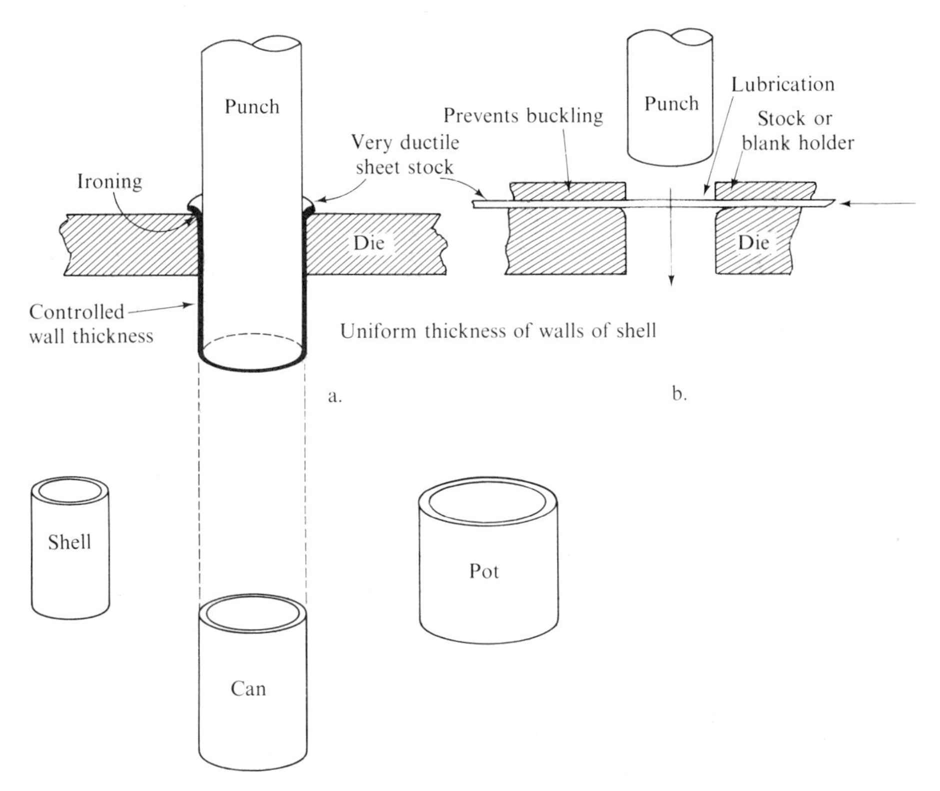

FIGURE 4–61 Typical deep-drawn parts showing effects of drawing.

continue on a collision course and parting of the metal occurs. In *c*, various types of cutting operations are illustrated.

Stamping Presses

The stamping industry is expanding its production capabilities to include heavier and more sophisticated parts. This capability is made possible due to larger and more powerful machines. Basically, presses have a common similarity in that biaxial tension and compression forces are exerted onto the parts to be formed. The bed and table of the press are, therefore, designed to handle the specific types of shaping operations.

Various assortments of dies and punches are available to accomplish piercing and blanking operations. Also, complex bending is performed whereby several bends are accomplished on the same metal. Properly equipped presses and brakes provide notching, piercing, curling, and folding operations on the same part. Such combined operations provide finished parts in various kinds of sheet metal such as aluminum, copper, brass, and steel. In automated lines, specialized machines are arranged in sequence. Stock is fed into the machines at rapid speeds, each machine performing its assigned task. Because the stock has an exceptionally long length, being unrolled from a coil, continuous operations are completed as the sheet moves from one machine into another. Lengths are sheared at the end of operations. Some machines

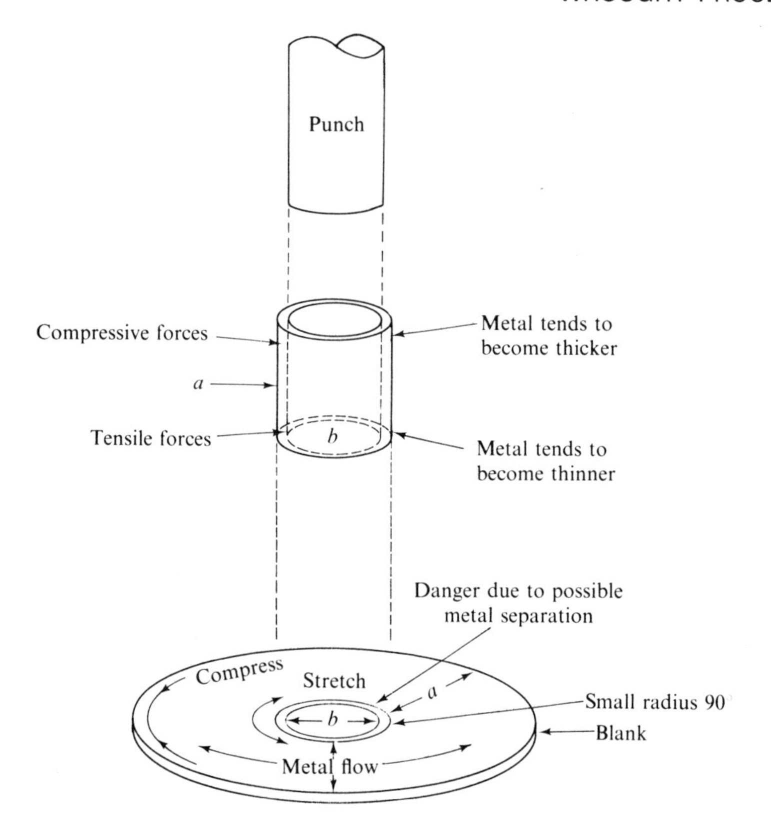

FIGURE 4–62 Metal flow during deep drawing.

receive stock at the rate of 2,000 inches per minute. Products from stamping presses include household and industrial shapes such as washer and drier parts, vacuum sweeper assemblies, refrigerator doors, exhaust vents, toys, and automotive, locomotive, and aircraft formed shapes. If a part can be pressed into shape, then dies can be cut in order to produce it. A typical stamping press is shown in figure 4–67.

Forming Processes

SWEDGING

When drilled or punched holes in thin sheets of metal are too thin to allow thread length, the hole's periphery is raised which gives height to the hole greater than the thickness of the original hole. In this respect, calculations require the hole to be produced at a smaller size so that when its edge is raised, the proper diameter will result for threading. Obviously, there is a minimum thickness of the sheet stock that will provide enough metal to allow thread roots in the circumferential surface of the hole, as illustrated in figure 4–68.

SPINNING

Spinning of metals is performed on hot or cold metal. A flat section of metal is held in a chuck so that the partially formed or flat stock spins around a horizontal, vertical,

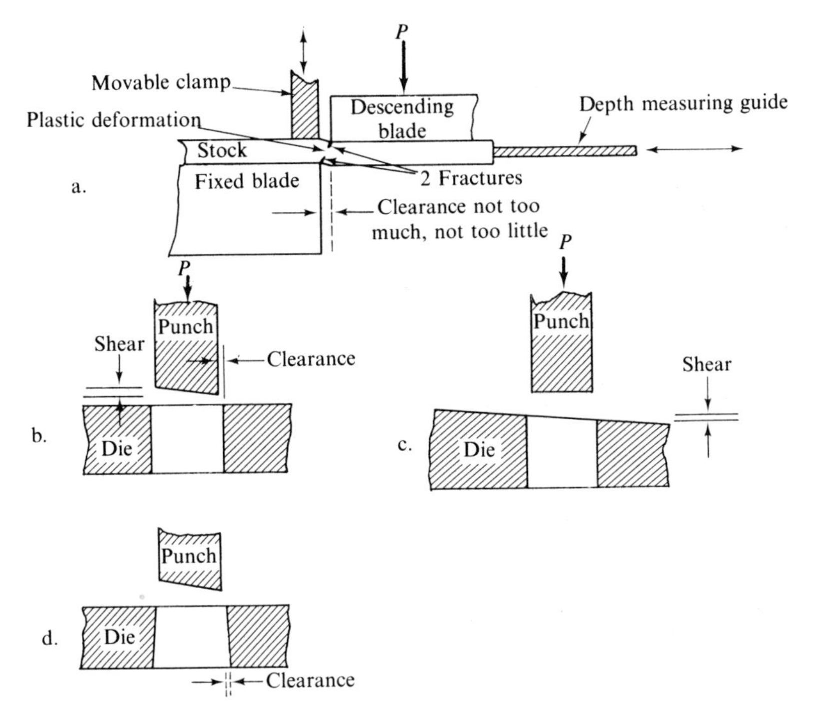

FIGURE 4–63 Shearing actions of blade and punch.

or angular axis. As the metal stock turns, it is slowly pressed against the attached backing plate or pattern so that it will take the shape of the pattern. The spinning tool is used to slowly force the metal into the pattern. Only that portion of the metal in contact with the tool is deformed (fig. 4–69).

Thick plates or sheets of steel are spun into shape while hot so that contact pressures will be lower and enable deformation to occur. More ductile and thinner sheets are spun cold. The governing factor concerning hot or cold working procedures is the deformation factor, keeping in mind the effects of the rotating stock. Metal along the periphery is compressed while inner sections of metal flow in a multistressed manner to include compression and shear, therefore, an overage of metal occurs in the compression regions, whereas shear forces cause sidewise plastic flow to compensate for the overage. In effect, plastic flow must occur smoothly and evenly in order to allow the proper thickness of metal in all regions of the circular shaped part. The skill of the operator is, consequently, an important factor in the successful production of spun parts when this operation is accomplished by hand. Once the technique is established, machine spinning can be subsequently performed.

Hot-spun products can be finished cold after proper cleaning. As has been pointed out, cold working increases tensile, compression, yield, and shear strengths in metals along with an increase in hardness and decrease in ductility. The increased hardness in any cold working procedure pertains only to the surface and subsurface metal affected by the deforming forces and not necessarily the whole cross section of the part. If hardness is only skin thick, then no appreciable total strength is increased because hardness and strength have comparable relationships. When large tank and

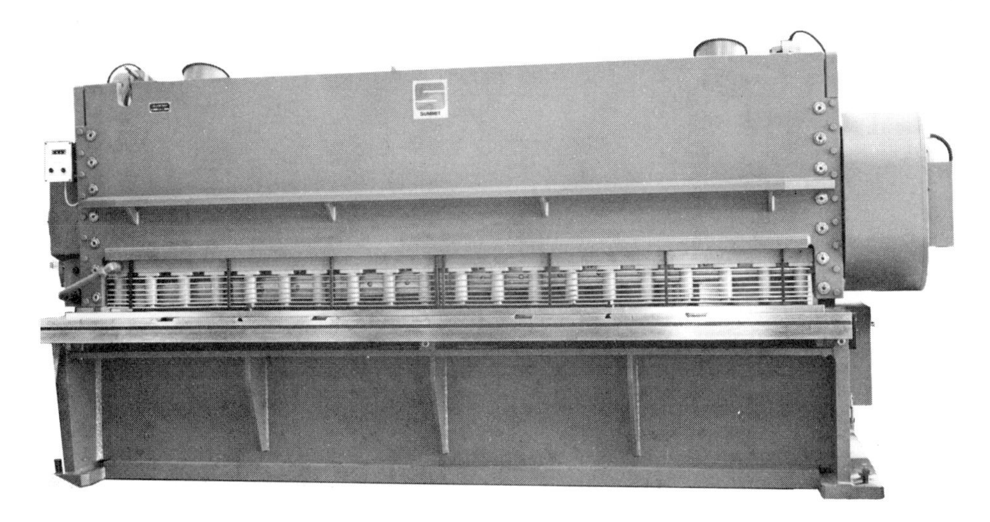

FIGURE 4-64 A metal squaring shear which provides powerful straight-line parting capability. As the cutting blade descends, observable behind the clamping bar, the bar securely holds the metal as metal separation occurs. (Courtesy of Summit Machine Tool Mfg. Corp.)

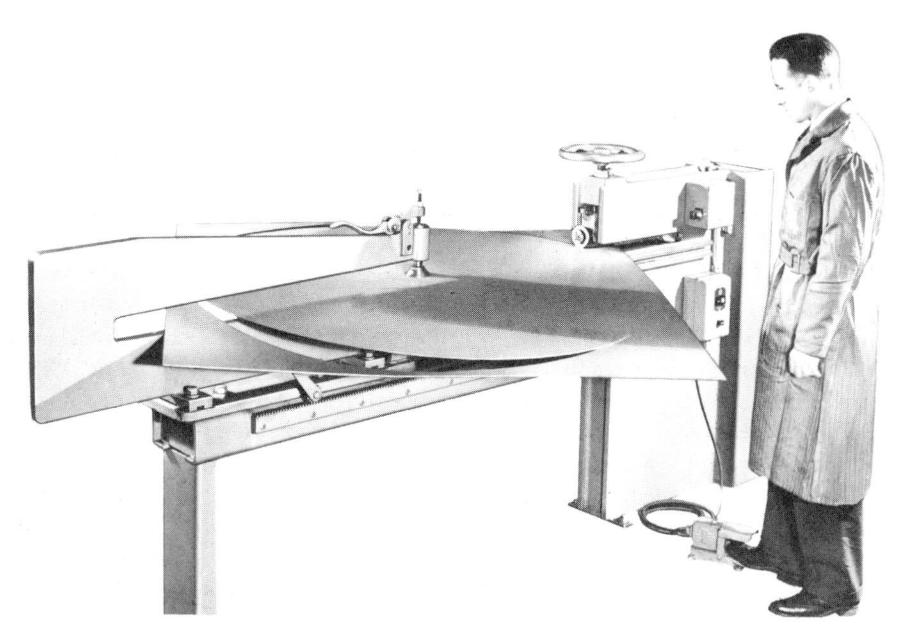

FIGURE 4-65 Cutting a circle with a ring and circle shear. (Courtesy of Niagara Machine & Tool Works)

cylinder ends are formed in this manner, caution is exercised to assure the attainment of the proper deforming stress.

Spinning is applicable to cylindrical shaped parts which are not economically produced by other methods. As an example, a single tank cover or high pressure cylinder

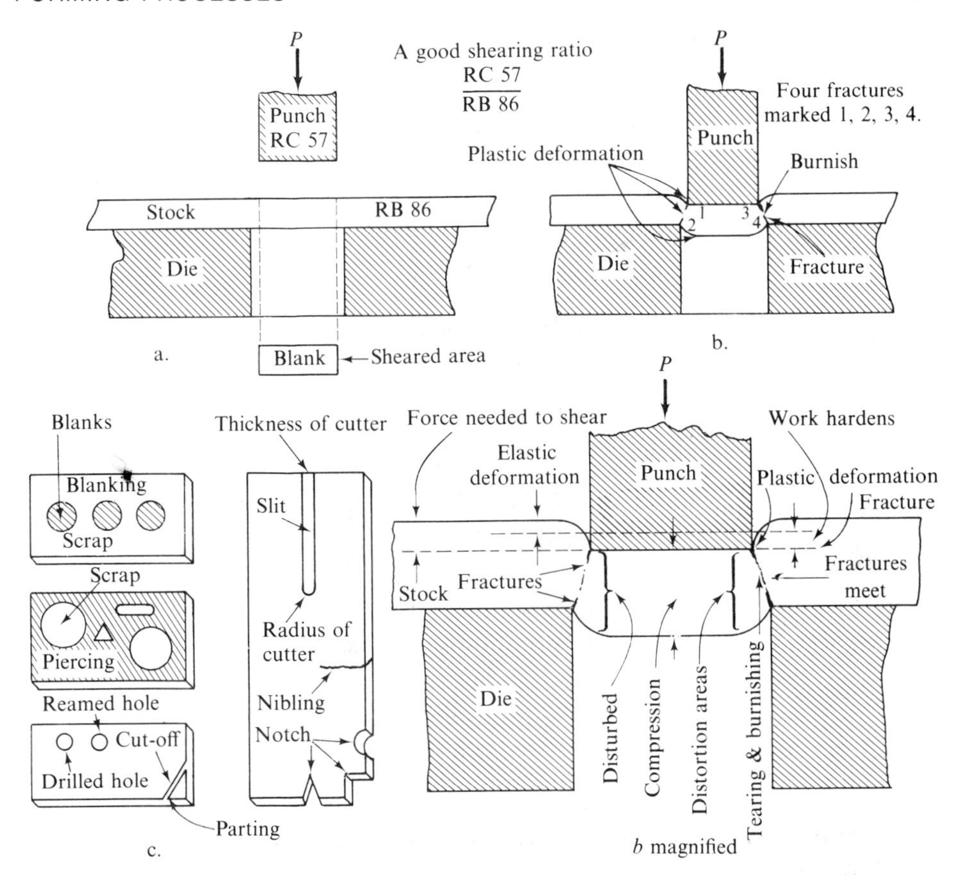

FIGURE 4–66 The shearing zones in sheared metal.

end is economically spun into shape, whereas a die for a stamping or pressing procedure would be too costly. Semiautomated spinning machines are available to shrink fit conical, circular, and cylindrical shaped parts such as light reflectors and horns.

SEAMING

Seaming is a folding operation peculiar to sheet metal work. It applies to any of the metals used for fabrication purposes. The purpose of a seam is to add to the appearance of the edge of the metal, add strength to the edge areas, provide locking joints, and provide curled edges for attachment to other parts. Different types of machines provide for capability in producing the several types of seams. Seaming and related edging operations are performed cold.

Some of the different types of seams include the simple edge seam, a modified simple seam, a wiring, a double curl, and different types of lockseams. Figure 4–70 illustrates these seams. Any of these seams are either used as folded or else the seam is exposed to solder. Solder provides a solid, fixed, and stiffer joint, being watertight and airtight. Products from the canning industry, along with numerous fabricated parts such as boxes, are locked in shape with specific types of seaming machines. When a

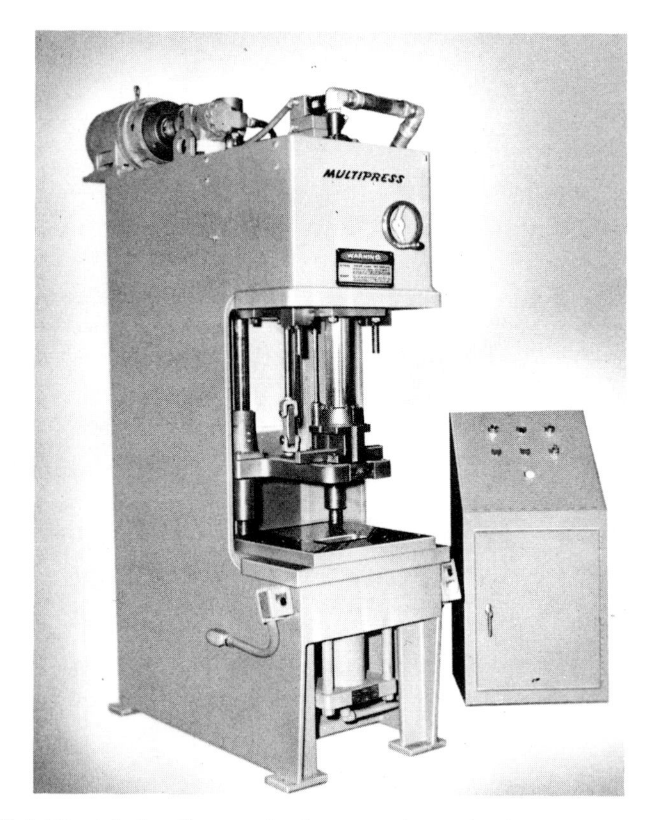

FIGURE 4–67 A hydraulic open-back automatic production press. (Courtesy of Abex Corp.)

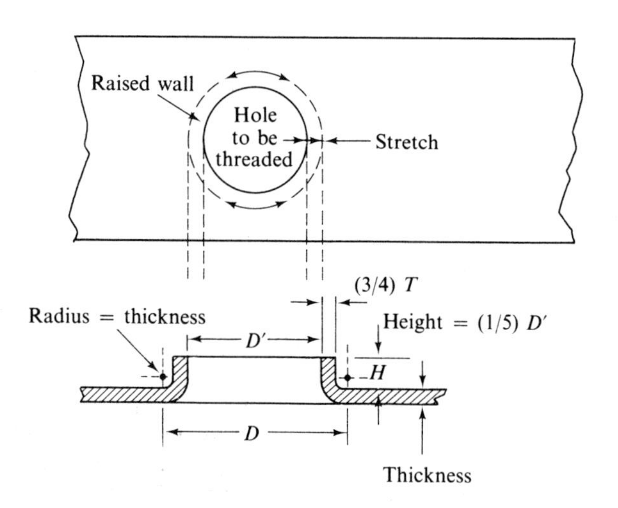

FIGURE 4–68 Swedging is stretch forming.

FIGURE 4–69 A metal spinning operation. (Courtesy of Eagleware Mfg. Co., Inc.)

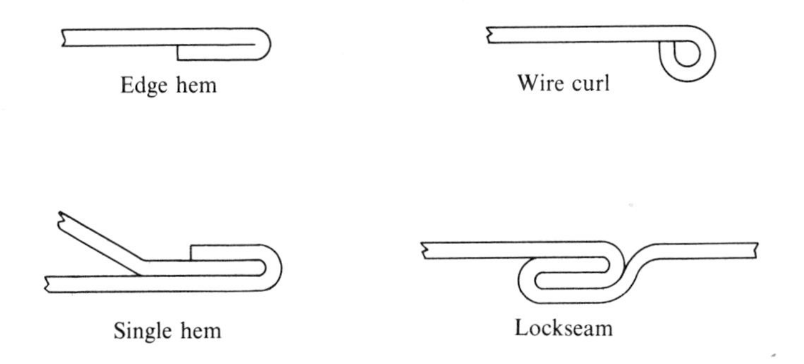

FIGURE 4–70 Typical types of folded seams.

sequence in production is established, parts with seams are produced by the thousands. Metals with hardness values in the Rockwell B scale are easily folded.

Seamed parts are provided with increased mechanical strengths because of the particular shape of the seam. Seaming is an extension of structural shapes. When a sheet of metal is turned back on itself, the metal is stronger in that particular cross-sectional area. Strength in this connotation refers mainly to stiffness. Some metal edges are merely folded 90° in order to eliminate the rough edge or to greatly increase the stiffness along the edge (fig. 4–71). The turned down edge then becomes a web in a kind of beam and automatically provides a section modulus factor peculiar to beams, as pointed out in dimension Z. This factor relates to bending resistance. When a load P is placed on the turned edge, less deflection occurs than if no turned edge existed. In this respect, should the lip be turned inward 90°, less stiffness would be provided than is presently illustrated. An examination of figure 4–72 shows that the part in *a* is

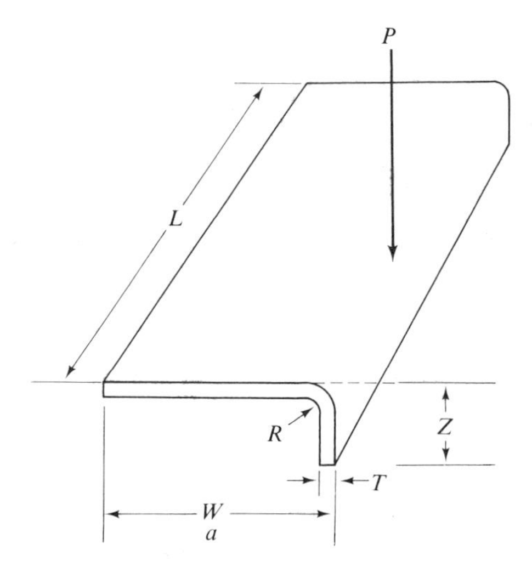

FIGURE 4-71 A turned edge of metal.

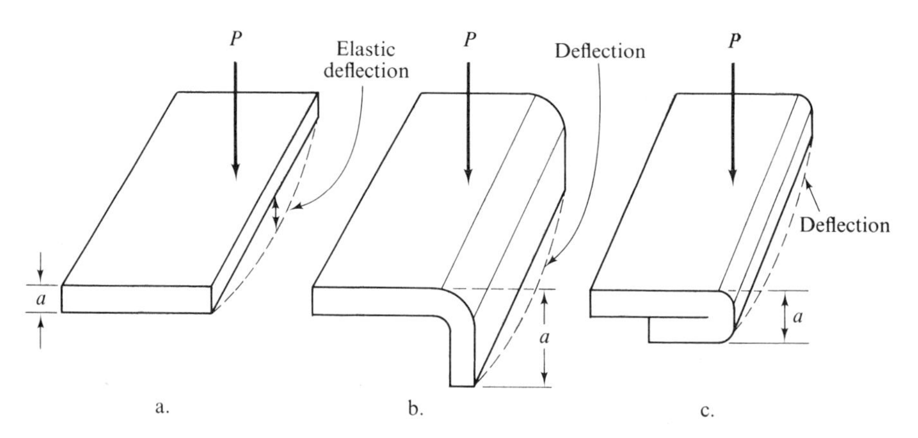

FIGURE 4-72 Strengths of edges of metal.

much weaker than the part in *b*. Should the edge in *b* be folded back on itself, as illustrated in *c*, the stiffness factor is greatly reduced, the edge then being equal to a strength factor between *a* and *b*.

As the shape of the part changes, other types of machines are required to produce the part, therefore, many edge shaping operations lose the seaming connotation and become known as bending, flanging, curling, crimping, bulging, corrugating, and beading. These examples are illustrated in figure 4-73. Accompanying stresses are pointed out in each example. As is illustrated, when a metal is bent, both tension and compression stresses are induced. Sometimes the seam or other shape is not at the edge of the metal but is in its inner areas.

When edges of metal are formed into configurations for purposes of adding strength to another part or assembly, the operation is called *ribbing*. Aircraft structural members such as wings, fuselages, control surfaces, and nacelles are

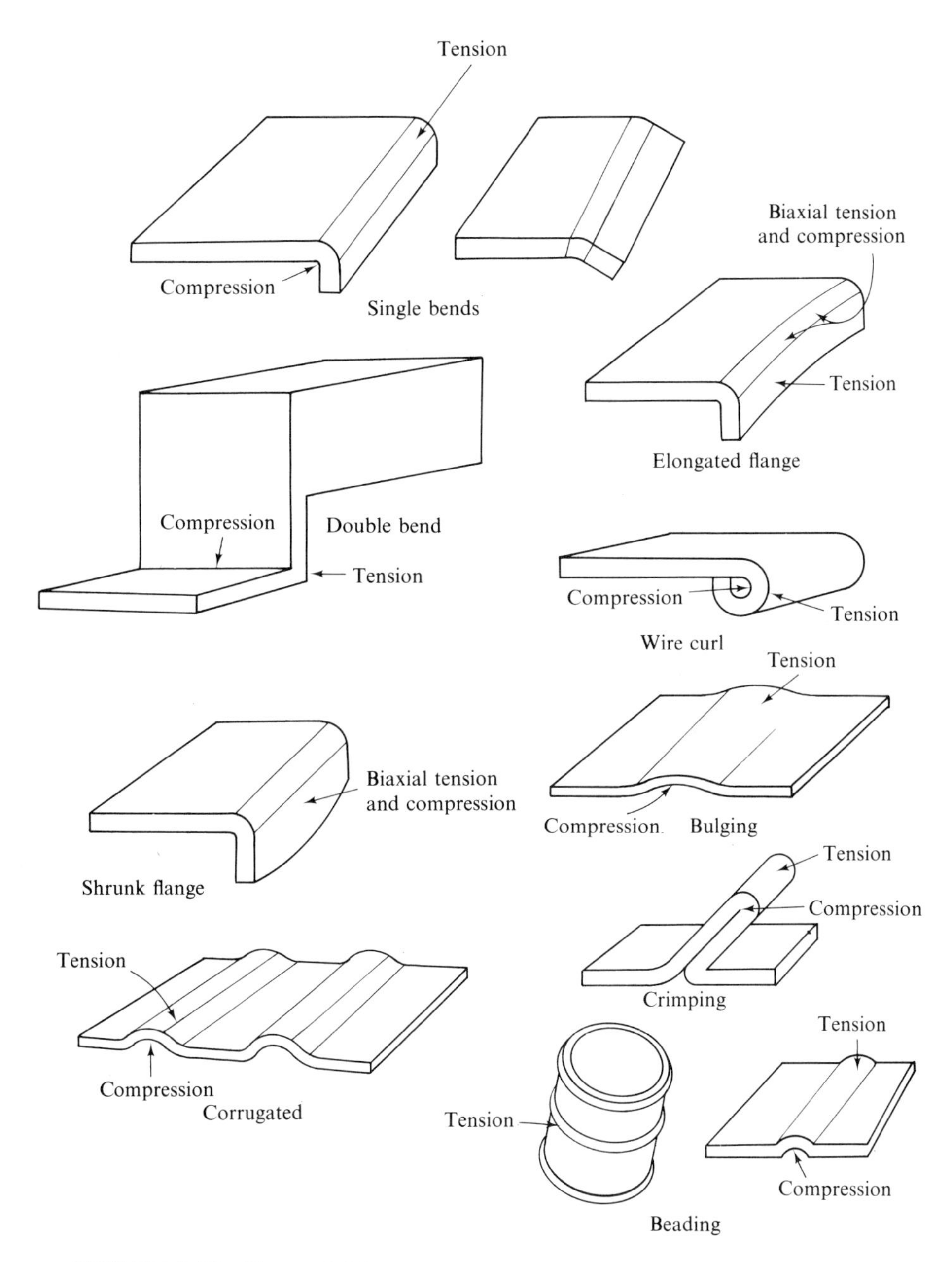

FIGURE 4–73 Types of folded edging.

reinforced with ribs, as illustrated in figure 4–74. Notice the several sheet metal operations necessary for completion of the rib in *b*. The rib is laid out on flat stock such as aluminum alloy and cut along the outer edges. In one operation a specialized press blanks holes, as shown, and then turns the edges of the holes and the periphery of the rib for stiffening purposes. Adjoining edges are notched when the holes are blanked.

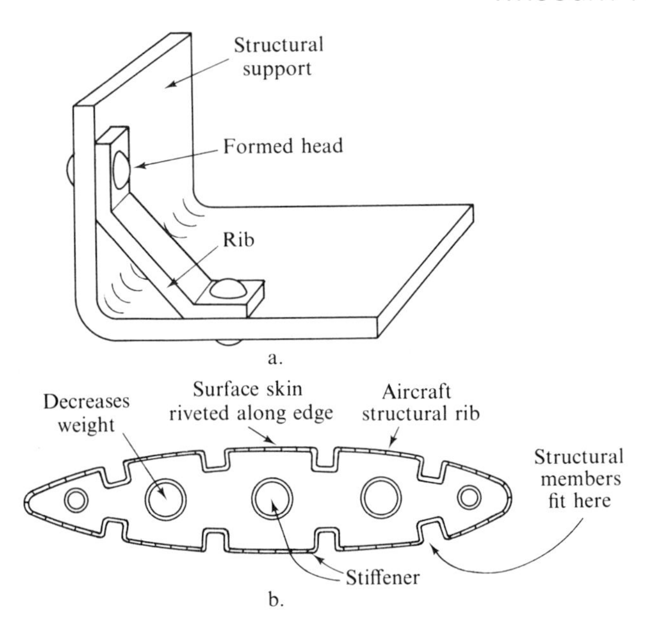

FIGURE 4–74 A supporting rib.

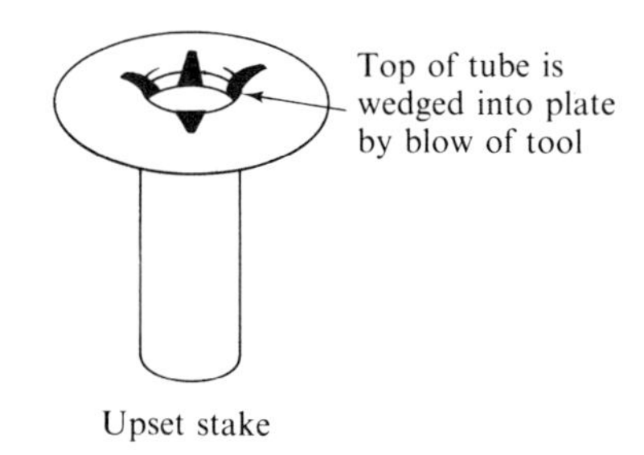

FIGURE 4–75 A staked fitting.

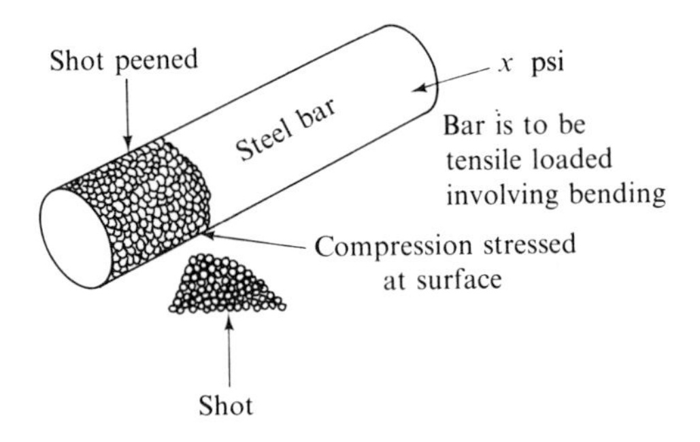

FIGURE 4–76 A peened surface.

When placed under multiloading conditions, the rib adds strength to the larger assembly. Straightedged holes and other edges are inherently weak. Curling at the holes and folding with generous radii provide for increased strength capabilities.

STAKING

Staking operations are cold forming operations that allow metal parts to be permanently joined (fig. 4–75). The stake connection is a depressed area of metal involving two pieces of metal. A punch, chisel, or forming tool is impacted against the area of metal to be joined. The top area of metal is compressed and embedded into the

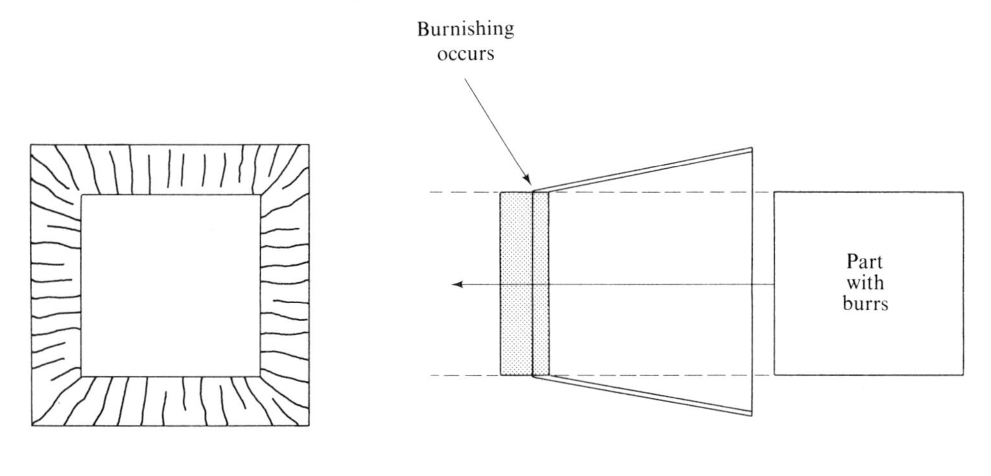

FIGURE 4–77 A burnishing operation. Part with rough edges and burrs is pushed through die and exits with smooth edges.

bottom sheet by deformation and cutting actions. Obviously, the strength of the joint depends on the kind of staking performed. Such a joint would not be exposed to heavy loads.

PEENING

The peening operation is performed on the surface of cold metal for the purpose of inducing compressive stresses at the surface. When a metal is exposed to a bending stress, the two outer surfaces become stressed at values higher than inside areas. If the stress is a changing one, then time under load with the changing load becomes a fatigue factor. When the bend in the metal causes stretching on one side as compression occurs on the opposite side, the tension area becomes subject to cracking. Peening with glass beads, metal balls, sand, or peening tools places impact compression loading on the surface and, in turn, compression stresses are fixed into the tension layer of metal. Therefore, when the part is tensile loaded, the residual compressive load is subtractive from the tensile, and a more fatigue resistant metal is produced. Peened surfaces are rough surfaces (fig. 4–76).

BURNISHING

Burnishing is sometimes used to smooth the edges and surfaces of metal parts. When a shaped part is pushed inside a slightly tapered opening, parts of the metal rub

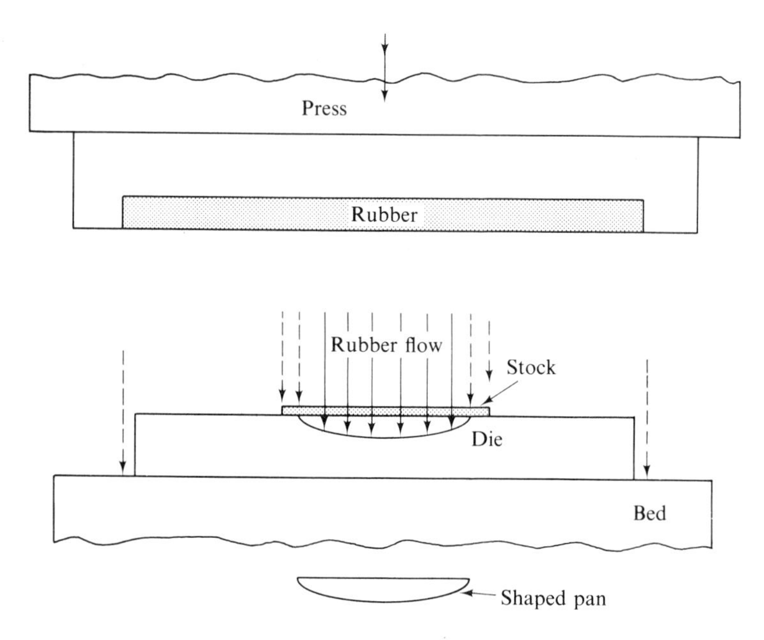

FIGURE 4–78 A rubber shaped pan.

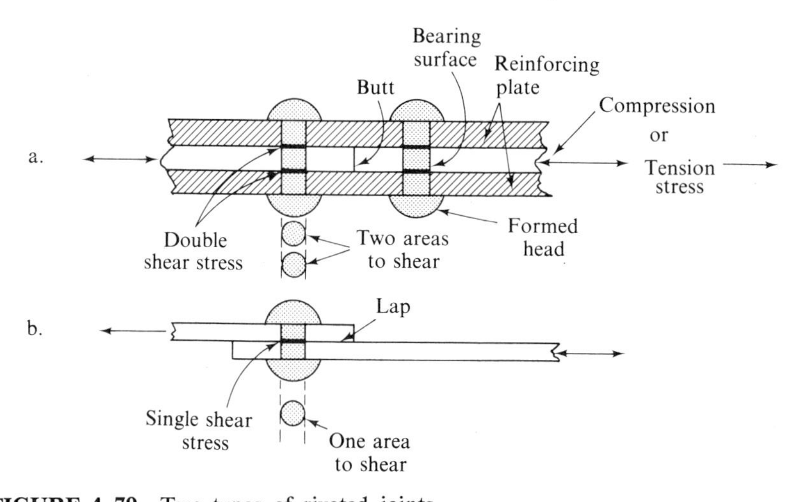

FIGURE 4–79 Two types of riveted joints.

against the sides of the die and are, therefore, smoothed. Rough edges and surfaces are removed by burnishing action. No metal is intentionally removed, but cold working occurs at the points of rubbing. A burnishing operation is illustrated in figure 4–77.

DEFORMING WITH RUBBER PADDING

Under pressure, rubber simulates a fluid condition, therefore, when a pad of rubber is substituted for one die, the enclosed sheet of metal is uniformly and smoothly pressed into the other die's cavity. The rubber forming process is a modified stamping process that reduces cost by eliminating one die. More intricately shaped parts at reduced costs are made by this process. Ductile metals are used in the production of all types of shapes while cold, provided sheet stock is used. Figure 4–78 illustrates the rubber forming process.

RIVETING

Riveting is a cold compression forming operation that uses a round metal rod to connect two or more sheets of metal. As illustrated in figure 4–79, the length of the rivet must be equal to the length of the head, the length of the thicknesses of metal, and the formed end. Rivets must be soft, as nonductile metals will split when deformation continues by squeezing or hammering action. When a proper sized rivet with the desired head shape is placed in a very slightly oversized hole, enough rivet shank must extend for *bucking* or forming. Usually, this is about 1½ times the diameter of the shank of the rivet. Bucking of the end is accomplished by holding a formed rivet set on the head of the rivet and applying a few light blows of the hammer on the shank's end. Skill is required to compress and shape the end to pull the metal sheets tightly together. As the end is compressed by shortening, the diameter increases. Consequently, a bulged end, or head, occurs on the rivet as the hole is tightly filled with metal to be used in shear. Keep in mind that when tension or compression forces are placed on the riveted sheets, the rivets are subjected to shear stresses. A rivet, then, is a shear substitute for a quantity of metal which was originally tension or compression metal.

As pointed out in figure 4–79, the joint in *b* is a lap joint in single shear, which means that one cross-sectional area is subjected to shear whereby tensile stresses are transmitted through the sheet to the rivet. In turn, the rivet transfers the stress to tensile stress in the other sheet. In *a*, two cover plates fasten the butt joint in double shear. Notice that two cross-sectional areas of the rivet must be cut if shear is to occur, therefore, this joint is stronger than the lap joint.

Rivets are used in numerous types of structural designs. The airplane is a riveted structure to a great extent. The outside skin of aircraft is riveted to internal stringers

FIGURE 4–80 This *F*-111 supersonic fighter aircraft is one of the world's most effective weapon's systems and its capability results from a combination of engineering materials, processes, and skills. (Courtesy of General Dynamics–Convair Aerospace Div.)

FIGURE 4–81 The *YF*-16 lightweight fighter incorporates many new types of materials and processes, including bonded surfaces. (Courtesy of General Dynamics–Convair Aerospace Div.)

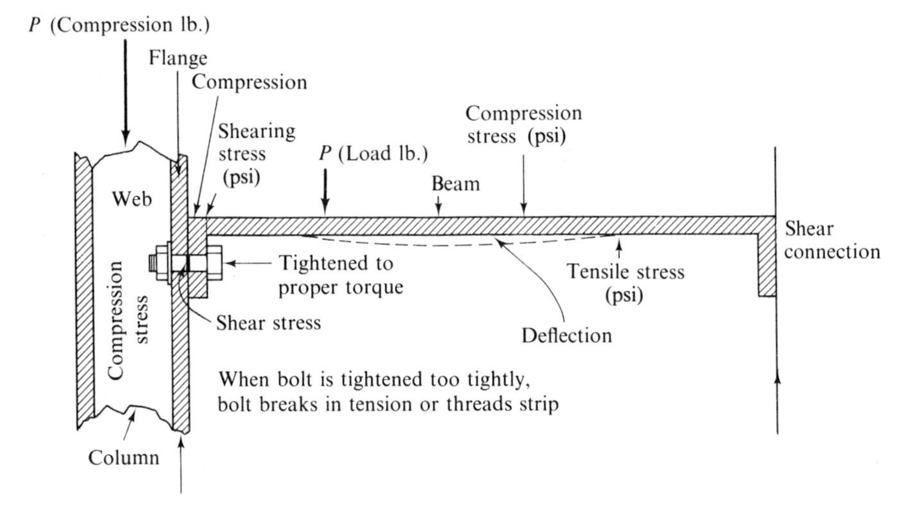

FIGURE 4–82 A bolted joint.

which, in turn, are riveted to ribs or bulkheads, and so on, to the main members. Flush rivets are used on the outside skin which is exposed to the air stream, while shaped head rivets are used in the interior. Most of these rivets are aluminum alloy. The *F*–111 (fig. 4–80) uses rivets throughout its structure. Several different types of joining methods, including bolting and adhesive bonding, are used in these and other highly stressed structures such as aircraft pictured in figure 4–81 and figure 14–10. In the bridge and building industry, steel rivets are frequently used, being driven while hot. However, much of this type of connection is transitioning into bolt connections (fig. 4–82). Rivets are designed to withstand shearing stresses, as illustrated in further detail in figure 4–83. Metals such as nickel-copper alloys, in addition to aluminum alloys, are used in highly stressed structures such as aircraft.

COINING AND THREAD ROLLING

When delicate impressions of symbols, wording, and other similar impressions are required on metal, coining is used. This process is automated whereby thousands of

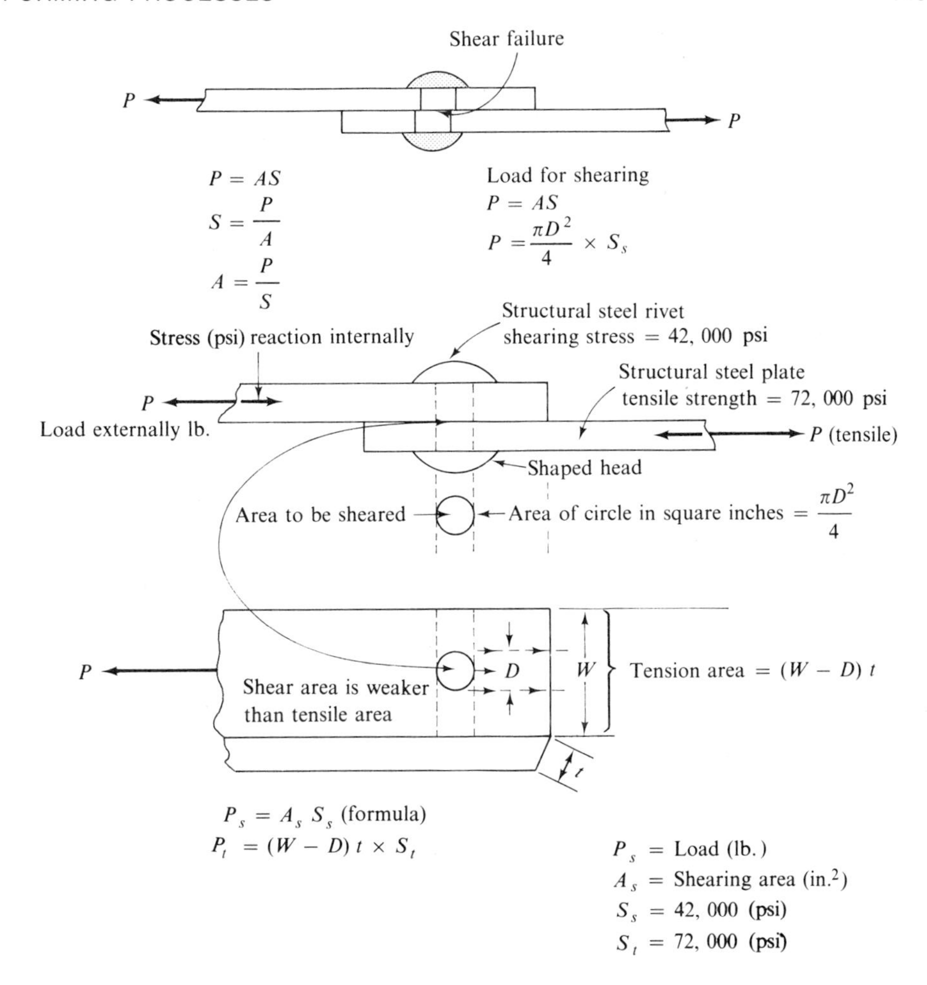

FIGURE 4–83 Load required to shear a rivet.

identically shaped pieces of metal are produced when a given quantity of a certain metal is placed between a very powerful punch and a die (fig. 4–84). Because of the precision of the process, the slug is carefully metered into the coining machine in order to assure proper functioning of the dies.

Coining is a cold working process that forms surface characters and increases hardness. In the same concept of hardening through coining, the threads on the bolt (fig. 4–84*b*) are cold rolled onto the shank of the bolt by actions of a sliding die. As the die moves along a line, the bolt spins, and thread grooves are rolled into the metal. Thread rolling is not coining, however.

EMBOSSING

The embossing process is a stamping process that forces the metal to take the shape of the dies (fig. 4–85). Metal slugs are placed in compression which, as in coining, forces designs onto the surface of the slug. The lack of precision differentiates this process from coining.

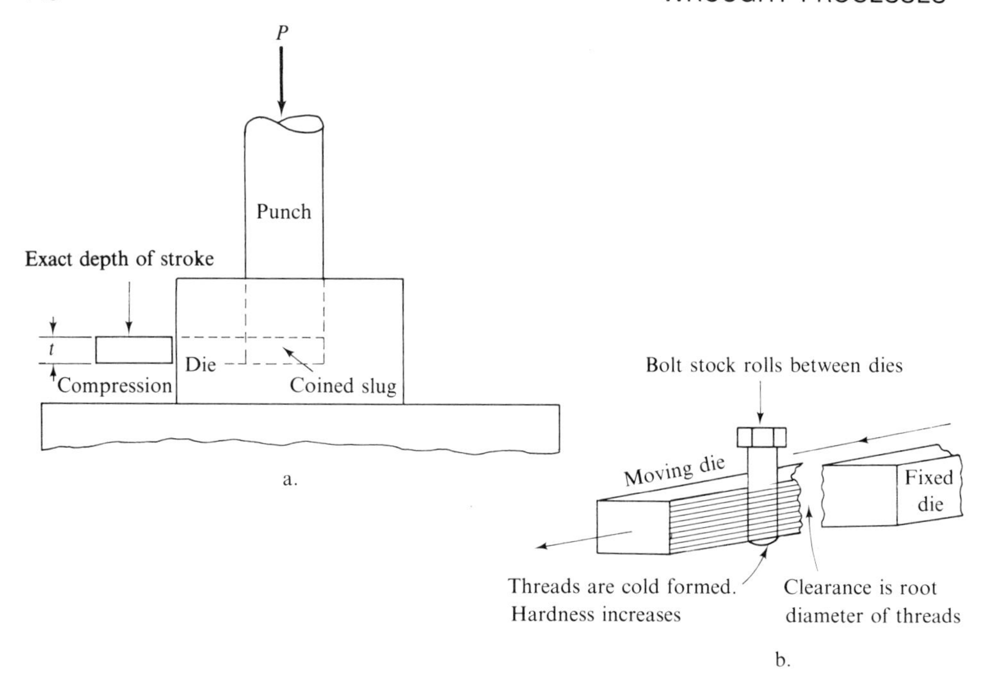

FIGURE 4–84 A slug being coined and a thread being rolled.

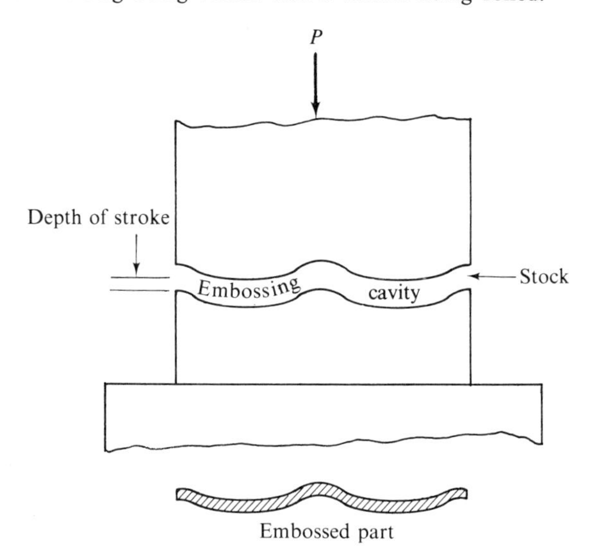

FIGURE 4–85 An embossed part.

Extrusion

Either hot or cold metal can be forced through an extrusion die, causing the metal to take the shape of the exit hole in the die (fig. 4–86). Nearly any shape can be extruded, including those parts having hollow spaces. Hollow shapes require *mandrels*, the extruded metal flowing around and along the mandrel. The inside shape takes the

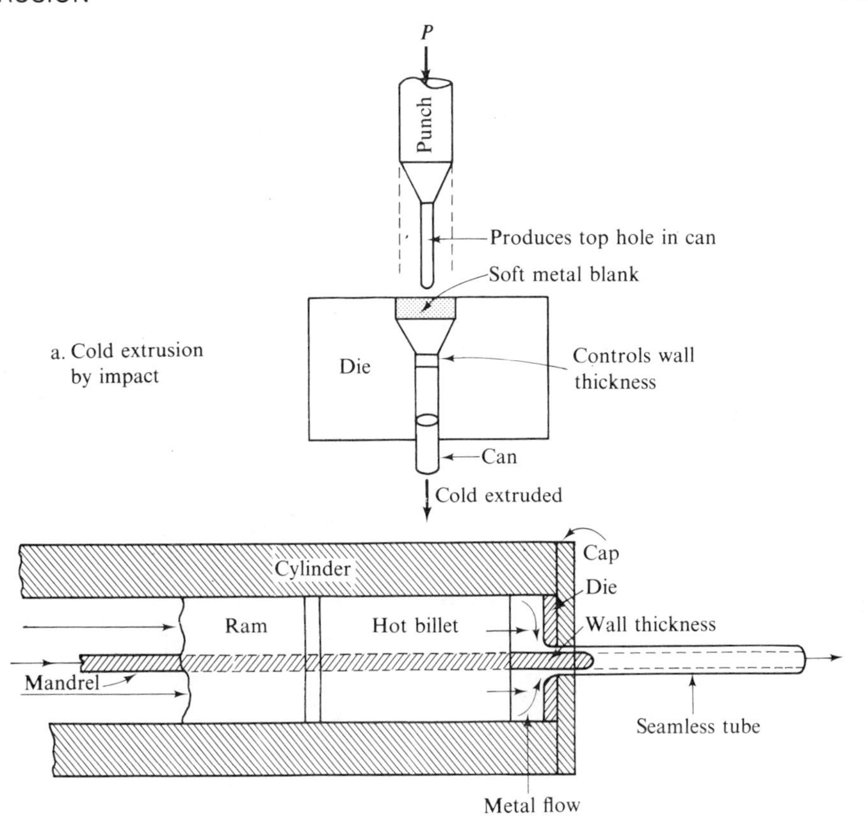

FIGURE 4–86 Extrusion processes.

shape of the mandrel. In figure 4–86*a*, a metal blank is cold extruded through a die by means of a punch, or mandrel. In *b*, a hot billet of aluminum is forced through the die by tremendous pressure from the ram. The tube is produced as the mandrel moves through the ram and blocks part of the die. As the extruded tube emerges, its shape is identical to the effects of mandrel and die. Figure 4–87 shows a billet being rammed into the extrusion chamber of a mill.

Most metals are extruded in sizes not to exceed a 30-inch circle. Possibly the main advantage of extrusion is the capability of forming shapes that are not feasible in any other method. However, many common shapes are rolled, as well as extruded, such as tubes and wide flange shapes. The most common metals which are extruded include aluminum and its alloys, copper and its alloys such as brass and bronze, magnesium and its alloys, and some of the lower melting alloys. Steel and titanium are also extruded, but with closer controls. Approximately five-inch diameter parts of these metals are limited to the extrusion press, as compared to 30-inch parts for aluminum. Limitations of sizes for steel and titanium refer to distortion ability and plastic flow of the metal. Because extruding is a multicompression stress operation, triaxial compression, that demands extreme ductility factors in the metal, deformation beyond the yield must allow replacement metal as needed to prevent tearing.

FIGURE 4–87 A billet being forced into the extrusion die by the ram of the extrusion mill. (Courtesy of General Extrusions)

Metals are literally pushed into a confined hole under tremendous pressures during the extrusion process. Just as paste leaves a squeezed tube in the shape of the tube's opening, metal leaves the extrusion die in the same manner. For example, a 10-inch diameter round section of aluminum 36 inches long is extruded into a 30-foot length of edging. The process is started with the billets. The billets are lifted into an automated conveyor system moving through a gas fired furnace operating at an extrusion temperature. As the hot billet leaves the long furnace, it is automatically lifted into position, resting in front of the ram. (Rams have various loading energies, some being able to exert millions of pounds of force on the billet.) Immediately, the ram moves into the completely enclosed billet. Only one way exists for escape of the metal, and it is through the die, as shown in figure 4–88. With a constant force of a million pounds, for example, moving the ram at a slow and constant rate, rapidly deforming metal moves into the tapered die opening and plastically flows through the die. The die's exit has the exact shape of the desired cross section of the part. Length of the original billet determines the length of the extruded part. Sometimes, metal must flow several inches before entering the die, consequently, most extrusions are produced while the metal is hot. However, some are produced cold, which requires higher energy input, but results in stronger shapes. Figure 4–89 illustrates the quantity capability of the extrusion mill, while figure 4–90 shows how some billets are handled as they enter the heating chamber.

FIGURE 4–88 An aluminum extrusion emerges from a hydraulic press. (Courtesy of Reynolds Metals Company)

EXTRUDED TUBING AND OTHER SHAPES

The extrusion press is also used for making hollow shapes such as seamless tubing. In order to produce the central hole or holes to form a tube, a mandrel or a series of mandrels is placed within the ram of the press and extended through the die to the exit side. (Figure 4–88 shows a beam being extruded.) The billet is squeezed by compressive forces between the ram's face, the die opening, and the walls of the die's enclosure. As the ram moves forward, escape of the compressing material is only through the die opening. Therefore, the deforming hot billet slowly flows through the die as it conforms to the shape of the die opening. An exact replica of the die opening is produced in the emerging part.

Other hollow shapes are produced by varying the ram, mandrel, and die designs. Square, rectangular, triangular, and oddly shaped solid or hollow configurations are extruded. Many extrusions, besides tubing, include structural channels and flange sections, window and door frames, kitchen edging strips, and numerous fittings, supports, and hardware brackets. Saws or shears follow the extruded length and part the shape at the desired length. When short lengths are cut from the fixed cross section of the extrusion, dozens of individual identically shaped parts are produced. As an example, a long brass hollow pinion wire with keyway is extruded and then

FIGURE 4–89 Numerous types of structural shapes are produced by this powerful extrusion mill. (Courtesy of Alenco)

parted in 1/2-inch intervals. Dozens of hubless gears are then produced. Such precision depends on the accuracy of the die.

Questions

1. Differentiate between cast and wrought processes.
2. Explain the difference between hot and cold rolling processes.
3. From the standpoint of a given metal and treatment, explain why forging is usually superior in strength compared to other shaping processes.
4. Describe the differences between stress and strain.
5. What is the difference between elastic and plastic conditions in a metal?
6. Describe the recrystallization process in metals.
7. Why can metals have differently shaped grains?
8. What does the modulus of elasticity measure?
9. Differentiate between yield and tensile strengths.
10. Name and describe the three main engineering forces.
11. During a forming operation, what is meant by pressure on a metal and stress in a metal?

FIGURE 4–90 Electromagnetic billet heater used to heat the aluminum billets prior to extruding. (Courtesy of General Extrusions)

12. Can rolling operations successfully close all gas holes in a cast metal? Explain your answer.

13. What is the relationship among ductility, malleability, and plasticity?

14. Why must springback be considered in the deformation of metals?

15. Describe how the angle of bite influences the rolling operation.

16. What is meant by seamless tubing?

17. Differentiate between an axial stress and triaxial stresses.

18. During metal fabrication, what is the meaning of plastic flow?

19. Explain what happens when a metal is cupped as compared to when a metal is sheared.

20. Surface hardness may not be indicative of the strength of the material. Explain.

Powder Metallurgy

Powder metallurgy is the process that produces parts by means of compressed and sintered powders. The process is not new; evidence of its existence has been established during pre-Christian times. However, this third method of producing parts did not become very active until the material needs of World War II sent engineers on the search for new ways of making parts. Closer tolerances in parts' dimensions were demanded by designers for specialized assemblies, and better ways of forming carbides into cutting tools were needed. Mass production of hard-to-shape parts was required. Because powdered parts had been produced to a limited extent, the potentials of this process were investigated. Immediately, new die designs and compressing techniques were discovered which led to a rapidly expanding technology.

Powder metallurgy is a simple process that involves the pressing and sintering of powders. Today, millions of parts are produced over short time periods by means of automated and semiautomated processes. The powder process is truly a third method of production, preceded in quantity output by casting and wrought processes. When metal powders are placed in a die, followed by compression under thousands of pounds pressure, a tightly packed briquette is produced. Heating the *green,* or raw, briquette to elevated temperatures causes *fusion,* or adherence, of the powders, and a strong and solid part results.

The Growing Need for a Faster Manufacturing Method

Even though tungsten carbide had been pressed and sintered into exceptionally hard cutting tools during World War II, there still remained the need to combine powders of other materials so that special mechanical and physical properties of each material would exist. As an example, self-lubricating bearings were required in specialized equipment. In this respect, it had been noted that copper and graphite powders could be compacted and sintered and used as bearings. Along with the need for the special bearings, there was also a requirement for close tolerances in these finished parts. Further, these bearings required mass production by automated processes.

Casting and wrought processes are limited in the kinds of materials and shapes that can be made, even though they produce the majority of parts for society. Also, much scrap metal is produced in these processes, especially in the machining aspects of the processes. When smaller tolerances and increased production were sought, the powder process was firmly established because it could provide identical parts by the thousands with tolerances of 0.001 inch.

Research was reflecting the tremendous potentials of powdered metals. Ferrous and nonferrous metal powders were mixed so that advantages of each would be provided. Even mixtures of hard and soft metals or metals and nonmetals were made. In fact, parts could be produced so fast and so exactly, that industry quickly accepted powder metallurgy as one of its major production processes. Today, nearly an endless list of parts is being produced by this process, but the powder method of making parts is not without limitations.

Principles of Powder Metallurgy

Because powder metallurgy requires the compression of fine powders in order to maintain a minimum density, special equipment is essential. Powders are compressed in a straight line effect, therefore, density is obtainable only where compression forces exist among the powders. These two factors limit the design of a part by powder methods. When flanges and shoulders, for example, extend beyond the straight line of compression, a density differential exists, and an unsatisfactory part results. When even slightly loaded, such parts will fracture at the line of demarcation between the two density regions. Density, then, is a governing factor which faces engineers, designers, and technicians.

DENSITY REQUIRES COMPACTION

In its simplest meaning, metal powder is tightly packed when compacted within a cylinder, but only that volume of powder will be compressed that is in line with the compacting ram (fig. 5–1). As the ram or punch descends, pressure is placed on the powder directly beneath the punch and in fairly close proximity to the punch. As pointed out in figure 5–1, powder at the bottom of the die will be less dense than powder adjacent to the punch. When an opposing punch is introduced from the bottom, however, powders become denser along the faces of both punches, but are less dense in the center (fig. 5–2). But when a die design as shown in figure 5–3 exists, there is absence of density in the part's flange regions, as illustrated in *a*. Solid ductile metal will plastically flow around corners when under high compression loads, but powdered metal will not. Bridges of powder will form at corners from high density to low density regions and, in turn, an unsatisfactory part will be ejected from the die. When a part design such as is shown in figure 5–3 is needed, dies must be built to provide compression loading of all the powder in the die, as illustrated in *b*. Obviously, this concept is also limited, but limitation is based mainly on the quantity of powder that can be effectively compressed. Opposing dies provide compression loading from each end which assures packing and loading of powders in the corners and in the center.

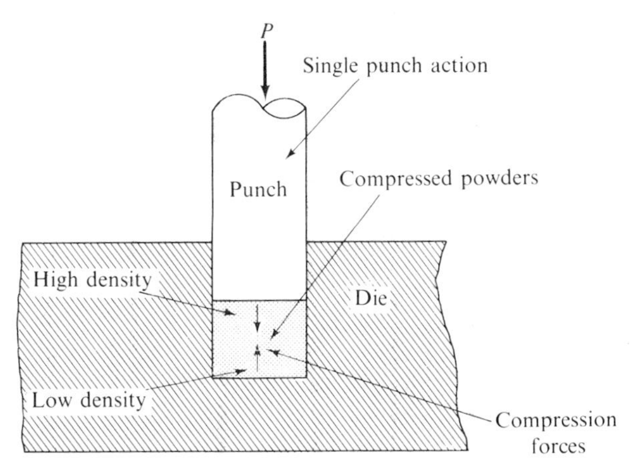

FIGURE 5-1 A one-punch design cannot provide for uniform density when length is involved.

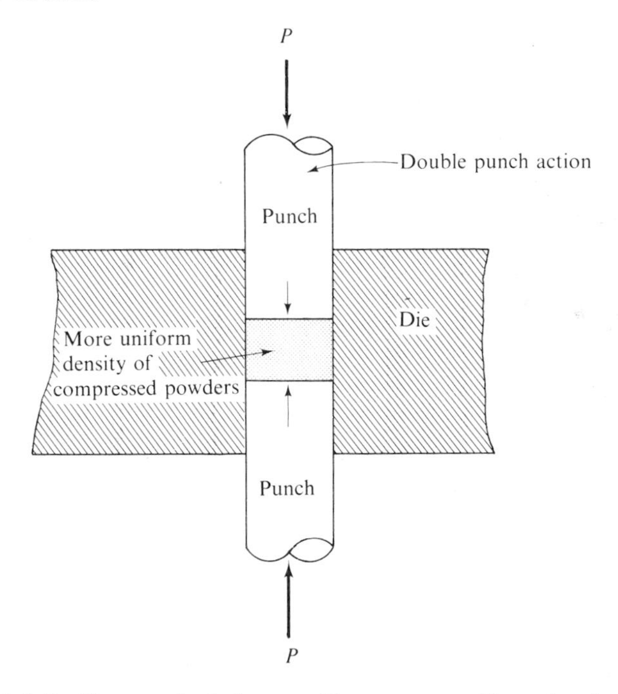

FIGURE 5-2 Two-punch design provides a more uniform density.

THE REQUIREMENT FOR UNIFORM DENSITY

The design of the part should provide for powder compression in all regions of the part. When holes, keyways, and related open areas exist in places other than parallel to the lines of compression, the design is unacceptable (fig. 5-4*a*). Compaction of powder will not occur below the hole. Often, however, compression can be performed at 90° to the original procedure and produce the holes and grooves (fig. 5-4*b*). When holes and grooves are required at angles where compaction cannot occur, these must

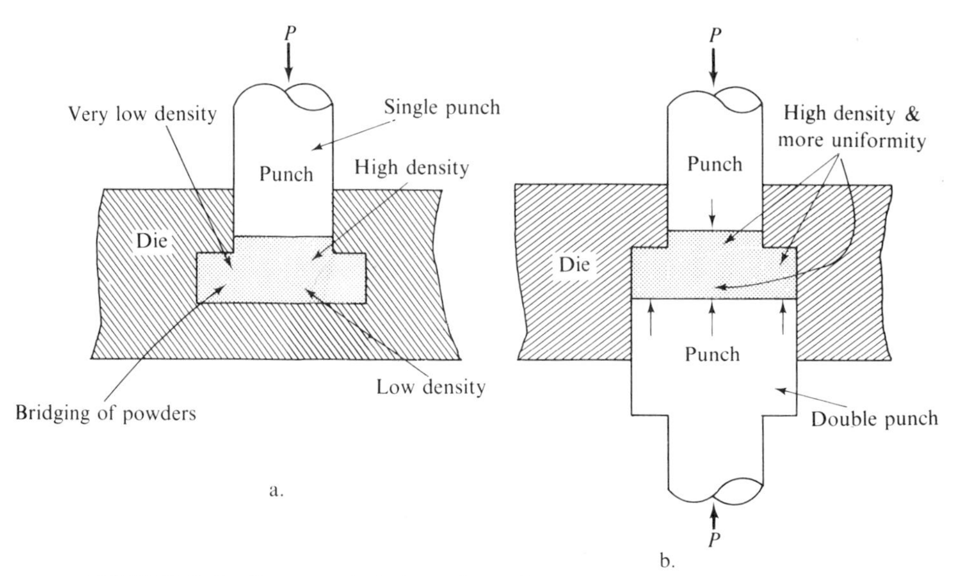

FIGURE 5–3 Two-punch design allows flanges to be made.

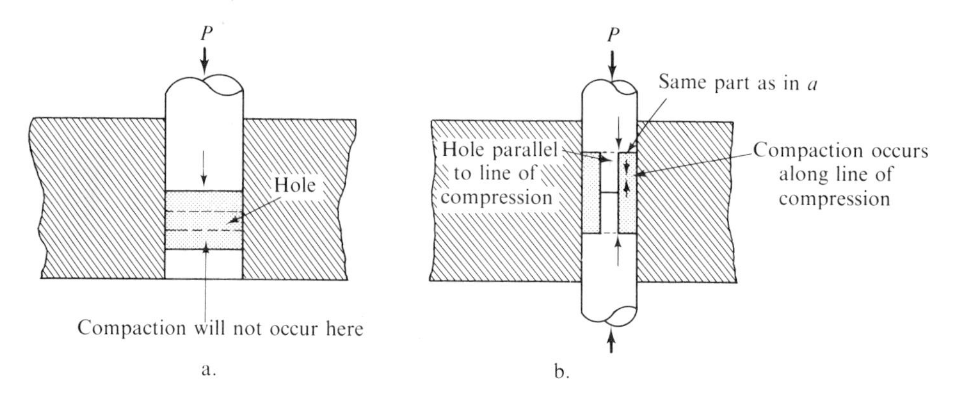

FIGURE 5–4 Hole in part turned 90° in order to produce.

be machined after a slight presintering, that is, shrinkage and dimensional change in powders occur as sintering begins and will not affect the accuracy of the machined threads or holes. Full sintering then follows machining, shrinkage having already occurred during the presintering.

DESIGN REQUIREMENTS

Threads, reentrants, undercuts, and related designs cannot be produced unless modifications in procedures are provided. Reentrants and undercuts can be produced when the line of compression is parallel to the vertical or nearly vertical sides of the angle, as illustrated in figure 5–5*b*. With respect to figure 5–5*a*, powders in those areas below reentrant angles will not be compressed. In *b*, the part has been turned 90°, enabling maximum pressure to be exerted across the faces of both punches. The angular areas are subject to shifting of particles, but will compact. Two punches are

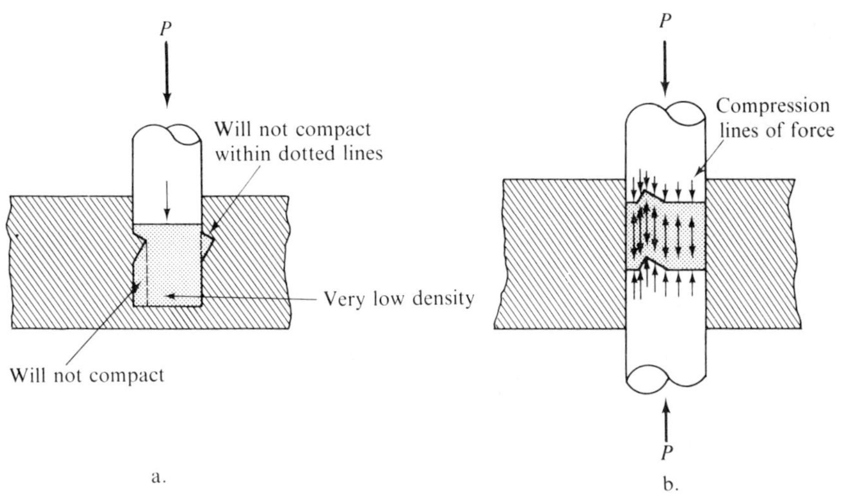

FIGURE 5–5 Compression forces are turned 90° enabling production of the part.

required. A theoretical "best" design is one which is symmetrical along the lines of compression, as illustrated in figure 5–2 and figure 5–4*b*. When cylinders are being produced, wall thicknesses should be uniform; if not, a smooth transition must be provided. For purposes of ejection, a draft of about 0.009 inch per inch is used along the wall. Another limitation is the thickness of flat parts or sections of parts. Thicknesses less than 0.030 inch become increasingly difficult to effectively produce. Feathered edges invite trouble. When a flange or lip is needed, secondary punches are required in the tool's design (fig. 5–3). Flat areas in contact with changing contours must be fitted with reasonable radii; sharp corners are potential fracture regions. If the corner is recessed within the part, care is needed to provide for essential compaction in the corner by proper procedure. Thicknesses and densities relate to the design of the cross-sectional areas of the part. When the design of the part calls for a thin groove, then the tool used in machining the die must also be thin. Therefore, tooling problems arise when length and diameter ratios become unreasonable.

There are many other limitations in design of the part when metal powders are used. As has been previously pointed out, powders do not carry stress patterns in the same way as does cast or wrought metal. Consequently, compression loading of powders extends outwards from the punch with an increasing reduction in density of material. This factor limits the length or depth of the part produced by powder metallurgy. Lateral dimensions are limited by die design with respect to feasibility of multiple punches and closeness of adjacent punches. Punches may be wide because pressure is exerted at 90° to the face of the punch, but when cross-sectional changes occur in the part, adjacent or opposing dies must have operating space. Unusually wide and conically shaped parts are feasible to produce when thickness remains within a three-to-one compaction ratio of the powders. However, such wide parts should be reinforced with some type of cast or wrought metal structure when placed in use and when heavy loading is expected. Extremely small parts are limited to the feasibility of die design. Very large parts are limited to acceptable density standards which depend also on die design and power of the press. Generally speaking, parts weighing up to 400 pounds are commonly produced by powder metallurgy.

DIAMETER-TO-THICKNESS RATIOS

When pressing ratios become greater than three-to-one, density differentials become rather high. In other words, when the unpressed powders are compacted to more than one-third of their initial volume, density differentials increase due to maximum density existing in the proximity of the face of the punch. Some thin parts, however, have been effectively compressed to as high as a ten-to-one ratio. The least length-to-diameter design ratios are best, as are the least length-to-wall thickness ratios. As pressures increase along the length axis, they do not necessarily increase laterally. This factor must be considered by technicians during design of the part as well as design of the process.

It has been found that parts having cross-sectional dimensions lying within a 12-inch circle with a depth dimension up to nine inches long or thick are satisfactorily produced. The length factor is then the density factor, while the lateral dimensions are limited by amount of pressure available, and this obviously depends on the size of the machine. When compaction ratios become greater than three-to-one, bridging of powders often occurs and density differentials result. Therefore, strength properties are reduced, and brittleness in lower density regions is present. This means that a close correlation exists between density and strength of the material. Consequently, uniformity in shape along the axis of compression is desired in order to acquire a uniform density in the part.

When density uniformness becomes a problem in a single-acting punch process, multiple punches are required. Multiple punches may consist of a punch emerging from the center of a punch (fig. 5–6). Inner punches move independently from parent punches, may reciprocate, may move at different speeds, and then may be synchronized to parent punches when all punches compress equally at the same time. Inner punches and multiple punches help fix a more uniform density in the powdered part. The faces of punches allow for designs other than flat.

TOLERANCES

Tolerances of pressed powders can be kept small, just as they are in the die casting process, because of the nature of the process. In die casting, molten or plastic metal flows under high pressure into all cavities of the die. When properly designed, loose powders compact to high densities inside a closed die. In such situations, close tolerances are expected and are obtainable. As an example, diameter dimensions can be contained within plus or minus 0.001 inch. Large parts can be produced within plus or minus 0.007 inch. Gear teeth, for example, must fit properly along the points of contact during the transfer of power. A smoothly moving mass of metal is required at all points of contact at the pitch circle. Often, plus or minus 0.0005-inch tolerances are obtainable. Powder metallurgy offers the technician the opportunity to reinforce the dimensions of gear and sprocket teeth so that greater loads can be transmitted. Powder metallurgy is playing an increasingly greater role in the gear manufacturing industry.

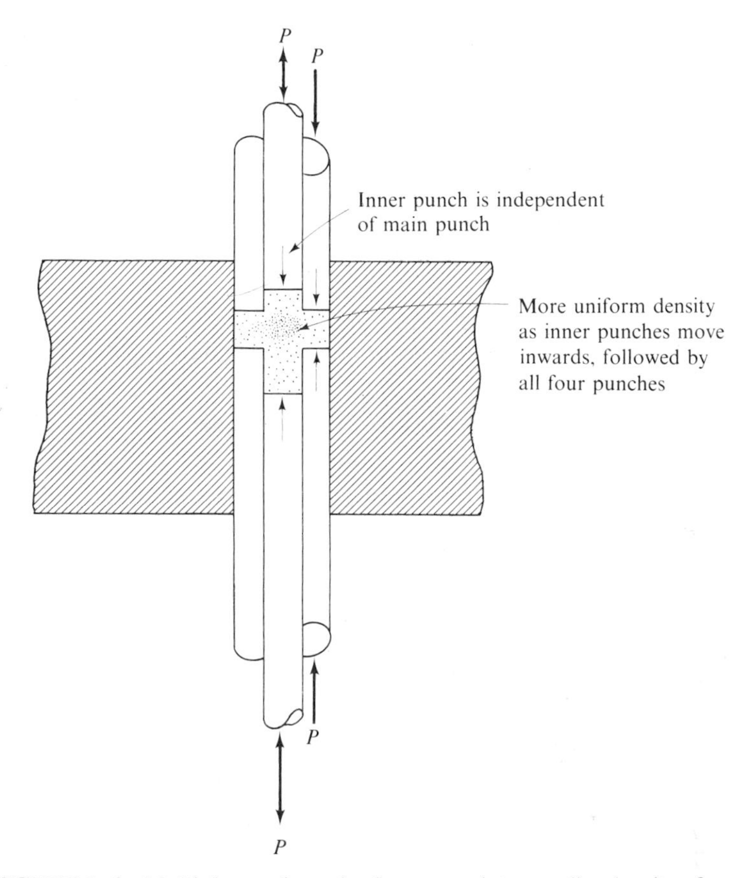

FIGURE 5–6 Multiple punches using inner punch to equalize density of powders.

Capabilities of Powder Metallurgy

Because new methods of powder production have been developed, the cost of powders is being reduced. Even though a given volume of powder may initially cost more than a solid mass of the same metal produced from casting or wrought processes, no scrap results in the production of parts. All of the powder is used during pressing, while scrap is produced during other manufacturing processes such as machining. Volume for volume of metal used, powder processes are more economical than cutting processes when the costs of scrap, labor for cutting, machinery, and labor for cleanup are considered.

Powders are available in the metallic and nonmetallic forms. Designs increasingly call for mixtures of metals and nonmetals, along with the need for many different alloys. In powder form, these mixtures are easily and very accurately prepared. Only three steps are required to completely make a part: mixing and blending of the powders, pressing of the powders, and sintering of the mass. As an example, the compound of aluminum and oxygen, aluminum oxide or *alumina,* is readily available in powder form at reasonable prices for the production of refractory material and cutting tools. It ranks near the top of hardness scales. As a comparison, the diamond and boron nitride are the hardest materials and are approximately equal in hardness, while only a few carbides such as boron, silicon, titanium, and zirconium are harder than alumina. As for geometry, shapes of powdered parts are limited only by design; the ease of shaping is fairly simple.

Two of the greatest advantages of powder metallurgy are the lack of waste material and the ability of the process to produce parts which cannot be produced by any other method. For example, when the soft metal aluminum combines chemically with the gas oxygen, the resulting compound alumina is harder than the hardest tool steels. Then, when alumina powders are compacted and sintered into a shape, a product exists that has little competition in regard to cutting tool capability.

PRODUCTION OF THE HARDEST CUTTING TOOLS

Powder metallurgy lends itself to the use of certain powders which can be cold and hot molded into shapes that will withstand various elevated temperatures up to several thousand degrees Fahrenheit. Cutting tools can be made that are many times more resistant to abrasion than the finest tool steels. Alumina powders produce cutting tools harder than the hardest martensite. As an example, RC 70 is convertible to a Knoop 972 on the hardness scale. Tests on aluminum oxide tools have provided a Knoop hardness of much greater hardness. Some carbides, as previously pointed out, produce Knoop hardnesses approximating the aluminum oxides.

EXCELLENT THERMAL SHOCK CAPABILITIES

When thermal shock is involved, powders are available to make materials that will effectively withstand alternating temperatures from below room temperature up to 4000° F. Unusual operating conditions such as those in the chemical, electronic, aerospace, and drilling industries are requiring materials capable of withstanding thermal shock, resisting acid and alkaline attack, resisting abrasion and compression loading, and resisting extremely high temperatures.

SIMPLICITY IN PROCEDURE

Basically, powder metallurgy processes require less costly machinery and supporting equipment than do several of the casting and wrought processes. Powders are requisitioned and stored under safe conditions which keep air from contacting the powders. Oxygen in the air contaminates powdered metals and often prohibits fusion during sintering. Mixing and blending machines are used to properly and uniformly blend a mixture of different materials, along with the various sizes and shapes of powders. Blended and measured powders are then placed in the die of the press and compaction occurs. A typical press is shown in figure 5–7. After compression of the powders, sintering occurs in a furnace similar to the one shown in figure 5–8. After sintering, the parts are either used as they are ejected from the machine, or they are sometimes processed additionally to conform with other requirements.

Presses are either mechanically or hydraulically operated, many presses being the mechanical type. A large press having a 100-ton capacity is considered more than

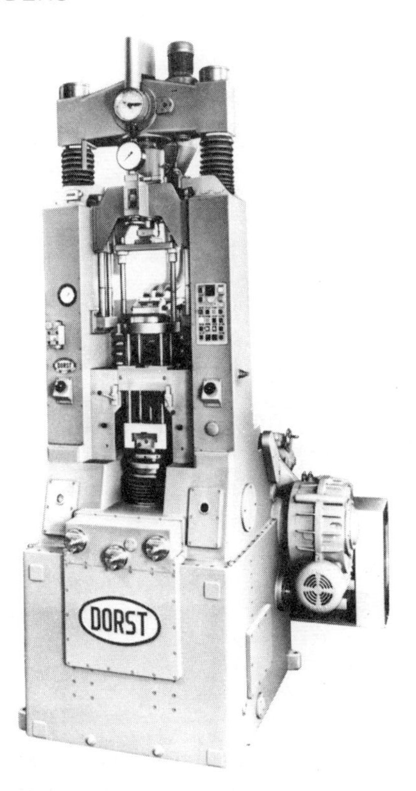

FIGURE 5-7 A 132-ton compacting press used in powder metallurgy. (Courtesy of AC Compacting Presses Inc.)

average in size. Up to 100,000 psi of cross-sectional area are required to produce satisfactory density factors, therefore, most pressing operations with smaller presses are limited to three or four square inches of area to be compacted.

Production of Powders

Metallic and nonmetallic powders are currently available for mass production of parts. Pure metals, alloys, compounds, and nonmetals include the common materials used in manufacturing processes. Much of the metallic and nonmetallic materials used in the casting and wrought processes are now available in powdered form. Just as the metallurgist varies the percentage values of different elements in a metal to be cast or rolled, the same is found in powder metallurgy. Exact analysis of each powder is known, and careful proportions of different powders are mixed. Some metals will not mix in the liquid state, but their beneficial characteristics can be obtained by mixing their powders. Many nonmetals and metals, hard-to-machine metals, hard-to-cast metals, and numerous difficult mixtures in conventional metallurgy are easily obtainable in powder metallurgy. Research has expanded the potentials in this field; its capabilities appear unlimited. Design possibilities are bringing new concepts and production techniques to the manufacturing field.

FIGURE 5-8 A continuous 12-inch mesh belt furnace for sintering powder metallurgy parts with heavy loading drive and connection cooling. (Courtesy of Drever Co.)

METHODS OF PRODUCING POWDERS

Some of the common methods used in producing powders include vapor condensation, shotting, granulation, precipitation, atomizing, thermal decomposition, grinding, oxidation–reduction, and electrolytic deposition. Some type of grinding operation follows the above processes in order to produce controlled mesh sizes. The size and shape of individual particles is important with regard to resulting physical and mechanical properties. From a given chemistry, several possibilities of mechanical properties are available; physical properties can also be varied by controlling the chemistry.

Reduction by Gas

Reduction of oxides is accomplished with the use of hydrogen or carbon monoxide. Oxides are first produced through chemical reactions such as are used in the production of alumina. Flexibility of this process lends itself to several procedures in the making of powders. Because oxides resulting from reduction processes tend to be porous, their shapes are irregular. This jagged surface contour provides for desirable characteristics in the matrix of the pressed and sintered part; interlocking of adjacent material occurs around the uneven sides of these powders. Even though this process is economical, chance of impurity contamination exists. Powders of molybdenum, iron, tungsten, and nickel are commonly produced by this method.

Atomizing

When a stream of molten metal is intercepted by a blast of gas, extremely small particles of solid metal result from the liquid. The practice must quickly divide the liquid mass into the smallest possible droplets and then allow these droplets to solidify. Inert gas prevents contamination of the metal. Those low melting point metals such as tin, lead, bismuth, antimony, aluminum, cadmium, zinc, iron, and some copper alloys like brasses and bronzes are powdered by atomization methods. Research is increasing the possibilities for other metals of higher melting points to be atomized. Atomizing causes extremely small sized particles to form, but particles vary in shape, size, and surface smoothness. Because variables are present such as gas pressure, temperature, shape of gas stream, kind of gas, and kind of metal, the resultant particles are directly affected. Many powders are ground to size following this process.

Figure 5–9 illustrates the shape of particles as they appear before and after pressing. A one-size spherical particle allows space between surfaces of particles. Smaller particles, in addition to large particles, fill these spaces. Jagged and elongated particles overlap and interlock with various shaped particles. Figure 5–10 illustrates the spherically shaped particle produced by atomization, the needle shaped particle sometimes produced by ball tumbling, and the porous sponge type produced by gaseous reduction. Depending on the type of process used in powdering metals, a work hardness factor is considered. Rates of cooling and work hardness influence plastic flow of the individual particle, therefore, many powders are annealed in a controlled or reducing atmosphere.

Grinding

One of the oldest methods used in powder production is the ball mill and some type of pulverizing. Powders are produced by crumbling and continuous size reduction. Harder metals and compounds such as tungsten-carbide tend to be irregular or jagged. Softer metals tend to flake. Jagged and uneven particles are used in pressing operations, but flaked particles do not compress to acceptable standards because too much bridging and work hardness occurs. Control of pulverization rests mainly with the type of grinding equipment and time. As time increases, both smaller and harder particles result, some being spherical.

Electrolytic Deposition

The flow of a metal solution from anode to cathode in an electrolyte is essentially an electroplating process. Anodes are produced in typical casting processes and may be of any of the metals normally used in this manner. Equipment consists of the electrolyte through which the current flows. When a direct current of sufficient throwing power and quantity flows from the anode, part of the anode is carried to the cathode. Consequently, pure metal is deposited at the cathode, similar to a cast structure, and is periodically removed. Softer types of metal produce more desirable particles than do the harder. However, all deposited metal which is considered usable is subsequently ground to the desired mesh sizes. A chief advantage of this method is the ability to produce pure metals. Grinding and atomization are mechanical processes, whereas deposition is electrical.

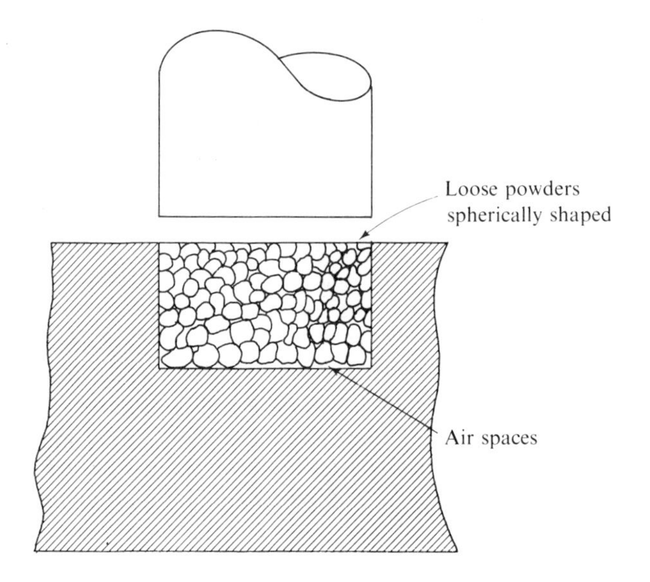

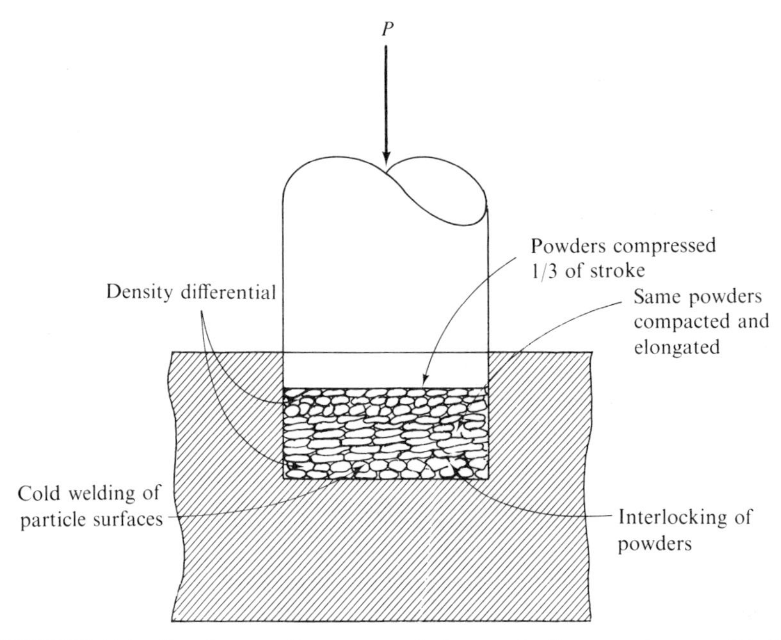

FIGURE 5–9 Powders as they appear before and after pressing one-third of punch stroke.

Shotting

When molten metal is poured through certain sized screens into cold water, coarse pellets are formed. These particles are subsequently ground to size. Such a process speeds up the grinding into powders by producing an initial small sized particle on which further grinding and pulverization is performed.

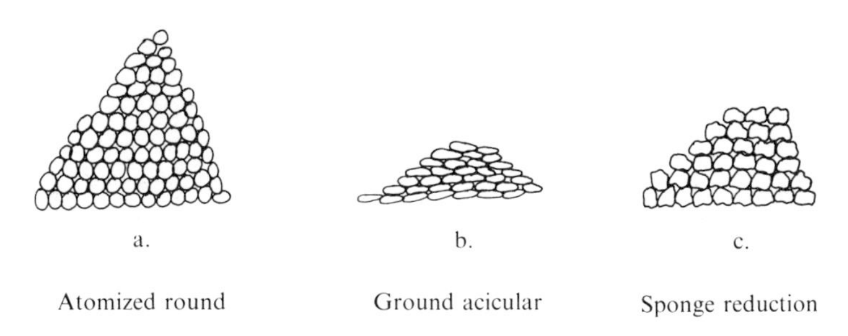

FIGURE 5–10 Three types of metal powders produced by different processes.

Granulation

When molten metal is agitated to a point where splashing occurs, fusion of colliding masses of metal is prohibited due to cooling. Because a reduction in temperature occurs, coarse particles of metal result. Subsequent treatments reduce these particles to acceptable powders.

Condensation

In condensation, a low melting metal is heated to its vaporization temperature. Being in contact with gas, individual dust particles are formed. Separation of metal particles is maintained due to temperature reduction. The very fine dust is gathered and run through screens to establish mesh sizes.

Precipitation

Precipitation is a chemical process involving precipitation of a metal from a solution of chemicals. The precipitated metal is subsequently processed to the proper shape and mesh size particle.

Kinds of Powders

Metal powders are being produced at an increasingly rapid rate, both in kind and in quantity. Two types of powders are available, the pure metals and compounds or alloys. Some of the kinds of pure metal powders are common to production needs such as aluminum, bismuth, cadmium, chromium, cobalt, copper, gold, iron, lead, manganese, nickel, niobium, platinum, rhodium, silver, tin, titanium, tungsten, vanadium, zinc, and zirconium. On the other hand, hafnium, thorium, and yttrium are uncommon, but have specific uses in the manufacturing of parts. Many pressed parts consist of compounds such as aluminum oxide, boron nitride, chromium carbide, molybdenum carbide, titanium carbide, and tungsten carbide. Also, graphite powders and other nonmetallics are available for mixing with various powders. The careful measurement and mixing of certain powders produces the alloys.

After pressing and during sintering, certain structural changes occur in the pressed powders such as precipitation or other results of heat treating. Some powders are mixed to cause a porous metal after sintering. The porous shape is subsequently saturated with oil. Such oil infiltrated metal is used for self-lubricating bearings. Another example of powder combinations is the physical mixing of metallic and nonmetallic powders such as copper and carbon. Any powders will mix physically, therefore, numerous combinations are available by mixing other metallic powders with each other or with nonmetallic powders. When pressed, a self-lubricating bearing results when the powders of copper and carbon are mixed and sintered. A main advantage of powders is their ability to be mixed among unlike powders and then used as a solid piece of material after sintering. Their size and shape are indicated on the manufacturer's container. When mixed powders are procured, the percentages of each mesh size are also indicated.

GRADES OF POWDERS

Basically, powders are graded by type of material, shape and size of particles, and their distribution in the mass. Many operations require round particles due to their slight flowability in the die while under increasing compression loads. Other operations may demand an interlocking type particle to resist friction loads. Either way, powders are available as blended mixtures of several sizes of particles, or several shapes and sizes, or whatever kind of powder is needed. The part's use determines the chemistry, size, and shape of particles. Shapes are selected by means of sieves and screens, much like the selection of sizes. When powders are agitated on screens having a designated number of openings per inch, a mesh size is established. When a square area of equal sized openings in a screen just barely retains a particle, that mesh is assigned to the particle's size. When different sizes or shapes are used, blending must be accomplished to assure equal distribution of all powder. When powders have different chemistries, thorough blending is also necessary. Regardless of the method used in production of powders, caution must be excercised to preserve the chemistry of the powder, oxygen being the chief contaminator. During and after production of powders, they must be protected under controlled conditions and sealed in a container. Two precautions must always be observed with respect to fine powder, that is, powders must not be breathed and powders may explode.

Molding Processes

Immediately prior to pressing operations, all powders must be properly blended in order to eliminate the possibility of finer powders being at the bottom of the container. The die is filled with the selected powder, excess powder is struck from the surfaces, and the die is closed. As previously discussed, the powder is then compressed to a desired density by means of the punch. However, density resulting from cold pressing is not the final density. Several additional factors influence final density, including particle size and shape, material, condition of material, rate of compression, length of ram stroke, compression force, presence of lubricants or gas, and sintering temperature and time. Length of punch stroke relates to the compression ratio of loose powder to compressed powder at a certain distance from the face of the punch (fig. 5–11). Sometimes a lubricant such as graphite is placed with the powders in order to facilitate a minute flow of powders. When other types of lubricants which will evaporate during sintering are used, the powdered part sometimes expands because of the expanded volume of gas being trapped among the powders.

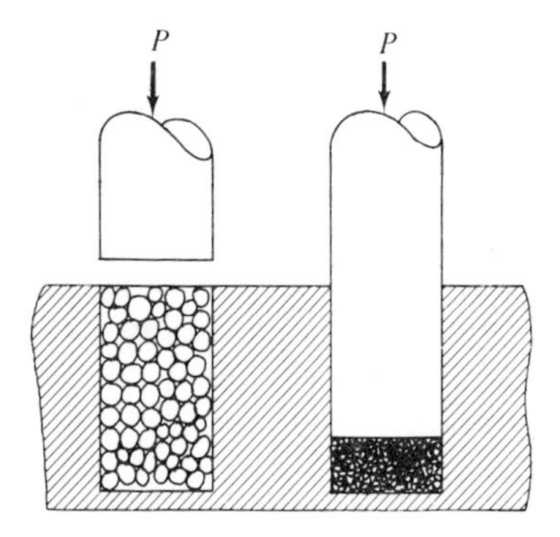

FIGURE 5–11 Metal powders before and after compaction.

The technician must assure proper die design, including die material. Design of the die is the basic design of the part. Because powders are abrasive during compression loading, inside faces of the die are subject to severe wear. When very abrasive powders are used, tungsten carbide inner faces of dies are required. Due to the high initial cost of dies, it is expected that several thousand parts will be produced before die replacement is necessary. Many presses using multiple dies produce several hundred parts per minute. Fully automated multiple die presses deliver as much as 100,000 parts in a few hours of operation. In order to profit from initial costs of equipment, as many as 6000 to 50,000 parts must be produced.

COLD WELDING OF PARTICLES

As pressure is exerted onto the many particles, they move together until particle-to-particle contact is made throughout all surfaces of all particles (fig. 5–9). In this respect, loose powder includes air among the numerous surfaces, whereas packed particles eliminate air. As pressure increases, some particles flow plastically while others flow slightly as whole particles. The result is a tightly compacted mass of solid material having a certain density at a certain point in the compaction or briquette. Pressures beyond 200,000 psi are often used to produce satisfactory densities. During pressing, thin layers of oxides are scraped off and metal-to-metal or other material contact is made. One particle is literally squeezed into its adjacent particle (fig. 5–11).

Briquettes are produced to satisfactory densities due to actions of single or multiple punches. A complex type of punch is schematically illustrated in figure 5–6. When length of part becomes a problem, an opposing punch is incorporated in the design. When a flange is required in the part, multiple punches are used, one punch extending from another (fig. 5–6). Pressing limitations are more concerned with length of part than width. As an example, 1½-inch thick parts are produced in 36-inch widths. Thickness and length dimensions are the same, and this factor relates to density. But width is not governed by compression ratio. Width is limited by feasibility of punch and die width, however, flow of powders is still a concern. As the cross-sectional area of parts increases, more power is required on the punch. Mechanical presses provide sufficient pressures up to 150 tons, but when greater loads are needed, hydraulic presses with capacities beyond 4000 tons are used.

HOT PRESSING

When extremely high pressures are required because of the kind of material in the particle or size of material, hot pressing is used. Powders are placed in the die and the die is closed as heat is applied. The punch compresses the softer and weaker material to desired densities because heat increases some flow of the powders. Being hot and often plastic, particles fuse together and close vacant spaces. Shrinkage in volume occurs, and a solid part results. Fairly large cutting tools are produced from hot-pressed tungsten carbides. Hot pressing eliminates the sintering operation in many procedures, the temperature reached being the determining factor.

AUTOMATION

Once a procedure has been established, the technician and engineer should move toward automated production. Observation of a slow moving hand operated pressing machine points out the types of actions with time intervals. By carefully designing press sequences and movements, automation can be obtained. Speed of output is geared to compaction time of the punch. Because compaction consumes time, multiple punches are sometimes used around a rotary. Hundreds of thousands of parts can be made when good design principles are utilized.

Sintering

Sintering is a diffusion process among the grains of powder which are in contact with each other. Because powder particles are very close due to high compression forces, surface coalescence bonds the many faces of each particle to its adjacent particles by diffusion. Temperatures used in sintering are high enough to cause diffusion, but in the majority of operations, melting of any constituent does not occur. Because critical stresses are frequently placed in the many particles due to pressing, recrystallization may occur. Also, temperatures above the metal's recrystallization temperature cause recrystallization in the applicable particles. In those particles stressed beyond their critical strains, both diffusion and recrystallization occur.

SHRINKAGE AND DIMENSIONAL CHANGE

From a given volume of pressed powders, dimensional change occurs as temperature is increased. The change is opposite to that occurring in solid materials. As a solid is heated above room temperature, its coefficients of expansion are predictable and at a given elevated temperature, an increase in dimension occurs. As pressed powders are heated, the dimension shrinks due to escape of vapors, the closing of tiny air spaces, and a small rearrangement of the compaction. The sum of these actions equals a tighter density, therefore, a three-dimensional change occurs. Fortunately, these changes are predictable and result in parts being pressed and sintered to within their very small tolerances, allowances being made for shrinkage. Parts being completed by sintering normally have good surface finish and are usually ready for use, unless some subsequent operation such as sizing or coining is required.

METALLURGICAL ASPECTS OF POWDERED PARTICLES

When metal powders are pressed and sintered, many basic metallurgical situations exist. Particles are measured in certain mesh sizes and are identified in microns in diameter. As the chemistry of the metal powder varies, so do the characteristics and capabilities of the finished part. As illustrated in figure 5–12*a*, loose powders as they

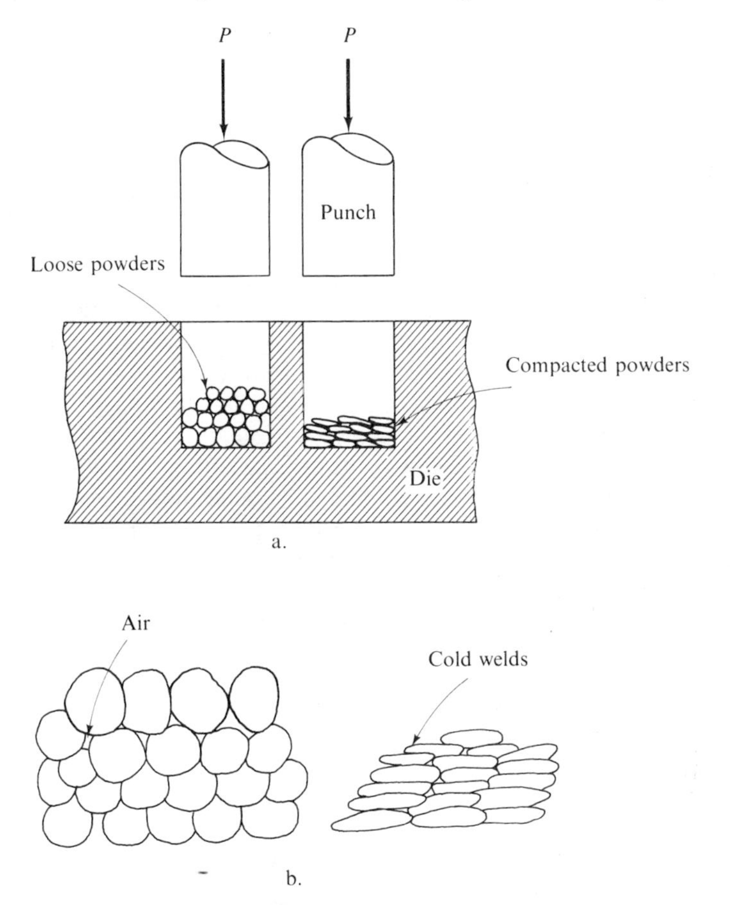

FIGURE 5–12 Spherical shaped grains compact, elongate, and cold weld under compression forces.

are placed in the die have physical arrangements very similar to small piles of sand. A cross-sectional view reveals the surfaces in contact at only a few points. Air or a controlled atmosphere envelops each particle. As pressure is applied to a given mass of loose powder (fig. 5–12*b*), compaction occurs, gases are mostly squeezed out, more surface contact is established among particles, plastic flow occurs within some particles, work hardening may occur due to compression stresses and abrasion, strains are established, and a type of cold weld results at the point of highest pressure. Many presses exert upwards of 200,000 psi on a given mass of powders, while other presses may apply only one-fourth this amount. As far as strength of the briquette is concerned, caution must be exercised when transferring pressed briquettes into the sintering furnace because they are weak and brittle.

Powder metallurgy processes are easily automated. Many presses release their briquettes onto a conveyor belt system which feeds into a continuous type furnace or batch type furnace. Because oxidation is present in a noncontrolled atmosphere, sintering furnaces must be equipped with protective atmospheres. Heat treatment occurs within the hot atmosphere, depending on the summation of previous processes, chemistry of the powder, particle size, pressing pressure, and density established. However, some materials do not respond to heat treatment. As temperature increases, the following changes occur within some of the many particles: softening, strain release, shifting, recrystallization, grain growth, contraction, coalescence, and bonding. Carbides and oxides do not soften.

SOLID PHASE SINTERING

In solid phase sintering, no melting of constituents occurs. Sintering temperatures range from 80 to 90% of the melting point of the major constituent. Consequently, due to a high atomic activity, diffusion occurs around the sides of the individual particles. As time and temperature increase toward maximum, changes occur in rapid succession. Grains within the metallic particles grow in size, seeking their compatible size at temperature. Grain boundaries collapse, just as in solids, and encroachment into adjacent particle material occurs (fig. 5-11 and 5-13). At higher temperatures, diffusion occurs rapidly. Particles of similar chemistry form homogeneous solid masses. Particles of dissimilar chemistry diffuse according to applicable physical factors of the heated chemicals. When powdered metallic alloys are sintered, mixtures and solid solutions execute their roles in microstructures. Metallics and nonmetallics contain their physical and mechanical properties, even though particles are hetereogeneous. The sintered part is cooled in a controlled atmosphere, often under controlled conditions. When cold, sintered parts are ready for use unless subsequent specialized processing is required.

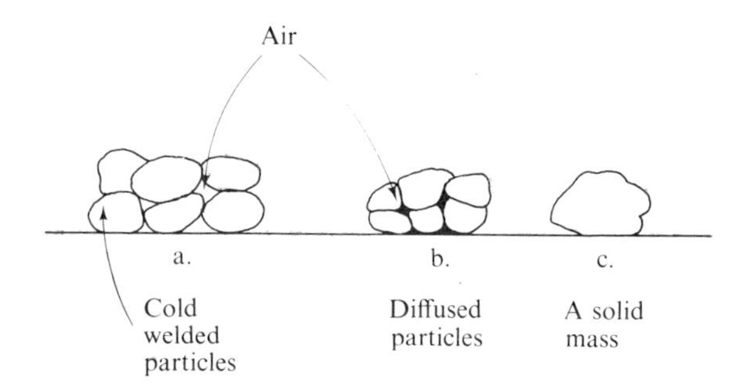

FIGURE 5-13 During sintering, particles diffuse into adjacent surfaces as air escapes.

PRESINTERING

Often, a part's design contains holes across the line of compression. These types of parts are presintered at the lowest possible temperatures and cooled, causing rigidity in the part along with shrinkage. Machining operations can then be performed because dimensional changes have occurred. Following machining, complete sintering is then accomplished, resulting in finished parts.

LIQUID PHASE SINTERING

Sintering powdered particles often involves the melting of a lower temperature melting constituent. At sintering temperatures, this lower melting point material liquifies and flows along the hot solid boundaries, closing crevices and pits, and cementing the main constituents in permanent place. Nickel is used, for example, to cement tungsten carbides into a solid shape. The toughness of the cementing nickel supports the abrasive power of carbide. These lower melting additives are deliberately added to the powder mixture for special purposes such as in the cementing processes.

Postsintering Treatments

SIZING AND COINING

When extreme accuracy and increased mechanical properties are desired in a powdered part, sizing and coining are accomplished. If a part has been cold pressed at 100,000 psi and sintered, further pressing is sometimes performed at slightly more than 100,000 psi for the purpose of obtaining more exact dimensions, causing plastic flow and increasing hardness and strength, and sharpening contour profiles and details. Sizing brings the previously pressed part to exact dimensions. Coining perfects profiles as it increases strengths.

ADDITIONAL POSTSINTERING TREATMENTS

Several postsintering processes are applicable. Following sintering, increased mechanical properties in some parts are attainable by causing liquid metal to flow into the partially porous surfaces of the sintered part. If a thin section of a specified metal is placed on top of a sintered part and allowed to melt, some of the liquid will infiltrate the porous structure. Mechanical properties are then changed, according to the treatment. On the other hand, sintering can increase porosity. When temperature drives off a volatile constituent during sintering, holes and crevices occur in the part, even though they are extremely small and narrow. When oil is forced into these porous parts, a semipermanent and self-lubricated bearing results. Another treatment involves heat treatment. Some sintered parts are heat treatable by case hardening through a carburizing and hardening process. Also, some powdered parts are heat treatable in the same manner as solid metals of the same analysis. Basically, powdered parts respond to many of the same treatments as solid metal parts, for example, electroplating is applicable, as well as machining.

Advantages of Powdered Parts

Repetition in powder metallurgy is assured because tolerances are close and procedures are such that automation eliminates the greater tolerances needed by an individual's skill. Parts are identical in both size and mechanical properties; uniform sintering conditions maintain uniform products. Another advantage is that the powder process is sometimes the only method available to produce some parts. Different metal powders are easily mixed. Also, metals and nonmetals are easily mixed, pressed, and sintered in the powder process, enabling the production of self-lubricating bearings when oil is absorbed in the open pores. Cemented carbides are produced by pressing nickel powder with carbides.

Many parts such as offset bosses and eccentrics require excessive and unusual machining operations. The powder process can often produce this complicated design in one stroke of the punch or punches, saving money and time. In this respect, all machining operations produce scrap metal which is costly. Once removed from a bar, pipe, or sheet, the metal is either a finished part or scrap, the cost of the scrap being frequently the same price as the cost of the material in the part.

Because powder processes are conveniently automated, mass production is feasible. Thousands of identical parts such as gears can be produced in a day. When multiple punches are designed into the press, increased quantities of parts are possible. Labor costs do not depend on skill of the operator in the same concept of skill of the machinist. A lathe operator must have skill to manipulate an ordinary lathe, but the press operator has no control over what happens inside the die once the machine is set to operate. Therefore, less skilled labor is needed in press operation than in the machine shop where men control the cutting operations. When very large quantities of parts are needed, the entire process is automated from the filling of the dies with powder to the receipt of sintered or coined parts.

Disadvantages of Powdered Parts

Metal powders are very abrasive and cause fast wearing of poorly designed dies and punches. Dies are larger because more material in their design is required to sustain the high pressing pressures. This disadvantage, however, is offset by the many advantages, including the increased life of properly designed dies. Pressed and sintered powders are frequently weak in mechanical properties when compared to solid metal, but new techniques are improving both the physical and mechanicl properties of powders. Compression properties run fairly high, but tensile and shear values sometimes are much lower than in the cast or wrought products. Lightly stressed designs do not usually involve powdered parts, but improved flow and bonding characteristics of the powders, along with improved techniques, are producing parts having tensile strengths up to nearly 200,000 psi. The metallurgy of the surface of the particles is currently being improved. A more solid and homogeneous association of particles promotes diffusion characteristics so that a nearly solid mass of metal is produced. Another disadvantage reflects from the limitation in design. Keeping in mind the fundamental principle of compression forces acting among loose particles, density is established only in the compression zones. This eliminates those designs having angular features which prohibit compression of material beneath a hole. Also, draft of edges and surfaces must be such that part extraction from the die can occur. Reentrant angles and undercuts simply will not work.

Typical Parts Produced by Powder Metallurgy

Powder processes are being used more and more as new ideas and techniques are discovered. Nearly an endless number of different parts are being produced by this process (fig. 5–14). Presently, the process is limited to small parts having cross-sectional compatibility with compression forces among powders. This means that thousands of common industrial and household parts such as bushings, bearings, pins, collars, washers, plates, gears, sprockets, splined fittings, hoods, supports, brackets, tubes, and special fittings in the automotive, garden, tractor, and small assembly businesses are being produced.

FIGURE 5–14 Typical parts produced by powder metallurgy. (Courtesy of Pacific Sintered Metals Co.)

Questions

1. Describe the meaning of powder metallurgy.
2. Why is the shape of metal powder significant?
3. What is meant by density of compaction? Why is a nonuniform density undesirable?
4. Name and describe the advantages of opposing punches.
5. Why does satisfactory compaction of powders occur only in a straight path?
6. What happens to the compaction during sintering?
7. Why can some metal powdered parts be heat treated for increased strengths after sintering?
8. What is the purpose for coining sintered parts?
9. Explain three main limitations in powder metallurgy.
10. Why can metals and nonmetals be compacted into parts?
11. Describe a main weakness in parts produced by powder metallurgy as compared to wrought products.

12. Why does the powder process ultimately limit a part's dimensions?
13. Why can tolerances be very small during the manufacture of parts by the powder process?
14. Explain a main advantage of powder metallurgy over conventional metallurgy.
15. What is meant by cold welding?
16. Why is automation very practical in a powder metallurgy factory?
17. Why do many powder parts shrink during sintering?
18. Differentiate between solid phase and liquid phase sintering.
19. Why is presintering sometimes performed?
20. Why is powder metallurgy a simple process?

Tooling for Manufacturing

A manufacturing plant depends on its tools to help produce its products. Good planning, preceded by scientific research, provides the means for a plant to become operational. Plans are transformed into blueprints, and blueprints initiate actions to procure the necessary tools to run the production line. Tools include all those machines, pieces of equipment, holding and handling devices, hand and machine tools, measuring tools, inspection devices, and standards needed to instigate production. The manufacturing plant uses tools pertinent to its finished products. Its production depends to a great degree on the types of tools, quantity of tools, and degree of automation of tools in use. In these respects, many machine tools are common to metals processing operations and include hacksaws and band saws, drilling machines, hand presses, cylindrical grinders, tool and cutter grinders, centerless grinders, pedestal grinders, straightening devices, lathes, production lathes and screw machines, various types of milling machines and planers, boring machines, gear and sprocket cutters, shaping and broaching equipment, shears and punches, presses and sheet metal support equipment, polishers, heat treating furnaces, welding and cutting equipment, materials handling equipment, conveyor systems, small power tools such as hand saws and drills, and precision cutting and measuring tools, gages, and instruments. With respect to the term *drilling machines,* it must be pointed out that any machine having the primary purpose of drilling, such as a drill press, is a drilling machine.

Primary tools include the major processing tools found in foundries, forge shops, machine shops, welding and sheet metal shops, and plastic shops. With respect to processing, producing raw materials has been discussed along with the casting and powder metallurgy processes. Therefore, subsequent chapters are concerned with the further processing of materials in other ways. Because manufacturing processes involve measurement, require holding devices to process materials, and require systems of power transfer, the material in this chapter is then concerned primarily with these areas of subject matter. Common characteristics of all tools are their associations with materials and the materials' holding devices. Basically, power tools require some means for material clamping and proper holding during processing by the machine. Consequently, several types of holding devices have been developed. The most common is a form of vise, while a great percentage of holding devices are more sophisticated and include the fixture and jig.

Tooling involves the pooling of processing machines and equipment, arranging them in a production sequence, providing connection systems between processing equipment such as conveyor systems, establishing inspection stations, and providing for final inspections and market consumption. A typical flow process in a manufacturing operation is schematically illustrated in figure 6-1. If the enterprise is not automated, then human hands and minds must support the various processes. To the engineer or technician, proper tooling is the foundation for the operation.

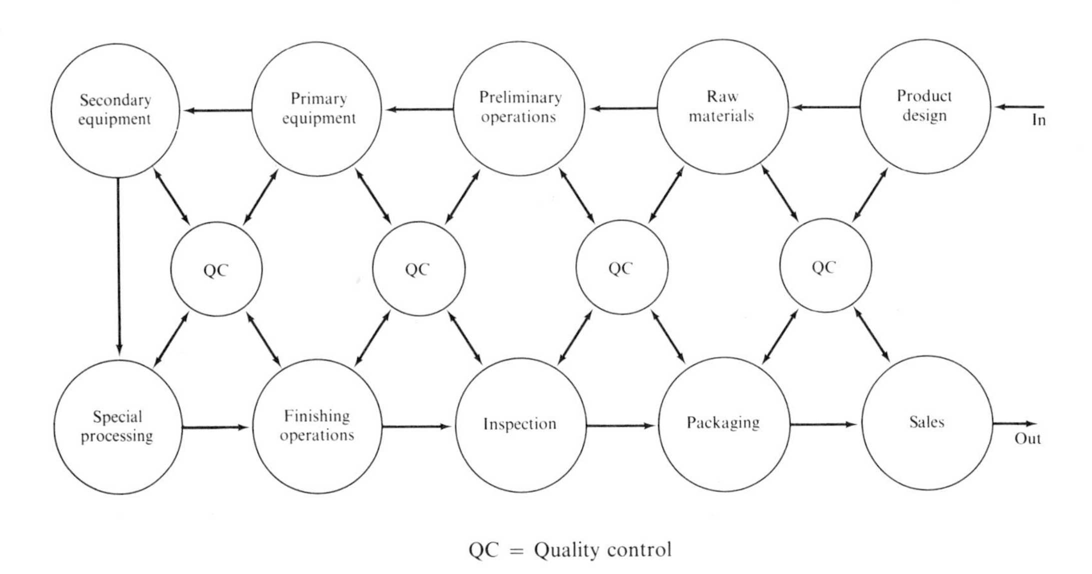

FIGURE 6-1 The flow process in a manufacturing operation.

Layout Processes

Many materials which have been produced through the casting, forging, and rolling operations require machining in order to produce parts having dimensions with closer tolerances. Examples include the machining of bosses on castings and the milling of bar stock to exact dimensions. Before cutting operations can begin, therefore, reference points must be established so that subsequent location dimensions can be obtained. In numerous cases, forgings are further processed by various types of cutting operations. At the same time, many castings require holes and flat spots for attachment to other parts. In the area of wrought products such as bar stock, thousands of cutting operations are performed over short periods of time. Before cutting begins, some means must be available to determine exactly how much metal or other solid material must be removed. It must also be established where this material is to be removed. Whether the metal removal operation is to be done by hand or by machine, reference points and dimensional location points must be established.

HAND LAYOUT

Many layout operations begin with the part or rough section of material resting on a table or surface plate. From this initial position, location dimensions can be fixed. Depending on the accuracy required, various tools and procedures are available. Once a base or horizontal point or plane has been established, the height or depth dimension can be located (fig. 6–2). When the horizontal plane is used as the reference, then the lateral or third dimension can also be located.

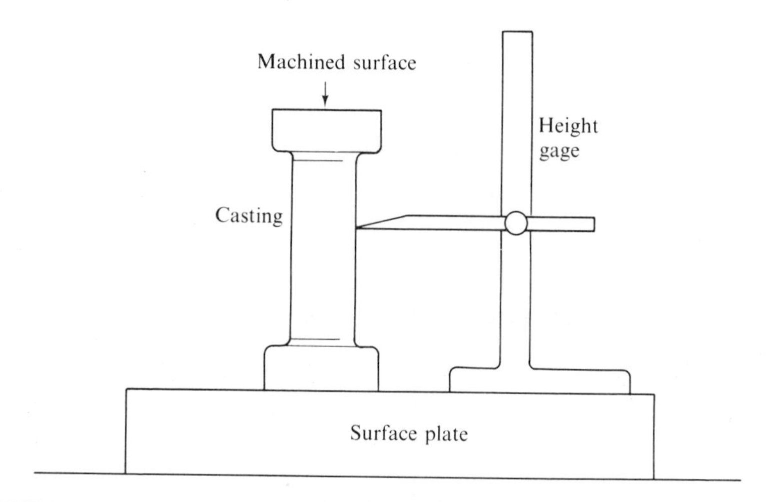

FIGURE 6–2 The height gage is used to transfer a measurement.

Layout Tools

Numerous tools are available for hand type layout processes, including rules and height and depth gages. Even though a technician may not be engaged in elementary layout processes, he must be cognizant of the process to assure accuracy in further production processes. Cumulative errors in dimensioning eventually cause scrapping the part, unless the errors are detected early in the process. An example of location dimensions where error can result is illustrated in figure 6–3. In order to locate point *C*, two preceding measurements must be taken to establish a direction or reference line, dimension *X*. Then point *C*, which is to be a hole, is found by measuring from point *A* along the line *AB*, using dimension *Y*. Another hole, point *D*, is also to be drilled and reamed. From point *C*, using dimensions *Z* and *Y*, the distance is measured to point *D* along a right angle line from line *AB*. In hand layout methods when reasonably clean and smooth surfaces exist, the layout is made either on the bare metal or scratched into a surface enamel or fast drying paint. A blue colored compound, *layout blue*, is often wiped onto the surface, allowed to dry, and then the surface is dimensioned with scratches and punch marks. With reference to figure

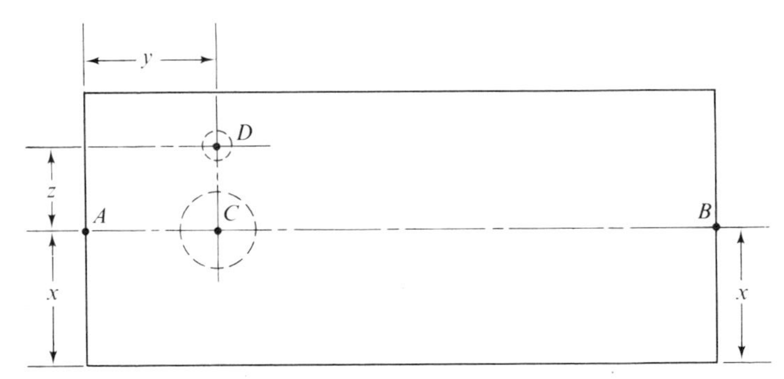

FIGURE 6–3 Hand layout of two precision holes.

6–3, points *C* and *D* are marked with a sharp pointed center punch, keeping in mind that the distance between the two holes is critical. When a small lead drill or center drill is used to penetrate through the punch marks, the beginning of the final holes takes place.

Skill vs. Error

At this time, it must be pointed out that no matter how carefully the punching and drilling of these two holes are accomplished, there is a great probability of error in the final dimensions. Review the two steps. First, point *C* was initially established within the essential tolerance. Then, point *D* was measured from point *C*. At the intersection of scribe marks, holes were punched, and then these locations were drilled. Chance of error at the first location existed during the punching step. Even though a sharp pointed punch was used, it requires a very high degree of skill to place the point of the punch in the proper place along line *AB* at dimension *Y*. Even scratch marks are 0.005 inch wide. Unless the punch is held vertical and properly placed on location, an error up to 0.015 inch is easily made. Obviously, when drilling into this conically shaped depression which may be slanted, chance of erroneous alignment of the drill exists. Should the drill lips bite into one side of the hole more than another, the resulting hole will be off center, unless the drill is mechanically guided. Upon completion of the lead drill or center drill operation for both holes, subsequent drilling brings the holes to within a few thousandths of an inch of their sizes. Reaming then completes the holes. Measurements between holes *C* and *D* are taken to assure proper fitting of the two pins which will move in these holes. If either dimension *X*, *Z*, or *Y* is off only a few thousandths of an inch, the part will be scrapped. This example illustrates the need for expertise, but at the same time, much time is expended. Such a procedure is justifiable if only a few parts are to be processed. Obviously, some type of mechanical device is needed which holds the material while drilling at fixed locations. Such a holding and guiding device would produce accurately spaced holes in a matter of seconds. But the manufacture of such a device would require much time and expense. Production quantities dictate the decision.

MACHINE LAYOUT

A faster and more accurate method of layout is available in certain types of machines. For example, the two holes in the above example can be more accurately machined in a jig borer or similar machine. Such a process immediately changes layout procedure from hand to machine and chance of error to no error. The accuracy of the layout rests with the accuracy of the machine. In this respect, the vertical milling machine cannot compete with the jig borer, but it is used in many machining operations. Once the part to be machined is securely positioned in a holding device, location dimensions are established by turning micrometer dials which control movements in different axes from a reference point or plane. The reference point is provided when the edge of the cutting tool touches the workpiece. Dimensions can then be located within 0.001 inch, and with closer controls, locations are obtained to 0.0001 inch. Many types of machines, both vertical and horizontal, have this dimension location capability. Some machines are capable of holding tolerances to plus or minus 0.0001 inch, and smaller tolerances are possible such as plus or minus one fifty-thousandths of an inch. Such close tolerances, however, are used only when a very high degree of fit and clearance are required.

Machine Control Over Layout

When equipment is set up with tape and computer controls whereby interfacing mechanisms or electromechanical devices drive the machines from control impulses, even greater accuracy is obtainable. Also, axes controlled machines such as the numerical control variety are capable of rapid production of parts, even in five axes, along with accuracy, repeatability, and closer tolerances. Layout is performed prior to cutting or forming operations by mathematical calculations or, in some cases, by operator control during the first operation when position locations are recorded on tape concurrently with the manual operation. A central reference point is established from which measurements along an axis are performed. Once the entire operation is programmed, processing of the part by controlled layout occurs rapidly and accurately, cutting or other forming being accomplished in response to multiple axes signals acting concurrently.

A PROGRAMMED HOLDING DEVICE

In many cases, layout of the part's dimensions is performed long before the actual cutting operation such as in drilling holes. A dimensional location device known as a jig is manufactured during the tooling phase of planning. With the aid of the jig, inexperienced operators can produce thousands of accurately drilled parts (fig. 6-4). In other words, layout of the part is accomplished within reference points of a special type of holding device for subsequent machining operations. When the part to be drilled is placed in the jig, it has only one position to assume in relation to the drill point. In order to perform such accurate drilling so rapidly, a toolmaker consumed many hours in positioning internal guides so that drilling accuracy would result.

LAYOUT TRANSFER

Another type of layout occurs during the transfer of dimension locations from a drawing paper or template to the part. When a part is drawn on paper at actual, or one-to-one size, its critical locations are transferred to the material by laying the drawing on the material and applying punch marks by hand or machine. Flat stock lends itself readily to this type of dimensioning layout. When templates, previously calculated formed layouts, are used, machining and layout occur simultaneously. Again, in

FIGURE 6–4 A rapid drilling operation due to a built-in dimension within the holding device. (Courtesy of The Cleveland Twist Drill Co.)

this process, time was consumed long before the forming operation so that time during forming would be saved. Consequently, when it is anticipated that large quantities of identical parts are to be produced, plans call for prior fabrication of devices such as jigs and fixtures. By having jigs and fixtures readily available, lesser skilled operators can produce the thousands of needed parts at a lower unit cost. It must be remembered, however, that highly skilled personnel such as toolmakers consumed many costly hours in producing the built-in layouts within some of the dimensioning devices.

Tooling

Once the "go" signal is given, manufacturing personnel, along with management, begin the "tooling up." This is a very broad term because it literally means the establishment of production facilities. Plans from books become three-dimensional objects, equipment, machines, fixtures, and materials. Once placed in sequence, the operation becomes alive when people move in and implement the process. Tooling in a foundry is quite different from that in a machine shop. Likewise, tooling in an aircraft plant differs tremendously from that in a ten-man repair shop. It is the kind, size, and purpose of equipment and tools that differentiate among the product outputs of manufacturing companies.

Tooling for a large factory may involve the purchase of conventional and special cutting machines and casting and forming equipment peculiar to the manufacturing operations. Such equipment may include melting furnaces and molds; power shears and assorted saws; forges and presses; pressure and heat forming autoclaves; specialized jigs and fixtures; lathes, milling machines, and grinders; screw machines and planers; parts conveyor systems; chemical tanks and controlling apparatus; dies; heat treating apparatus; welding and sheet metal facilities; straightening equipment; testing and inspection machines; materials transfer systems; computers for controlling processes; and hundreds of common and special cutting tools, gages, and measuring instruments. Mass production demands a large percentage of automated equipment, therefore, interfacing systems are installed on selected equipment. Such interfacing between signal impulses from controlling devices and mechanical actions of machines requires skills in the electromechanical field of energy transfer. The cost of all tools runs into many thousands of dollars and even into millions. Therefore, care must be exercised to provide the correct kind of processing equipment, being especially aware of machine capability and machine structural stability. Automation and advanced mechanization require many specialized machines arranged in a continuous pattern, while hand operated equipment is arranged in a manner which will result in maximum efficiency during processing operations.

Integration Supports Production

Engineers and technicians begin manufacturing processes with the materials of industry. It is the raw materials which are fabricated into usable items that attract primary and initial attention in the development phase of planning for manufacturing. Responsible personnel also see new machines becoming worn within a given time period, even though some equipment lasts for twenty or more years when properly maintained and used. On the other hand, new equipment frequently becomes worn out when handled improperly. Also, new equipment sometimes becomes obsolete when better machines become available. Over a period of time, machines must be replaced, therefore, funding for equipment replacement becomes part of the tooling plans. In this respect, 10 years is sometimes used as a life expectancy of a production machine. Once the tooling operation has been formulated, details are completed with reference to process procedures and standing operating procedures. Some companies write specifications and set up a series of standards to control processes. In turn, both skilled and unskilled labor benefit through assistance and guidance of written procedures. Better quality control and product assurance result.

Holding Devices

Machines which are used for cutting and forming materials depend on the holding device to securely hold the material while cutting or forming is being accomplished (fig. 6–5). Should the material slip, danger exists to both operator and machine. Danger to the operator exists during a drilling operation with a drilling machine,

FIGURE 6–5 A special drilling operation. (Courtesy of Clausing Corp.)

especially when large drills are used to drill thick plate. There is a possibility of the metal slipping as the drill point digs into the metal. Even greater danger exists when the drill begins to cut through the plate (fig. 6–6). Once the metal slips, it begins turning with the drill, endangering the operator's hands, arms, and body. Severe accidents have been caused from such an apparently simple drilling operation, therefore, holding devices must be securely bolted to the proper part of the machine. The metal must then be tightly clamped in the device (fig. 6–7), but not so tight as to cause springing of the metal. Overtightening is a common cause of part scrappage (fig. 6–8), because after forming or cutting, the overtightened part springs back to its non-stressed condition, often containing a convex or concave surface (fig. 6–9). Even a very small misalignment may cause results sufficient to warrant scrapping the part.

Most manufacturing processes require materials' holding devices. One of the most common implements used for securely holding a material for machining or forming is the *vise*. Depending on the operation to be performed, vises vary in size and purpose. The bench vise is used for hand type operations which require little accuracy in bending and cutting procedures (fig. 6–10), however, precision operations such as machine slitting require precision vises and other fixtures. Machine vises are precise in their

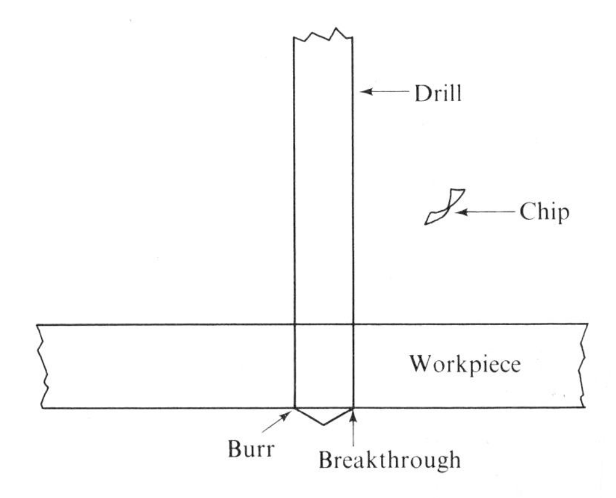

FIGURE 6–6 A hole being drilled showing drill breakthrough.

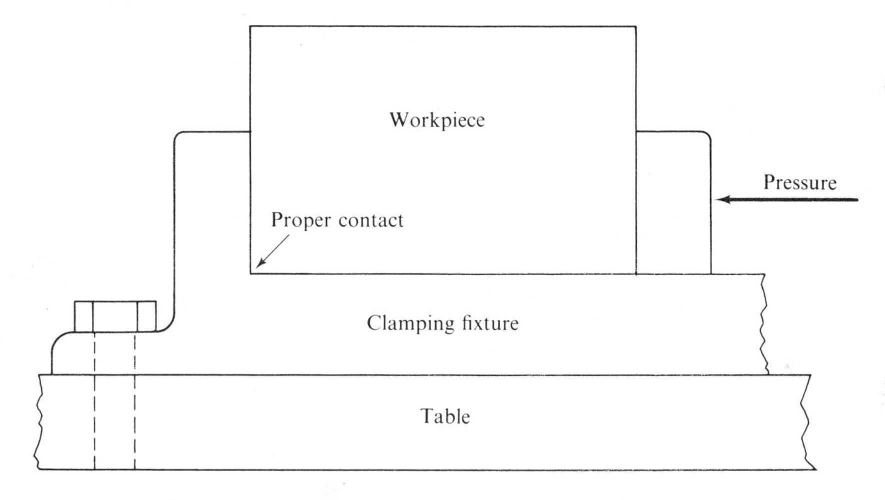

FIGURE 6–7 A part securely clamped in fixture.

functions and allow materials to be processed within very close tolerances (fig. 6–11). Vises or other fixtures are used with machine shop machinery. In the other metals fabrication industries, various types of clamping fixtures hold the work for forming, welding, and cutting.

ECONOMICS AND INTERCHANGEABILITY

The output from production equals its input. If an unskilled operator drills holes in plates which are secured with special precision holding devices, the output is greater than that from a nonprecision holding device by a skilled operator. In this respect, it must be remembered that many hours of skill go into the manufacture of the precision holding device which enables the unskilled operator to maintain a high production output (fig. 6–4). Such a procedure is justifiable only when thousands of operations are to be performed. On the other hand, should the operator have to stop and lay out his work for each part, his production will be low (fig. 6–12). The two holes shown in figure 6–12, or possibly a dozen or more, can accurately and quickly be

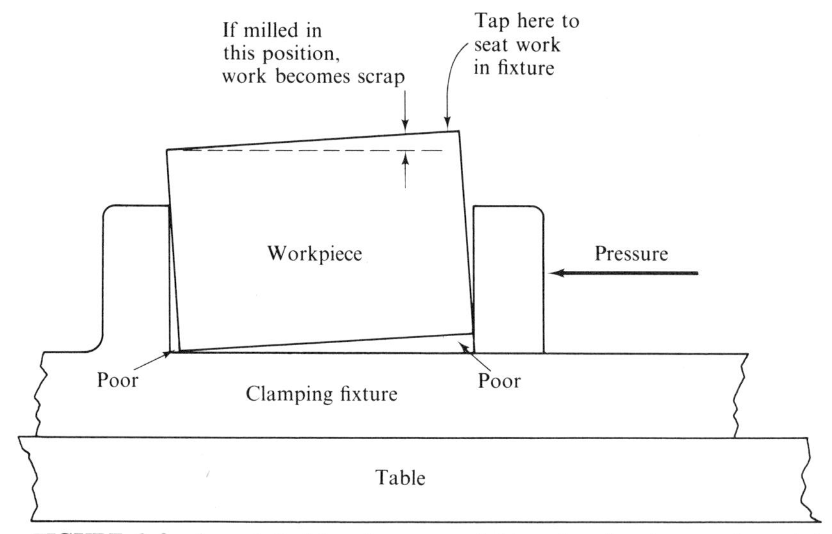

FIGURE 6–8 A part that has been over tightened and not properly seated.

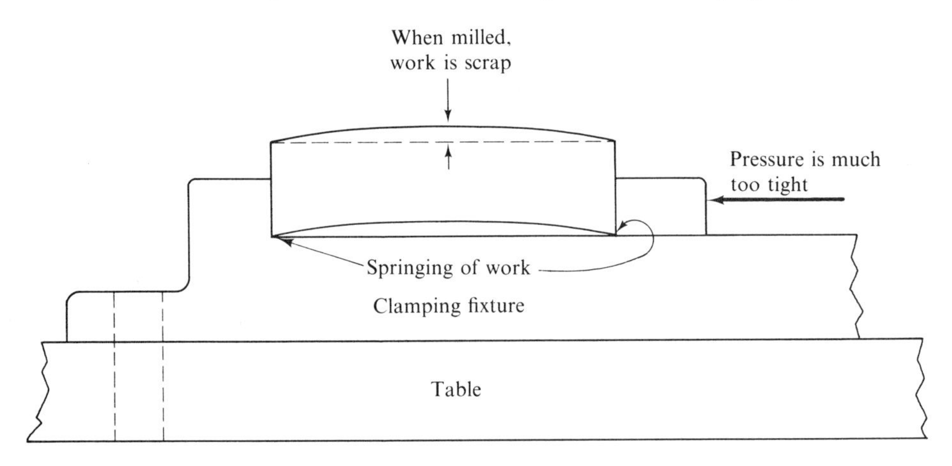

FIGURE 6–9 Work is tightened too tightly causing springing of work.

drilled in a plate jig by an unskilled operator, while a skilled operator lays out and drills only the two holes in one plate. A typical plate jig is illustrated in figure 6–13.

When holding devices can be attached to the machine and the materials secured properly so that rapid production exists, then lesser skilled operators can be hired to perform the operation. As an example, when a drill jig is used, little skill is required to drill the hole correctly (fig. 6–14). When no jig is used, however, a more highly skilled operator is required to locate the hole and then drill it properly. Economics dictate the use of precision holding devices when mass production quotas are to be maintained.

Along with the requirement for precision holding devices is the requirement for an extent of interchangeability of the device. Holding devices must have some types of general applications, as well as specialized, however, most holding devices other than the vise are more specialized than generalized. In other words, the holding device

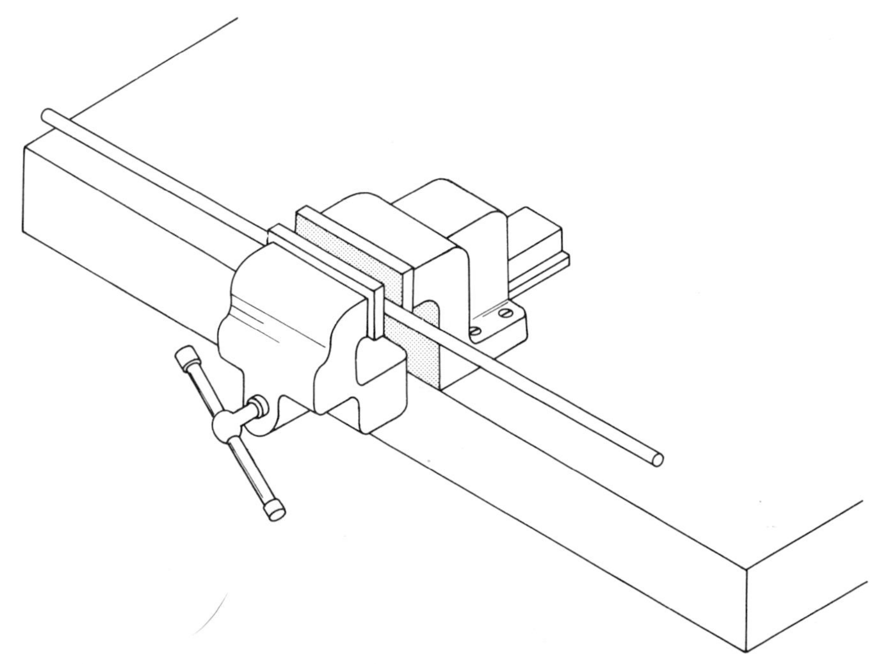

FIGURE 6–10 A bench type vise.

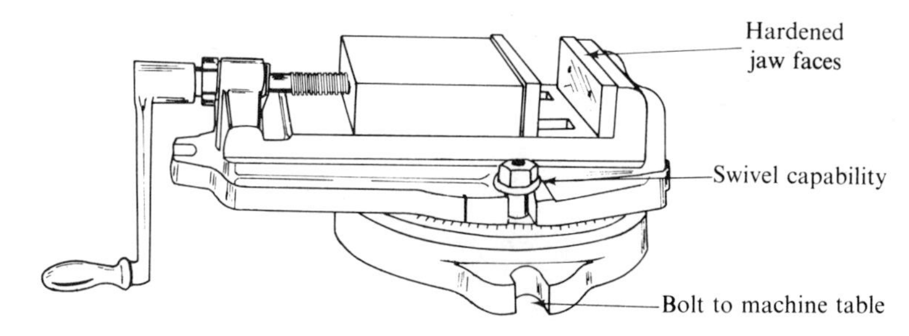

FIGURE 6–11 A precision vise which is used in lieu of fixtures during some machining operations.

should be able to be used with different machines, but within the capacity of the holding device. Also, the device must be quickly adaptable to the job both in clamping power and release capability.

VISE VS. JIG

When location dimensions for machining the workpiece which is held in a device are required, the specialized *jig* is used. The greater the accuracy and capability of the holding device, the less skill is required for completing the specified operation. The jig holds the work as it is being machined, as does the vise, but the jig is also arranged to guide the cutting tool. The vise has no guides to provide an exact location for machining the workpiece. In other words, the jig (fig. 6–15) provides for built-in precision, while the vise's precision depends mainly on operator skills (fig. 6–11).

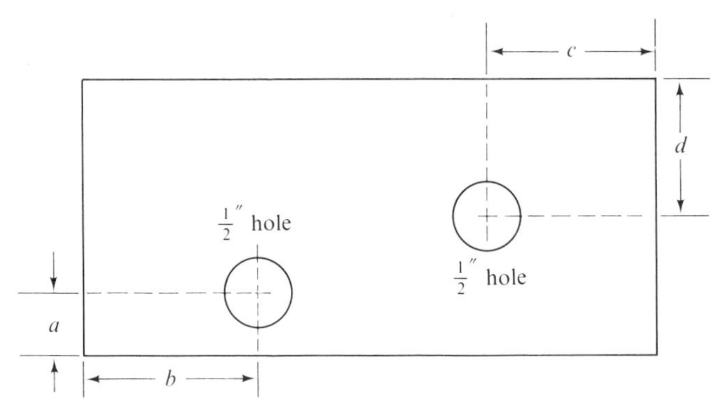

FIGURE 6–12 The layout of two one-half inch holes.

FIGURE 6–13 A plate jig. (Courtesy of National Acme)

FIGURE 6–14 A box jig. (Courtesy of National Acme)

When many parts are to be produced, the jig or fixture is used. The drill enters the bushing at point *A* (fig. 6–15). Point *B* on the part will not be drilled accurately and repeated in subsequent parts unless a constant location dimension exists where the drill enters and where it contacts the part at *B*, therefore, some means of material holding must be available.

Holding of the part illustrated in figure 6–15 is accomplished with the handle of the jig which moves in one upward swing, locking the part in place by contact buttons *C*. Guiding buttons 1, 2, 3, 4, and 5 force the part to take only one position in the jig box.

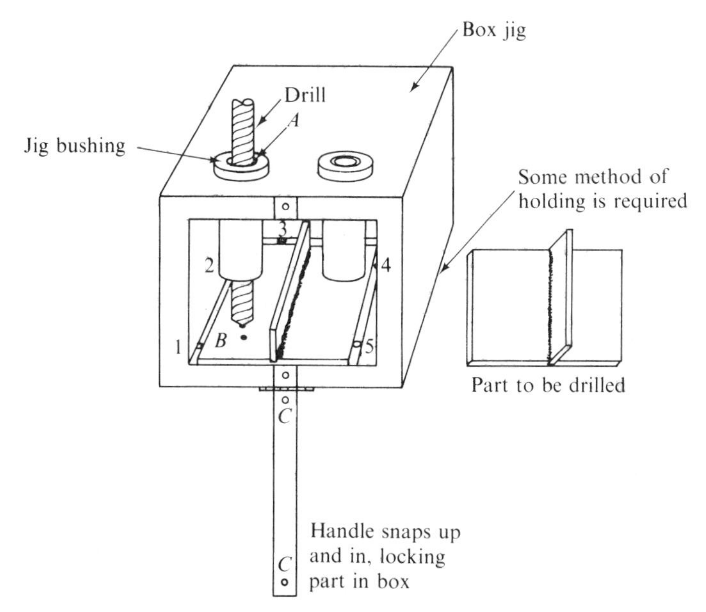

FIGURE 6-15 A T-shaped part being drilled in a drill jig.

The drill enters at point A and rotates within the very small clearance along the bushing's wall. Three rapid operations occur: the part is slid into the box while buttons fix the part's position, the handle is snapped into place, and the drill is lowered into the part at point B while the chips fall through a slot in the bottom of the box. This sequence is accomplished by an unskilled operator because skill and accuracy have been built into the jig. Outside dimensions of the part to be drilled are previously inspected for proper tolerances.

The above manual processes can be automated by simply adding five programmed operations which are assisted with mechanical means: (1), the part is pushed into the jig; (2), the handle snaps into place; (3), the drill is lowered a given distance ànd raised; (4), the handle is lowered; and (5), the drilled part is ejected. The process is repeated as parts are mechanically fed into the jig.

FIXTURES

A fixture is a device that holds a solid material for purposes of forming, welding, and machining. Such a general definition fits most situations in the metals processing industries where materials are tightly secured for some type of dimensional change operation. Because the primary purpose of a fixture is to hold a part or material during processing, the fixture must be capable of effective and repetitive performance.

A fixture is designed to do a specific job. It is not a general holding device such as a vise, but is patterned for use during closely related machining operations such as holding automotive engine blocks for specific milling operations. Parts of the fixture can be moved to continue milling on other surfaces of the same blocks. Just as a vise holds a part which is to be sawed, the fixture clamps and holds the work for similar operations. While a vise is limited in its capabilities, both in capacity and type of holding power, the fixture is more universal in that it can be moved about, adjusted for varying capacities in more than one plane, and is able to fit into close places where forming or machining is critical. However, such a description does not place the fixture in the same general use category as is the vise.

Design of Fixtures

Fixtures securely clamp work so that precision machining can be accomplished. In this respect, the fixture must allow the thrust of the load to be against its fixed structure and not on an adjustable part, unless this adjustable part is structurally capable. Also, a fixture must be secured to the work in such a way that the work is not sprung an excessive amount, especially not beyond its modulus of elasticity. Contact points of the fixture, with reference to the work, must be sufficient to allow stress dispersal in the part being altered to the fixture and then into the structure of the machine. Improper clamping and overtightening of a fixture often throws the work out of alignment, resulting in scrapping of the part. When fixtures are tightened, the loads should be against supports, otherwise, excessive springing of material results.

With regard to structure of the fixture, heavy sections should be placed where stresses will be exceptionally high. Fixtures used in machining operations must allow for.passage of chips and coolant application. Ease of clamping, access to the machine and part, and holding power are primary qualities of a fixture. The fixture is a common part or attachment to a machine (fig. 6–16). This is especially true for

FIGURE 6–16 A part being machined while being held in place by a fixture. (Courtesy of Cleveland Twist Drill Co.)

milling machine operations; the setup for milling often includes the fixture. The setup for forming and welding also includes fixtures. In other words, many welding, machining, and forming operations depend on the fixture for holding the material, and fixtures are usually attached to the machine or assembly. Fixtures keep the part from bending during processing and also maintain proper alignment of dimensions.

General Nature of the Fixture

Production fixtures are unlimited in both design and function. A simple machining operation, for example, may require only *L*-shaped clamps (fig. 6–17). Most machines

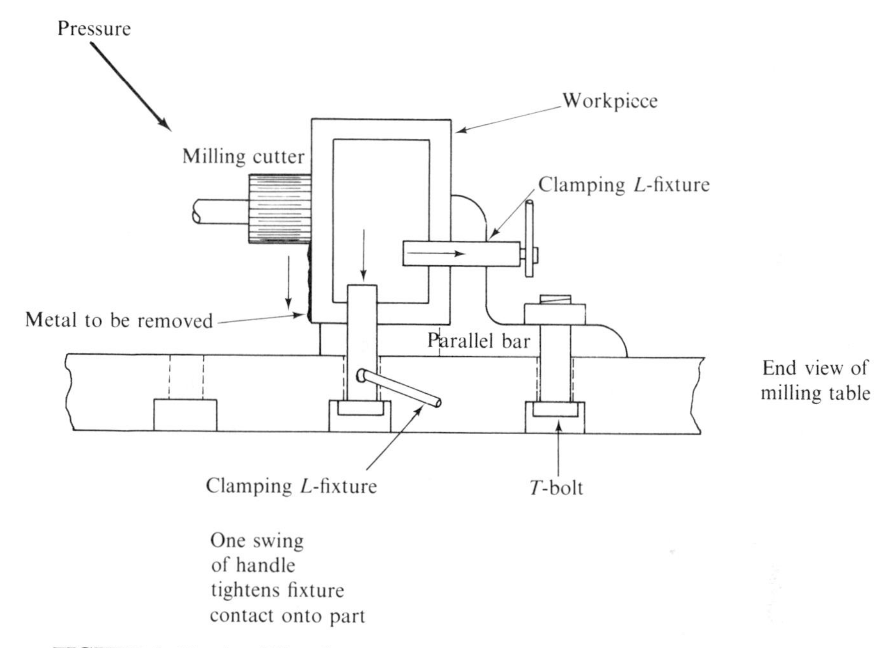

FIGURE 6–17 A milling fixture and a *T*-bolt holding a part during machining.

have *T*-slotted facilities in their tables, or they are sometimes equipped with dovetails (see fig. 9–9 and fig. 9–24). Matching bolts or dovetails fasten the fixture to the machine's table. According to the functions of the fixture, it has no special arrangements for guiding the cutting tool. In figure 6–18, however, there are guides for the drills, these being part of the jig which is clamped to the fixture. The fixture is merely a clamping device which holds the part or other holding device securely against the high stresses imposed into the material during cutting or other processing.

A double holding device is illustrated in figure 6–18. The base of the fixture is equipped with a dovetail to allow accurate and easy sliding along its base. The base is provided with slots so the *T*-bolts can be used to bolt the fixture to the machine's table. The top of the fixture is also equipped with *T*-slots so that additional fixtures or jigs can be securely attached and locked with levers. In this example, two drills are lowered into the part which is fitted into the jig, resting against buttons. Part insertion is rapid and accurate, followed by accurate drilling. Coolant is used as needed, while chips are released at the bottom. This type of holding device, the jig, is specialized and is patterned to fit precut parts which exactly fit into the jig. Clamping levers swing on

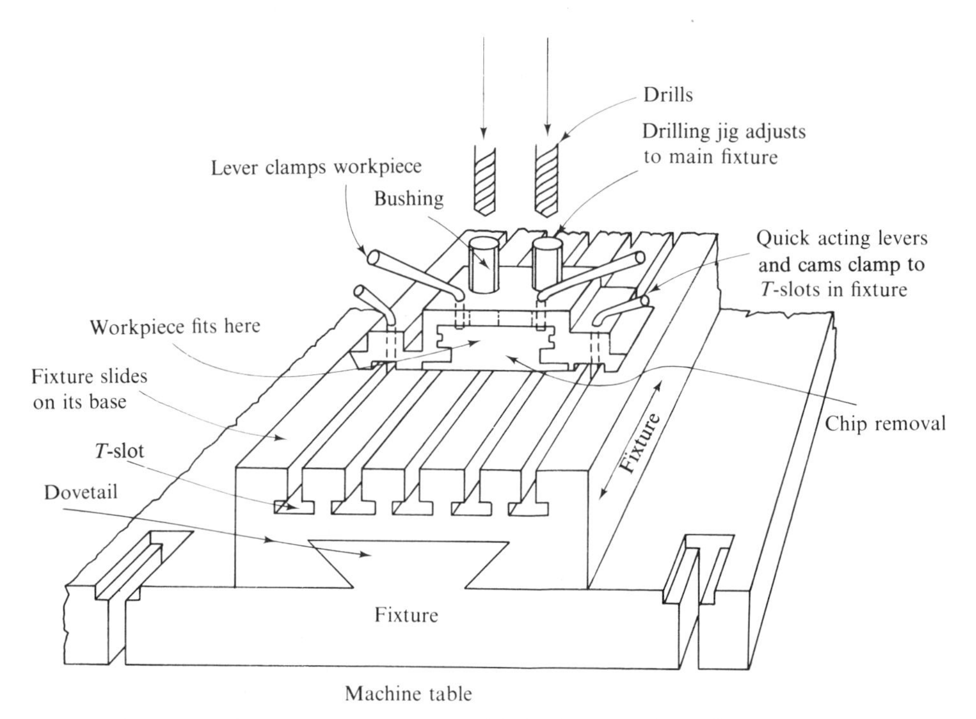

FIGURE 6–18 A fixture equipped with *T*-slots and dovetail.

cams and securely hold the work. When the top holding device is removed, the bottom fixture is available for many other specialized and generalized operations.

A more sophisticated fixture is illustrated in figure 6–19 where several points of attachment to the material are required. When large quantities of identical parts are to be produced, the fixture becomes more specialized to the holding needs of the part. When multiple use capability is built into the fixture, however, it becomes more universal.

Fixture and Material Relationship

The association of the fixture to the material and the machine is critical. For example, the base of the fixture is frequently bolted or clamped to the machine's table. The least number of individual operations reduces time for this operation, therefore, consideration must be given to quick clamping levers and other attachment means. A study of the basic holding and alignment factors, pressure points on the material, direction and magnitude of stresses, clearance from cutters and forming tools, ease of material insertion into the fixture, speed of clamping, number of clamping points, and strength of the fixture is needed to produce a fine fixture in relationship to the material. Basically, fixtures, especially those used in machining operations, are more massive in structure than are jigs. Some important points to be observed in regard to design are to keep the material as close to the table as is possible in order to reduce bending stresses, to position clamping points over established bearing points of the fixture to help reduce the possibility of vibration with its resultant chatter between material and cutter, to have a clamping arrangement that allows stress transmittal

FIGURE 6–19 Face milling a part while part is held in a circular milling attachment. (Courtesy of Cincinnati Milacron)

directly into the fixture's base and then into the machine's table, to have clamping device handles that are limited in length to eliminate the danger of excessive torque and resultant overpressure at the point of contact between fixture and material, to have as few clamps as are practical with each clamp having holding or releasing power within one swinging arc of the handle's clamp, to be able to mount the fixture to the machine quickly and easily, and as important as any aspect, the fixture must be as simple as is practical.

JIGS

The most common and fastest layout procedure for drilling is accomplished with the use of a jig, either by manual or automated process. A prime requirement of the jig is that it must automatically establish a location dimension, as has been pointed out, that is, perform layout. The jig has built-in features which provide fixed location points whereby a part will only fit in one manner within the jig. Even though the jig's primary requirement is to assure consistency of location with regard to a position on a part, it also guides the drill or other cutter during machining. A drilling jig, for example, locates the point where the hole is to be drilled, holds the part, and then guides the drill during the operation. Because jigs are used for several different types of operations, the particular kind of jig must be specified when describing jigs. For example, a diameter jig is illustrated in figure 6–20.

FIGURE 6–20 A diameter jig. (Courtesy of National Acme)

Fixture and Jig Compared

One basic difference between fixtures and jigs is that the fixture is tied to the machine because it must hold the part while work is being done. The jig is not necessarily fastened to a machine. The jig is more closely related to a specific type of operation; the fixture is common to the machine and not the operation. A drill jig is designed to drill a hole at a given location; a fixture is designed to hold material so that various types of operations can be performed. Keep in mind that the jig may or may not be attached to a machine.

Accuracy of the Jig

The jig is a precision device that automatically provides for dimension locations on a part. This means that the jig is portable in many cases and that it can be attached to a machine or independent of a machine. There is a fixed relationship between the jig and the operation, however, once the location is established. When a jig holds and guides the working tool, care must be exercised in the manner in which the jig is held.

The jig is designed with many of the same features that the fixture contains. Because dimension locations for inserted parts are inherent with the jig, all other features support this requirement. When the jig also holds the material (a drilling jig, for example), strength of the jig's structure must assure that twisting of the structure will not occur. Other requirements include speed of part insertion, the operational speed of work performed, speed of part extraction, a method of clamping the work, a method of guiding the drill, a provision for chip removal, and repeatability of precision performance.

The drill jig is equipped with bushings placed within bored holes by exacting methods, such as with a jig boring machine. When the bushings are inserted and the part is locked within the jig, the drill being inserted into the bushing can find only one location on the part to drill. As drilling progresses, drill guidance is also dictated by the close proximity of the bushing's wall. This type of drilling is illustrated in figure 6–21.

TYPES OF FIXTURES AND JIGS

Even though fixtures are basically different from jigs and are more generalized in use than are jigs, some common relationship exists between them. Both devices are

FIGURE 6–21 The drill jig enables fast and accurate hole drilling in machined parts. (Courtesy of Cleveland Twist Drill Co.)

associated with manufacturing operations, primarily in the metal machining industries. Each has certain functions to perform in relation to work being done on a part, and each promotes speed of production. Both fixtures and jigs are divided into different types, dependent on work to be performed.

Fixtures are used to hold work while welding is being performed or while forming or cutting processes are accomplished. The fixture is a very common attachment or part to many different machines. As an example, the milling machine uses many different types of fixtures, each being dependent on the processes to be performed. Boring and broaching operations require fixtures (fig. 6–22).

Jigs also are available in different types, but all have a common characteristic, dimension location with repeatability. Some common jigs are listed as channel, box, universal, plate, and diameter. In each type, there is a common relationship between the part's position in the jig and the location of the work to be done on the part. A common leaf jig is shown in figure 6–23, while a channel jig is shown in figure 6–24. With regard to dimensioning and position of dimensions, accurate measurements are essential in the layout and production of fixtures and jigs.

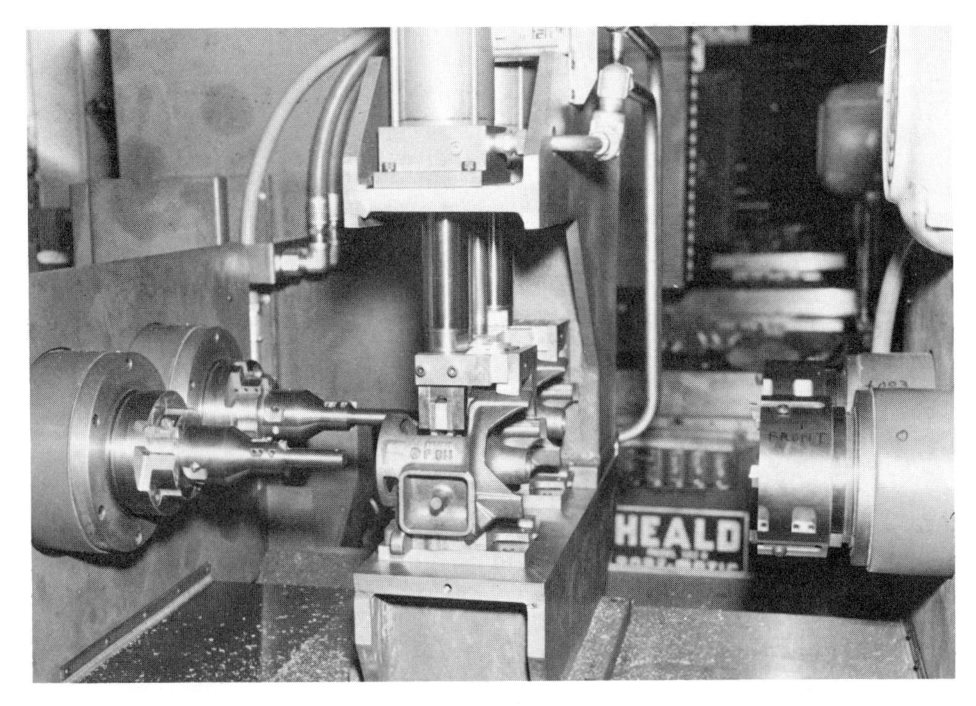

FIGURE 6–22 A boring operation. (Courtesy of Cincinnati Milacron)

FIGURE 6–23 A leaf jig. (Courtesy of National Acme)

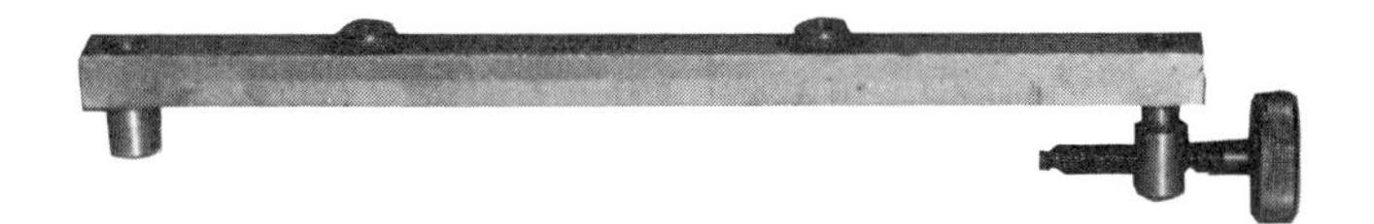

FIGURE 6–24 A channel jig. (Courtesy of National Acme)

Measurement

Measurement is a means of establishing a quantity of some particular thing such as a dimension indicated on a part. The dimension may indicate a size as in length, breadth, or thickness. Also, the dimension may indicate the location of the measurement. With respect to measurement and dimensioning, the designer draws objects in three-dimensional display so that the fabricator can produce the object in accordance with the drawing. The dimension has an indicated size, for example, two inches, but the indicated size may be an ideal dimension obtainable only under ideal conditions. During production, the ideal conditions often do not exist. In other words, the resultant measurement of an object may be anywhere from just a little less than two inches to just a little more than two inches. Specifically, the dimension may be indicated as two inches with a tolerance of plus or minus 0.002 inch. The object is considered acceptable from a dimensional aspect when it measures anywhere within the above measurement. When a closer tolerance is desired, the dimension may be indicated as two inches with a tolerance of plus or minus 0.0002 inch. The smaller the tolerance, the more difficult it is to obtain unless special provisions are made so that production will not be reduced. Finally, it may be worthwhile to limit the tolerance to millionths of an inch. When a dimension is established in such precise exactness, production quantities are reduced, increased manual skills are required unless special production tools are used, and the cost per unit output goes up. Basically, the acceptable tolerance of a dimension should be no smaller than is necessary for the proper use of the object. The tolerance, then, depends on many factors, all relating to the use of the object.

MEASURING TOOLS

The measuring tool verifies the accuracy of the finished part. When a large tolerance is specified, rules are used wherein the eye guesses the location of the measurement by comparing a pencil or prick punch mark on the part with a line on the rule or by measuring the diameter of a round rod by comparing distances between locations of lined locations on the rule. Often, tolerances up to plus or minus 0.015 inch are acceptable when using this method, but when the part has precision qualities attached to it, such as a crankshaft, the micrometer is used. Measurements having tolerances of plus or minus 0.0002 inch, for example, are quickly made with the micrometer. A comparison with a precision rule verifies the greater accuracy of the micrometer (fig. 6–25).

As the geometry of the part changes, different kinds of measuring tools are required. A hole, for example, is measured in depth with a depth gage, the telescope gage being used to measure the inside diameter of the hole (fig. 6–26). Length is measured with precision rules, verniers, and tapes, and diameters are measured with calipers or micrometers, either inside or outside (fig. 6–27). When a diameter increases appreciably, the measuring tool also increases in size. Parts having an outside diameter of less than one inch, for example, are measured with a zero to one-inch micrometer. This precision tool (fig. 6–25) is made with capabilities up to several feet whereby a large turbine shaft can be measured with the same accuracy as can a small

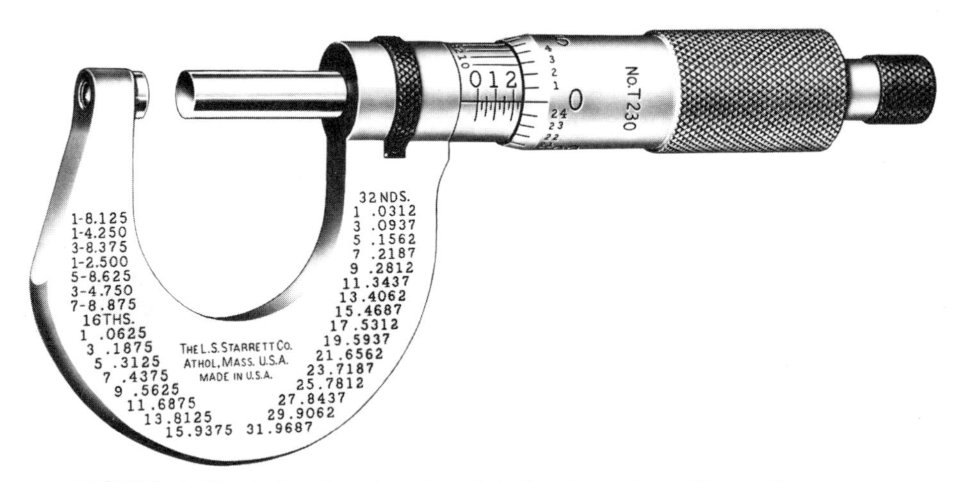

FIGURE 6–25 A 6-inch rule and a 0 to 1 micrometer with ten-thousandth capability. (Courtesy of The L. S. Starrett Co.)

bolt. Small micrometers are available in increments of one inch with either the thousandth or ten-thousandth measuring capability.

In other geometrical situations, the measured surface (a thread, for example) may not be uniform. Special measuring tools such as thread micrometers are used whereby the anvils of the tool fit the shape of the thread (fig. 6–28). Ball-point anvils measure the thickness of wall tubing, the ball resting against the inner wall surface. The main difference, then, between the rule and the micrometer is the accuracy factor of the measuring tool, the eye being relied upon more by the rule than the micrometer. Even the size and shape of cutter angles, gear teeth, and splines are measurable by using the proper type of instrument. Table 6–1 lists some of the important hand measuring tools. Faster measurements are accomplished with gages and instruments.

TABLE 6–1 Hand Measuring Tools

Micrometers	Calipers	Gages	Rules
Bench Inside Outside Internal groove	Outside Inside Vernier Dial Vernier Vernier gear tooth	Micrometer depth Height transfer Vernier height Vernier depth Telescope Pitch Thickness Thread Wire Radius Taper Center	Steel Steel squares Universal protractors Dividers

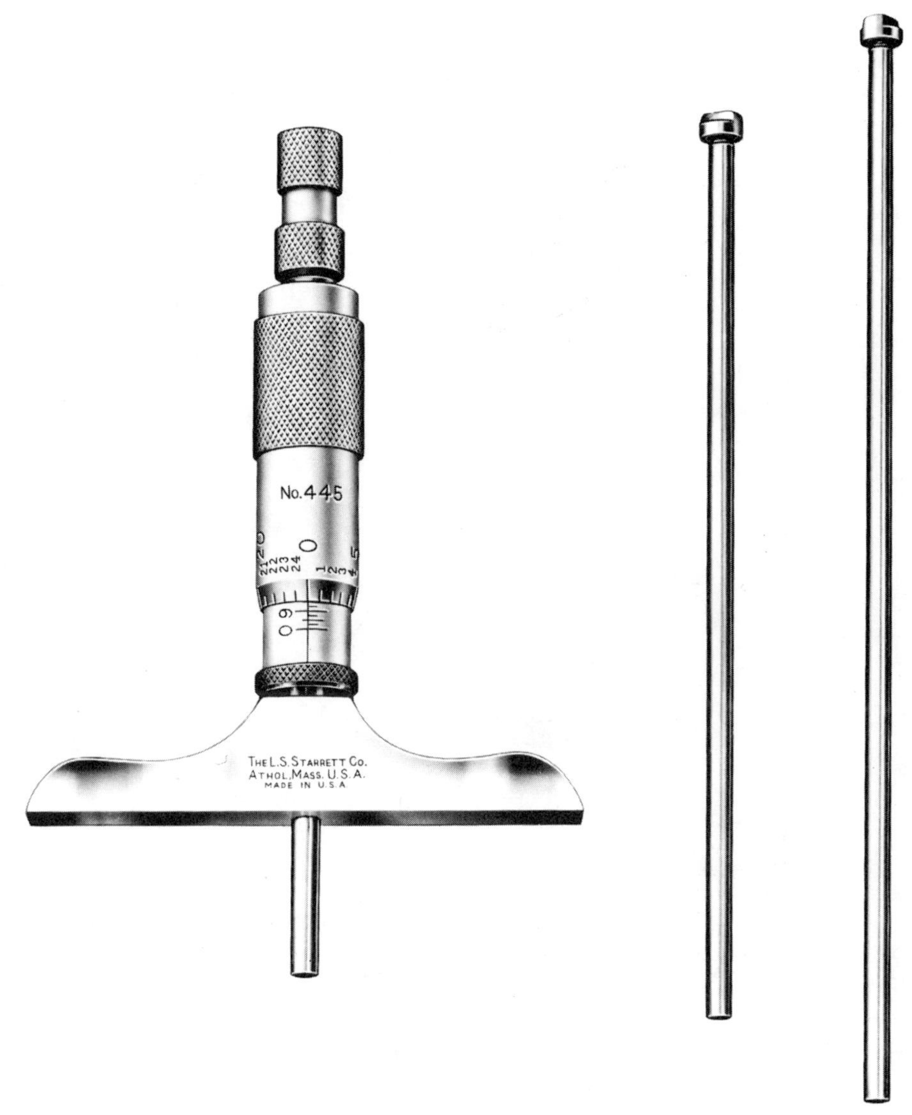

FIGURE 6–26 A micrometer type of depth gage. (Courtesy of The L. S. Starrett Co.)

GAGES AND INSTRUMENTS

Many instruments are equipped with dials which show each division of measurement. A dial indicator, for example, has many uses. When a part is pressed against the plunger of the gage, the pointer turns to a point which indicates the measurement. When the indicator is attached to a fixture (fig. 6–29), parts are inserted and extracted in rapid succession, each reading showing the dimension of the part in thousandths of an inch. From this concept of rapid measurement, a type of fixture has been produced that allows a quick measurement to be taken of differently shaped parts, whether the part has holes or grooves, or rounded or flat shapes. Table 6–2 lists several of the different types of gages.

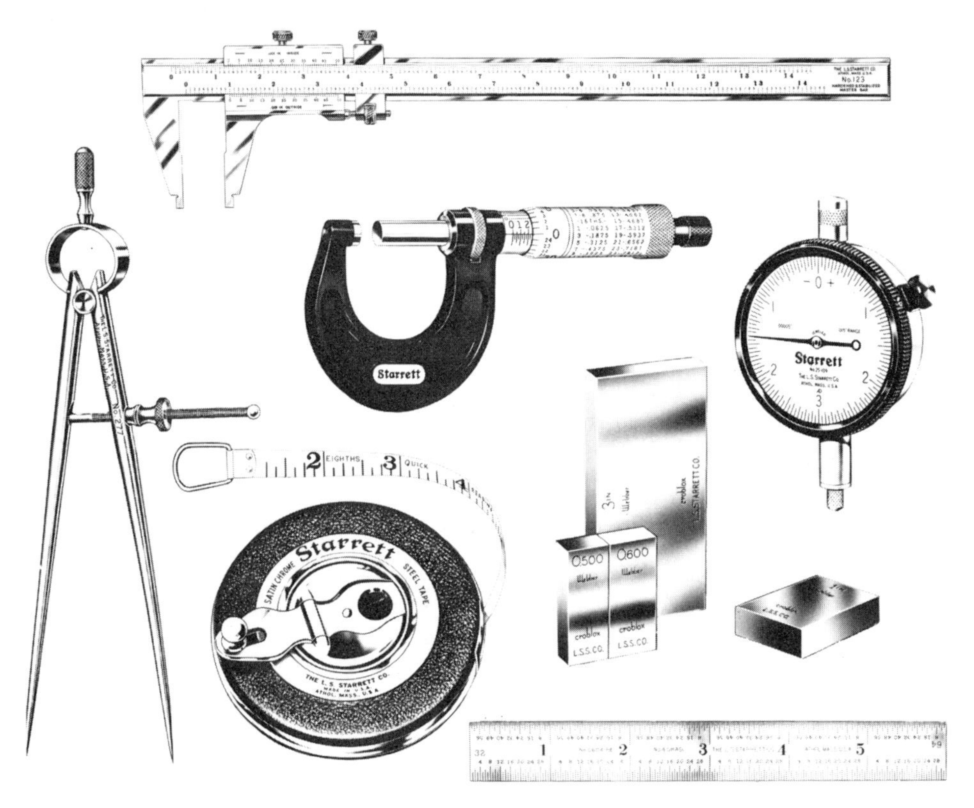

FIGURE 6-27 Precision measuring tools. (Courtesy of The L. S. Starrett Co.)

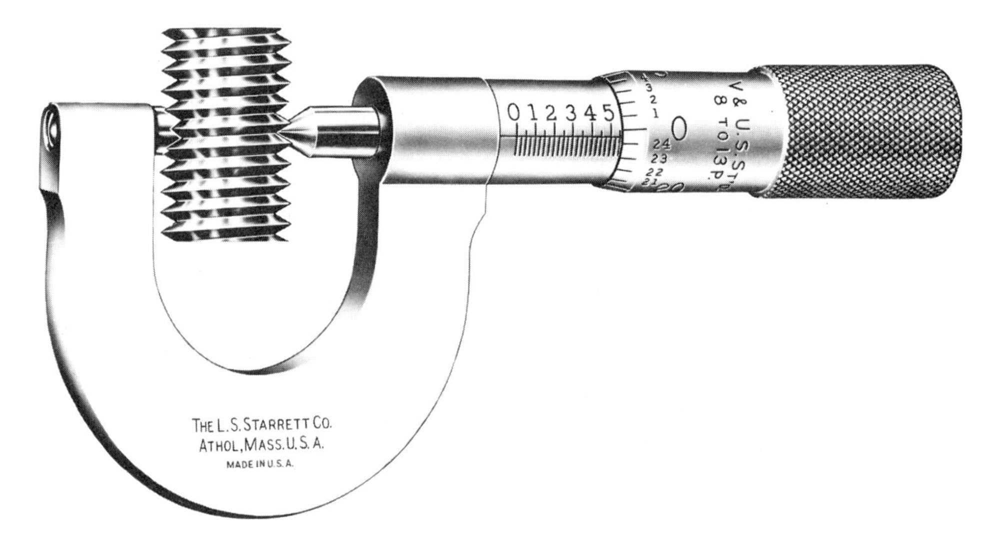

FIGURE 6-28 A special measuring micrometer for measuring threads. (Courtesy of The L. S. Starrett Co.)

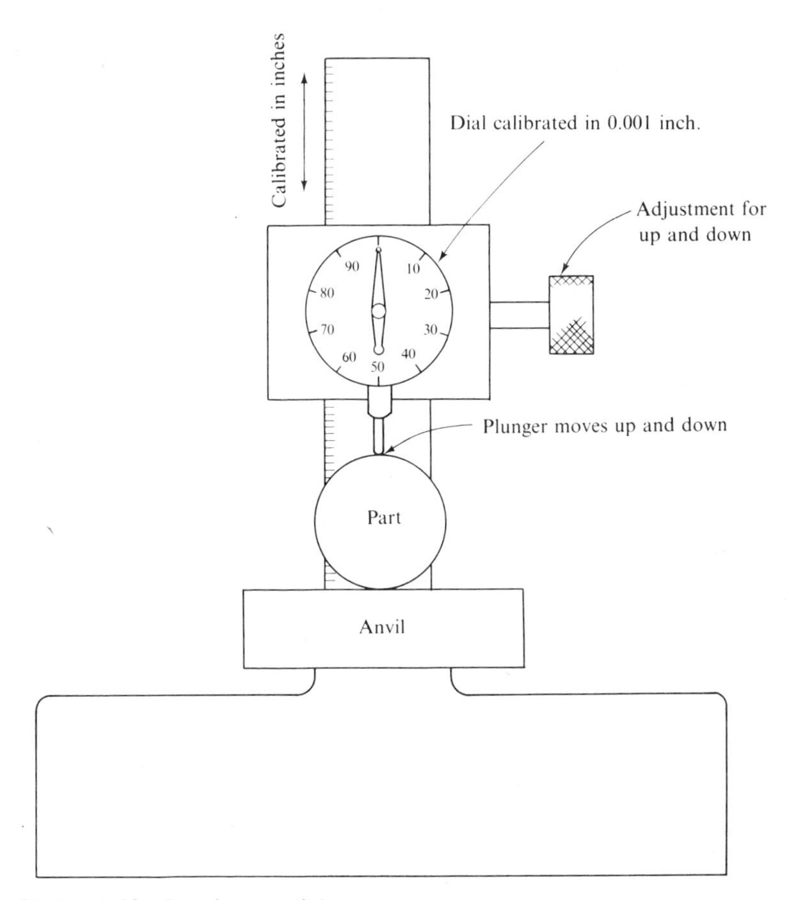

FIGURE 6–29 Bench type dial gage.

Special gages such as the plug gage are available for gaging tapered holes. The ring gage measures outside tapers. "Go" and "No Go" gages are available for rapid·verification of acceptable parts. Built-in measurements within the tool allow rapid insertion of a part between the contacts of the tool. Parts beyond the tolerance are rejected.

GAGE BLOCKS AND STANDARDS

Some means must be available to assure the accuracy of measuring tools, instruments, and gages. When production requires parts having close tolerances, there should be a standard measuring block on hand to verify the accuracy of the measuring tool. A gage block (fig. 6–27), for example, is used to verify the accuracy of a micrometer. When the measurement differs from the standard block, the micrometer is adjusted to conform with the block, or compensation must be accounted for. Gage blocks are available from tool manufacturers in various sizes. Because manufacturing type measuring tools must be very accurate, all measuring tools must be checked periodically against standards to assure accuracy of measurement. In this respect, measurement is indicated at a specified temperature. When small parts are held in

TABLE 6–2 Indicators and Instruments

Indicators and Instruments
Snap gages
Groove gages
Out-of-roundness gages
Cylinder gages
Vibrometers
Sheet gages
Bore gages
Strain gages
Test indicators
Comparators

the hand for a period of time, the part expands due to an increase in temperature caused by the human body. If machining follows measurement of parts with very small tolerances, the parts will frequently be scrapped due to their undersize as shrinkage occurs at room temperature.

Power Transmission

The engineer and technician are concerned with the means of transferring power from the source to the working tool. Once the proper quantity of power has been established, the means for delivery must be worked out so that work can be performed. Power originates from an engine, generator, turbine, motor, pressure system, or other source and is delivered by electrical, mechanical, or fluid means.

Electrical power is distributed so that motors or devices will actuate a mechanism. Electricity may be direct or alternating current and may originate from batteries, generators, or a power line. When a fabrication machine is connected and energized by an electrical motor, the mechanisms of the machine operate, making the machine available for production. In other words, electrical energy is transferred to mechanical energy through the electromechanical power transfer system. Basically, electromagnetic lines of force, or flux lines, create a turning action in a motor and, in turn, the shaft of the motor transfers power through mechanical means to the working tool.

POWER TRANSFER SYSTEMS

Mechanical transfer systems are of several types, the gearing system being predominantly used. Trains of gears are connected from the first driver gear to the final driven gear. Shafting attached to the driven gear creates torque, and work is performed. Gears are produced in various kinds, including spur, helical, bevel, and worm gears. Figure 6–30 illustrates the power transfer characteristics of several types of gears. Most machines contain gears, the gearing being activated from a power source such as an electric motor. For example, the push of a button causes the chuck of a lathe to turn, the chuck being driven by a powerful electric motor which is connected to the gear train.

Gearing

Gears illustrated in figure 6–30 perform work in accordance with the designs of the gear teeth. Helical gears (*a*) transfer power at their points of teeth contact. Two right-hand helical gears transfer power from shafts at 90° to each other. The helix of the teeth allows smooth pressures from teeth to teeth. In *b*, a lefthand mated with a right-

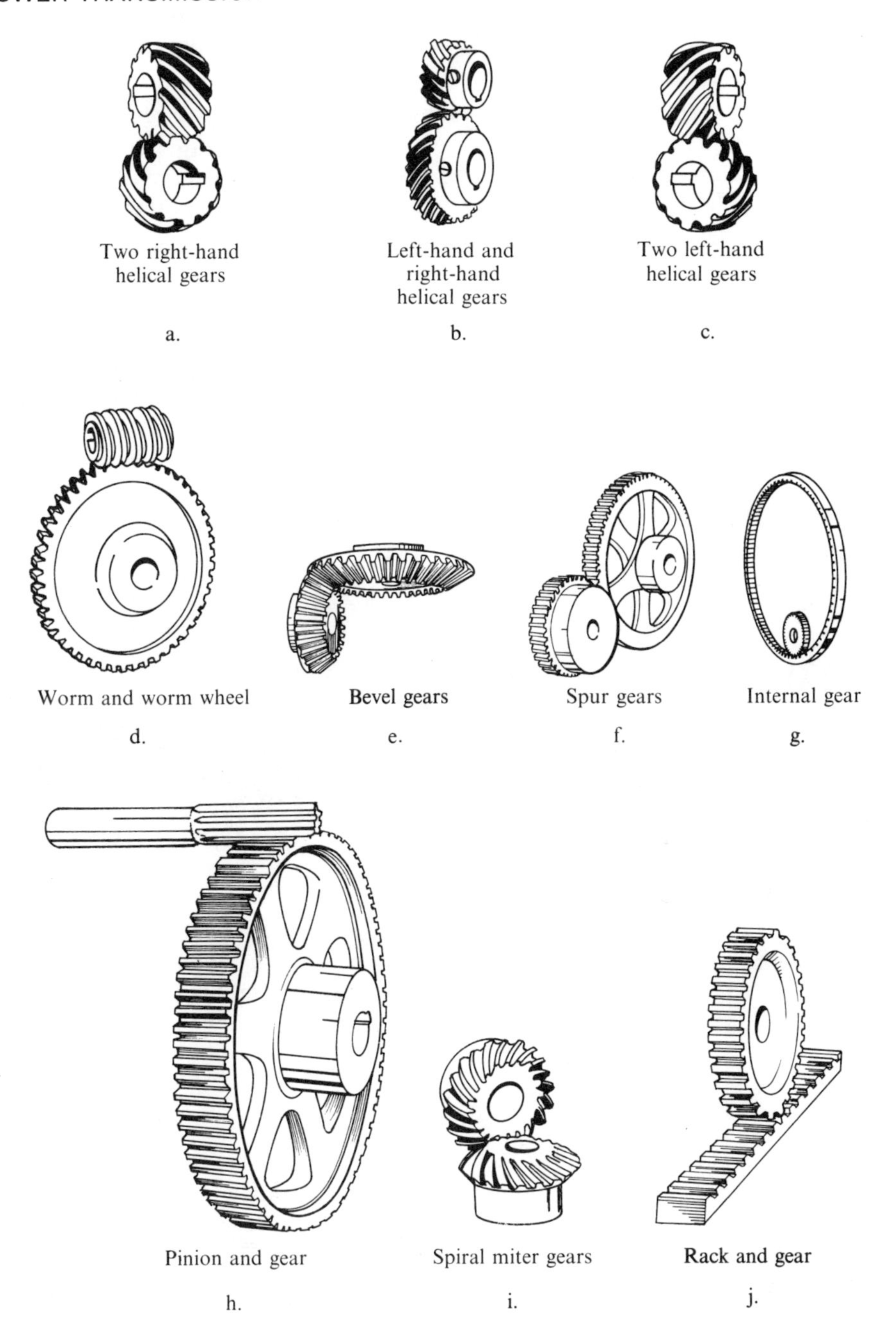

FIGURE 6–30 Typical gears used in the transmission of power.

hand helical gear provides power transmission along parallel shafting. When two left-hand helical gears are mated, right angle power transfer also occurs (*c*), but opposite to the relationship as shown in *a*. When a worm is associated with its mating wheel (*d*), right angle power transfer occurs. Many steering mechanisms use this system of power transfer. In *e*, the two bevel gears also transfer energy at 90°. The simplest of gears is the spur gear illustrated in *f* whereby power is transferred to parallel shafting. In a few instances, an internal gear moves inside another, causing circular transfer of power, as shown in *g*. In *h*, the spur gear and pinion drive apply to many of the common shafting arrangements in mechanized production lines and in machines. In *i*, a relationship similar to that in *e* exists whereby the spiral miter gears transfer power at 90°. When vertical motion is required in a hand press or in reciprocating machines, for example, the spur gear and rack (*j*) are used. There are other gears such as herringbone in electric machinery, but the above gears occupy an important area in power transmission and subsequently in the manufacturing operation. Manufacturing and mechanical technicians use these gearing arrangements in materials handling equipment, conveyor systems, feeder mechanisms, and related power drives.

Typical Drive Systems

Other types of driving mechanisms are used to transfer power. Some of these are illustrated in figure 6–31. The sprocket and chain arrangement shown in *a* is a typical

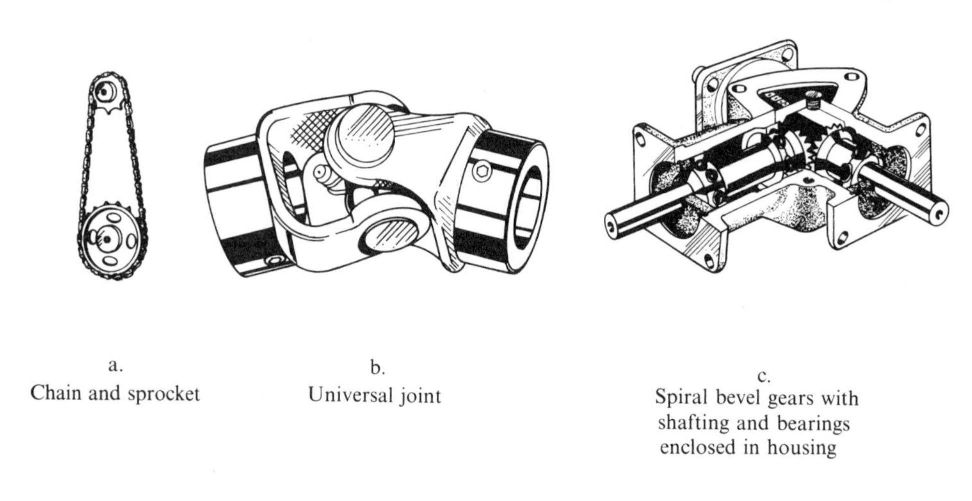

a. Chain and sprocket

b. Universal joint

c. Spiral bevel gears with shafting and bearings enclosed in housing

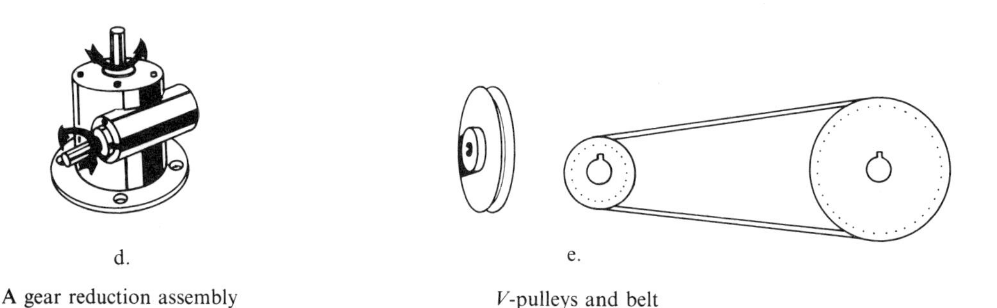

d. A gear reduction assembly

e. *V*-pulleys and belt

FIGURE 6–31 Typical drive systems and power transmission.

system of power transfer, a common example being the bicycle. In industry, many sprockets and chains are used in numerous types of machine operations, both inside and outside of the machines. The universal joint in *b* is used when shafting is not in complete alignment or when universal motion is required. Then, in order to provide an operational system of power transfer, necessary components must be assembled. A typical power transfer system is illustrated in *c*. In this illustration, right angle power is transferred from a shaft by means of the bevel gears. Notice the placement of bearings to absorb the loads and guide the shafting properly. Torque stresses from the shafts move into the bearings and, subsequently, into the cast housing. The shafts have key-ways for positive connection to the gears and drive capability with mating parts. When it is essential to reduce the speed of a shaft or change its direction of rotation, a gear reduction assembly is used. Such a device is illustrated in *d* whereby a 90° power transfer is performed along with rotation reduction. When transfer of power is required in a manner similar to the sprocket and chain, the belt and pulley system is used (*e*). Pulleys may also be of the variable diameter type which help regulate the revolutions per minute by expanding and contracting their diameters. When heavier loads are imposed, several belts are arranged in a group. Each of the above power transfer systems is common in industrial applications throughout a manufacturing plant.

Other power transfer systems include belting, push rods, levers, cams, and fluid power mechanisms. The use of belting allows power transfer over a large gap of space and eliminates the necessity for gear trains. Belt drives are smoother than gear drives and have built-in safety features. A belt will slip or break when caught under excessive load, while a gear train is a positive drive, even though it is often protected against sudden overloading by shear pins.

Fluid power systems include hydraulic and gas systems. When a fluid such as oil or air is forced into a closed cylinder, pressure is generated. This pressure actuates a piston. When the piston is attached to a push rod or lever, a mechanical power transfer system exsists. Actuation of the piston moves a rod and, in turn, work is performed. The cam, for example, is used to cause different types of work, as illustrated in figure 6–32. The lobes on the cam shown in *a* raise and lower the push rod which, in turn, causes a reciprocating motion of the rod, the energy transfer being at an extended distance from the cam. In *b*, the circular motion of the wheel causes reciprocating work in the extended rod.

POWER TRANSFER RUNS THE PLANT

Power transfer systems are part of the manufacturing plant. Machines are energized by some form of power, therefore, engineers and technicians must be cognizant of power capabilities and work performance in regard to machines and related equipment. Each machine is a potential work source. Its structure and function, much like the human body, enables it to do work through its many dozens, even hundreds, of parts. In many types of machines these parts include electrical switches, power drives, guides, stops, rods, clutches, brakes, gears, cams, shafts, and numerous other working mechanisms. For instance, when a spinning wheel must be stopped immediately, electromechanical brakes are used, some governed by a momentum reduction mechanism. In turn, clutches grab the shaft in a mechanism and quickly accelerate it

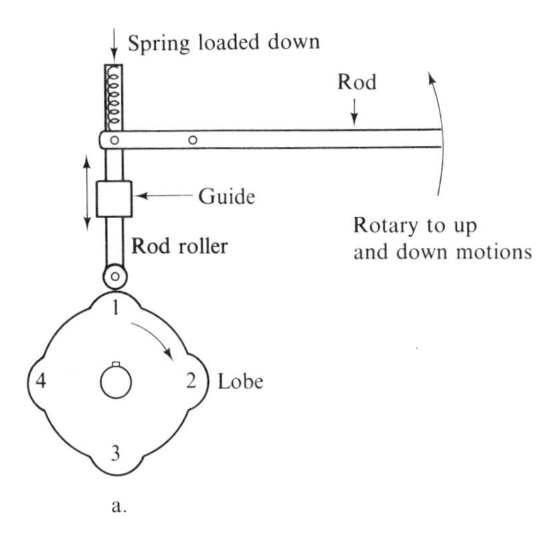

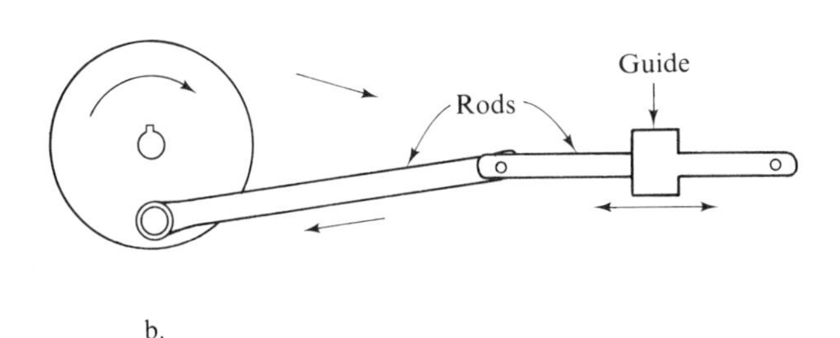

FIGURE 6–32 Operation of typical mechanisms for power directional changes.

to the desired r/min. When an exact moment is required for stopping the mechanism, the brake is applied. Slow turning mechanisms are stopped immediately in order to provide an exact dimension or to coincide with a specific reference point.

Because vibration is detrimental to machines and parts produced, great care is essential in verifying the mechanical balance in the system or in the part. Rotating parts vibrate when out of balance. A quick way of obtaining balance in a whirling part is to drill a small, sometimes predetermined, amount of metal from the heavy side. Obviously, some means must be available to balance the mechanism in order to determine where imbalance exists. When balance is established, operation is smooth. All vibration is not out-of-balance vibration, however. Machine functions such as punching, shearing, and stamping cause vibrations. These are natural and must not be interpreted as abnormal. Also, many machines use levers and cams whereby eccentric motions of parts create jerks and vibrations. Again, these functions are normal. On the other hand, when the pulley of a belt driven lathe is out of balance, the machine must not be used until balance is restored.

Once the capability of the machine is known, it needs only a control system to put it to work. The machine may cut a part, bend it, hammer, punch, squeeze, cup, heat, stretch, or perform many other tasks. The control may be manual or mechanical. The operator may be skilled or nonskilled, and the mechanical control may be tape or computer. Keep in mind, however, that automated machinery requires skilled people to set and limit mechanical capabilities.

Machine Control

Fabrication machines vary in both size and function. A small press used in the cake pan industry, for example, may turn out several thousand formed pans a day as a result of pressing operations. Because such a process lends itself to automation, the production line is arranged so that each operation occurs in a sequence. A large coil of aluminum strip, mounted on a roller and having the necessary width and thickness is arranged so that the end of the strip is pulled onto a roller type conveyor system. The strip is fed into a power shear which operates from a programmed tape. Equal lengths of metal rectangles are sheared at equal increments of time, the coil of metal steadily unrolling to feed the shear. The rectangular sections then move into the pressing die. The ram moves downward on command, forcing the punch smoothly onto the ductile metal. In a reciprocating motion of the ram, the punch is lifted while the shaped pan is ejected and moved farther down the conveyor line into the trimming and edging machine. Each finished pan is rhythmically ejected, cleaned, and subsequently arranged into a stack. Periodically, the stack is moved to final processing.

One programmed tape controls the entire operation, while the services of an observer are required periodically to monitor the operation, replace the coil of metal strip, move the stack of finished pans, and assist the process, if necessary. A push of the correct button controls the entire operation. Such an operation is especially designed to fit the needs of this particular process. Each machine is equipped with special fixtures, fittings, lubrication points, guides, power transfer mechanisms, and other parts to move the metal, hold the metal while it is being formed, and move the finished part to its destination. A nonautomated assembly line would require the services of several skilled operators, increasing the cost of each pan a significant amount. Each machine would require separate controls, and many manual operations would be essential. After the initial cost of machinery and equipment is absorbed, cost per part produced reduces the costs on previously produced parts.

Testing and Inspection During Manufacturing

Because manufacturing processes use materials in the various sequences of operations, some means must be available to assure product reliability upon completion of the last operation. Obviously, any testing performed on the product must be of a nondestructive nature in order to prevent damage to the product. Methods of product inspections to verify a satisfactory product include the penetrant or magnetic particle processes, ultrasonic or eddy current procedures, and radiographic inspections. Often, all five methods are used on one part. Details of these inspections are pointed out in a previous text. According to the manufacturing flow process (fig. 6–1), testing and inspection are performed continually. These procedures are essential in maintaining quality control. Basically, the raw material is exposed to one or more nondestructive tests prior to any further processing. Should unsatisfactory conditions prevail inside the metal, such as a hole, crack, inclusion, or segregation, further processing of the material into a part may be useless. As previously discussed, the presence of oxide on the surfaces of holes prevents closure of the metal during hot working. Also, inclusions are elongated and often result in cracks during heat treatment. Consequently, the casting or ingot is tested by radiographic or ultrasonic processes prior to further processing. Disclosure of unacceptable discontinuities justifies scrapping of the incomplete part at this point. Should these internal conditions be overlooked or disregarded, serious consequences may result, even disaster to equipment and personnel. Such a decision to accept or reject the material naturally includes the economics of the process and the resulting profit. However, economic considerations are also involved in instances where lawsuits occur as results of material failure and disaster. Such legal actions tend to focus attention on the initial soundness of the material.

As the manufacturing process continues to shape the part, periodic nondestructive tests are made to disclose any process irregularities. Keep in mind that birth defects in the solid material often remain with the material throughout its lifetime. Secondly, processing may tend to increase the birth defect to near failure of the material. Naturally, these circumstances must be detected, therefore, testing and inspection are performed all through the manufacturing process, even on the final product. It must be pointed out that nondestructive tests are not the only tests accomplished during the manufacturing processes. In conjuction with nondestructive testing, the several types of destructive tests are performed on samples of the material. Such examinations include hardness, impact, ductility, tensile, shear, compression, and microscopic tests. Destructive tests sample the lots of parts in support of nondestructive tests on the parts to be used. The end products of quality control are, then, quality assurance and product reliability. Manufacturing plants which produce any types of stressed parts use the nondestructive and destructive testing processes. The need for material soundness and safety and dependability of the final part forces manufacturers to comply with the safety standards established by law and good practice. The OSHA, Occupational Safety and Health Act, established by the United States Congress, is detailed in its objectives to assist manufacturers in promoting safety in the material, safety for personnel during manufacturing, safety of the final part in service, and, consequently, safety for the consumer.

The processes of nondestructive testing are being used more extensively due to the demands of designers for sound materials so that less material can be used for their designs. Further, manufacturing costs are being reduced when unsound material is discovered at the beginning of the manufacturing process. During fabrication or other processing operation, tensile tears in the metal may be found, whereby further processing is useless. Also, parts which have been in service, for example, train wheels, aircraft parts, pipeline welds, and nuclear components, must be periodically tested to assure soundness of the metal. Nondestructive testing is rapidly accelerating in both importance and necessity.

Questions

1. Differentiate between a jig and a fixture.
2. Describe the major differences between a milling machine vise and a milling machine fixture.

3. Why must fixtures be constructed of heavy materials?
4. What is the purpose of the drill bushing in a jig?
5. Describe how layout is constructed into a jig.
6. Describe the need for tolerances in dimensions.
7. How does the temperature of a material relate to its measurement?
8. Describe three methods of power transmission.
9. Describe an advantage of the worm and wheel arrangement in power transfer in relation to a pair of spur gears.
10. Why is a belting system of power transfer sometimes used instead of a gear drive?
11. What is meant by "tooling up" during the preparation phase of manufacturing?
12. Why is part interchangeability very important during manufacturing processes?
13. How is part interchangeability guaranteed during manufacturing?
14. Explain two layout procedures.
15. How is layout which is built into a machine a great advantage?
16. Why must the technician be cognizant of the principles of power transmission?
17. Why must strength of materials be considered during the design of power transmission machinery and mechanisms?
18. What is the purpose of nondestructive testing?
19. List the five main methods of nondestructive testing.
20. Describe how economics relates to soundness of material and product reliability.

7

Conventional Machining Operations

The machining of metals is basically a chip removal process, while its purpose is to shape a material to its required dimensions within tolerances of plus or minus a very small quantity. The smaller the tolerance, the more difficult it is to obtain the desired dimension. As an example, the tool grinder removes chips so small that a fraction of a ten-thousandth of an inch within the desired dimension is obtainable. On the other hand, rough machining accomplished by the lathe may leave the dimension at 0.007 inch greater than the desired dimension.

Chips are removed by knifelike cutting tools such as rotating cutters, reciprocating cutters, linear type cutters, abrasive cutting wheels, and special process cutting apparatus. Regardless of the cutting method, materials are reduced to their desired dimensions in the shape of parts and objects. Machining is necessary because most forming processes are incapable of producing parts to the shapes and close tolerances which are obtainable in machining. Powder metallurgy and die casting processes, however, produce parts with tolerances comparable to machining. On the other hand, parts produced by sand casting, forming, and forging vary considerably within larger tolerances. Frequently, these parts require machining to obtain a smooth surface, a flat surface, or a dimension at some point with a closer tolerance than the "as cast" or "as forged" condition. Such machining operations remove only a small quantity of material. In parts having undercuts, holes and recessed areas, the machining sometimes removes up to one-half and even more of the original material. Consequently, two factors exist as a result of machining—a quantity of scrap material in the form of chips and a finished part. Machining, therefore, is a necessary process in an industrial society because no known method is capable of replacing it.

Materials Requiring Machining

Most metals and some plastics are at some time machined into a shape. These materials include mainly the ferrous and nonferrous metals and the commercial plastics. Plastics are usually softer than metals and have machining characteristics different from the metals. Also, many of the nonferrous metals such as aluminum and brass have machining characteristics much different from the ferrous metals. Several groups of nonferrous metals such as the alloys of cobalt, chromium, and tungsten machine under different circumstances than does low carbon steel. Basically, each material has its own set of machining qualities. These qualities must be considered prior to the cutting operation.

RATE OF METAL REMOVAL

The chemistry and physical structure of a material allows its material to be removed by cutting at certain rates. The rate is a combination of factors, including the cutting speed, the rate of feed, and the depth of cut into the material. Depending on the tool, horsepower of the machine, and rigidity of the total equipment, the depth of cut is selected. Generally, the operator uses as heavy a cut as is practical within the limitations of the machine. With respect to tool life, the depth of cut has the least adverse effect. As for the feed rate, the same factors listed above must be considered, in addition to the type of surface finish desired. Also, the relationship between the depth of cut and feed rate must be considered. Fast metal removal requires the greatest practical feed rate with a given depth of cut into the material. Lastly, the cutting speed is selected, usually from charts which have been prepared from results of experimentation. The cutting speed is listed in feet per minute or surface feet per minute. This is the measurement of material in feet that passes a set point in one minute.

REVOLUTIONS PER MINUTE

Because the machine is built to operate in revolutions per minute, some means must then be available for converting cutting speeds into revolutions per minute. The common formula for conversion is

$$\mathrm{r/min} = \frac{12\,CS}{\pi\, D}$$

Spindle speeds in revolutions per minute equals r/min. *CS* is the cutting speed factor which is obtainable from charts, while *D* is the diameter in inches of the work or tool, whichever is turning. When the cutting speed is desired, the following formula is used.

$$CS = \frac{\mathrm{d} \times \pi \times \mathrm{r/min}}{12}$$

CUTTING SPEEDS

Cutting speeds are influenced by the machinability of the material. For example, it has been determined that AISI B1112 cold-drawn steel has a machinability rating of 100%. The chemistry of AISI B1112 is carbon, 0.12%; manganese, 0.85%; sulfur, 0.20%; phosphorous, 0.10%; silicon, 0.10%; and the remainder is ferrite. However, the cold-drawn condition is different than any other structural condition; it is somewhat harder than the annealed. Those metals which are harder to machine have machinability ratings less than 100%. Those metals having ratings higher than 100% are easier to machine. One revolution of the work in a lathe, for example, moves around a circle a distance of $\pi D/12$. Metal is removed at the rate of depth of cut x feed x cutting speed x 12.

A variable such as metal structure will cause reasonable deviations from the standard. Annealed low carbon steel is softer than the cold drawn and deforms more readily at the tool's cutting edge, therefore, this steel machines slower than the cold drawn which is less ductile at the cutting edge of the tool. Machining deviation in a metal means that AISI 1040, for example, which machines by turning at approximately 90 feet per minute (table 7–1) using a high speed steel cutting bit, may or may

TABLE 7–1 Cutting Speeds in Surface Feet Per Minute.

1520 Metal	Cutting Speed	
	Condition	sf/min
Low carbon steel	Cold rolled	100
Free machining steel	As drawn	200
Mild carbon steel	Cold rolled	90
High carbon steel	As rolled	70
Alloy steel, low	Annealed	80
Tool steel	Annealed	70
Alloy steel, low	Normalized	55
Cast gray iron	As cast	90
Bronze	As rolled	95
Brass	As rolled	220
Aluminum	As rolled	600

*NOTE: For lathe operations and drilling using high speed bits.
For thread cutting, use approximately 30% of above speeds.
For milling operations, reduce speeds approximately 10%.
For finishing, add approximately 20%.
For alloy inserts, add approximately 50%.
For carbide and oxide inserts, add approximately 300%.

not perform at the indicated speed because of the large structural variable present. In other words, the particular steel may be annealed and cold drawn or cold drawn and annealed. Performance is influenced by metal microstructure which is reflected from forming operations and heat treatment. Such factors as grain size and shape, arrangement and shape of the carbides, and amount of cold work influence the cutting speed of a metal. The arrangement of the carbides is greatly influenced by heat treating. Also, all calculated revolutions per minute are not available on a machine, so the next lower available machine setting is often used. Coupled with the variables of metal structure and the available machine setting, appreciable deviation exists from the theoretical calculation. Table 7–1 lists some typical cutting speeds, but these are approximate. Trial and error experimentation is suggested. Personal experience and existing variables influence the cutting speed factor which, in turn, determines the rate of metal removal. Incidentally, when a carbide cutting tool is used instead of high speed, metal removal is, theroretically, three to four times greater.

When alloy steels are machined, the presence of additional carbides reduces the cutting speeds. As an example, AISI 3140 has a cutting speed of approximately 80 when high speed bits are used in lathe turning and is calculated to be 240 when carbide tools are used. Gray cast iron and cast steel may be machined at approximately the same cutting speeds, whereas type 321 stainless steel is reduced approximately 50%. Aluminum and its alloys, on the other hand, have cutting speeds during turning operations of more than 600 sf/min when a high speed tool is used. The carbide tool provides three times the metal removal ability, while the ceramic tool has an even greater capability. When annealed high speed steel is turned by the high speed steel cutter, the cutting speed factor is reduced to 70, or slightly less. High speed (*HS*) steels contain numerous hard carbide particles imbedded in the matrix of metal, and these particles slow down metal removal. On the other hand, the carbide cutter, not a steel, enables the annealed high speed steel or other highly alloyed steels to be cut at a cutting speed of more than three times that of the *HS* steel. As the type of machining operation varies, for example, from turning to milling, the cutting speeds may also vary somewhat. Appropriate charts in engineering handbooks should be consulted, therefore, to ascertain the correct cutting speed for a given material. In practice, the speed of machining depends primarily on the type and kind of cutting tool being used, machinability of the material to be cut, rigidity of the setup and machine, and the power capacity of the machine.

Metal Preparation for Machining

As the hardness of a material increases from a certain value, the volume of material removal usually decreases, because the machinability of the material is usually reduced. But very ductile steel, AISI 1020, for example, has approximately only 50% ease of machining compared to AISI B1112 which has 100% or a *CS* value of approximately 200. Therefore, too ductile a material may also decrease machinability. The AISI B1112 is significantly harder on the Rockwell B scale than the AISI 1020. Some metals which are too ductile and plastic cannot be machined as rapidly as a slightly harder metal because of plastic flow interference at the cutting edge of the tool. A high carbon alloy tool steel, because of its inherent complex carbide structure, is machined at a lower rate compared to the AISI 1020. In this example, the difficulty in cutting and breaking through the brittle carbides in a soft matrix of metal is greater than in avoiding the tearing of the softer low carbon steel. In other words, there is a "best" metallurgical condition for machining purposes, and this condition varies with the metal.

Metals which are properly prepared for machining enable faster machining processes to be performed. In some cases, coarse-grained steels machine faster than the fine-grained, while the slightly cold-worked steel machines faster than the hot-worked. Low carbon steel, when quenched in water from austenite, may frequently machine better than when it is in the rolled or annealed condition. A fully annealed high carbon steel has better machining characteristics than when the carbides are spheroidized. In this respect, plates of carbides in pearlite allow faster removal rates than carbides in spheres resting in a very soft and plastic ferrite matrix. Pearlite is a microstructure resulting from slowly cooled steel from the red-hot or austenitic condition.

As the quantity of carbides and other constituents in addition to the hardness of the metal increase, lower cutting speeds are established due to increased interference at the tool's cutting edge. Surface finish is also influenced by varying the cutting speed factor. As the hardness increases, cutting speeds can often be maintained by using carbide and ceramic tools. A steel with a reading of RC 52 is machinable with a ceramic tool, while a carbon or low alloy steel cutter will be destroyed on contact. Basically, steels in the hardness vicinity of Rockwell B 90 have very good machining qualities. Generally, any of the metals in the Rockwell B scale of hardness are considered machinable. Metals in the Rockwell C scale usually decrease in machinability as the C number increases. Again, ceramic tools allow these harder metals to be cut. Those steels below the Rockwell B 90 value usually machine with less ease. Those steels above this value also machine with less ease, unless carbide or ceramic tools are used. On the other hand, the nonferrous metals aluminum and brass have much higher cutting speeds than the steels, even though their hardnesses are lower than the steels. This condition in nonferrous metals relates to cutting edge actions. With respect to hardness, the effects of heat treatment or forming have varying results on machinability.

Cutting Tools

The type of cutting tool is chosen when the kind of machine is known. As an example, the single-point tool bit is used in lathe, shaper, planer, and slotting operations. Milling operations, along with drilling, reaming, and tapping, require differently shaped cutters. Sawing operations, as well as broaching, require linear toothed cutters. Whether the work is moving or stationary makes no difference when the type of tool material is chosen or when calculating cutting speeds, because the cutting speed is relative. In lathe work, the material turns, whereas in drilling work, the cutter turns. Either way, the tool bites into the material and chips are moved away.

In order for a material to be cut, the tool doing the cutting must have a higher hardness value. Metals in the Rockwell B scale, for example, cannot be cut by a tool which also measures the same hardness value. Therefore, as the hardness difference between the cutting tool and the work increases, the cutting efficiency also increases. A cutting tool that is at least three times harder than the material being cut performs an excellent job. A cutter which reads Rockwell C 47 is no longer a cutter. Machine cutters have Rockwell C values in excess of C 60 for steels, mostly in the range from RC 62–RC 67. Material having such a high hardness value is also brittle, therefore, caution is used in using tools with very high hardnesses. Excessive overhang of a tool from its holder and vibration cause tool damage and even breakage.

CARBON AND ALLOY STEEL TOOLS

Cutting tools are made from high carbon steel, high carbon-high alloy steel, nonferrous alloys, high speed steel, carbide-tipped steels, and ceramics. These tools came into use because of need. Even today, many fine cutting tools such as drills are made from high carbon steel, AISI 1095. When properly heat treated to an RC 61–63 hardness value, this steel will perform very well when cutting materials in the RB scale, but not in the presence of heat.

Because machining operations create heat, carbon and low allow steel cutters must be protected, usually with a liquid coolant. Temperatures generated in the tool in excess of 430°F. cause softening of the tool with subsequent loss of cutting edge. Therefore, lower cutting speeds and a coolant are used with carbon steel cutters. Alloy steel cutters with high carbon contents have similar characteristics to the plain carbon steel and must also be kept cool while cutting. In this respect, many fine twist drills are produced from carbon and alloy steels, but the high speed steel drill is used in production work. During use, however, once the "blue" heat is absorbed into the carbon or low alloy steel cutter, it is ruined. During use, a cutting tool generates heat which, in turn, may influence the microstructure of the tool. With respect to steel, a temperature of approximately 500° F. will cause the surface to turn blue in color. This color is surface iron oxide which appears as different colors when the surface is clean. Accordingly, as metal temperature varies from around 350–650° F., the surface color varies from light yellow to gold to purple to blue and then to gray. During machining operations, when metal is removed too fast, the tool becomes very hot, causing softening and edge dulling. Grinding will restore the edge but will not restore the hardness to the sharpened edge as it will immediately be folded over when again used. High speed cutters, on the other hand, are not affected by the blue heat color due to their stability up to approximately 1050° F.

Alloy Tools

Special alloys of chromium, cobalt, and tungsten are cast and ground to the shape of the cutting tool. These tools are used at cutting speeds up to 2,000 sf/min whereby red temperatures are sometimes generated, and the nonferrous alloy does not soften. The great difficulty in using this alloy is the need to cast its shape.

High Speed Steel Tools

During the early part of this century, the high speed steel came into use. Experiments showed no loss of hardness in this metal when subjected to 1000° F. The tungsten and molybdenum types of high speed steels are now extensively used not only as cutters, but also as parts where heat is involved. High speed steel cutters and drills are marked with *HS* to indicate their chemistry. During machining, the RC 65 hardness value is maintained, even after exposure to 1000° F. Temperatures in excess of 1100° F., however, cause softening of the metal. Usually, these steels are used with a coolant to minimize stresses in the tool.

Carbide Tools

The need for more effective cutting tools during World War II introduced a product of powder metallurgy, the carbide tool. Carbides of tungsten, titanium, or tantalum were sintered after pressing into desired shapes, resulting in hardness values in excess of RC 70. Consequently, cutting speeds from 300 to 1100 sf/min were made available for cutting many materials. Temperatures in the red heat range do not soften these carbides, therefore, many steel cutters are tipped with carbide inserts, forming a tool capable of removing large volumes of metal. Tipping is accomplished by brazing or by the use of set screws.

Oxide and Diamond Cutters

Aluminum oxide ceramic, when pressed and sintered, provides the hardest mass produced cutters. Cutting speeds beyond 2000 sf/min are used during some machining operations. Temperatures up to 2000° F. have had no adverse effect on hardness. As a result, these cutters are often used dry.

Another cutting material is the diamond. Diamonds are harder than the aluminum oxides, but are costly. Therefore, diamonds are not used in large production quantities in the same manner as other tools. Diamonds used in manufacturing processes are of the commercial type and usually have a brown or dark color. Their cost is only a fraction of that of the blue-white diamonds.

Chip Formation

When a wedge-shaped tool having a hard and sharp cutting edge is forced into a much softer material, compression and shear forces induce plastic deformation in the metal, and metal failure occurs along a certain plane of the softer material. Cutting tools are available in many shapes, each having its specific purpose. In order for the tool to cut, several factors must be present in the tool—a very high hardness; a specific shape, depending on purpose; a keen cutting edge; clearance angles from the cutting edge; the need for a minimum of power; and a means of being supported during the cutting operation. When enough power is applied to either the work or the tool while the cutting edge advances into the work material, a chip will shear away from the softer material. The size and shape of the chip will depend on the geometry of the tool and chemistry of the material being cut.

ORTHOGONAL CUTTING

A two-dimensional illustration of chip removal is used so that the shear plane can be easily described. Figure 7–1 illustrates the cutting of a section of metal in which two dimensions provide a plane suitable for illustrating shearing actions. The area of the chip to be removed is shown as width b and depth a. The tool has a depth of cut

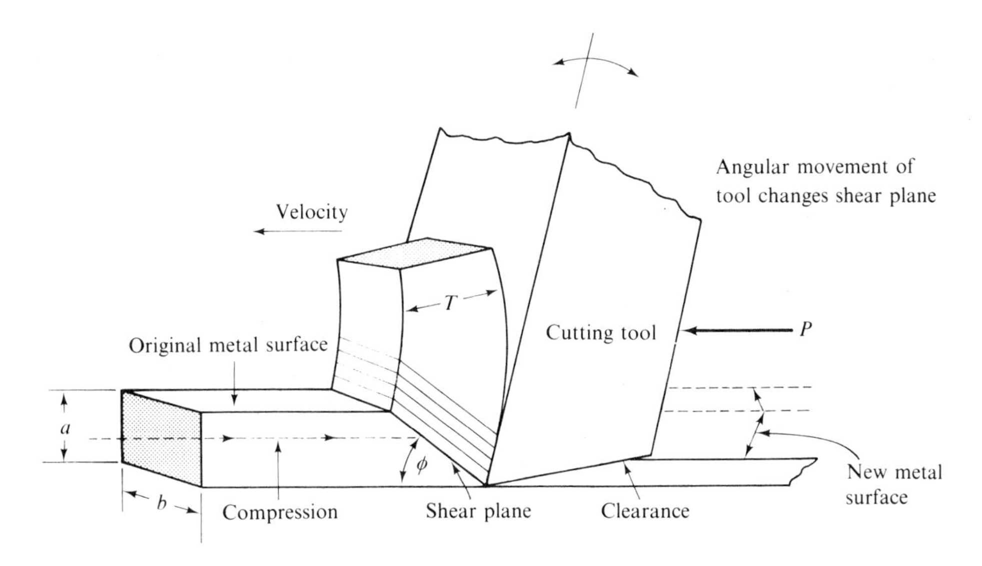

FIGURE 7–1 A two-dimensional illustration of chip removal.

according to *a*, causing the chip to be sheared at the shear plane. As the tool advances at a specified velocity driven by the force *P*, the chip slides plastically across the face of the tool, leaving the original metal at a new dimension resulting from the loss of the thickness of the chip. Because the tool moves in a straight line or constant radius against a resisting metal, compression forces are quickly generated in the metal immediately ahead of the tool or cutter. As the moving force, or load *P*, increases the compression stress ahead of the tool, a stress factor accumulates in the metal until the shear failing factor is attained. Because the failure plane, the dimension between the flat face of the metal and the edge of the cutting tool, has theoretically no dimension, being less than 0.001 inch, shear occurs. When the shearing stress builds up to a factor greater than the shearing strength of the metal, immediate plastic slippage of metal occurs across a shear plane whose dimensions are indicated in figure 7–1.

Because time is a factor in this moving situation, and because the moving tool is intercepted by the resisting metal, the failing plane of metal separates from the parent metal. Because the tool or work is moving, a series of failing planes at the angle ϕ results when stress becomes greater than strength in each plane and the chip remains continuous. When the shear angle ϕ is small, a thick chip results because the shear plane is long compared to a large shear angle. In this respect, the longer shearing plane requires more power to cause failure. On the other hand, when the shear angle is large, the shear plane is short and the chip is thin with a constant width. Less power is needed to cause failure. Obviously, control over the shear plane determines the chip formation. Relationship of the tool to the metal controls the shear plane. You can see that the chip thickness ratio is a variable and is equal to the depth of cut (*a* in fig. 7–1) divided by the thickness of the resulting chip.

TYPES OF CHIPS

Materials vary in their chemistry, tools vary in their cutting capacities and shapes, and shearing angles vary. Due to these and other variables, three main types of chips are produced. The long and continuous chip from the lathe or similar operation moves away from ductile materials such as those produced from aluminum and low carbon steels (fig. 7–2*a*). Depending on the tool, the chip may spiral, coil into a helix, or wind randomly. These chips are dangerous, especially the steel chips. Being hot and often razor sharp, they are hazardous to the operator. Consequently, care must be exercised to guide the chip in an appropriate direction or break the chip in some manner. The cutter may often provide a chip breaker to eliminate the long chip condition.

Brittle materials produce short chips (fig. 7–2*b*). These discontinuous chips are due to material chemistry and structural condition of the material. Hard metals contain little plastic flow ability and, therefore, chips are segmented in intervals. Soft and brittle metal such as gray cast iron, as well as lead and sulfur in steels, also produces the discontinuous chip.

When a ductile metal has been machined for a period of time, a built-up edge appears on the leading edge of the cutting tool. Such a buildup changes the tool's geometry, and, eventually, harmful cutting actions occur (fig. 7–2*c*). The buildup consists of cut material being transferred to the tool where it strongly adheres due to pressure and temperature. The modified continuous chip with the built-up edge is a common occurrence. Built-up metal should be ground away.

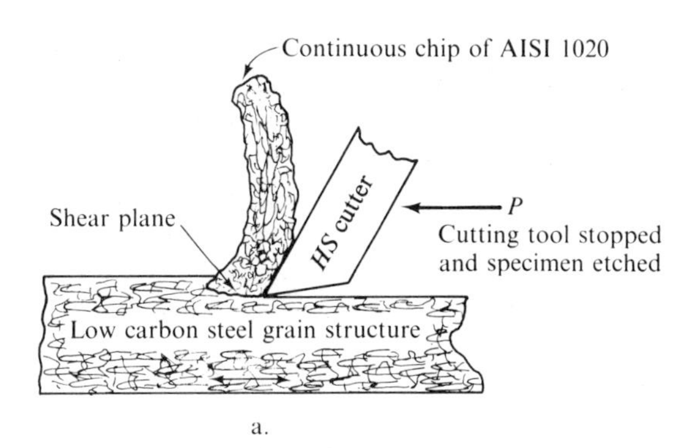

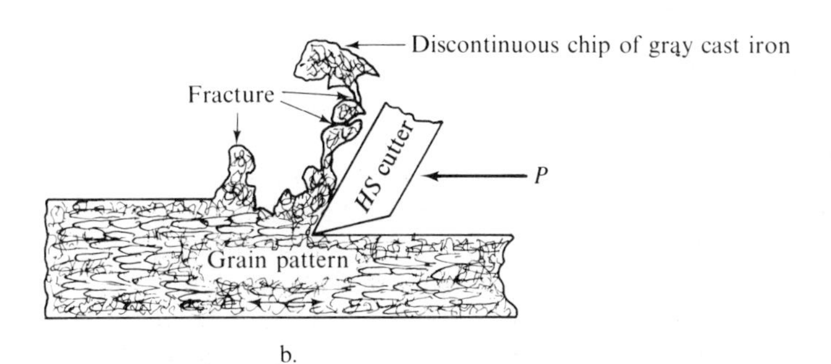

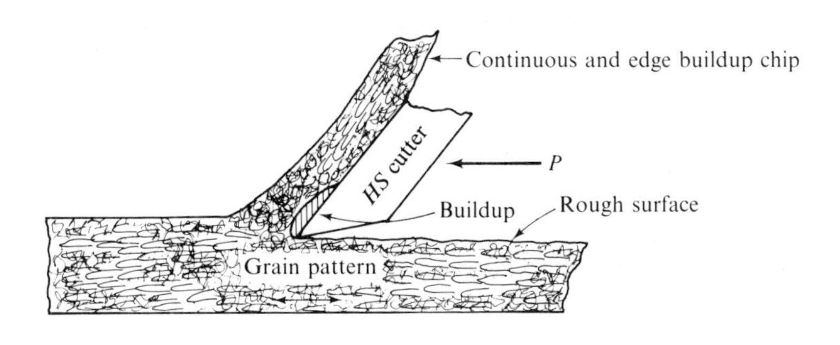

FIGURE 7–2 Types of chips in metal removal operations.

As illustrated in figure 7–2, the grain pattern of the metal is severely disturbed at the shear plane, the disturbance increasing as the chip moves up the face of the tool. Periodic shear planes occur, slippage results, and the deformed metal becomes distorted throughout its volume. As a result, some work hardness results in the ductile metal, while fracture results in the more brittle metals. Fracture results from either severe work hardness or the presence of a nonmetallic material such as graphite or a very soft metal such as lead. The continuous chip in *a* may remain until the end of the cut. The discontinuous chip in *b* continually fractures against the face of the tool. Because of loss of angle contour, the built-up edge in *c* produces rough surfaces due to the tool pushing a different material ahead of itself without proper clearances. Tools must be cleared of buildup.

Regardless of the type of chip removed from the material, shearing action produces the chip and, at the same time, induces a small hardness increase in the new surface of the cut metal. Depth of hardness penetration is shallow and is often insignificant. Chemistry and structural condition of the material influence the chip removal process. In a material such as metal, mechanical properties are variable. Hardness, for example, is directly related to tensile strength in steels up to approximately RC 55, while tensile strength is related to the lesser shear strength. Therefore, as hardness increases for any reason, the shear strength will change accordingly and will require a greater power factor for chip removal. Consequently, metallurgical conditions of the metal must be considered when calculating revolutions per minute of spindle speed or work speed. Incidentally, hardness-to-tensile conversions are only approximate.

CHIP REMOVAL AND TOOL LIFE

The tool life is the length of time the tool is used between sharpenings. When tools are out of service, production is reduced. Usually, this time factor is listed in operating hours or minutes at certain conditions. When metal is to be removed by machining, caution must be used in regard to volume of metal removal. As an example, when the feed is increased, the sf/min factor must be reduced for equal tool life. When carbide-tipped cutters are used while doubling the feed factor, up to a one-third reduction in speed must be provided to maintain an equal tool life while removing more metal. When the depth of cut is increased up to 100%, approximately one-fifth reduction in speed is necessary for equal tool life. Keep in mind that only the tool's cutting edge is cutting, therefore, the depth of cut has less effect on tool temperature change. In practice, the operator will choose the optimum cutting speed.. This is the speed that will give the lowest machining cost.

Geometry of Cutting Tools

As discussed, a cutting tool has knifelike characteristics. Regardless of the shape, the tool has a certain point or edge which is designated as the cutting edge. Single-point cutters such as lathe and shaper bits have an edge on the leading side which performs the cutting. Because the tool can be ground to various shapes for cutting flats, curves, forms, and threads, certain relief angles must be provided in order that the tool will cut. As an example, the lathe, shaper, or planer tool bit has a lip angle which results from the back rake angle and an end relief angle (fig. 7–3). Also, to prevent rubbing of the tool on the work during cutting, a side relief angle is provided. The nose angle on the face of the tool where the chips move across has an end cutting and side cutting angle. Along with the back rake angle, there is the side rake angle. The tool's face provides the surface for chip flow and is made as the result of the combined back and side rake angles. As illustrated in figure 7–3, the cutting tool is provided with a sharp cutting edge which is the metal remaining after clearances are formed by removing metal behind the cutting edge. In this respect, a small rake angle, the angular relationship between the face of the tool and the surface of the metal being cut, induces high compression stresses in the metal being cut. High compression stresses induced in the workpiece require more power for chip removal because the

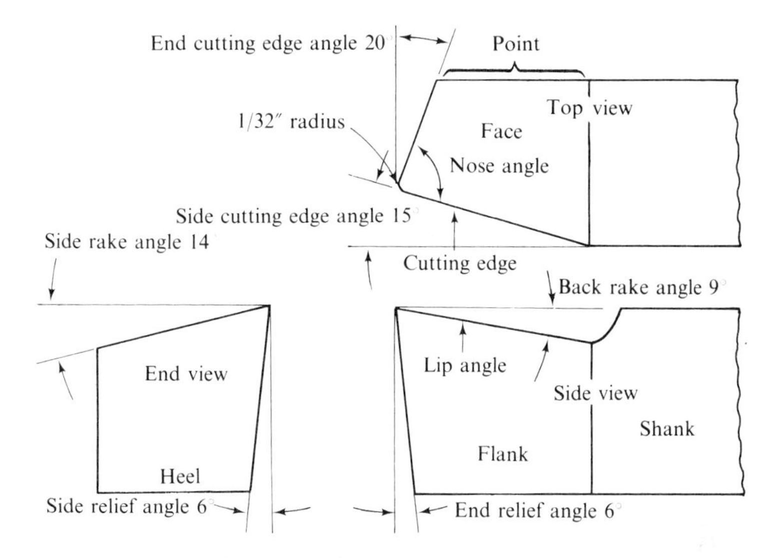

FIGURE 7–3 Angles and nomenclature of single point cutting tools.

amount of metal forming the chip tends to increase as the rake angles decrease. Consequently, as the rake angles increase, especially the back rake angle, the amount of metal in the chip is reduced due to the reduction of stresses in the workpiece. With respect to tool geometry and chip removal, it must be strongly pointed out that the relationship and angle between the tool's cutting edge and material are both important factors in effective machining operations. Standard 16½ degree tool holders (fig. 7–4) help fix the angular relationship between workpiece and cutting tool. As the cutting tool material or material to be cut changes, this angle is sometimes changed.

Single-point cutting tools are held securely in tool holders. Usually, three holders are required. The right-hand turning holder holds the cutter so that cutting can be

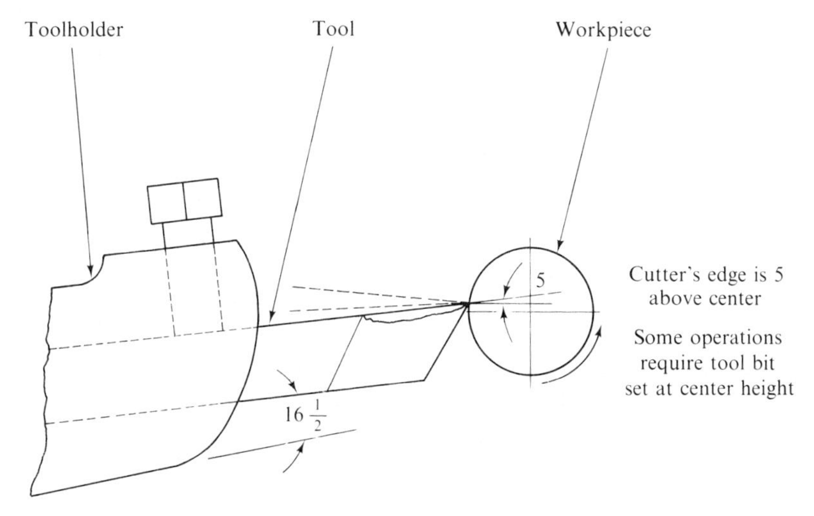

FIGURE 7–4 A 16½ degree toolholder.

accomplished close to the tailstock. Left-hand turning holders hold the tool bit for cutting close to the headstock. Straight holders allow cutting in either direction. Figure 7–5 shows some of the common cutting tools and holders. Cut-off holders hold the cutter so that parting of the workpiece can be accomplished. The boring and threading tools are also held in special holders for boring holes and cutting threads on the lathe. Insofar as the cutter bits are concerned, a right-hand tool cuts to the left, while a left-hand tool cuts to the right.

As the number of cutting edges increase on the tool, differently shaped edges and different relief angles are necessary. A drill cuts with two edges concurrently. Therefore, each cutting edge must be balanced by the opposite side. In order for this type of cutting to occur, each edge must have sufficient clearance at the periphery. Reamers, taps, dies, broaches, drills, saws, and milling cutters have their cutting edges arranged so that efficient cutting occurs according to the shape and use of the tool.

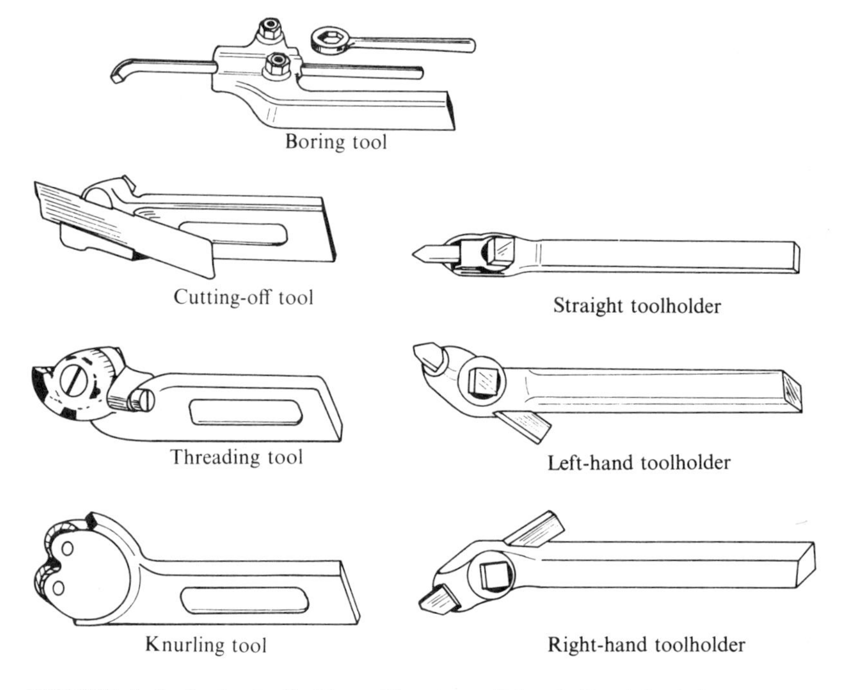

FIGURE 7–5 Lathe toolholders. (Courtesy of South Bend Lathe)

Coolants

All machining operations generate heat. Many steel tools are greatly affected by heat due to softening of their martensitic structure. Carbon and alloy steels, other than high speed, must not be allowed to generate more than 430° F. High speed cutters often generate up to 1050° F. without adverse effects. In order to assure cool cutting tools, a liquid coolant is used in generous quantities on the cutting edges of steel tools, especially in milling operations. The coolant washes away chips as it cools the cutter and helps lubricate the cutting actions. Because lathe operations vary in magnitude, the water based liquid coolant is sometimes replaced with a cutting oil, but often no coolant is used. Lard oil, for example, promotes cutting action. Several types of lard and mineral oil solutions are used when high production of parts is required. The soluble oils, when mixed with about twenty parts of water, make a good cutting lubricant in that the thin liquid wets the metal much more than oil. However, certain chemicals such as chlorine and sulfur added to the liquid increase cutting efficiency. For gear cutting, engine oil and mineral oils are used. Prepared coolants are available under many trade names.

During the cutting of a metal, much of the energy used is converted to heat. Because the power of the machine pushes or pulls the cutting edge of the tool into resisting metal, very high compression stresses are generated in the metal immediately ahead of the cutting edge. As the stress builds up and the cutting edge induces a shear plane in the compressed metal, slippage of metal occurs, and a chip slides off the parent metal. Heat is thus generated as thousands of pounds of stress per square inch of metal area are induced and subsequently reduced to zero as the metal fails. Coolants increase tool life by helping to prevent chipping of the tool and wearing of the cutting edge.

Conventional Machine Shop Machinery

A machine shop which performs general type machine operations may have on hand band and hacksaws, engine and toolroom lathes, single and multiple spindle turret lathes, automatic screw machines, plain and universal horizontal milling machines, vertical mills, special milling machines such as the planetary and multiple spindles and the rail and bed types, boring and broaching machines, slotting and special cutting tools, tool and cutter grinders, as well as cylindrical and centerless, surface grinders, drilling machines from the bench type up to and including the radial drill, hand presses, straightening equipment, special purpose gear and spline hobbing machines, small power tools such as drills and impact tools, and a large assortment of measuring tools, standards, and gages, along with numerous machinist tools and cutters. Some machine shops do their own heat treating and, thus, maintain certain heat treating furnaces, quench tanks, and testers. Even a microscope is required.

When a general machine shop moves to automation or semiautomation, many of the above machines are placed in the production line and equipped with automatic handling mechanisms and equipment. The machines are controlled by the tape or computer, thus eliminating the need for the specialized operator.

TYPICAL MACHINIST TOOLS

When a machine shop technician or machinist uses the different pieces of equipment in the pursuit of general machine operations, he needs certain general tools. Table 7-2 lists many of the tools which are normally used in the machine shop. Unless the shop is specialized, much of the machine operator's time is spent in planning and layout, using information from specifications and drawings. In this respect, draftsmen and design technicians are responsible for producing completed drawings. Too often, the machine shop technician or the machine operator has to query the engineer or foreman with regard to material for the part, radius and clearance dimensions, type of

TABLE 7–2 Typical Machinist Tools

Tool	Size
Boring Bar	Standard
Caliper	Hermaphrodite 6 in.
Caliper	Outside 6 in.
Caliper	Inside 6 in.
Card file	Regular
Center gage	60°
Center punch	Set
Center drills	Set
Combination set	Set 12 in.
Dial indicator	0.0001 in.
Diagonals	Cutter
Drift	Drill
Files	Assorted
Hammer	Ball peen
Hammer	Lead
Hammer	Rawhide
Honing stone	Regular
Jack	Machinist
Level	Machinist
Micrometer	Outside 0–1, 1–2
Micrometer	Inside 1½–2
Parallels	Small assorted set
Pliers	Slip joint
Pliers	Side cutting
Radius gage	Set
Rule	6 in. solid
Rule	12 in.
Screwdriver	Assorted
Scriber	Regular
Square	6 in.
Telescope gage	Set
Thickness gage	Standard
Thread gage	American standard
Tool bits	Set
Toolholders	Set
Wrench	Adjustable 6 in. and 12 in.

fit, and many things which are not his responsibility. The drawing, therefore, must be clear and complete so that machine operators will only have to follow instructions.

As listed in table 7–2, many tools and instruments are used by the machinist. The most commonly used tools are the ball peen hammer, rawhide hammer, six- and twelve-inch measuring rules, diagonal cutter, center gage, slip joint and side cutting pliers, several assorted files and a card file, a set of center punches, adjustable wrenches, a combination measuring set, a thread gage, several screwdrivers, a radius gage, micrometers, telescope gages, inside and outside calipers, hermaphrodite caliper, level, scriber, center drills, and a honing stone. Attachments and accessories to the machine then enable machine operations to be performed.

HOLDING DEVICES AND TAPERS

The precision vise is used in many machine shop operations. It is a common attachment to a milling machine (fig. 7–6). This vise has holding lugs on its bottom sides so it can be securely bolted to the table of the machine. The shape and size of the vise

FIGURE 7–6 A milling machine vise holding a part being milled by a gang type cutter operation. (Courtesy of Cincinnati Milacron)

(some are plain while others swivel) provide adequate holding power for most machining purposes. On the other hand, a bench vise is sometimes needed for bench work purposes such as layout and tapping processes. This type of vise is not precision in its structure; the accuracy of the work depends on the capability of the machinist. Operations such as hand drilling, reaming, tapping, punching, and hand sawing are performed with the use of this holding device. As the size of the manufacturing operation increases and the greater quantity of production warrants the use of special holding devices, the fixture and jig are used.

Tapers

Several types of tools require tapers to provide attachment to another tool or part. As an example, the tailstock of a lathe requires a tapered hole in order that quick holding and quick releasing can be accomplished when a tapered shank drill, a center, or drill chuck is inserted. Large drilling machines require tapered sleeves so that the drill can be inserted and released quickly. The sleeve is a coupling device that closes the distance between two tapers which are of different sizes. The outside taper fits the larger taper and the sleeve's inside taper fits the tool's taper. Many spindles and sockets have internal tapers. Tapers hold tools securely, yet the tools are quickly released when the operator needs to change tools or parts. When properly snapped into the tapered hole, a tight, accurate, and strong attachment occurs. The only easy way to release the self-holding type tapered tool is to provide a pushing pressure at the center of the small end of the tapered insert. Therefore, the taper is used as a quick means of attaching one part to another. Friction holds the two tapers together. An example of the self-holding taper is the plug gage illustrated in figure 7–7.

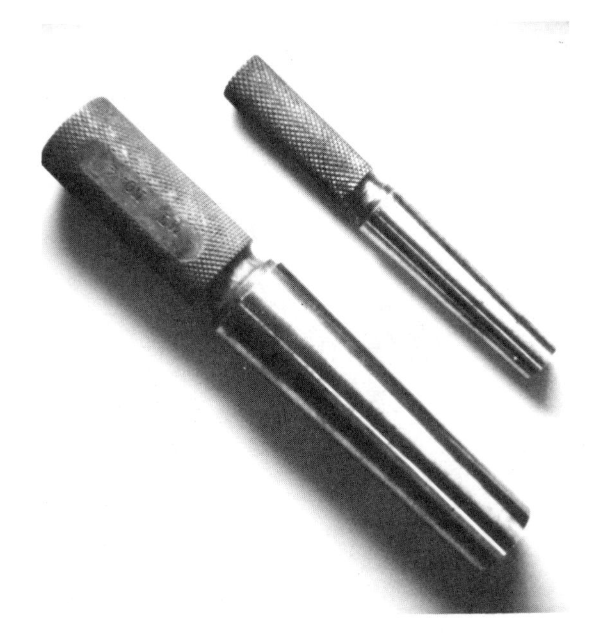

FIGURE 7–7 Plug gages.

The self-holding taper is a small taper having from two to three degrees of taper. The Morse, Jarno, American Standard, and the Brown & Sharpe taper are self-holding. Most of the Morse tapers (there are several sizes) have approximately a five-eighths inch per foot taper. The general purpose twist drill has an American National Standard taper on its shank when the shank of the drill exceeds one-half inch diameter. However, straight shank drills are available for sizes larger than one-half inch, the shank being machined to one-half inch diameter. Cutting tools such as reamers and mills have Brown & Sharpe tapers which are approximately one-half inch per foot, but a common taper is the American National Standard. This taper consists of twenty-two sizes, the tapers varying from approximately one-half inch per foot up to three-fourths inch per foot. The Jarno taper per foot is six-tenths of an inch on the diameter and is available in several sizes, just as are the Morse and Brown & Sharpe. American Standard includes Brown & Sharpe, Morse, and others.

Some tool shanks and spindles have Morse, Jarno, Brown & Sharpe, or American Standard machine tapers which are all self-holding, while some spindles have a steep machine taper and are not self-holding. The steep machine taper is used on some machine spindles, such as a milling machine. When this self-releasing type of taper is used (a larger taper such as three and one-half inches per foot), a key is needed to eliminate slippage between the mating parts when torque is applied. In general, each of the above types of tapers is available in various sizes, the Morse tapers ranging from approximately one-fourth inch at the small end of the male taper to two and three-fourths inches. Morse has eight standard tapers, number zero being the smallest and number seven being the largest.

Chucks

A common holding device for both moving and stationary parts or tools is the chuck (fig. 7–8). The chuck has three or more movable jaws which uniformly and securely clamp the part or tool while its other end is attached to a spindle. The spindle is either tapered or threaded. In other words, one end of the chuck holds the tool while the other end is held by the spindle of the machine. The Jacobs chuck which fits in the tapered or threaded spindle of the drilling machine is a common type of chuck. The Jacobs chuck is also used in the tailstock of lathes to hold drills and other cutting tools. Several sizes of Jacobs chucks are available.

One of the most popular chucks is the three-jaw universal which is attached to the spindle of a lathe by either a key-drive, cam-lock, or thread type arrangement. A chuck wrench opens and closes the three stepped jaws simultaneously in a manner similar to the Jacobs chuck. Sometimes, the four-jaw independent chuck is used in the lathe to secure square stock or to provide eccentric turning. Most all other chucks provide concentric turning. Three- and four-jaw chucks are available in several sizes.

When polished drill rod is to be machined, the collet chuck is frequently used. This chuck collapses only a small amount, approximately 0.005 inch, because of its split face. These collets are available in many hole diameters, each size accommodating standard sizes of rods. When placed in the spindle of the lathe and tightened onto the workpiece by a draw bar, the split face shrinks uniformly and clamps the work.

Questions

1. When the shear angle is small with reference to a single-edged cutting tool, why is the chip thick?
2. What factors influence machinability and the resulting cutting speeds?
3. What is the relationship between revolutions per minute and cutting speeds?
4. Describe three types of chips.
5. How does hardness of material to be cut influence machinability?
6. Describe the principle of the self-holding taper.
7. List some advantages of carbide and oxide tools over conventional steel cutting tools.

(Questions continued on p. 228.)

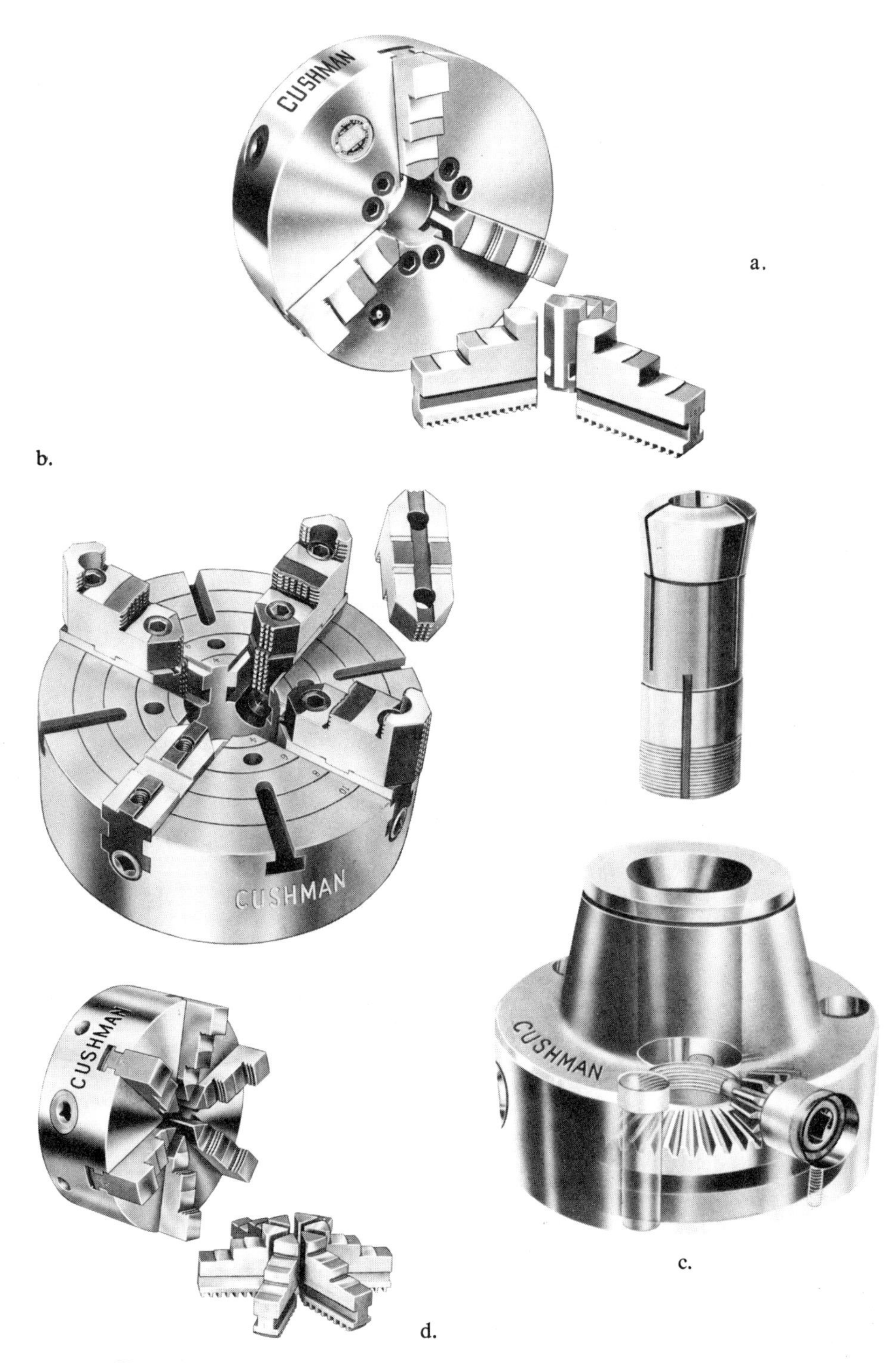

FIGURE 7–8 Some typical holding devices: *a*, 3-jaw self-centering chuck; *b*, 4-jaw independent chuck; *c*, collet, chuck, and collet; *d*, accroset chuck with 6 self-centering jaws; *e*, 3-jaw power chuck; *f*, tri-action drawdown chuck; *g*, indexing spacer. (Courtesy of Cushman Industries Inc.)

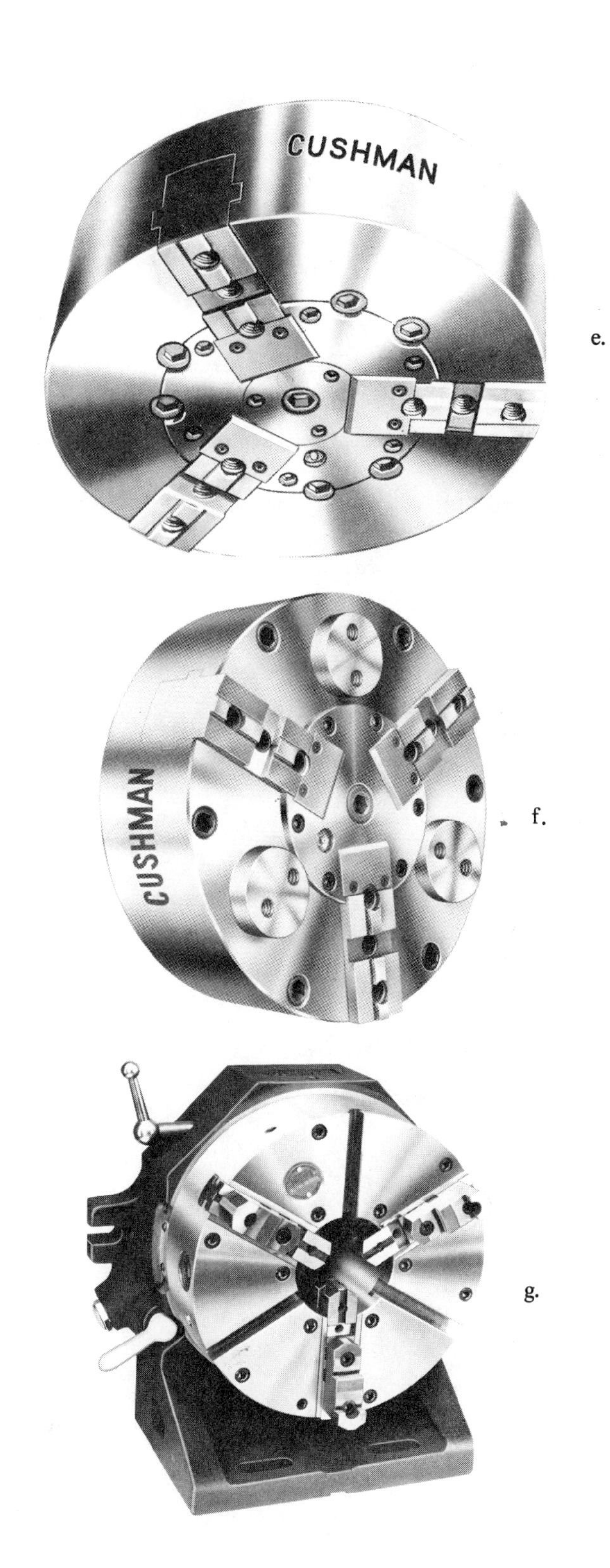

FIGURE 7-8 (continued)

8. Why do brittle materials produce small chips?
9. A ratio of three-to-one of the hardness of the cutting tool to the metal to be cut is considered satisfactory. Why?
10. What constitutes the rate of metal removal?
11. Machining operations always produce two things during machining, an asset and a liability. Describe each.
12. How does the microstructure of a metal influence machinability?
13. Why are alloy steels harder to machine than plain carbon steels?
14. In Rockwell hardness values, what is the minimum acceptable hardness for a cutting tool such as a reamer?
15. Why is a steel having a RB 67 hardness value more difficult to machine than one having a value of RB 89?
16. Why is the chip thickness ratio a variable?
17. Describe the buildup on a cutting tool and how this influences cutting.
18. Why does the depth of cut have the least effect on the temperature of a cutting tool?
19. What is back rake angle?
20. Name three reasons why coolants are used.

Basic Machine Tools

Whether a shop is general or is production oriented, the saw, drilling machine, and lathe are all basic tools used in the first machining operations of the part. Raw materials such as bar stock require sawing before any other operation is performed in order that a length can be established. For example, cutting tools such as reamers begin as short lengths of bar stock. Subsequent lathe operations accomplish the turning requirements, while the mill machines the flutes, and the grinder completes the tool, including its outside dimensions and the sharpening of the cutting edges. Many cast parts and plate stock, for example, require holes before such operations as threading are performed. Therefore, the various types of drilling machines are available to drill and ream holes. When production is forecast to very large quantities of specific parts, however, the automatic screw machines are used. Such machines change the general type machine shop into a production shop. But the presence of screw machines does not require the elimination of other machines, because many differently shaped parts are produced in other machines, and the quantities are frequently less. The screw machine changes piecework into mass production.

Power Saws

The power saw is a basic machine tool in the shop because sawing is usually the first operation to be performed on a material during processing. Power saws function in the same manner as hand saws, except that the power saw has greater capability because it is a machine. The principle of metal removal by sawing is similar to that by a single-point tool because each tooth in contact with the material removes a chip. Varying pressure on the blade causes the size of the chip to vary. As the blade passes through the material and causes a cut, or *kerf*, friction and heat are generated. In order to reduce friction, the teeth are set in patterns which cause the kerf to be wider than the thickness of the blade. Chips move freely down the kerf and between the teeth and then away from the material.

SAW BLADES

Metal cutting saws predominantly use the wave and raker tooth forms, however, some of the nonmetallics use the straight form of tooth. Figure 8–1 shows the basic forms of

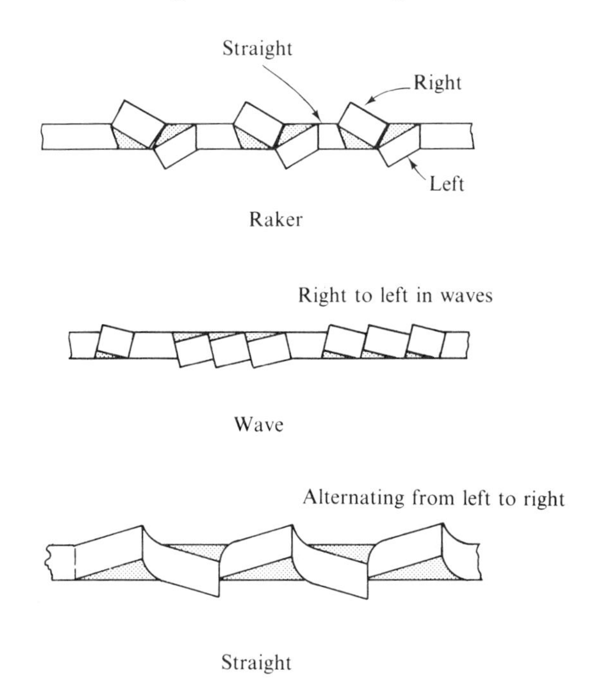

FIGURE 8–1 Basic forms of set in saw teeth.

teeth called *set*. Saws are available having fine or coarse teeth, and at least two teeth must be in contact with the work during cutting in order to prevent stripping of the teeth. Because pressure is exerted onto the blade during cutting, the mass of metal in the blade must be great enough to allow the blade to efficiently cut without twisting. Blades which are too thin or too narrow often twist and produce irregularly shaped cuts. Widths of blades vary, therefore, depending on use.

Pitch refers to the spacing of teeth on the blade. Hacksaw blades are available in pitches of 3, 4, 6, 10, 14, 18, 24, and 32 teeth per linear inch of blade. Pitches larger than 14 are usually used in hand type blades. Because each tooth cuts and drags a small chip from the kerf, the saw accomplishes fast severing of metals. Then, too, as the speed of cutting increases, the number of teeth passing a fixed point in a given time increases; more chips are removed, and the rate of cutting accelerates.

TYPES OF SAWS

Saws remove metal by either continuous cutting or intermittent cutting. The band saw which uses a band or loop type blade provides teeth on the work at all times as it moves in a single direction. On the other hand, the hacksaw provides teeth on the work in alternating strokes. Both types of cutting actions, continuous and intermittent, are used in sawing flat and round stock and tubing. In order that straight and angular sawing can be performed, both kinds of sawing machines are equipped with swivel vises which hold the material to be cut.

Power Hacksaws

Several types of power saws are available. The most common is the hacksaw and is available in either the wet or dry types (fig. 8–2). Wet type saws direct a liquid coolant

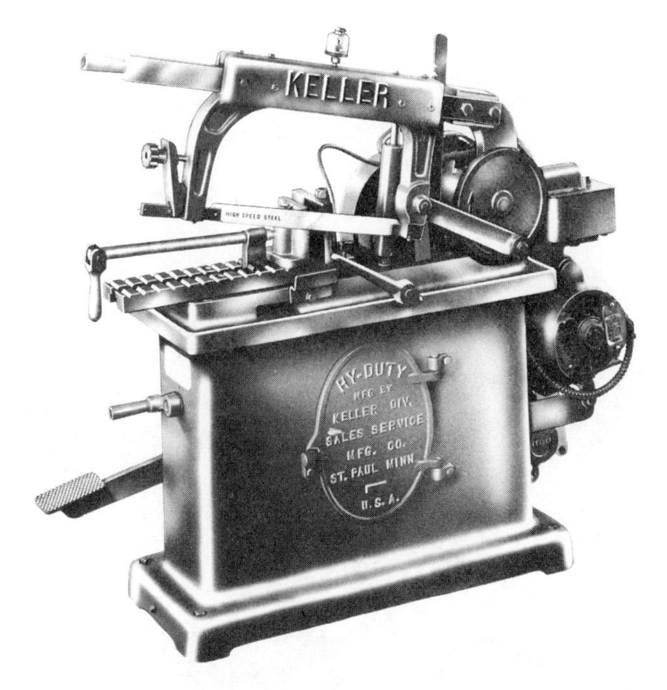

FIGURE 8–2 A typical power hacksaw of the automatic blade lift type. (Courtesy of Keller Division of Sales Service Mfg. Co.)

onto the work at points of cutting. Chips are washed from the kerf, the blade is lubricated, and heat is reduced. Thin sections of metal, however, are sawed dry.

Hacksaws operate on a reciprocating principle, that is, the blade cuts on the power stroke as feed is applied and lifts slightly on the return stroke in order to reduce damage to the offset teeth while sliding backwards. Feed, or pressure on the blade, is governed by the machine setting and is automatic. The number of strokes per minute varies according to the kind and size of material being cut. Softer materials cut faster than harder ones. Keep in mind that either low alloy or high alloy steel is used in manufacturing saw blades. When the hardness of the metal being cut enters the Rockwell C scale of hardness, machinability is reduced, along with a reduction in tooth life. Most metals received from the supplier will have a Rockwell B scale reading which indicates soft metal.

When harder metals are to be cut, adjustments to the machine are made along with a change in blade. Many power hacksaw blades are 10 to 14 inches in length, having usually from four to 12 teeth per inch of blade. Large hacksaws have blades longer than 14 inches. The softer the metal to be cut, the coarser the pitch; the harder the material, the finer the pitch. Wet cutting saws operate usually from 35 to 140 strokes per minute, depending on the kind of machine. The harder the metal, the fewer the strokes per minute. Dry cutting machines often are available with 70 to 100 strokes per minute. When the sawing operation is completed, the more expensive saws will raise the blade and holder automatically.

Saws produced by numerous manufacturers vary in their size and capability. All operate on a common cutting function, however. Saws for small shops usually have a four-inch capacity vise with a swivel jaw so that angular work may be cut. When larger sections of metal are to be cut, more massive machines are used. When properly aligned, the band or hacksaw will produce thin wafers of round stock which may be several inches in diameter. The hacksaw is used for cutting lengths of round steel bar stock to be used in the manufacture of bolts. If a lathe's chuck is to hold the workpiece, then the stock must be cut the needed length plus a few inches which will be in the chuck. The saw will cut many shapes of stock, in addition to round, to size.

Band Saws

Continuous cutting is produced with the band saw because the band blade is a loop and travels in one direction (fig. 8–3). Pressure is applied to the top of the blade,

FIGURE 8–3 A horizontal band saw operation. (Courtesy of DoAll Co.)

pressure being automatic in the more expensive saws. Unless the blade is of sufficient width and is held securely in special guides, the saw blade will cut in curves and angles. Heavy-duty and appropriately guided blades will perform in the same manner as hacksaws, some exceeding the work of the hacksaw because of the continuous cutting action. The same general rules apply to band saws as to hacksaws in regard to the cutting processes. Instead of strokes per minute, however, the band saw cuts at the rate of approximately 50 to more than 2,000 surface feet per minute. Some saws vary considerably from this cutting factor (fig. 8–4 and fig. 8–5).

FIGURE 8–4 The band saw cuts through tubing at a slower speed than several other types of sawing operations. (Courtesy of Simonds Saw and Steel)

Band saws are produced in the horizontal and vertical types. The horizontal type competes with the hacksaw. High quality saws are provided with hydraulically governed feed mechanisms. A vise with a swivel base securely holds the work in both straight and angular cutting. As with hacksaw blades, a band saw blade breaks now

FIGURE 8–5 A high speed band saw blade is rapidly cutting the large bar of steel while a coolant is used. (Courtesy of Simonds Saw and Steel)

and then. The hacksaw blade is discarded, whereas the band saw blade is welded together, reestablishing the loop. Because the blade periodically bends around driving and guiding wheels, the back side of the blade is soft, allowing the hard teeth to cut yet operate within a reasonable radius of flexibility. Speeds and feeds are manually set by the operator.

A vertical type band saw has more use capability than the horizontal, possibly because it is vertical (Fig. 8–6). Based on some of the same general principles of operation as the horizontal, the vertical saw has universal usage in that it can perform contour cutting, as well as inside cutting or plain cutting. Vertical saws are equipped with tables for the purpose of operator control of the cutting operation. Because the band saw also performs a different type of work than the hacksaw, teeth pitches are available up to 32 teeth. Thin sections of metal up to several inches thick can be sawed with these saws. Some have automatic control, but most are used manually. The saw blade can be easily removed and welded if broken. A fully equipped vertical saw of good quality has a welder attached to the machine. When a blade is broken, the two ends are ground squarely, placed in the welding and annealing attachment on the saw and welded, followed by annealing to reduce brittleness resulting from fast cooling. The blade is then ground again by a grinder attached to the machine, allowing proper thickness and sliding action between the guides.

Such a welding capability allows inside cutting to be performed. In order to remove metal from inside a plate, for example, a hole is drilled larger than the width of the saw blade. The blade is broken, pushed through the hole, and welded together. With care, the work is secured to the table while the blade is placed on its drums. After proper tension is applied to the blade and proper adjustments of the guides have been made, the r/min is selected, depending on the cutting speed of the metal. Any shaped

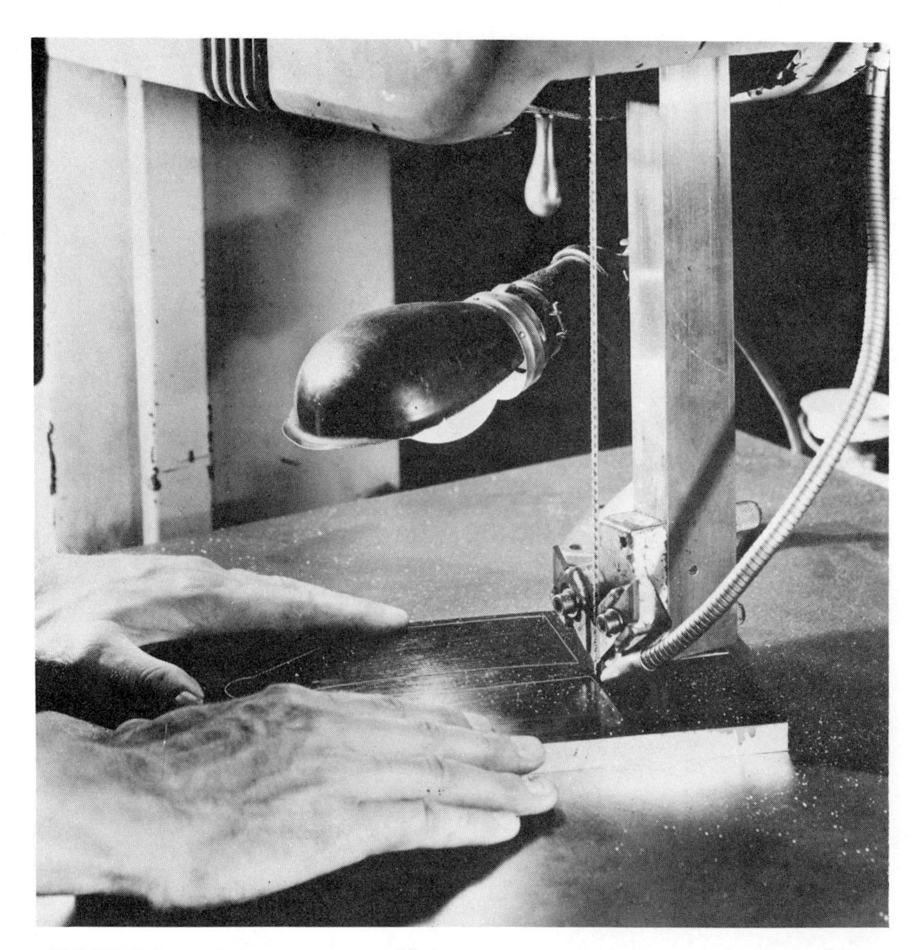

FIGURE 8–6 A hand sawing operation using a vertical band saw. (Courtesy of Simonds Saw and Steel)

hole may be cut. When finished, the blade is again broken, removed from the hole, and rewelded.

Many types of sawing operations are practical with the vertical saw; both hand and automatic operations are performed. Guides on the table allow straight line sawing. Radius fixtures provide circular sawing capability. Ease of blade control is accomplished by a variable pulley which provides the proper cutting speed setting. Figure 8–7 demonstrates the versatility of the vertical band saw.

When very hard metals are to be parted, the blade must contain dull teeth or, possibly, very small teeth. When the sf/min increases up to speeds of 14,000 and higher, the blade-to-metal contact causes instant friction, and metal melting temperatures are generated. Therefore, as the fast moving blade moves into the material, melting or near melting occurs, and the swiftly moving blade snatches the very weak material from the heated zone, resulting in a cut. This type of friction cutting will part hardened steels, just as the abrasive parting wheel, but produces a coarse cut. Keep in mind, heat changes the metallurgical structure of hardened steel,

FIGURE 8–7 A vertical band saw operation. (Courtesy of DoAll Co.)

so when the structural factor of metal is important, only the abrasive cut-off wheel should be used while coolant is generously applied.

Circular Saws

A circular type of cold saw is frequently used to part all kinds of soft and hard materials, both metals and nonmetals (fig. 8–8). The cold saw may contain inserted teeth of carbide material whereby it has universal usage. Steel bars, as well as clay bricks, are cut with it. When large masses of metal are to be parted, the large circular saw is used with a surface cutting speed beyond 20,000. These cut-off saws are used in steel mills and supply warehouses. No teeth are required on some blades because friction creates the heated trench of metal which is instantly removed by the fast moving periphery of the saw. A special type of sawing operation is shown in figure 8–9.

Files

Filing is a modification of sawing (fig. 8–10) in that many more cutting edges are in touch with the workpiece at a given time interval. Individual band saw files are attached on a band. On the downward, or cutting, direction, they overlap each other to form a single and fast moving file. Types of cuts are those available in the hand file series, the single and double cut being very effective. The file is also classified as to its

FIGURE 8–8 A metal cutting saw. (Courtesy of Simonds Saw and Steel)

smoothness or coarseness while filing speeds are indicated on the machine's data card. Filing bands replace the saw blades as some finishing operations are performed with the file band.

In addition to the file band, there are reciprocating filing machines which provide a speed indicated in strokes per minute (fig. 8–11). These small machines are effective in the fast removal of metal in small and selected areas. Another rapid metal filing machine is the disk file (fig. 8–12). These radial toothed files are shaped in the form of disks. They cut on the downward stroke. Great care must be exercised when using these machines due to the tendency of the fast moving cutting edges to grab the work and possibly the fingers. Disk files are similar to disk sanders.

Several miscellaneous cutting operations are accomplished on the vertical band saw. Sanding belts are available for the fast finishing of metals. These belts are impregnated with commercial diamonds, carbides, and other hard materials and perform like file belts.

FIGURE 8–9 A special type of sawing operation. (Courtesy of Simonds Saw and Steel)

Drilling Machines

The drilling machine (fig. 8–13) is also a basic machine tool in a metal shop because of its many common capabilities. Just as the power saw provides some of the initial overall dimensions of a workpiece, the drilling machine provides drilling, tapping, polishing, and even operations like filing and milling. The drilling machine is a basic machine because many machined parts require holes. The saw and drilling machine are normally found in most metalworking shops, regardless of the work being performed in the shop. The drilling machine allows an operator to machine holes deep into a part, the holes normally being 90° to the table of the machine. When holes at other angles are required, the table may be swiveled, or special vises may be used to tilt and swivel the workpiece.

A unique feature of the drilling machine is that it removes more metal from a part per pound of its weight in a given time interval than most other machine tools. Such an operation is based on the geometry of the twist drill and its functioning and the principle of the machine. The cutting edges of the two-fluted drill, when forced into a material, shear the metal, and chips emerge in a helical pattern from each flute

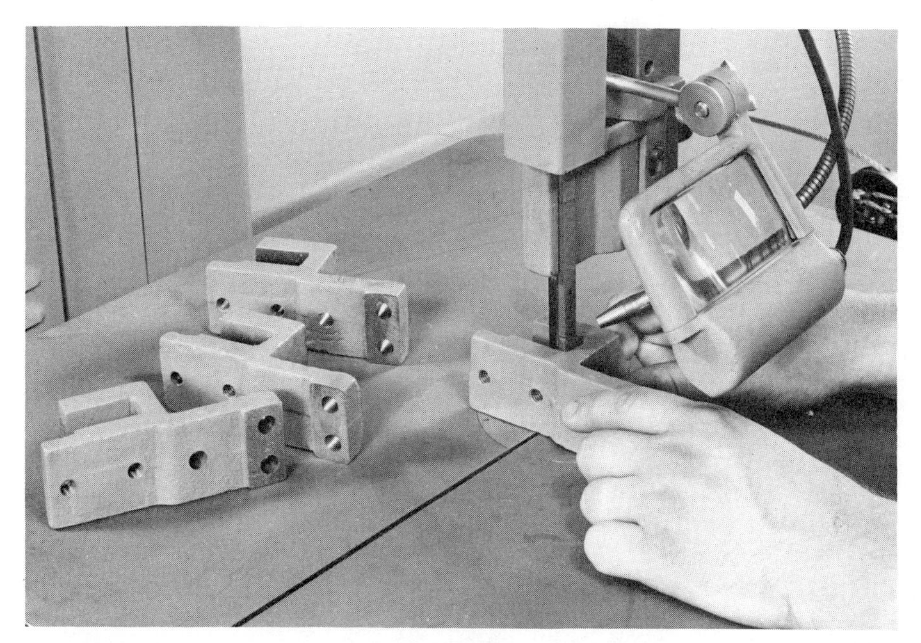

FIGURE 8–10 A filing operation in a vertical saw. (Courtesy of DoAll Co.)

FIGURE 8–11 A filing machine which can also be used for sawing. (Courtesy of Oliver Instrument Co.)

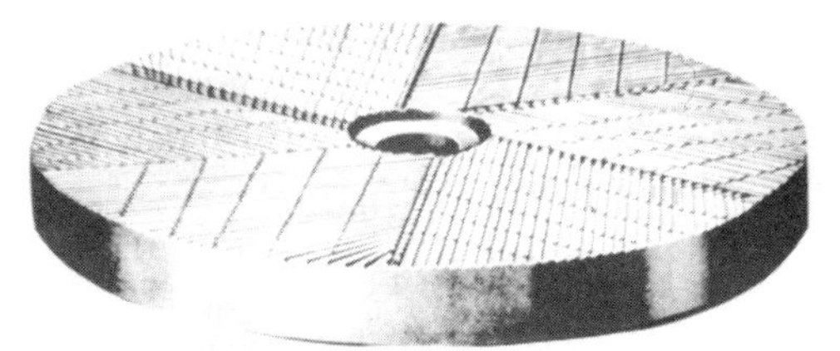

FIGURE 8–12 A filing disk, part of a filing machine. (Courtesy of Severence Tool Industries Inc.)

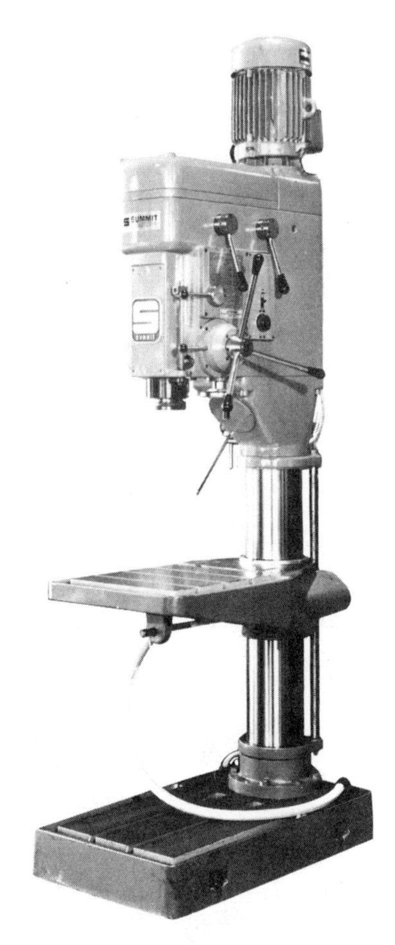

FIGURE 8–13 A heavy duty automatic precision vertical drill. (Courtesy of Summit Machine Tool Mfg. Corp.)

(fig. 8–14). A constant shearing action occurs as long as pressure on the drill exists. As pressure, or feed, is reduced, the chip's thickness is reduced and finally breaks.

TYPES OF DRILLING MACHINES

Drilling machines are available in several types and sizes. According to type, the hand operated is the most common. Often, it is called the sensitive type of machine because

FIGURE 8–14 Drilling in the lathe showing the two helical chips being formed by the drill. (Courtesy of Clausing Corp.)

the hands can feel the presence of drilling actions (fig. 8–15). As an example, a small diameter drill must be moved into the material with a slow feed while exercising

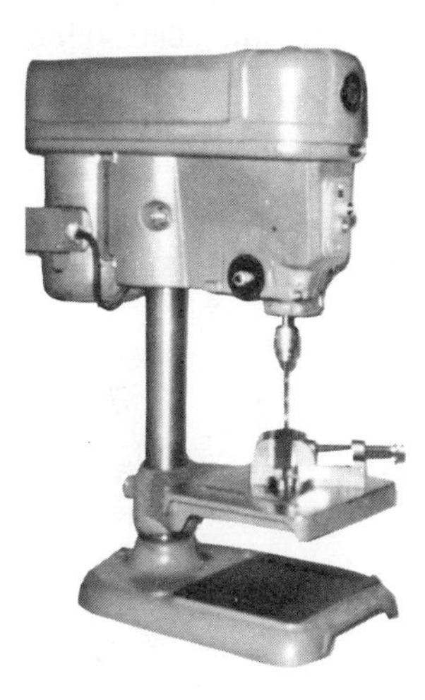

FIGURE 8–15 A bench type drilling machine.

caution so as to prevent breaking the drill in its hole. The operator is sensitive to this operation.

Hand Operated

The sensitive drilling machine, being hand operated, is produced in both the bench and floor types. Three dimensions of the drilling machine are critical—the maximum size drill which it can take in its chuck or spindle, the throat distance or distance between the center of the drill to the edge of the column, and the distance of vertical adjustment on the column other than the few inches of handle movement. Drilling machines are equipped with the appropriate chucks, the 1/2-inch Jacobs chuck which holds straight shank drills being a common type. However, chucks of larger capacity are available. The raising and lowering of the spindle allows for drill feed. This quill portion of the machine is movable within the machine's head, thus, the spindle moves upward and downward on command from the feed lever. The quill is springloaded up, eliminating the additional necessity of removing the drill from the workpiece by hand.

Power Operated

Another type of drilling machine, the vertical floor model, is merely an extension of the sensitive, but has an increased drilling capability. A power feed is available, along with a larger drilling capacity accompanied with the needed increase in power. When the entry point on the securely held workpiece is determined, the r/min and feed of the drill are calculated. When turned on and properly set, the machine drills automatically, relieving the operator of manual procedures (fig. 8–13).

Because larger drills require a different holding device in the spindle socket, special tapers are provided to obtain friction holding. Drills larger than 1/2 inch have tapered shanks, being the Morse taper. When the drill is snapped into the female taper of the spindle, a firm friction grip is established between the two tapers. The spindle's taper and the drill's shank taper are sometimes of different sizes and do not touch one another. In this situation, a drill sleeve is used in the spindle. The sleeve's two tapers, outside and inside, allow a perfect fit between a small Morse taper and a larger Morse taper. The sleeve contains a tang at its small end which provides a positive drive in the event of improper contact and slippage.

The machine can be operated by manual feeding or by an automatic system. Automatic feeds vary from 0.002 inch per revolution of the spindle to usually about 0.025 inch per revolution. When the air-hydraulic feed system is placed on larger drills, a variable feed arrangement is available whereby increased feeds are used when needed.

Radial Drilling Machine

When special types of drilling are required and when larger drills are needed, the radial drilling machine is used (fig. 8–16). Instead of the conventional center column arrangement, the radial has its drilling mechanisms mounted on the overhanging arm. The arm's length varies from two feet up to several feet. The arm is counterbalanced to equalize stresses in the machine's column. A radial drilling machine uses a semifixed anvil or vise on its base for holding the workpiece. The drill head is then

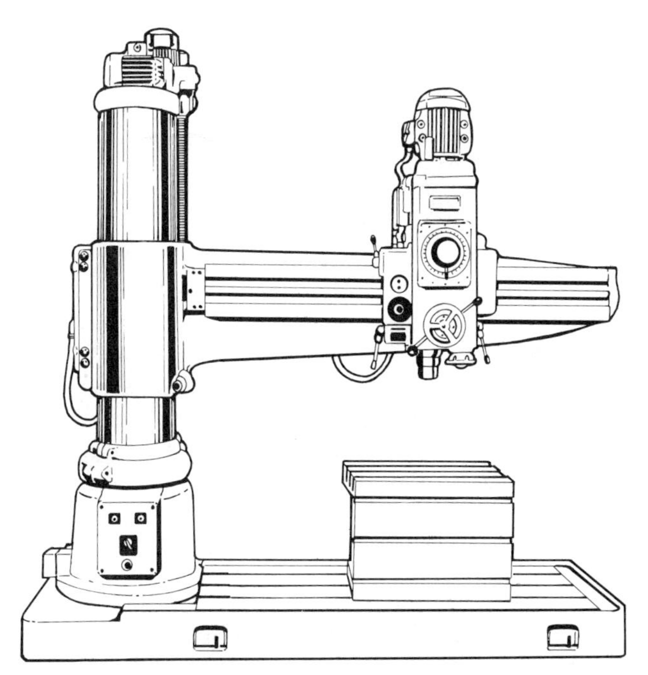

FIGURE 8–16 The radial drill has numerous machining capabilities due to its versatility, size, precision, and power feeding ability. (Courtesy of Summit Machine Tool Mfg. Corp.)

moved outwards along the arm by power feed to a point over the work. Conventional machines require movement of the vise, whereas the vise is usually stationary in radial machines. Due to the size and power of radial drills, all movements are power driven, however, a manual control is available in setup procedures.

Radial drilling machines are capable of driving three-inch drills with ease. Such a metal removing operation requires strength and stability in the work holding device. All work to be drilled in this machine and all other drilling machines must be securely tightened to the vise, anvil, or base of the machine. The vise is the common holding device. When angular work is to be drilled, special vises are available. When drilling holes larger than 1/4 inch, the vise should be bolted to the table of the machine. Smaller holes can be drilled with machinist vises which have no hold down lugs, however.

Special fixtures are also available for many drilling purposes. These fixtures have *T*-slots or dovetails in their bases for quick attachment and release from the table of the machine. Upon completion of drilling, drills are removed from the tapered spindle or sleeve by means of a drift. Insertion of the pointed end of the drift in the spindle's slot accompanied with a sliding snap of the drift's handle, releases the taper of either drill or sleeve. This procedure is also used to remove drills from the larger machines. When specialized drilling operations are needed requiring automatic or multiple spindles, the production model is used. These machines may contain many spindles, each being automatically controlled to perform a specific operation.

SPEED

The speed of the spindle and drill refers to the revolutions per minute. Speed is influenced by the cutting speed of the metal or the number of feet per minute considered as good machinability for that metal. Standard charts are available for determining this factor. Because machines are geared to revolutions per minute, a means must therefore be available for converting *CS* into r/min. The following formula is used to calculate r/min.

$$\text{r/min} = \frac{CS \times 12}{\pi \times \text{diameter of drill}}$$

The drilling machine spindle is turned by a gear or pulley system. Power is released from an electric motor having a given r/min. This fixed r/min is changed to the needed r/min through a geared system of speed changes or through a step or variable pulley system. Bench type and small drilling machines are produced with r/min capacities from a few hundred to several thousand. Larger machines have reduced r/min capacities, but some have up to 3000 r/min or less than 100. Most machines have gear or pulley settings which provide a given r/min. The radial drill, being larger, more sophisticated, more universal, and more powerful, has the most speed settings which provide r/min settings from that needed for the largest drill to the maximum for smaller drills.

Because r/min are fixed in the machine according to placement of certain levers, the closest setting to the calculated r/min is used. A quick r/min calculation is obtained by using the cutting speed of the metal (using carbon steel, high speed steel, or carbide drills) multiplied by four and divided by the drill's diameter.

FEED

Feed of the drill means the penetration of the drill in one revolution. Feed is determined by several factors—the material being drilled, the drill's material, the revolutions per minute, the diameter of the drill, and the stability of the setup in the machine. As a general rule, the larger the drill, the slower the r/min. Length of drill must be considered when extra long drills are used. When several factors pertinent to the drilling operation are known, another factor can be found. Should the sf/min be needed for a setup when r/min and diameter of drill are available, the following formula is used.

$$\text{sf/min} = \frac{\text{r/min} \times \pi \times D\ \text{(drill)}}{12}$$

Many machines provide machine settings on a data card which is attached to the machine. However, in its absence, speeds and feeds can be calculated. If the feed rate (inches per revolution) is needed, use the following formula.

$$\text{feed rate} = \frac{\text{feed rate in inches per minute} \times \pi \times D}{12 \times \text{sf/min}}$$

The feed rate in inches per minute equals the feed rate in inches per revolution multiplied by r/min. As in sawing operations, drilling is usually accomplished with the use of cutting fluids or a special cutting oil.

TWIST DRILLS

Most drills are two fluted and helical, having 118° lip angles at their cutting edges (fig. 8–17 illustrates the parts of a drill). When a lip clearance of 8° to 12° is ground at

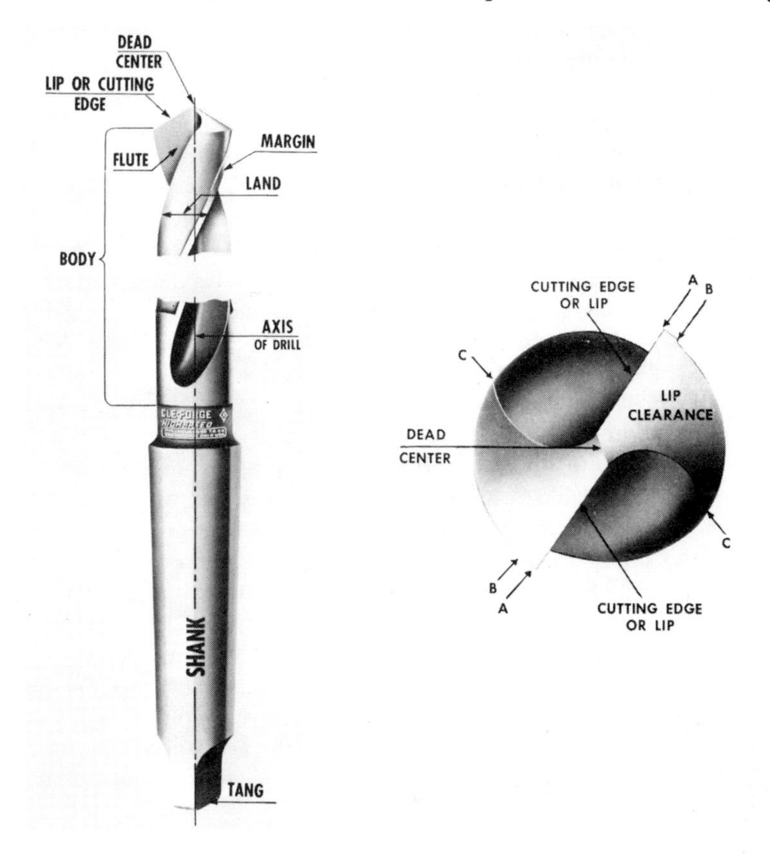

FIGURE 8–17 Parts of the twist drill. (Courtesy of Cleveland Twist Drill Co.)

the end of the drill, a chisel point is established which forms an angle of 120° to 135°. The *flutes* which wind around the thin backbone of the drill form a web. When the conical point is ground on the drill, a cutting edge is formed at the lip of the drill. Space is then provided for chip removal and coolant entry. The lip clearance provides freedom of rotation for the drill whereby no rubbing of tool occurs behind the cutting edges. This is a constantly increasing angle from point to the trailing edge at the periphery.

The angle of helix for general purpose twist drills is 24°. The main rake angle is formed in connection with the relationship between work and helix angle. This is the angle formed between the margin's leading edge and the axis of the drill. When the flutes are machined, the body surface remaining between the flutes is known as the *land*. The leading edge of the land includes a narrow width of metal which is the *margin*. Provision of the margin allows body clearance for the drill due to part of the land being ground away. Limited flute depth allows the central axis, or *web*, to provide strength and rigidity to the drill.

Other types of drills are multifluted, single-fluted, high helix angle, center drill and countersink, bit shank, straight shank, core, and taper shank, Single-fluted drills are used for deep drilling where drift of the drill is less likely to occur. A multifluted drill provides more space for chip passage, and it is usually used for some special operation. The core drill is multifluted and cuts along its chamfered edges, machining along the sides of the cored hole produced by forging or casting processes. The precast or forged hole is machined to size with the core drill. Drills having more or less than the 24° helix angles are special purpose drills which are used on harder or softer than average materials. The bit shank drill fits a special drive drill. Center drills are used to start a hole because their construction helps eliminate the tendency for drill drifting when entering a punch mark or a flat surface. Their 60° bevel terminating at the major diameter provides a tapered hole for use with lathe centers and for other purposes. The tapered hole also affords guidance to larger drills entering the hole. Center drills are available in several sizes to accommodate the large size differences between lathe centers which are used in lathe tailstocks for holding workpieces.

The shank of the drill is the driving end and its sides are either parallel or tapered to employ friction holding. Diameters beyond 1/2 inch normally require friction through taper action due to its much greater holding power as compared to the chuck's grasp of straight shanks. Chisel-pointed drills must be pushed into the work due to their large negative rake angle. Such force on the drill creates pressure and heat. Consequently, many small drills are constantly being broken due to overpressure. When a special procedure is used to grind a spiral point on the drill instead of the chisel point, breakage is reduced, because approximately 50% of the negative rake is removed, thereby increasing drill entry efficiency. The spiral points cut easier and even reduce the tendency of the drill to drift. Chisel-pointed drills can be machine or hand ground whereas the spiral-pointed drill is machine ground. Caution must be exercised in hand grinding clearances on drills due to the great possibility of leaving the point off of center. Off-center drills perform unsatisfactorily, even drilling oversized holes.

Twist Drill Sizes

Because drilling is one of the most important machining operations and because standard size hardware such as bolts is used to fit the drilled holes, the drill is produced in all standard sizes. A systematic method is used in drill production; fractional size drills are produced in increments of 1/64 inch up to three inches. Drills which fit in between the fractional sizes are produced in a size from a few times larger than a human hair to nearly 1/4 inch in increments of just a few thousandths of an inch. These are drills numbered from 80 to 1, number 80 being the smallest. Drills are also produced in increments of a few thousandths of an inch to accommodate sizes larger than numbered drills but between fraction size drills up to 0.413 inch. They are designated by letters *A* to *Z*. Drills larger than four inches are available.

The drill holding device used in drilling is often the *chuck*. The chuck is available in sizes greater than 1/2 inch, but the 0–1/2 inch is the common size for use in small factories and schools. It is most often attached to the spindle by taper action. Larger drills are equipped with the tapered shank, the machine's spindle holding the drill. Differently designed collets also act as quick holding devices for drills. In addition, the quick change drill chuck is used when many different sizes of small drills are used.

SPECIAL DRILLING MACHINE OPERATIONS

Many machine operations require bolt heads to be either flush with the surface or below the surface of the metal part. Because the bottom side of a bolt head is flat, the top portion of the drilled hole must be counterbored to allow the bolt to fit flush. The squaring of the bottom of a hole is performed with a *counterbore* (fig. 8–18). The pilot at the end of the cutting tool fits into the drilled hole and acts as a central guide so that concentricity of the two holes results. Counterbores cut square corners at the bottoms of holes as opposed to conical holes produced by drills. Multifluted counterbores are available in several sizes to fit standard machine operations.

The *countersink*, either 60°, 82°, or 90°, is used to provide a beveled edge at the top of a hole for the purpose of receiving rivets or screws. The countersinking tool, as well as the counterboring tool, fits into the spindle or chuck of the drilling machine. The countersink may or may not be equipped with a pilot drill, therefore, the workpiece must be securely bolted to the table while the tool is lowered into the exact center of the hole. For best results, the pilot drill type should be used. This operation, if done without bolting the workpiece to the table in addition to having a rigid setup, may often result in scrapping the workpiece. Other special type cutting tools are also shown in figure 8–18.

Reaming

On many occasions, drilled holes must be reamed in order to obtain a smooth interior wall and an exact dimension. The drilling operation leaves a hole's wall in a series of helical shaped scratches, but in most situations, the "as drilled" wall is satisfactory. However, when a spindle is to be inserted in the hole, for example, the wall's surface must be smooth in addition to being dimensionally accurate within a usually close tolerance. To accomplish this, the reamer (fig. 8-19) is used to remove a small quantity of metal from a hole, the resulting hole equalling the diameter of the reamer. Holes to be reamed are to be drilled to sizes which enable the slightly tapered end of the reamer to enter, that is, the drilled hole must be within 0.005 to 0.030 inch of the desired size. Reaming then removes small ribbons of metal. To maintain hole accuracy, reamers must not be tapered, except at the very end. The tapered reamer, however, is deliberately tapered for specialized operations. Also, reamers must measure to within 0.001 inch of the required dimension when precision is required. A special reamer is shown in figure 8–18.

Reamers are produced as hand, machine, shell, and adjustable types. The reamer may be straight or tapered and either straight fluted or helically fluted. Hand reamers are turned with a wrench and may be guided with the spindle of the drilling machine. The machine reamer is chucked and moved into the drilled hole at a specified r/min. When large holes are to be reamed, the shell reamer is used. An arbor holds the shell reamer which may have high speed steel or carbide-tipped edges. Adjustable reamers are equipped to provide a small range of adjustment, while the expansion reamer

FIGURE 8–18 Special purpose cutters: *a*, helical flute shell reamer; *b*, straight shank counterbore (a pilot fits into the end of the counterbore); *c*, adjustable shell reamer with inserted blades; *d*, four flute center reamer. (Courtesy of Cleveland Twist Drill Co.)

FIGURE 8–19 Removing a small quantity of metal from a hole with the machine reamer. (Courtesy of Cleveland Twist Drill Co.)

expands less than the adjustable reamer. Due to the cost of large reamers, the adjustable reamer provides for expansion by salvaging the reamer when its blades are worn and must be ground. Figure 8–20 illustrates machine reaming with a coolant.

Tapping

Other machining operations such as tapping or threading can be performed in the drilling machine by hand manipulation. Tapping is a cutting operation which produces threads in a hole. The tap is turned in the hole, the resulting thread being determined by the tap thread size. When a small center finder is placed in the chuck and engaged into the center hole of the tap, guidance from the spindle helps align the tap while hand turning with the tap wrench is performed. When machine tapping in the lathe, for example, a specified r/min is used due to the nature of the operation and the cutting speed of the metal.

FIGURE 8-20 High speed reaming and turning which necessitates the use of coolant. (Courtesy of Cleveland Twist Drill Co.)

Boring and Circle Cutting

Boring and circle cutting are quickly accomplished by the drilling machine. The boring tool is placed in the chuck with the cutting edge extended the exact radius of the hole. A downward pressure of the spindle enlarges the drilled hole to specification requirements. When a circle cutter is used, the rapidly moving circular saw removes chips at a swift rate. Again, the diameter of the saw must be exactly calculated, keeping in mind the thickness of the saw blade and whether the inside or outside piece of work is required.

Another device used to cut large holes is the fly cutter. A fly cutter uses a small drill to enter the metal first and act as a guide for the cutter. The cutter, similar to a lathe tool bit, is adjusted to the proper radius. As the spindle is lowered, the drill guides and maintains the periphery cutter at a constant radius. Usually, a soft piece of metal or wood is placed beneath the plate of metal being cut to allow the slowly rotating cutter to completely penetrate the workpiece. Great caution must be exercised during this operation because a whirling tool cannot always be seen.

Lathes

The lathe is used if any form of turning is performed. Saws and drilling machines are needed in most metal shops, while the lathe is needed only if the type of work done in the shop justifies it. The lathe is capable of any type of machining which involves rotation of the workpiece. Straight and taper turning, facing or squaring, radius and form turning, threading, knurling, drilling, tapping, grinding, sanding, reaming, boring, and several other operations, are performed on the lathe. No other machine can do as many different machine operations as the lathe. With special attachments, the lathe becomes a universal tool; it can perform the work of most other machine tools such as milling, grinding, and slotting. Production shops, however, are equipped with special lathes and do not resort to using attachments.

LATHE TYPES

The most common lathes are the engine and toolroom lathes. They are used in all types of lathe operations and are the workhorses in industry (fig. 8–21). The toolroom lathe is more accurate in its structure than the engine type. Engine lathes are used in most industrial shops, other than automated shops, while the toolroom lathe is used

FIGURE 8–21 The lathe equipped with the four-jaw chuck, follower rest, tool post holder, steady rest, and tailstock center. (Courtesy of Summit Machine Tool Mfg. Corp.)

for precision and special work such as in tool and die making. When the kind of work is similar and repetitious, the production lathe is used. This machine is more heavily constructed and has a larger electric motor in its drive mechanism. Special attachments on this lathe such as power turrets allow numerous parts of the same type to be made in short periods of time. The production lathe may have a gap bed in order to swing a very large diameter workpiece. Automated and semiautomated lathes having special cutting tools, turret functions, and sequential operations are used by the thousands when large masses of parts are required. The controlling computer or the tape is programmed to operate the machine hour after hour. There are various modifications of each of these four basic types, each modification being more specialized in its function or size.

LATHE SIZES

The lathe has two main dimensions which give it its size. A lathe swings a workpiece of a given diameter and length, therefore, its size is stated in the number of inches in diameter of a workpiece it will swing and the number of inches there are between the headstock and tailstock centers or the length of the workpiece it will contain. These dimensions refer to distances between centers and between the center's point and the ways multiplied by two for the diameter. Because a compound rest slides along the ways, the height of the saddle and compound rest must be taken into consideration when calculating the diameter of work swing and determining the size of machine to use. Maximum diameter work will rub the carriage if the work has much length. The radius of working swing is a more common dimension of capacity, as it is the maximum clear dimension between headstock center and working conditions.

MAJOR PARTS OF THE LATHE

Lathes are variable in their construction, depending on the make and model. Because the lathe is a complex machine tool, and because the movements of its parts control many types of operations, a brief description of the more important parts is given. As pointed out in fig. 8–21, the lathe *bed* rests on its legs or on a stand. The bed is one of the six main parts of the lathe. The bed, being the long and thick base of the machine, supports the sliding *ways* along its top part. Some machines have *V*-shaped ways while others have a combination, that is, one flat and one *V*. The ways, which act as guides for the moving carriage and tailstock, are aligned along the longitudinal axis of the bed and must remain in alignment for effective performance. Because the ways are frequently used, their surfaces are usually hardened. The headstock is permanently mounted on the left end of the ways and bed. When the headstock, carriage, and tailstock are properly aligned on the ways, precision and fine work can be produced. When placing the lathe on the concrete floor, care must be taken to see that no twists occur in the bed and that the lathe is properly leveled.

At the left end of the lathe is the *headstock* which houses the power source and most of the control mechanisms. An electric motor is usually mounted just below the headstock. As the size of the lathe increases, additional power is required, therefore, the horsepower rating of the motor is a large variable. *V*-belts and pulleys connect from the motor to the driving mechanisms. The lathe's spindle is driven by either a belt drive or gear drive, the driving power source being the motor. When belting is used from the gear train to the spindle, a smooth and quiet transfer of power occurs. When gears are used, the power transfer is often noisy. Gear drives are direct drives, while belt drives will slip on the pulley when the spindle's work is jammed. Step pulleys in the belt drive system allow for r/min change by shifting the belt from one pulley to another (at both ends of belt loop). Usually four pulleys are available in order to help provide several spindle speeds. Shifting levers on some machines also help provide the desired r/min in conjunction with the pulleys. The gear driven spindle is turned by means of a series of levers for producing the desired spindle speed.

Lathe *spindles* are hollow so that long rods or tubes may be held in the chuck while the rod's length is extended through the spindle. With this setup, a cutoff operation in the lathe saves a sawing operation. The spindle's nose extends slightly from the head-stock, its inside diameter being tapered with the large end on the right. The outside diameter is sometimes threaded to receive and hold the various kinds of chucks, faceplates, or other holding devices. A modification of the threaded spindle is the steep tapered spindle with a locking key. Another type of spindle nose includes the cam-lock. When chucks are not used, the drive plate drives the workpiece because the dog's tail is connected to the driving plate. The *dog* (fig. 8–22 and fig. 8–28) is

FIGURE 8–22 Lathe accessories. Included are faceplate, toolholders, 4-jaw chuck, wrenches, draw bar and collets, jacobs chuck, drill and countersinks, live center, and dog.

securely attached to the workpiece which is centered in the spindle by means of the headstock center. A dog is a connecting device used to connect the workpiece to the driving mechanism. The dog slides around the workpiece and is tightened by a set screw. The bent portion, or tail, fits into a slot in the drive plate, resulting in a positive drive between plate and dog. The collet chuck also fits into the tapered spindle and holds small rounds securely.

Another major part of the lathe is the *carriage* which carries the toolholder and tool. Mounted on a sliding *saddle*, the carriage moves between the headstock and tailstock, carrying the *compound rest* and cutting tool. Control mechanisms are mounted on the *apron* which is part of the front of the carriage. The carriage moves along the longitudinal axis of the lathe either by hand or machine power in either direction. Many lathes are equipped with a *taper attachment* which, when operating, is connected so that lengthwise movement of the carriage provides a taper along the length of the part being turned. Movement of the cutting tool which is perpendicular to the axis of the spindle is controlled by the cross slide mounted on the saddle. This movement is also hand or power moved in either direction. The compound rest provides the toolholder and tool swiveling action up to a 360° turn. The cutting tool is held securely in the toolholder. Cross feeds are controlled by micrometer dials. Threading is controlled by a chasing dial attached to the apron.

On the right-hand end of the lathe is the *tailstock*. This device serves the purpose of supporting a piece of work extending from the headstock. Its spindle is also tapered to provide a means for holding a center, a drill, a chuck, or some other tool. In order that accurate work is performed on the lathe, the headstock center and tailstock center must be in perfect alignment and at equal height. Any misalignment will result in tapered work. Sometimes, however, a taper is deliberately turned on the lathe by offsetting the tailstock from a straight line position.

The lower portion of the headstock contains a quick-change gear system. Power is transmitted through a series of gears to both the lead screw and feed rod. Gear boxes allow for several settings of the gears to accomplish feeding and threading. Spindle speeds are converted into needed r/min of the feed rod and lead screw which, in turn, moves the carriage and energizes carriage controls. Lead screws provide for thread cutting operations, while the feed rod is used in power drives.

LATHE ACCESSORIES AND ATTACHMENTS

The basic lathe illustrated in figure 8–21 cannot be used unless additional equipment (fig. 8–22) is mounted such as a chuck and a tool post holder with the cutting tool. When a round bar of steel, for example, is secured within the 3-jaw universal chuck, machining operations can be performed as soon as the proper tool is placed in position. The type of machine operation, then, determines the kind of accessories or attachments required.

A 3-jaw chuck is universal as all three jaws move at the same time. The 4-jaw chuck, however, holds square stock and also provides a means for turning eccentrics, and the jaws move independently. A chuck is used in the tailstock spindle to hold center drills and twist drills. Both the headstock and tailstock provide for the taper holding of centers, either live or dead. The live center's point turns with the workpiece, whereas the dead center must be lubricated with a heavy lubricant such as a paste. Collet chucks, draw bars, and face plates with driving dogs fit the spindle for holding work.

The tool post holds the several types of toolholders and may be considered as a single turret. The regular turret, either hand or machine operated, is mounted on the carriage instead of the toolholder. Because different cutting operations are performed on the lathe, differently shaped holders are available, as well as different purpose tools. Lathe toolholders are illustrated in fig. 7–5.

Cutting tools shown in figure 8–23 are capable of providing most of the cutting operations on the lathe. Notice the different cutting and relief angles required in the several operations. Basically, the tool should not overhang the holder more than is necessary to perform the operation because of the danger of breaking the tool or inducing chatter marks on the workpiece.

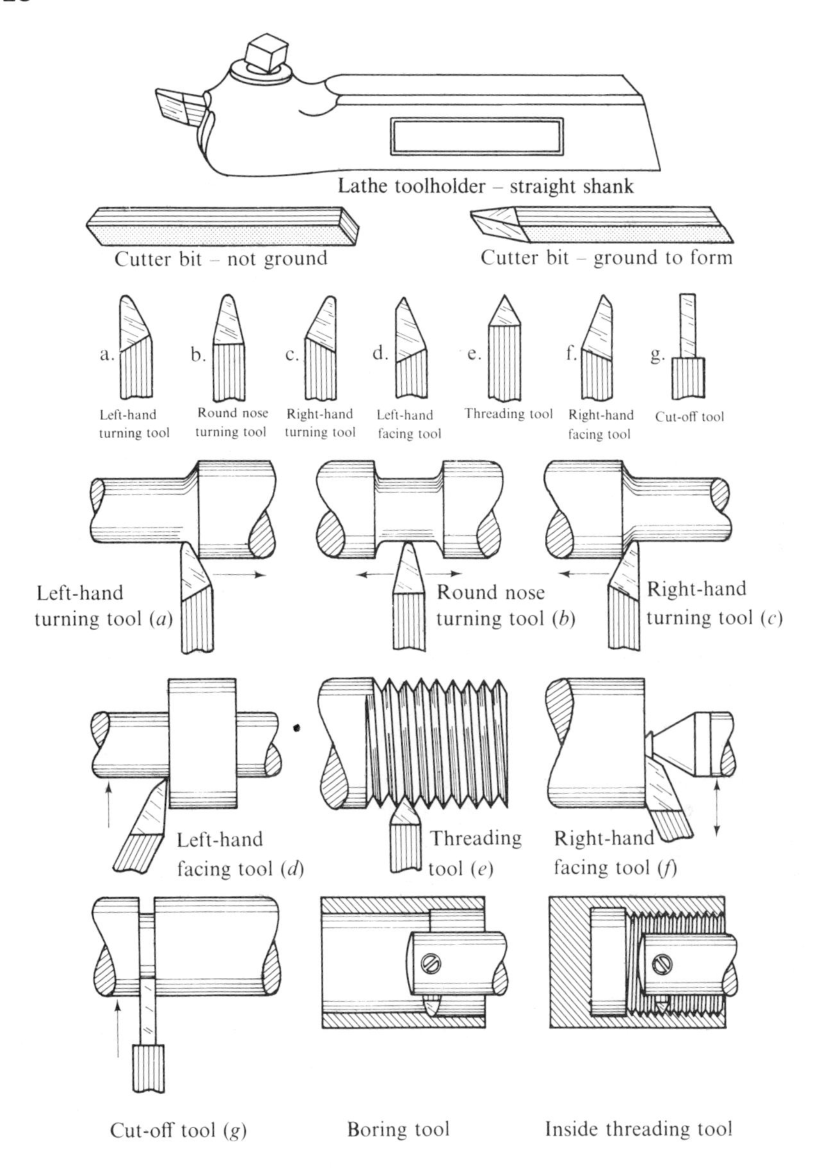

FIGURE 8-23 Common lathe cutting bits and typical uses. (Courtesy of South Bend Lathe)

Other attachments include those for work support. When turning a long and thin rod, for instance, the follower rest or steady rest is used to prevent bending of the rod as the tool cuts along its length. Either of these rests maintains a straight line axis of the workpiece and prohibits bending and erroneous dimensions of the part.

CUTTING TOOLS

The chip formed from a workpiece during the lathe cutting operation results from failure along a shear plane, as has been previously pointed out. As the tool is fed into the revolving work at the rate of a fraction of an inch per revolution, power is consumed because of the metal's resistance to being cut. Consequently, the chip slides across the face of the tool, being guided by the angular rake. Specifically, as the revolving work moves downward into the path of the tool's projected shear plane, the upper layer of metal slips along the plane and moves away from the work as a tangent to the circular movement of the work. The speed of rotation and the thickness and width of the chip, along with the cutting speed of the metal, determine the power required to produce the chip at a constant rate. The chip, then, is formed due to the shape of the cutting tool as the revolving work collides with the knifelike edge of the tool. Therefore, the importance of these angles must be understood.

The cutting edge of single-pointed tools has been shown in figure 7–3. Notice that reduction of tool material in certain planes provides an edge at the intersection of two edge relief planes. Small relief angles are used for the cutting of hard materials while larger relief angles reduce the nose angle. Too small a nose angle will result in a broken tool, however, small nose angles allow the continuous removing of chips from softer metals. Relief angles are ground into the tool for the purposes of providing clearances between tool and work and these angles vary from 3° to 20°. This angle variable depends on the kind of material being machined and its hardness, the type of cutting tool used, the relationship of the work to the tool, the physics of the cut, and the overall setup of the work and tool.

The actual operating tool relief and cutting angles vary according to those angles ground into the tool and those angles between the tool's position in the toolholder and the work. The toolholder may be moved in the post by rocking the holder to the desired angle. The side rake angle forms a plane across the tool's face. Cutting ability of the tool is strongly influenced by this angle due to the direction of the shear plane. The back rake angle rests between the tool's face and a line at 90° to the center of the work at the tool's cutting edge tip. Variable positioning of the toolholder and the type of holder affect the relationship of this angle with the workpiece. The side cutting edge angle varies from a small angle up to approximately 25°. An angle of 18° gives a good cutting condition. The end cutting edge angle works with the side cutting edge angle to produce the cutting edge and its relation to the work. Approximately 25° provides satisfactory cutting actions for the end cutting edge angle, this angle varying a large amount. When the end cutting and side cutting angles are formed, the nose angle results. It is at the tip of the nose angle and along the leading edge of the tool that the cutting occurs.

As soon as the chips begin to form and slide away from the cutting edge, a continuous or noncontinuous chip is produced. In materials such as aluminum and ductile steel, the chip will be continuous, and often is dangerous. This long chip can be broken in equal length increments by grinding a chip breaker into the path of the moving chip. The sudden change in direction of the moving chip causes immediate fracture.

LATHE OPERATIONS

As has been mentioned, the lathe can perform more different types of operations than any other machine tool. The more common operations include facing, turning, knurling, parting, threading, filing, polishing, grinding, drilling, boring, reaming, and special operations such as template operations. All of these operations are classified into external and internal, facing being external and drilling internal.

Before any machining operations can be performed, the cutting speeds of metal (table 7–1) must be determined in order to set the proper spindle r/min. Much of the same criteria existing for drilling are used in lathe operations. In drilling, the work is stationary, while in the lathe, the work turns. If cutting speed of the metal is known, the rpm can be calculated. If the rpm is known, the *CS* can be calculated. As an example, low carbon steel is turned at approximately 100 sf/min speed. The r/min setting then is found by multiplying this factor by 12 inches and dividing the total by πD, D being the diameter of the workpiece in inches. As discussed, most cutting speeds of small diameter work are posted in various charts and are easily obtainable. When the r/min is known and no *CS* chart is available, multiply the r/min by πD and then divide the total by 12 to change the value to feet. Calculated values for *CS* are amended as the type of cutting tool changes. In this respect, consider the different ratios used in drilling operations when different types of drills are used.

Another factor which must be considered prior to machining is the setup, that is, the relationship of the tool to the work, including stability of the relationship. Many operations are performed by hand feeding while others are machine controlled and power fed. Prior to metal removal, an inspection should be made to assure clearance of all moving parts and that the power feed is not engaged. Should the revolving chuck hit the edge of the carriage, serious damage can occur. If the chuck handle is not removed from the chuck prior to starting the spindle rotation, very serious injury to the operator and machine may occur. As the chuck moves toward the bed on the operator's side, the wrench may be thrown into the operator's face. Besides checking for proper clearance of work, check for tightness of the tool in its holder and the tightness of the work in its chuck.

Depending on the operation being performed, the tool's cutting edge is positioned at the best cutting position. This position indicates the tool's angular relationship to the axis of the work and the height of the cutting edge of the tool compared to the center of the work's diameter (fig. 8–24). Notice the adjustment of the tip of the tool to the center of the work for taper turning and most other operations including straight

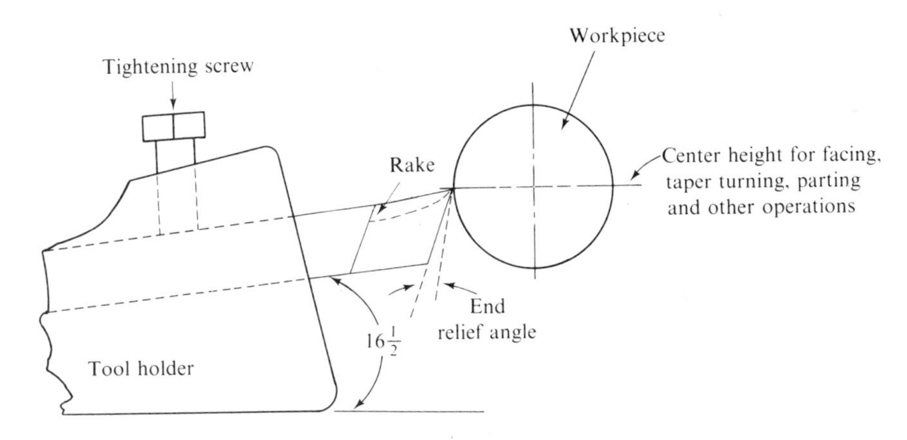

FIGURE 8–24 Relationship of cutting edge with work.

turning. For straight turning, the tip can be adjusted for a small degree above center height (fig. 7–4). When all these variables are determined, machining operations can commence. Variables change as the operation changes, therefore, a special setup is required for each operation.

Some words of caution in regard to most lathe operations must be given. If something should slip while machining is in operation, quickly turn off the spindle and power feed. Do not try to grab or hold anything. You will only fail in your attempts and will probably get seriously hurt. Also, remove rings and watches, roll up sleeves and tuck in loose clothing. Long hair and machine operations are incompatible. Then, make a final visual inspection to assure the safety of the setup and that proper clearances prevail. The power feeds should be off prior to turning on the machine. Turn on the spindle for proper rotation and proceed with the operation. Face shields are used in some operations when hard and short chips are flying from the work.

Facing

Facing is one of the first operations to be performed on the lathe following the initial setup. Because the original sawing operation is incapable of producing final finishes to most parts, the lathe provides finishing operations to exact dimensional requirements. Any holding device is suitable for facing, such as the chuck. As shown in figure 8–25, the right-hand toolholder is used while the tool's cutting edge is placed at

FIGURE 8–25 A facing operation in the lathe. (Courtesy of Clausing Corp.)

exactly center height. If not at center height, some of the material's center will not be removed. A reasonable depth of cut is adjusted, the carriage lock screw is tightened to prevent longitudinal slipping, r/min is established, and cutting commences either by hand or power feed. Deep cuts remove metal fast, but leave a fairly rough surface. Light cuts usually provide smooth machine finishes when the proper cutting tool is used. Feeds are established in accordance with the quantity of metal to be removed in one revolution, therefore, all variables must be taken into consideration. Feeding is accomplished with the micrometer dial of the cross slide. The facing operation is performed to bring the workpiece to exact length dimension and provide the desired machine finish. Coolant is sometimes used in situations which require deep cuts and fast metal removal when using steel cutting tools.

Facing prepares the end of the workpiece for subsequent operations such as center drilling, drilling, reaming, tapping, or boring. During facing, the tool moves from either the center towards the surface of the workpiece or the opposite, depending on the cutting tool. The left-hand tool has its side cutting edge on the right and, therefore, cuts to the right. The right-hand tool has its side cutting edge on the left and cuts to the left. These tools are illustrated in figure 8–23. A typical example of facing is illustrated by the initial machining operations on a reamer. After cutting the reamer stock to the desired length, one end of the workpiece is then faced. This operation squares the stock's end with relation to the longitudinal axis. A center drilling operation may follow. When the opposite end is faced, the exact length dimension is established. With respect to facing, only a small length of the workpiece should extend from the chuck to prevent wobbling of the workpiece.

Straight Turning

Turning is used to remove metal along the longitudinal axis of the lathe. A combination turning and reaming operation is demonstrated in figure 8–20. For straight turning, the tool's cutting edge is set at either center height for soft and small diameter materials or up to 5° above center for slightly harder materials or larger diameter materials. Coolants are used according to the need. Often the right-hand tool is used in turning to allow the cutting thrust of the heavy load to be against the headstock. Metal which is close to the chuck or other holding device can sometimes be better removed with the straight holder, otherwise, the holder is used which best suits the job. When working close to the tailstock, the left-hand tool is often used. Again, depth of cut, r/min, and feed depend on the setup, material being cut, and the kind of cutting tool. The r/min is derived from the cutting speed of the metal being cut. Roughing cuts, 0.060 inch for example, are heavy cuts to remove the material as quickly as possible. Finishing cuts up to around 0.016 inch leave the material in a smooth condition within exact dimensions. When long pieces are to be machined, the steady rest is used to prevent the workpiece from bending away from the cutting tool, while better still, the follower rest follows the tool and supports the piece all along the longitudinal axis rather than at one point. Steady rests are clamped to the lathe's ways, while the follower rest is clamped to the saddle. Because turning may involve different types of operations, specific lathe tools are available for any specified operation. A typical turning operation uses a left-hand holder.

Examples of straight turning operations include the machining of gear blanks for subsequent tooth cutting operations, the lengthwise machining of the reamer workpiece to bring it to outside diameter dimensions prior to fluting, the turning of a bushing or spacer to an exact dimension, the turning of a bolt's diameter in preparation for subsequent threading operations, and the machining of a long spindle to be used in a lathe's headstock. After the workpiece is placed in the chuck and held securely on the right-hand end with the lathe's center, the turning operation is ready to commence. The right-hand cutting tool is placed at center height of the workpiece, being slightly past the right-hand end, r/min and feed are established, and the longitudinal power feed is engaged which causes the tool bit to move from the right end of the workpiece towards the headstock. The depth of cut produces a chip as the cutter moves smoothly to an established dimension near the chuck. When the scribed line of the workpiece is reached, indicating the length of the part, the power feed is stopped, and the cutter is returned to a position just past the right-hand end of the workpiece. With the aid of the cross-feed micrometer dial, another depth of cut is arranged, the power feed is again engaged, and the workpiece is reduced in diameter. The finishing cut is a smaller depth of cut at a slower feed so that a smooth machine finish will result. By zeroing the micrometer dial when the tip of the cutting tool touches the workpiece, diameter control is established by feeding in the required number of thousandths of an inch.

Taper Turning

The main difference between straight and taper turning procedures is the placement of the tool's cutting edge at center height (fig. 8–24). Many lathes are equipped with a taper attachment. When the degree of taper is known, the attachment is adjusted to provide this taper and then tightened to the lathe's bed. As the saddle moves longitudinally, the attachment moves the cross slide and cutting tool in a taper plane to provide the needed taper per foot according to the adjustment made on the attachment. The attachment's upper swivel slide includes a guiding rod and mechanism and causes the cross slide mechanism to provide a direct and constant taper ratio between diameter and length of workpiece (fig. 8–26).

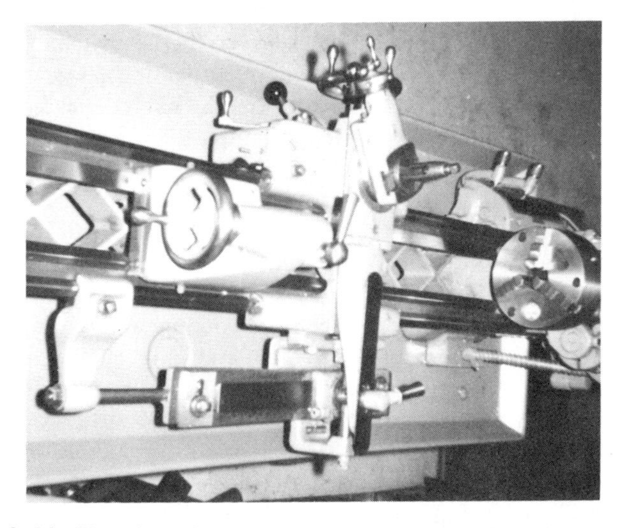

FIGURE 8–26 Top view of a toolroom lathe showing 3-jaw chuck, compound rest with toolholder, tailstock, and taper attachment.

Another method of taper turning includes the offsetting of the tailstock from its center point, but excessive loads may be placed on both the headstock and tailstock centers when using this method. When the length of the workpiece in feet multiplied by the taper in inches per foot is divided by two, the offset of the tailstock is known. This method is not recommended unless no other is available, because it consumes time in realigning the two centers to their proper alignment. Realignment is accomplished with the use of the dial indicator, that is, when the tailstock center has been zeroed properly, the indicator will show no deflection from zero when moved along the length of a workpiece. If the tailstock is off center only a very small amount, an unwanted taper will occur during other lathe operations.

A fast and efficient taper operation is also accomplished by means of the compound rest. When the rest is adjusted to the desired angle of taper, the tool is merely fed into the work by hand, turning the knob of the compound rest slide, not the cross-feed knob. Only small and steep tapers are turned by this method, such as the 60° taper on lathe centers. Straight turning and taper turning are related, the difference being the constant taper ratio existing during taper turning.

Knurling

The knurling operation is a cold forming lathe operation which forces the surface of the metal into pointed or lined patterns according to the type of knurls used (fig. 8-27). When the turning workpiece is properly set up to resist heavy loading, the

FIGURE 8-27 Knurling in the lathe. (Courtesy of Clausing Corp.)

knurl is lightly pressed into the surface whereby a pair of knurls are equally touching the metal. Then the knurling tool is moved to clear the right end of the workpiece. A depth of approximately 0.020 inch is set with the cross feed dial, along with the longitudinal power feed setting of approximately 0.025 inch. The r/min is set in back gear or very slow r/min. With this slow turning, the longitudinal power feed is engaged to allow the knurls to pass along the needed length toward the headstock, using oil as the lubricant. The machine is then stopped at the appropriate place, the tool remaining engaged. The longitudinal feed is then reversed as the workpiece again turns and is stopped just beyond the right-hand end of the workpiece or at a designated dimension. The cross feed is again increased approximately 0.015 to 0.020 inch at exactly the same time that the power feed is engaged. The knurling operation is again stopped at the left-hand dimension. These operations are repeated until the desired pattern is completed. Very soft metal will deform into shape at a single pass of the knurls. An increase in depth of cut can be made on the return cut.

With regard to any power feed operation, a word of caution is given for both lateral and horizontal feeds. Do not allow the moving carriage and tool to crash into any part of the machine. When the spindle is turning and when the power feed is on, unless a stop is provided, a crash will occur and the machine may be seriously damaged. This warning applies to all power feed operations on all kinds of metal cutting machines.

Parting

The parting operation is a cut-off operation. When a thin-faced cutting tool is fed into the workpiece with the cross slide, the carriage being locked to the ways to prevent side movement, the diameter of the workpiece is gradually reduced to zero. The cutting tool moves 90° to the longitudinal axis. Parting is frequently performed as a means of bringing a part to its proper length, with facing occurring at the same time. The parting tool mounts in the toolholder in the same manner as the knurling and turning tools do. Clearances in front of the tool beneath the cutting edge and on both sides are required, as well as a slight back rake. The tool's cutting edge is square and narrow, similar to a wedge operating in reverse, to allow quick severence of the workpiece. Unless the tool's edge is at center height of the workpiece, parting will not result. An example of parting is the final lathe operation on the previously mentioned reamer. After all lathe operations have been completed, the reamer workpiece is cut off at the desired length by action of the parting tool, the remaining stock being held by the jaws of the chuck.

Threading

Threading is a common operation performed by the lathe (fig 8–28). Because many types of threads exist, the threading tool is ground to the thread's shape. A thread is a helix around a core, the advance of the thread's angle being uniform along the axis of the threads. There are single threads and multiple threads. The double and triple threads provide an increase in the lead as compared to the lesser lead of the single thread. A double thread on a bolt will cause a nut to move faster along the length of the bolt, the lead being twice the pitch. The lead of the triple thread is three times that of the single thread. Details referring to screw threads are available in appropriate handbooks and should be consulted prior to thread cutting. Such a definition as lead, which is one divided by the number of threads per inch, must be understood in order

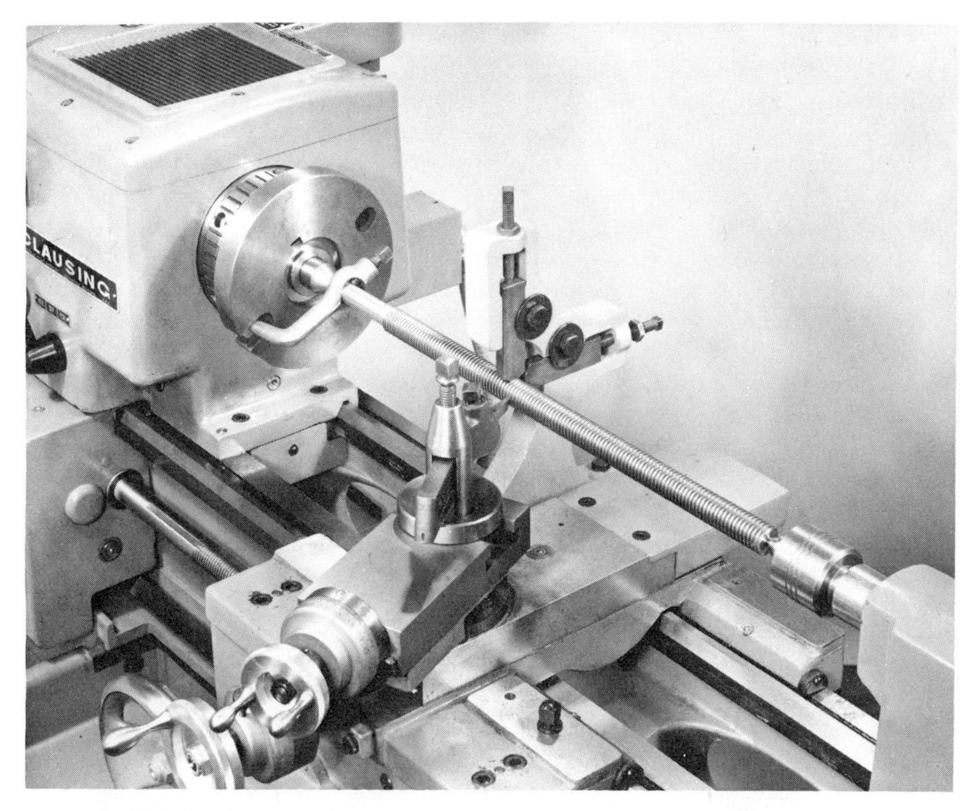

FIGURE 8–28 Threading in the lathe. (Courtesy of Clausing Corp.)

to calculate essential thread setups. Besides the regularly used American Standard Unified thread, there are square, Acme, and worm threads. Threads are also left- and right-hand and either internal or external.

The American Standard thread is used in six common pitches to hold mating parts in most industrial applications. A pitch is a distance from a point on one thread to the corresponding point on the next thread. When the number one is divided by the pitch, the number of threads per inch is determined. If the number of threads per inch is 20, then the pitch is 1/20. The best method of determining the number of threads per inch is to use a screw pitch gage. The pitch diameter is an imaginary dimension that divides the thread widths and spaces into equal spaces. This dimension allows equal strength of mating parts such as a nut on a bolt. The bolt's threads and the nut's threads hold each other equally. The three more common pitches are national coarse, national fine, and national extra fine. As the diameters of bolts decrease, the number of threads per inch increase. Such a system provides safety in shear loading on the threads while tensile or shear loading exists at the root diameter of the bolt. The other three pitches are used in special applications such as in the pressure vessel industry.

Threading tools are carefully ground to exact angles and form to correspond with the type of thread to be cut and the cutting procedure. The use of gages is very important in the proper preparation of the tool to produce exact clearances. The workpiece is machined to the indicated major diameter of the thread. Specific thread cutting procedures pertain to the particular thread to be cut, however, there are several general procedures. The tool's cutting edge is set at center height of the workpiece and at an angle which provides the type of cutting required. For example, in the cutting of a right-hand external *V*-thread, the compound rest is secured at an angle of 29° while the threading tool is prepared to move toward the headstock. Back gearing is engaged for slow r/min. The cutting tool's point, being at center height, must fit into the 60° notch of the center gage while the gage rests against the longitudinal axis of the workpiece. Depth of thread to be cut is performed by turning the knob of the compound rest, however, another procedure uses only the cross feed. Feed is selected according to number of threads per inch required, and the levers are adjusted accordingly. The knob of the compound rest is turned and brings the tool in contact with the work, and then the dial is zeroed. The tool is moved past the right-hand end of the workpiece and fed in by the compound rest about 0.003 inch. The thread chasing dial is then engaged according to the lathe's directions, causing the carriage to move to the left automatically. A trial thread results along the longitudinal axis of the workpiece.

After ascertaining the proper setup and correct pitch, the needed depth of cut is calculated by dividing 3/4 by the number of threads per inch to be cut. The compound rest screw is again turned inwards only a few thousandths of an inch so that the tool will move into the turning workpiece when the thread chasing dial is again engaged. A thread is then cut, the tool following the path of the previously cut thread. The thread chasing dial is relied upon to accurately chase the previous thread by a split nut action onto the longitudinal lead screw. As the tool reaches the marked left end of the workpiece, the cross feed dial is quickly turned to the left and the chasing dial is disengaged. Again, the tool is returned past the right end of the workpiece and the cross slide fed inwards to zero. Another depth of cut is provided by turning the knob of the compound rest the desired amount. Several cuts may be necessary to remove all but a few thousandths of an inch which is removed by turning the cross feed screw, causing polishing of the right side of the thread. Threading requires a cutting oil to assist in the cutting by lubrication. Other types of threads are machined according to specific procedures with a differently formed tool. Typical examples of other threads are the worm, Acme, and square.

When internal threads are required, a hole must be drilled into the workpiece. The hole provides the entryway for the threading tool which cuts in a similar manner to the boring tool. Threading is accomplished by dependence on the machine's dials and mechanisms, keeping in mind that the depths of cuts are provided in an opposite manner to outside threading procedures. Threading is done either by tapping or by a thread cutting machine such as a lathe. When depending on the tapping procedure, several limitations exist because of the unavailability of many tap sizes. Taps are manufactured according to standard sizes, whereas many internal threads have the same number of threads per inch, but the diameters of the threaded holes vary according to need. Indeed, the holes may be any diameter. Only specially produced taps can be used for specific diameters.

Threaded parts have clearances between the mating threads. These clearances are called *fits*, there being the loose, medium, and close fit for thread purposes. Most bolts and screws are machined with the medium fit or tolerance so no movement can be felt when parts are being tightened. The loose fit applies to commercial parts having no special tightness factor. A looseness can be felt while tightening parts. The close fit has its tolerances greatly reduced and a wrench is required for tightening. Machined threads are inspected with a thread gage.

Often; threads can be hand cut on a rod by means of the die and stock. Dies have thread sizes etched into one side. When the proper size rod is obtained, that is, the major diameter of the thread, the die is turned onto the rod with the assistance of the die stock. Care must be exercised to maintain 90° between the face of the die and the axis of the rod or the thread will be slanted, and the part will become scrap. A cutting oil is used as a half turn is made, then a slight reversal is made, followed by another half turn, and so on, until the thread is completed. The half turn and backup procedure help prevent tearing of the thread on some metals by the die.

Filing and Polishing

The filing operation in a lathe is similar to hand filing. The type of file is chosen and held by a handle at the opposite end with a slight pressure as the file is moved across the work, no coolant being used. Extreme care is used in positioning the file onto the revolving workpiece. Even though filing is a dangerous operation, it is sometimes used. The left hand grasps the handle while the right hand holds the end of the file. Small quantities of metal are removed as the file bites into the revolving workpiece by the action of the forward movement of the file.

Small amounts of metal, measured in thousandths of an inch, are removed by filing, the surface finish being regulated by the type of file used and the skill of filing. The 10-inch mill file produces excellent results as the work turns at a speed of several revolutions per file stroke. Sharp corners and burrs are also removed by filing. When cross filing is done, the file alternately is moved from left to right during the strokes so as to produce a more uniform surface appearance. Many filing operations require polishing in order to bring the surface to an acceptable appearance. With the use of a grade 90 emery cloth in the shape of a strip looped around the workpiece, pressure is applied as the piece rotates. Adjusting the pressure of the cloth against the workpiece results in the desired surface finish. Sometimes an oily lubricant is placed on the cloth to accelerate the polishing operation.

Grinding

An electrically driven grinding machine, the tool post grinder, is manufactured to fit onto the carriage of the lathe (fig. 8–29). Such a setup gives mobility to the grinder because the carriage movement is the same as with conventional cutting tools. Sharp angles found on lathe centers, clearances along the cutting edges of tools such as reamer and milling cutters, and straight and other taper grinding are made by the combination of grinding machine and carriage. As the workpiece rotates, the rapidly rotating grinding wheel is cautiously brought in contact with the workpiece. Dry or wet grinding can be accomplished.

Prior to grinding, a cloth must be laid over the lathe's ways to prevent grinding wheel dust from lodging in the precision parts. Also, great care must be exercised not to take too heavy a depth of cut or feed to eliminate the danger of local overheating of heat treated parts. Lathe centers, cutting edges of carbon, or low alloy steel cutters must not be heated beyond 430° F., indicated by a pale straw color. Temperatures higher than this result in immediate softening of the surface, which subsequently

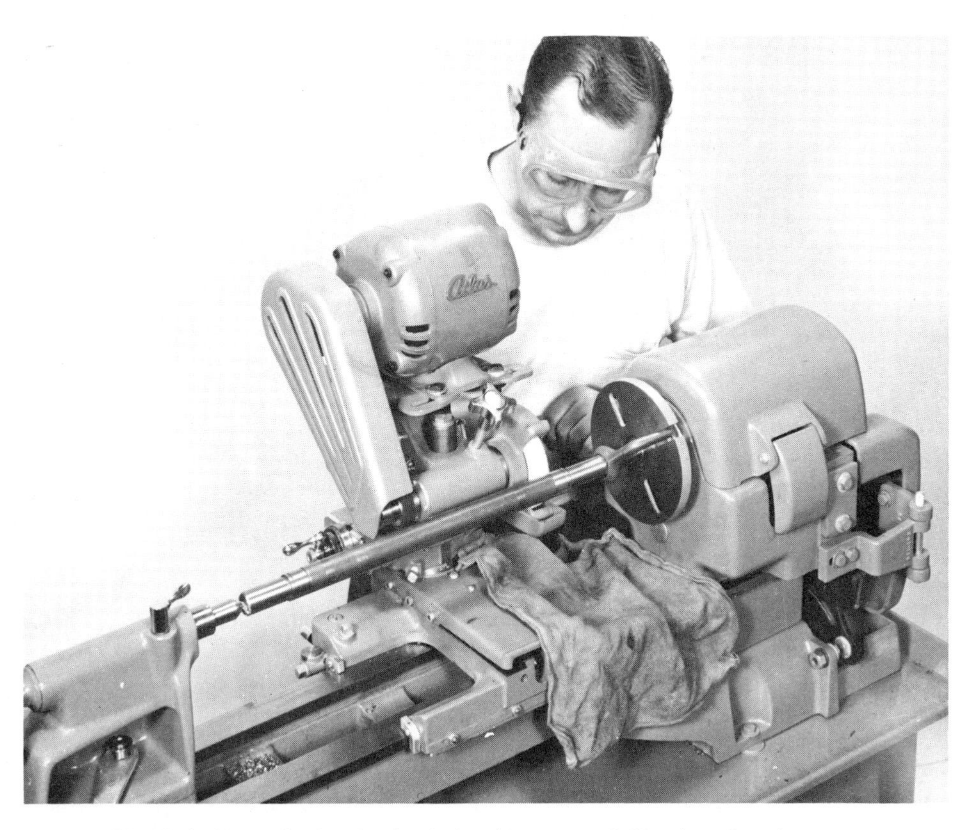

FIGURE 8–29 Grinding in the lathe. (Courtesy of Clausing Corp.)

reduces the part to scrap. On the other hand, high speed steel cutters will withstand temperatures up to 1000° F., but such temperatures are rarely needed. All kinds of small grinding operations can be performed with the tool post grinder.

Drilling

Drilling in the lathe is a common operation. Drilling is necessary to provide a working area for boring, for internal threading, for a means of reaming, for tapping, and for holes needed as finished holes. Cutting speeds for drilling in the lathe are similar to those used in the drilling machine because the relative speeds are the same, that is, stationary drill and revolving work in the lathe or stationary work and revolving drill in the drilling machine. Also, coolants are used, usually an oil. Often, the facing operation precedes drilling. In order to assure proper centering of the drill, the center drill and countersink is used, unless only a small twist drill is required. Normally, drills are held in the tailstock while the rotating work is held in a chuck or spindle. When twist drills up to 1/2 inch are required, the chuck is used in the tailstock. Larger drill shanks are tapered and fit into the lathe's tailstock spindle. Sometimes, however, the work may be held in a special pad by the tailstock while the chuck holds the rotating drill. Figure 8–14 illustrates drilling in the lathe.

The drill and countersink, or center drill, is available in several sizes and is used to provide a tapered hole for holding one end of a part such as a reamer workpiece while it is being machined. In this respect, the lathe, mill, and grinder are equipped with centers which fit into the tapered holes of the spindles whereby proper alignment and holding power exist. Also, the drill and countersink assures accurate centering of the initial hole and guidance during subsequent drilling. Its sturdy construction assures alignment as drilling proceeds. Drilling stops when half the tapered surface has penetrated the hole or when the desired diameter is attained.

Twist drills are used in all sizes during the many different machining operations. Usually, a large drill is preceded with a smaller drill to help assure concentricity of the hole along the longitudinal axis. Hand feeding of the part or drill is performed by turning the tailstock spindle handle. Care must be exercised, however, when using long and thin drills due to the tendency of the drill point to wander off the longitudinal axis. An example of drilling in the lathe is the drilling of the hole in a gear blank.

Boring

Boring in the lathe follows most of the general rules of turning. The main difference is the internal operation requiring a special boring tool. Being internal, the cutting tool is brought into the wall of the workpiece by a reversal of direction of the cross slide. It must be kept in mind that boring can be performed only when a previously drilled or cored hole exists. The hole must be large enough in diameter to enable the cutting tool's edge to enter while, at the same time, there must be allowances for clearances at the back side of the tool. The specially designed tool is fed into the hole using similar turning speeds and feeds as in straight turning (fig. 8–30). The main difference

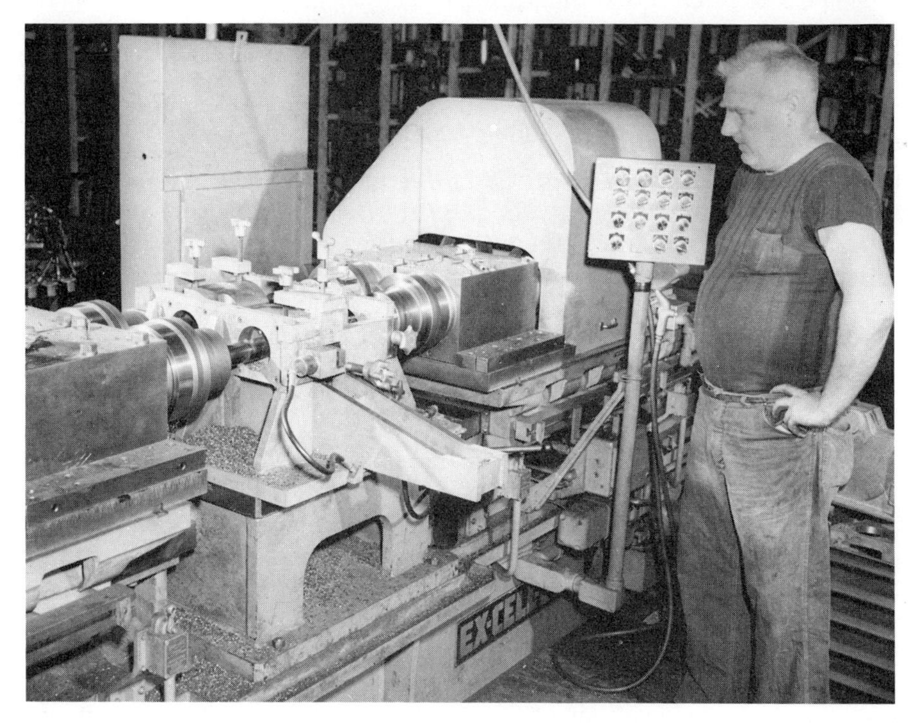

FIGURE 8–30 A boring operation being performed in a special type of boring machine. (Courtesy of Clausing Corp.)

between boring and turning is the possible reduction in feeding and depth of cut due to the long overhang of the cutting tool. As the cutting edge of the tool increases in distance from its holder, the tendency to spring away from the work also increases, therefore, boring has a depth limitation. Tool spring away from the work leaves a bell-mouth hole. Cutting on the way in and allowing cutting on the way out with no increase in depth of cut helps to eliminate the taper potential. An advantage of boring is the ability to machine concentric holes, as boring is independent of the drilled hole's walls. Sometimes, the drill, if exceptionally long, wanders from the horizontal axis; boring the hole will remove the error. An example of boring is the boring to size of the drilled hole in a gear blank. Boring tapered holes such as the inside taper of a drill sleeve is also performed in the lathe.

Reaming

The purpose of reaming is to provide a smooth hole having precise dimensions. Reaming follows drilling because the reamer will remove only a small amount of metal, possibly from 0.002–0.030-inch depths. Reamers are turned into the drilled hole resulting in a thin ribbon of metal being removed. Holes may be straight or tapered; reamers are manufactured to finish the sides of both straight and tapered holes, including the several types of tapers. The reamer may be straight fluted or helically fluted, while the number of flutes varies according to the size and purpose of the reamer. Reamers are available in standard sizes similar to the fraction size drills. Also, reamers are produced in special sizes to fit the growing needs of industry. (Hand reaming has been previously discussed). When reaming in the lathe, the reamer is held in the tailstock spindle, either in the chuck or in the tapered spindle, and hand fed into the revolving hole.

Milling Attachment

The milling attachment gives the lathe a nearly universal usage. The attachment fits on the lathe's carriage, while the workpiece is held in a special vise which provides vertical feed of the workpiece. The milling cutter is mounted in the lathe's spindle. With proper manipulation of both the vertical feed of the attachment and the cross feed of the carriage along with its longitudinal capability, many milling operations can be accomplished. Machine attachments such as the milling attachment are not used in production operations, but some small shops do use these additional facilities of the machines.

Automatic Screw Machines

Due to the continuous need for faster producing machines, especially those which could produce threaded hardware such as screws, the automatic screw machine was developed. The principle is simple. A bar of metal is fed into the machine, and completed hardware or parts result in a fast and automatic series of production cycles. Modern machines produce numerous types of parts other than threaded. These machines are equipped with a sequential response controlling mechanism which causes planned machining on the bar stock as it is automatically fed into the machine.

Several types of screw machines are available such as the automatic chucking and the single- and multiple-spindle types. The single-spindle machine rapidly processes the bar stock into finished parts as the stock moves through the spindle. Each of the main types of machines is further divided into several specialized types, depending on the need for differently shaped parts. One brand of machine functions with an automatic turret which houses different tools. Automatically, each tool moves into position sequentially, resulting in identically shaped parts being produced within small tolerances. A series of cams and chucks provide positioning and movement of the stock. Multiple-spindle machines decrease the production time of finished parts by the spindle's movement around a turret. The automatic chucking machines can perform most of the work of the multiple-spindle types and employ automatic chucks on their spindles for holding the forging or casting while machining occurs. These machines do not process bar stock in the same manner as the multiple-spindle machines.

While the lathe is used to produce parts singularly by hand or by automatic processes, the screw machines increase production capability many times. But the cost of these machines is high, and their use is only justifiable when many thousands of identical parts are to be produced. When 100,000 or more parts are needed, then this particular type of machine should be placed in operation. An analysis of the part's design dictates the type of machine needed. Such things as length-to-diameter ratios, tapers, threads, grooves, shoulders, and special machining such as knurling all require study in order to procure the best type of machine for the job to be done. As as example, a single-spindle machine can chuck up to eight-inch diameter work for very rapid processing, whereas some of the multiple-spindle machines easily handle 12-inch stock. Basically, if a part can be made in the lathe, then it can be made in the screw machine. All types of work such as collars, spindles, shafting, threaded parts, and most round shapes, both small and large, are processed in these machines. Tolerances are normally established as plus or minus 0.002 inch for one-inch parts and plus or minus 0.005 inch for parts greater than two inches in diameter.

Questions

1. During lathe cutting operations, how does control over the shear plane control the formation of the chip?
2. Describe three ways to produce a taper on a lathe.
3. What is the relationship between the cutting speeds in the lathe and in the drilling machine?
4. What is the difference between reaming and boring in the lathe?
5. During facing in the lathe, why must the tool bit be placed at center height?
6. What factors must be present to prevent a horizontal band saw blade from cutting a wavy or slanted edge?
7. What forms of saw teeth are predominantly used in metal cutting saws?
8. Why must a minimum of two saw teeth always be in contact with the workpiece as sawing progresses?
9. Compare the advantages of the vertical and horizontal band saws.

10. What is meant by pitch of a tooth?
11. Why do twist drills larger than one-half inch usually have tapered shanks?
12. List several advantages of the radial drilling machines over other types of drilling machines.
13. What is the relationship between cutting speeds of the drilling machines and the lathes?
14. How can a seven-inch hole be produced in a steel plate with the use of the drilling machine?
15. Describe the principle of the automatic screw machine.
16. Name three main types of screw machines and indicate an advantage of each.
17. When should a general type of machine shop consider the purchase of a screw machine?
18. Name four kinds of threads which can be machined in the lathe.
19. How does the feeding rate influence the metal's surface while turning in the lathe?
20. Why are clearances required along the edges of cutting tools?

Special Operations and Automation

Machined parts which have been produced on the lathe are often processed further by means of milling and grinding. The reamer which was cited in the previous chapter must be milled after the lathe operations to provide flutes and cutting edges. The lathe provided the overall dimensions by facing and turning operations, but the milling machine is required to provide the flutes. Then, it is the grinder that establishes the final diameter and places cutting edges along the flutes. The milling machine is capable of producing flat and angular surfaces, as well as formed surfaces such as splines and teeth. Precision grinders provide flat or cylindrical work and are also used in numerous cutter sharpening operations.

Other machines such as the shaper and planer are also capable of providing flat and angular surfaces, as well as grooves, but the grooves must be in straight lines due to the reciprocation of the cutting tool or table. Because milling cutters turn, the milling machine can perform the operations of the shaper and planer, but the shaper and planer cannot accomplish all the work which the mill is capable of doing.

When various shaped holes are required in a part, the *broach* is often used. Basically, the broach is a series of single cutting teeth along a straight-line axis of the tool's length. The last group of teeth shape the resulting hole. However, a basic requirement for use of the broach is that a hole be present so the leading part of the cutting tool can be inserted through the hole. Nearly any shaped hole can be produced with this tool as long as the axis of the workpiece continues in a straight line and no change in shape of the hole occurs.

Machine shops vary in their size and purpose. As the shop grows and its output substantially increases, consideration is frequently given to mechanization and automation. These processes are actually exchanges of precision machines for skilled operators. As the built-in skill in the machine increases, skills of operators may decrease. When mechanization requires no human assistance, the process is automated.

Milling

Even though the lathe is a universal type of cutting machine, it cannot perform all the essential operations in industry. Basically, milling machines are capable of machining flat and angular work, form work such as gear teeth and grooves, and operations involving slotting. The cutter is a multiple-tooth cutter, as opposed to the single-point cutter of the lathe and the two-point cutter of the twist drill. A milling cutter may have 15 to 20 cutting teeth around its periphery, each cutter removing a small amount of metal which results in many small chips and a large quantity of metal removal. Milling is a procedure which produces a part in an exact shape and dimension as a result of the material being moved progressively into a rotating cutter. In effect, the milling machine's spindle, like that of the drilling machine, turns the cutter, whereas the lathe's spindle turns the work. The lathe usually has two axes; the milling machine has three, the third providing the vertical capability. A horizontal milling machine is shown in figure 9–1 which has performed a conventional up-milling operation.

FIGURE 9–1 A horizontal type milling machine showing a part that has been up-milled. (Courtesy of Cincinnati Milacron)

TYPES OF MILLING

Milling is a material reduction process that involves a relationship between moving work and a rotating cutter. The relationship has two possibilities, that is, the cutter can rotate against the incoming workpiece as in *up milling*, or the cutter can rotate in the same direction as the feed as in *down milling*. Up milling is the conventional method of milling (fig. 9–1). The cutter tends to force the workpiece upwards due to the cutter's movement. A chip is removed whereby the first dimension is larger than zero and the last dimension is the largest, thus, maximum chip size is produced as the cutter leaves the workpiece along with maximum stress in the workpiece. This action tends to lift the workpiece from the table, therefore, parts must be securely tightened in the holding device. Up milling is used on most of the common metals.

Down milling (fig. 9–2) allows the cutter to begin with a large chip and end with a

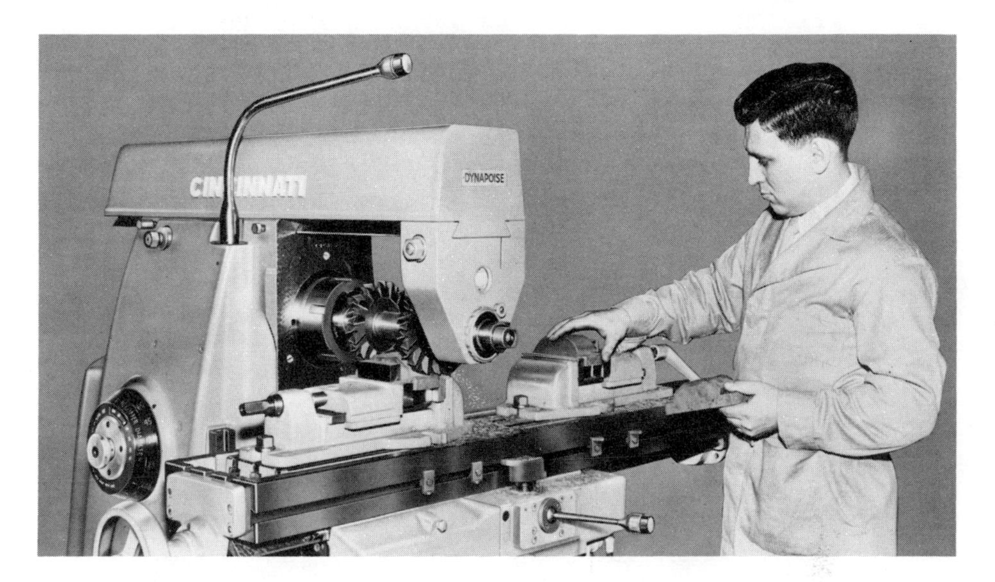

FIGURE 9–2 Down milling in a milling machine. (Courtesy of Cincinnati Milacron)

chip approaching zero dimension. Such a cutting action forces the workpiece into the holding fixture while, at the same time, the work is being pulled into the cutter, but is held back by the feeding mechanism. Down milling is also called *climb milling* due to the cutter's attempt to climb onto the workpiece. Again, the workpiece must be securely clamped in the holding device to avoid shifting of the workpiece and subsequent damage to the machine such as springing of the arbor. In both types of milling, cutters must be sharp, r/min must be appropriate to the task, feeds and depths of cut must be carefully calculated, and the whole setup must be reviewed prior to the milling operation. Again, r/min is derived from appropriate cutting speeds (table7–1).

Milling operations are also classified into the types of work performed. Two main classes are distinguishable, *face milling* and *peripheral milling*. Work produced by face milling processes is flat, because the rotating cutter removes material by both its side and periphery teeth as it moves across the part's surface. Face milling removes metal or other material which is parallel with the cutter's face and 90° to the cutter's axis. End mill cutters are typical examples of face milling cutters as they cut on the bottom and side surfaces (fig. 9–3 and fig. 9–4). Both horizontal and vertical milling

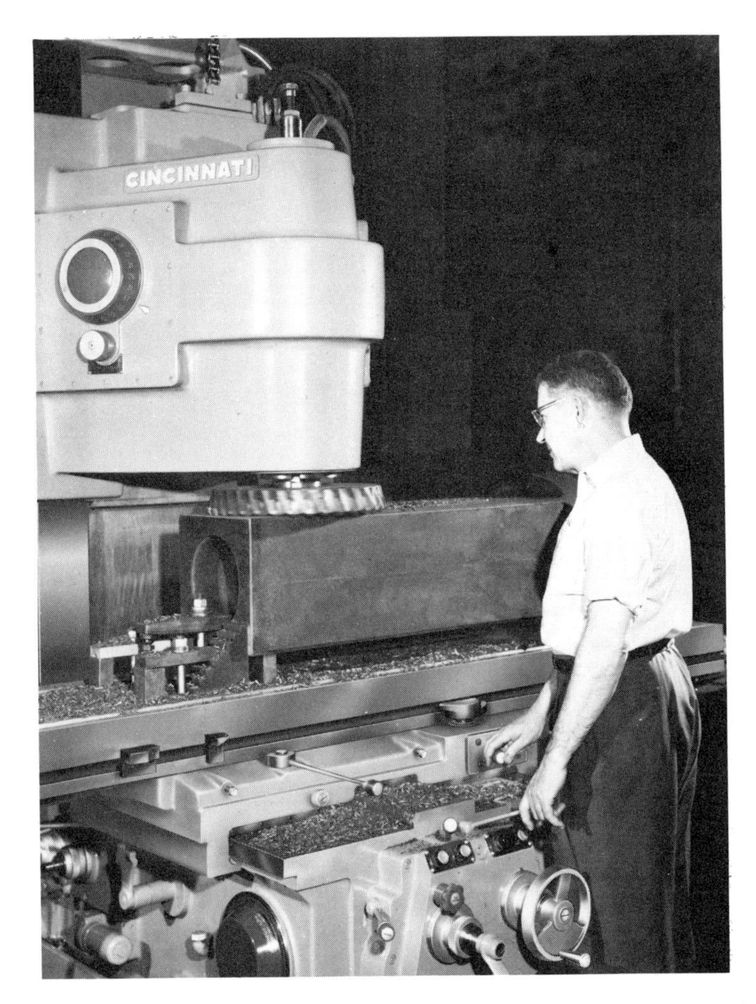

FIGURE 9-3 A face milling operation. (Courtesy of Cincinnati Milacron)

machines will perform face milling. Peripheral milling is commonly known as *plain milling* due to the part's surface being parallel to the cutter's axis. Peripheral cutters cut on their outer edges (fig. 9-5). Flat and formed surfaces such as the sliding ways of a machine or grooves in a plate are produced according to the shape of the cutter (fig. 9-6). When the horizontal type of milling machine is used for any operation, cutters can be mounted singularly or in gangs (fig. 9-6).

MILLING MACHINES

Due to the complexity of the milling machine's operation, many specific types are manufactured. However, the *column and knee type* is the most popular, versatile, and commonly used. The other type is simply the *special type* due to its deviations from the column and knee type. Many bed type, planer type, and other production types fit into the special type of milling machine category. The column and knee type is

FIGURE 9–4 An end mill machining operation. (Courtesy of Cleveland Twist Drill Co.)

manufactured in the plain, universal, and vertical models. Each machine has a sturdily constructed column which provides structural support and mechanical assistance to the powerful knee. As shown in figure 9-7, the table rests on the knee.

Plain Type

The plain type milling machine (fig. 9–1) is a horizontal mill having its spindle horizontal and at 90° to the column. The spindle is internally tapered to hold cutter arbors. Plain mills have three axes, that is, the table and workpiece move horizontally a given amount, crosswise a given amount, and vertically a given amount. The combination of the X–Y–Z axes provides three-directional milling within the capability of the machine. Amount of movement along an axis is measured in inches, the inches movement and total measurement capacity relating to the size of the mill, indicated by a number from one to six. Many machines are built which differ from

FIGURE 9-5 A peripheral milling operation. (Courtesy of Cincinnati Milacron)

this general numbering system. The longitudinal table travel relates to the size of the machine more than other data. Plain mills are both hand and power operated, being driven by electric motors slightly larger than those used in the lathes. Possibly more sophistication exists in the milling machine than in the lathe. Examples of special parts in the milling machine are automatic trip dogs, a *T*-slotted table, three axes table movement, and an arbor which can rotate numerous types and shapes of cutters.

Universal Type

The universal type milling machine has all the arrangements and facilities of the plain mill, but has a swivel table capability which enables a helix to be machined. Tables on plain mills do not swivel; the tables of universal mills are mounted on a block which provides swiveling at the center of the saddle. Plain mills cannot machine a helix along the horizontal axis without special attachments. While universal machines cost up to 25% more than the plain mills, the increased capability is well worth the difference in cost if universal milling is used in the shop. Many tools and parts have helices along their lengths such as reamers, twist drills, and cutters, and the helical

FIGURE 9–6 Angular milling cutters arranged in gangs to accomplish special surface contours. (Courtesy of Cleveland Twist Drill Co.)

gear is a standard gear used in machinery. Universal machines are sometimes more heavily constructed than the plain mills

A helix is machined along the longitudinal axis of a reamer when the dividing head is connected with a gear train to the machine's table lead screw. A dividing head is a device that arranges for the division of round objects into any number of equal divisions within the limits of the device. By swiveling the table a few degrees, the cutter is maintained in a normal relationship with the slowly turning workpiece. As the table moves the workpiece into the turning cutter, proper depth of cut having been provided, the gear train provides turning motion to the chuck of the dividing head and workpiece. The ratio of longitudinal travel of the table to the turning workpiece is provided in the gear train ratio between the driver and driven gears. In other words, as the table moves the workpiece into the rotating cutter, the gear train which connects the table with the dividing head causes the workpiece to slowly turn, or rotate, as it moves into the cutter. The result is a helix in the surface of the workpiece. An odd number of gears provides the same directional rotation of the driver and driven gears. Other tools such as some taps and milling cutters, twist drills, and some gears have helices. The helix provides a smoother turning motion both for cutters and gearing.

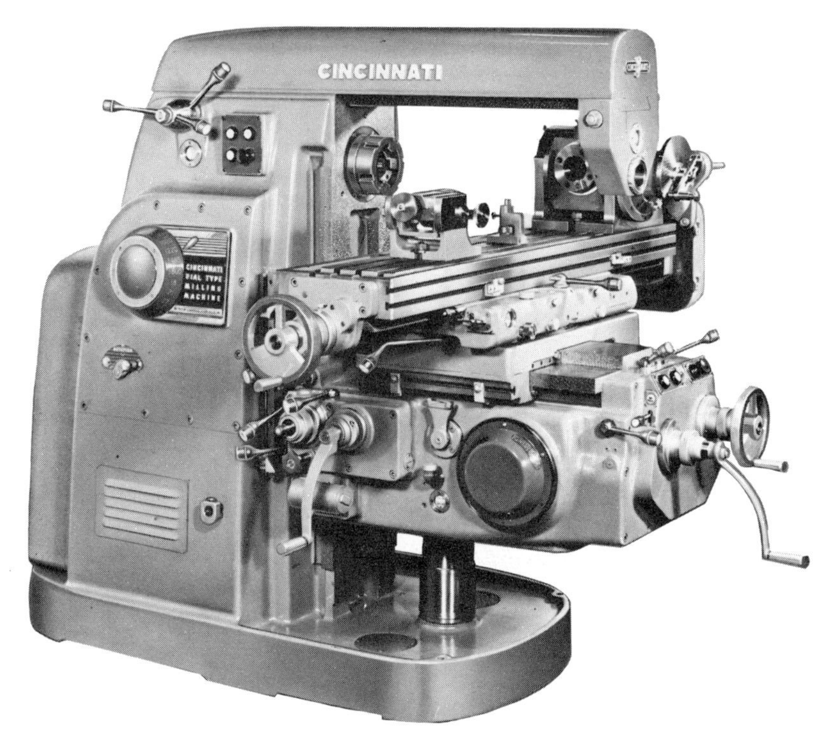

FIGURE 9–7 A horizontal universal milling machine. (Courtesy of Cincinnati Milacron)

Vertical Type

The vertical milling machine has characteristics of the drilling machine, but has additional capabilities (fig. 9–8). The spindle is vertical and is mounted in the head which provides up and down movement. Vertical mills have the same general capabilities as the plain mill with regard to the table movements and control functions. Because of the exceptionally high torque existing during milling operations, the column and spindle head are produced in more massive designs than in the usual types of drilling machines. Vertical mills can perform drilling operations in addition to their primary function of milling. The vertical mill has its place in industry, as does the universal horizontal and the plain mill, but the vertical mill uses different types of milling cutters. Vertical and horizontal mills are capable of performing some of the same milling operations such as the milling of a flat surface. Precision drilling and boring are exceptional qualities of the vertical mill, along with its end milling capabilities. Some manufacturers build the vertical mill with the tilting head for the purpose of angular milling operations, in addition to conventional milling.

Special Type

Special type milling machines vary according to need. As an example, the *bed type mill* is rugged and is production oriented for heavy and continuous milling operations. Bed mills have longitudinal movement of the table and vertical movement of the spindle. Spindles are available in one or more units for the purpose of increased metal removal capability. Bed mills are available in several models (fig 9–9), some

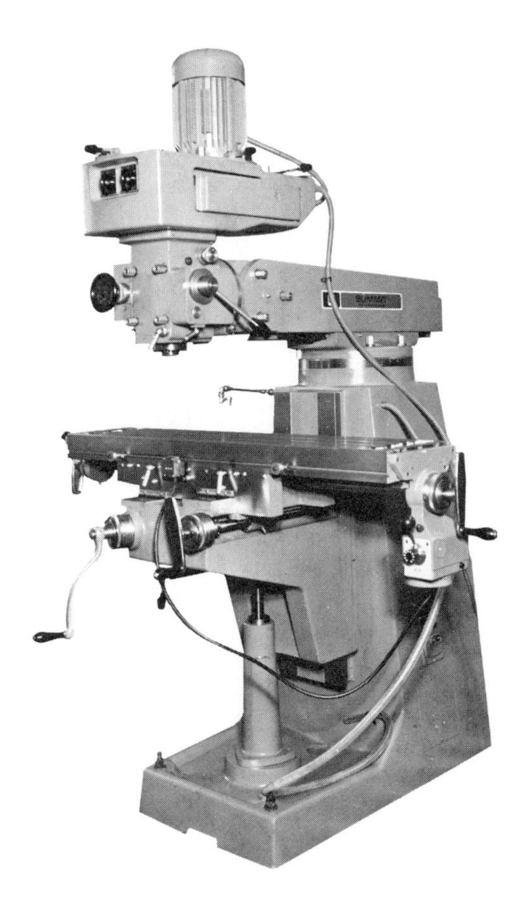

FIGURE 9–8 A vertical milling machine. (Courtesy of Summit Machine Tool Mfg. Corp.)

incorporating the advantages of column and knee, and transverse motion, along with massive rigidity. This type of mill is also called the manufacturing mill. The workpiece is clamped in a fixture. Once the setup is completed, the operation becomes automatic.

Planer type mills, another special type, provide milling operations similar to operations of the planer in that the work-holding bed reciprocates beneath the rotating end mills of vertical heads (fig. 9–10). When equipped with very powerful motors, and when designed to withstand high stresses, the planer mill will remove large quantities of metal in a short period of time, often in one pass across part of the surface of the metal being milled.

Other special mills include the *profile* which is used to duplicate parts. Profiling mills often have more than one spindle and are capable of both internal and external machining. Locally produced templates act as guides during the profiling operation. The *duplicating mill* is used for three-dimensional duplication such as in the die making industries. Profiling operations cut in two dimensions; duplicating mills function in the three dimensions. Additional special production type milling machines are the *rail type* for milling long and flat surfaces, the *automatic rise-and-fall miller* which allows the cutter vertical movement to skip across obstructions along the

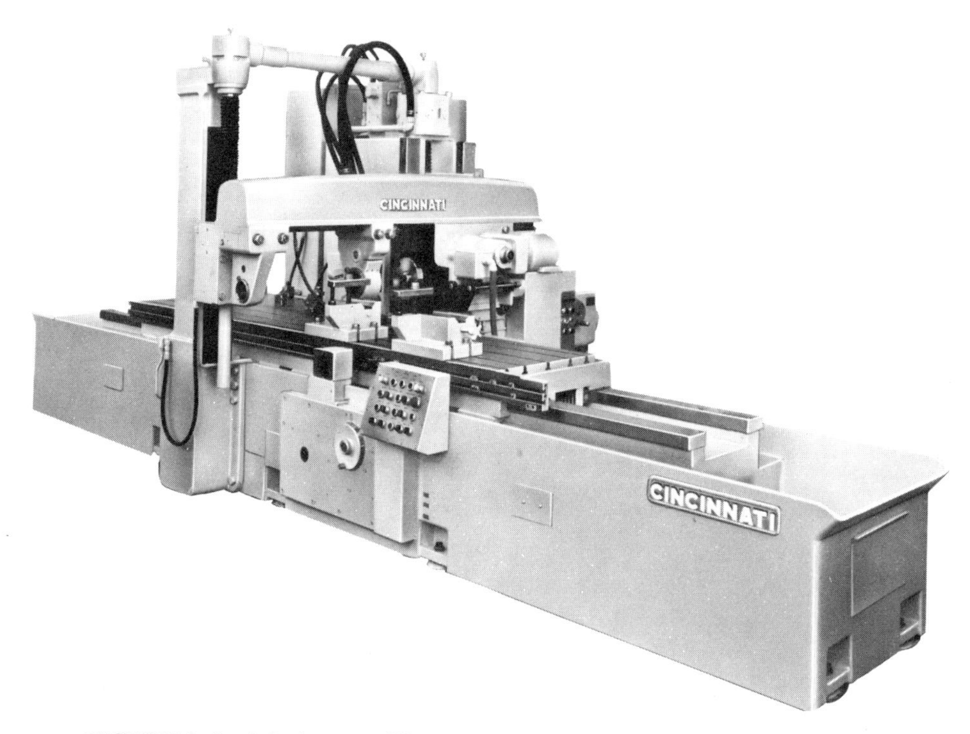

FIGURE 9–9 A bed type milling machine. (Courtesy of Cincinnati Milacron)

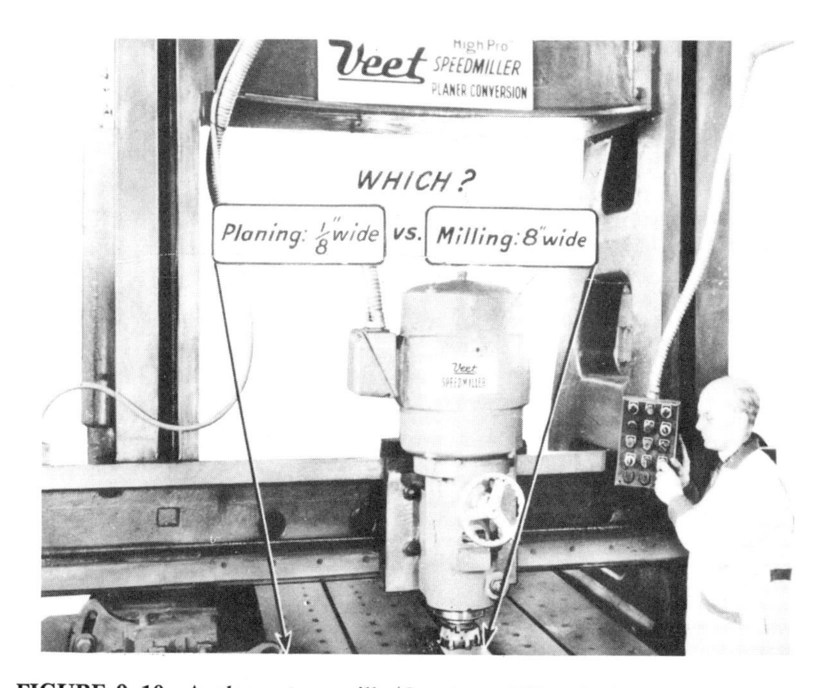

FIGURE 9–10 A planer type mill. (Courtesy of Veet Industries)

line of milling, and *rotary millers* for both slab work and contour machining. Basically, the machine can be designed with parts to accomplish the needed machine work, for example, gangs of cutters properly spaced for profiling or the machining of splines in a special mill.

Milling Machine Attachments

Most milling machines are equipped to handle some types of attachments in order to make the machine operational or more versatile. All milling machines have their tables equipped with *T*-slots for the purpose of securing attachments, accessories, and fixtures. The basic machine without some attachment is incapable of machining operations; at least a vise is necessary to hold the workpiece. Several types of vises are available, in addition to numerous types of special fixtures. When the proper cutter is placed in its proper place on the spindle, machining commences when the cutter engages the workpiece which is held on the table by some type of holding device.

Vertical Attachment When the vertical milling attachment is installed on the horizontal mill, vertical milling can be performed. Usually, attachments do not have the massive structural designs as do other similar parts of the machine. For production type work, changes in functions of the basic machines such as the use of the vertical attachment are not normally made. Decisions must be made prior to purchasing equipment as to the kind of work to be accomplished in the shop. General machine work often justifies the vertical attachment for the horizontal mill. However, if a reasonable quantity of vertical milling is to be done, consideration should be given to the vertical mill and the universal mill. Two different machines provide constant work capability, while changeovers require loss of production time. One changeover is actually two, because of the need to convert back.

Slotting Attachment Slotting attachments change circular motion into reciprocating motion, giving the mill an additional capability which is unlike other milling operations. When installed on the horizontal mill, the cutting tool reciprocates inside a hole, cutting a predetermined slot such as a keyway, spline, or shaping the lands on a spark plug wrench. The tool is similar to the shaper tool or lathe tool, except the cutting edge at its end is shaped exactly as the required shape of the slot. The tool produces a chip as it pushes through the metal at a specified depth and width. Several entries into the hole or along an external surface are required to produce the desired size.

Universal Attachment The universal milling attachment is installed on plain mills to give them universal capability. Angular milling can be accomplished such as the machining of helices. Again, consideration should be given to the need of this attachment before final purchase of the machine. When the cost of this attachment is added to the plain mill's cost, possibly the universal mill may be the better buy.

Dividing Head When the part to be produced requires machined divisions of equal measurement around its surface, the dividing head is required. This device is precision made and complex in operation. When the workpiece is held in the dividing head, being supported on the other end by a footstock, divisions around the circumference or along the part's length, such as a helix or helices, can be machined. An example is the straight-fluted reamer or the tap. When the tap has been turned to its proper length and outside diameter on the lathe, including the proper threads, subsequent machining operations include squaring the end and fluting. When flutes are needed, then the chuck of the dividing head holds one end of the partially finished tap while the cutter cuts one flute. If four flutes are required, then a 90° turn of the part provides the proper dimension for flute number two. Two additional 90° turns allow four flutes to be machined by the tap-fluting cutter. Each flute is exactly divided around the circumference of the part.

The purpose of the dividing head is to divide a dimension into any number of equal parts so that subsequent machining can be performed. As shown in figure 9–11, the

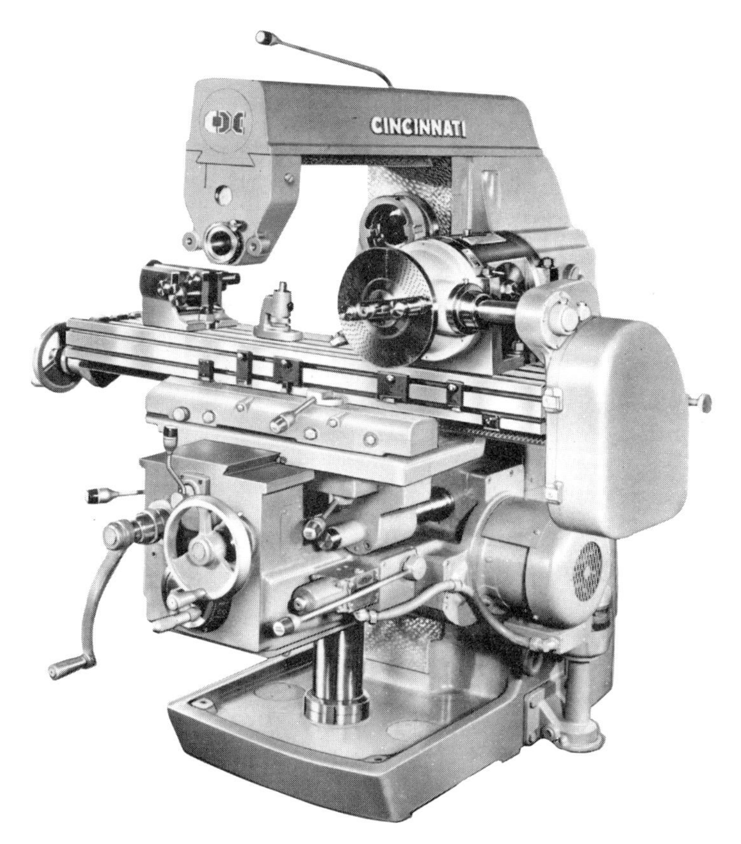

FIGURE 9–11 A typical dividing head mounted for a milling operation. (Courtesy of Cincinnati Milacron)

left side of the head contains the spindle screw on which can be mounted the face plate and center or the chuck. In this setup, the opposite end of the workpiece requires support, therefore, a footstock and center are provided. Both the footstock and dividing head are securely attached to the milling machine's table (fig. 9–11).

The right side of the dividing head contains the drive mechanism which causes rotation of the workpiece. The mechanism can be driven by the power of the table's lead screw in helical milling, or by the hand operated crank in the front of the head if division of the workpiece must be performed prior to milling. In other words, the dividing head not only holds a workpiece while it is being milled, but it also has division of surface capability, and can even rotate a workpiece concurrently with machining, as in helical milling.

The capability of the dividing head results from its construction. A gear ratio provides one revolution of the spindle when the crank is turned 40 revolutions, this being a worm gear reduction. Off centering of the spindle is also practical due to the tilt ability of the spindle. The 40:1 ratio then indicates that the desired number of divisions is a fraction of 40. For example, if 20 divisions, or teeth, in a spur gear are needed, then 1/20 of 40 equals the number of crank turns for each tooth, or two turns per tooth. Then, the cutter is allowed to pass through the gear blank and returned. Again, two turns of the crank are accomplished, and the cutter passes through the blank causing the first tooth to be completed. When the second tooth is finished, the pitch indicates that 20 teeth will be cut and all teeth will be equally spaced around the circumference of the gear blank. When divisions such as 18 are required, then the fraction becomes 1/18 x 40, or 2 2/9, turns of the crank per division. The 2/9 fraction of a turn is made possible by the index plates attached behind the crank.

Index Plates There are three basic indexing systems, one being the *direct plate* which is mounted on the spindle's nose for direct indexing of divisions within the capability of the plate. The other division capability, or *plain index plate*, is mounted behind the crank. Each plate contains a system of holes which indicates the number of equal divisions in a given circle. Different manufacturers arrange the concentric series of holes around the face of each of several plates in different divisions. That is, one manufacturer will use a system of holes whereby each circle has one more hole than the inner circle. Several plates are required to provide for the large number of possible divisions. Another manufacturer will provide a circle of holes having a still different spacing with holes being on both sides of the plates. If several plates are available within a given system, there is a good chance that the remaining fraction of the turn can be made by finding the circle that has the number of holes needed to exactly complete the turn.

Caution must be used in indexing to remove backlash in the system and to not overshoot the last hole with the crank's pin. The pin should gradually slide up to the hole and then drop into the hole. When backlash has been removed, the required number of divisions will be evenly spaced. Achieving the 2/9 of a turn in the previous example is a matter of attaching the proper plate to the head prior to milling. Any circle divisible by nine will then provide the fraction of the turn. The 18-hole circle, if used for 18 divisions, will require two complete turns of the crank plus four holes. More complex indexing requires a differential system which also includes a system of gears in conjunction with the plates.

MILLING CUTTERS

A milling cutter is a multiple-tooth cutting tool functioning similarily to the single-point cutter, except many chips are removed in one revolution of the cutter. Milling cutters are made of the same material as other cutters, being predominantly carbide or ceramic-tipped steel, high speed steel, or a cast nonferrous alloy. Cutters are formed to have a positive radial rake along with ample clearances for effective cutting and chip removal.

Some of the common cutters are shown in figure 9–12. Milling cutters are so numerous in function and shape that several classifications exist. However, if they fit a stan-

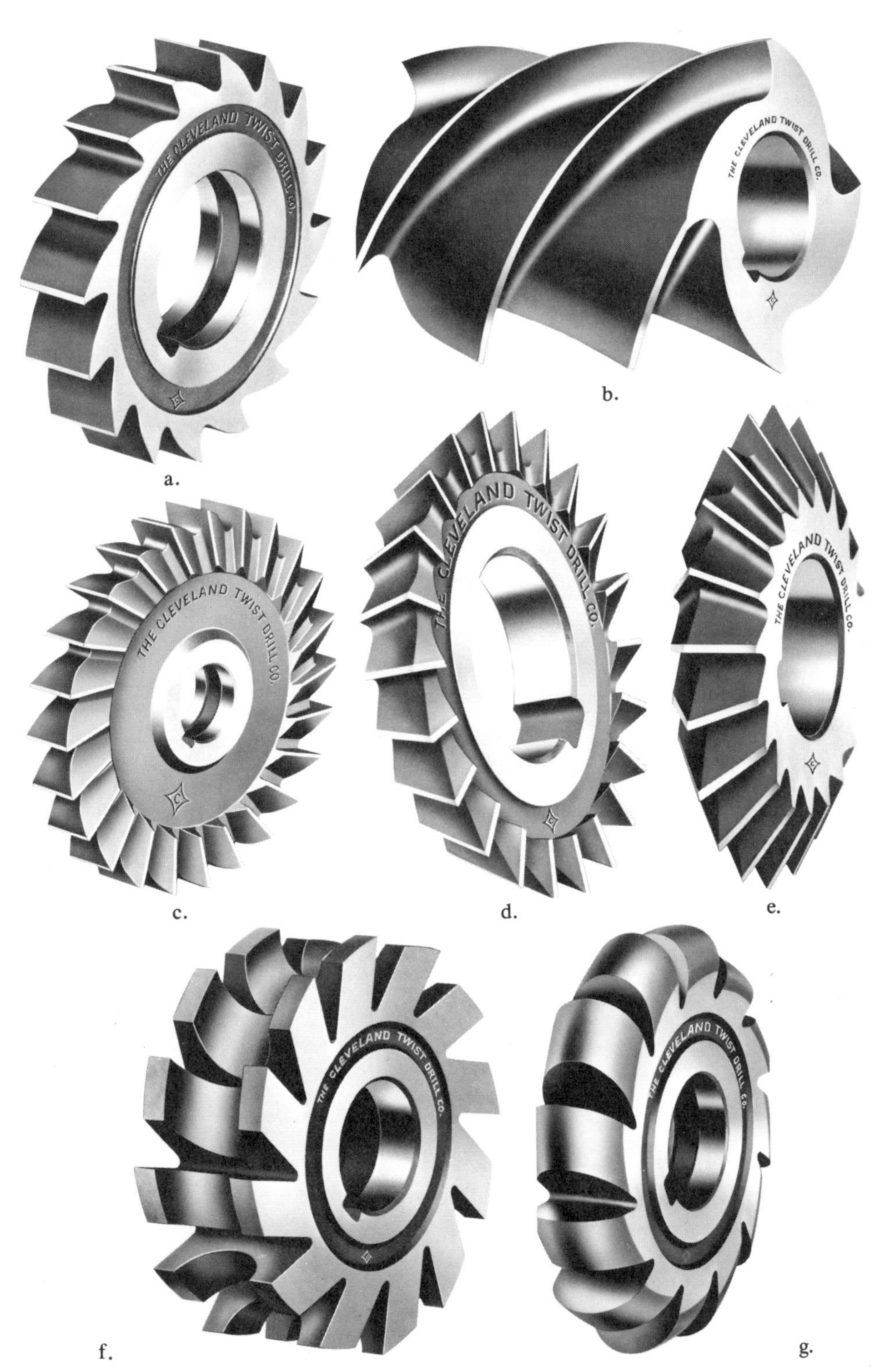
THE CLEVELAND TWIST DRILL CO.
THE CLEVELAND TWIST DRILL CO.
THE CLEVELAND TWIST DRILL CO.
THE CLEVELAND TWIST DRILL CO.
THE CLEVELAND TWIST DRILL CO.
THE CLEVELAND TWIST DRILL CO.
THE CLEVELAND TWIST DRILL CO.
a.
b.
c.
d.
e.
f.
g.

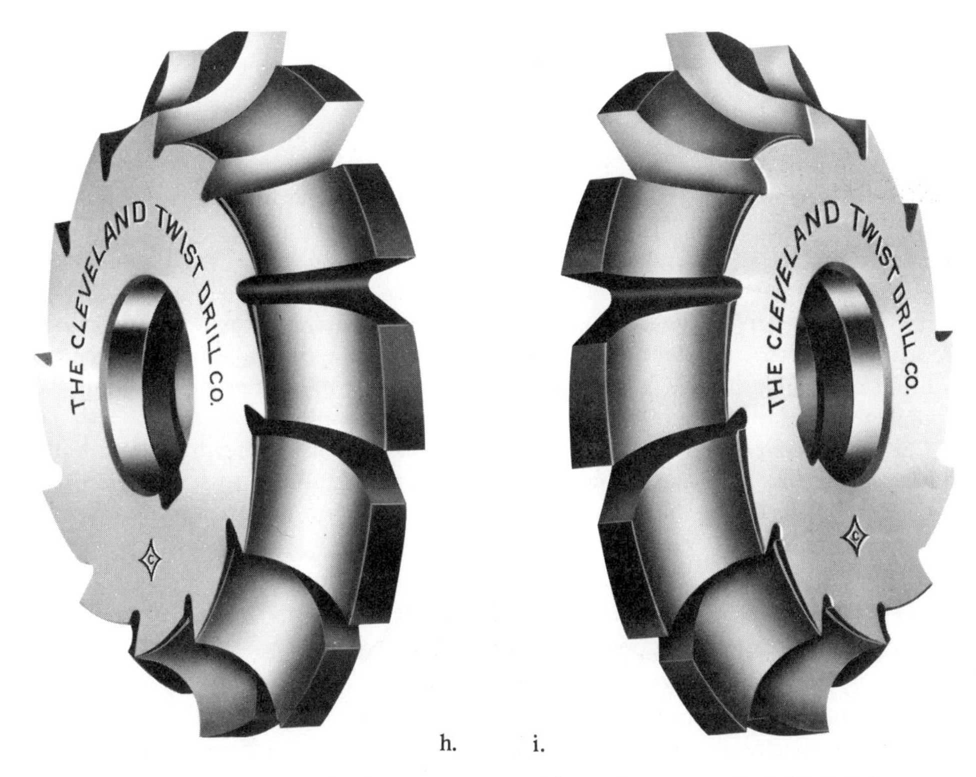

FIGURE 9–12 Typical milling cutters: *a*, plain milling cutter; *b*, helical plain milling cutter; *c*, side milling cutter; *d*, right-hand single-angle milling cutter; *e*, double-angle milling cutter; *f*, concave milling cutter; *g*, convex milling cutter; *h*, left-hand corner rounding milling cutter; *i*, right-hand corner rounding milling cutter. (Courtesy of Cleveland Twist Drill Co.)

dard measurement system to include such measurements as outside diameter, hole diameter, and width, they are known as *standard cutters*. All other especially made cutters are *special*. According to purpose, cutters have several names. Common cutters include form, plain, angular, and face.

The *form cutter* mills an irregularly shaped profile such as a round corner, gear tooth, tap and reamer flute, and even the fluting cutter which is to subsequently do the cutting. The *plain cutter* has either helical or straight cutting teeth on the circumference and is used to cut flat surfaces parallel to the axis of rotation. The width of teeth may be narrow and straight with many teeth such as those used for light duty milling operations. Large *slab cutters* usually have helical teeth and remove large quantities of metal. When viewed through the cutter's hole, the right-hand helix curves in a clockwise direction, which means that it cuts to the left and is a *left-hand cutter*. The *side milling cutter* cuts along the sides as well as the periphery. Many types of these cutters are available, and they are used for several types of milling operations. *Arbor cutters* may be reversed on the arbor to reverse their cutting directions. The *angular cutter* mills a V-groove, serrations, dovetails, and performs operations like notching. The *angular cutter* cuts angular surfaces at different angles to the cutter's axis of rotation. The *face mill* is used for milling flat surfaces which are 90° to the axis of the cutter. The face mill is similar to the end mill but may be larger. Many of these cutters have inserted teeth. *End mills* cut on the ends and the sides such as in grooving. The two-flute end mill will cut its own hole and then extend the round hole into a groove, for example.

Other cutters include *T*-slot, dovetail, slitting, keyseat, and fly, although there are many more special cutters. The *T-slot* cutter machines the table surface of a mill with a series of *T*-slots which hold *T*-bolts. *Dovetails* allow evenly sliding parts which have been mated to each other. *Slitting saws* perform parting operations and grooving. *Keyseat cutters* perform milling operations such as semicircular keyseats whereby a key can be inserted in the slot for purposes of locking one part to another. A *fly cutter* will cut a circular hole in a steel plate, the radius of the cutter determining the diameter of the machined hole.

Cutters are also classified as to arbor and shank types. *Arbor cutters* have holes through their diameters and include the various saw and form cutters, side and plain cutters, and inserted teeth and angular cutters. The *shank cutters* include keyseat, *T*-slot, end mill, and fly cutters.

CUTTING SPEEDS AND FEEDS

Because the cutter turns in milling, it is the length of the cutter's circumference that assists in calculating cutting speeds and surface feet per minute travel. Cutting speeds must suit the milling operation. Slow speeds can become too slow and will damage the cutter's edge, whereas cutters which turn too fast will not cut efficiently, and severe damage to the cutter may result. Obviously, surface speed is related to r/min. Each term has its own meaning, but a relationship exists in that r/min influences sf/min. With a given r/min, the larger the diameter of the cutter, the greater the sf/min speed.

Cutting Speed

Cutting speeds are influenced by the metal's ability to be cut, that is, the machinability concepts pointed out in lathe operations. Again, the AISI B1112 steel is used as a standard from which to compare cutting abilities of other metals. If the B1112 is rated as 100%, then a low carbon steel would be rated less, approximately 50%, due to its extra ductility factor. Cutting speeds for metals are available from manufacturers' literature and some are listed in table 7–1. Cutting speeds increase as the machinability of the metal increases. When coolants are used, the cutting speed ability also increases. Cutting speeds are influenced by the type of cutter being used, use of coolant, kind of cutter and its function and shape, depth of cuts to be made, the material to be machined and its condition, and any other related variable. As an example, annealed tool steel has a cutting speed of approximately 70 sf/min, whereas bronze may be machined with a greatly increased cutting speed. Aluminum machines very well and its speed may be six to ten times that of a low carbon steel or most of the other steels.

Cutting speeds can be calculated just as in drilling or turning on the lathe:

$$CS = \frac{\text{r/min} \times \pi D}{12}$$

With a given r/min setting, the cutter's diameter in inches is multiplied by 3.14 and then by r/min. This value when divided by 12 equals the metal's cutting speed. Usually, however, the r/min is needed. This calculation is as follows:

$$\text{r/min} = \frac{\text{cutting speed in feet x 12}}{\pi D}$$

Machines have fixed r/min settings, consequently, the r/min is used which is closest to the calculated value, the cutter's value being noted in inches.

Feed

As the work moves into the rotating cutter, metal chips are removed. Feed, then, is the rate at which the work moves into the cutter and, in turn, the quantity of metal removed can be determined. As the workpiece collides with the cutter, the total chip removed indicates the width and depth of cut, in addition to the length of metal which slipped into the cutter's path, or the *linear feed.* Cutting speeds and feeds are related. The feed rate should be as great as the setup and machine will satisfactorily allow. Too fast a feed produces poor finish, while too slow a feed results in cutter damage. Small machines are hand fed by sight and feel. More expensive machines operate automatically. In other words, as the cutter size increases or spindle speed increases, the feed rate increases. These feeds range from a few thousandths of an inch per spindle revolution to approximately 1/4 inch. Also, feeds are regulated on many milling machines according to inches per minute and are separate from the speed of the spindle. Feed rate in inches per minute can be calculated by multiplying the product of r/min and number of teeth by the feed per tooth per revolution. No positive method will provide the exact feed. Due to the many variables in both the cutter and the material, only a best setting of the machine can be made.

MILLING OPERATIONS

Once the operator has completed the setup, the milling operation is started. The cutter is placed on the arbor at about the midpoint of the workpiece, spacing collars being used on both sides of the cutter to enable tightening on the arbor. Arbor rotation must allow cutting teeth to rotate in the proper direction. If milling a flat surface along the longitudinal axis, the depth of cut is adjusted by bringing the work up into the rotating cutter. This is accomplished by cranking the knee upwards. In this regard, many column and knee machines have power vertical feeds, especially the production types. When set for automatic feed, the workpiece moves horizontally into the cutter, leaving a smooth machine finish. The width of cut is adjusted, as well as the depth of cut, by adjustments of the vertical and transverse feeds. Passes across the surface are repeated until the exact dimensions are obtained. Often, only one pass is necessary to produce the required dimension. Plain milling may use the narrow cutter with straight teeth or the slab cutter with helical teeth for wider cuts. An example of plain milling is the machining of a flat surface on a casting where a part is to fit properly. Automotive cylinder heads are milled flat on one side in order to match the faces of the adjacent blocks. Parallels are machined flat on four sides. Flat areas on forgings are machined for attachment of bolts and other fittings. With respect to flat surfaces, many angular surfaces are also milled, the flat surfaces being at angles to the main part. The six sides of a cube are easily milled, for example, when properly set up.

Side milling operations are used to produce machined surfaces along the edges of metal. As an example, the square end of a hand reamer or tap is machined with the side milling cutter. This operation is performed by causing the workpiece to move upwards into the path of the turning cutter. In this respect, only a calculated portion of the rounded surface is machined flat as it moves past the cutter. Along with the use of the dividing head, three additional flat surfaces are machined which result in the square end. Straddle milling is also performed, whereby two or more cutters are brought together or separated to produce the desired shape such as ways for a machine. When grooves in a flat surface are required, the end mill is used. This cutter cuts its own round hole and then can be moved along to form a slot equal to or larger than its own diameter. End mills cut on their ends as well as their sides. Spur gears are cut with the form cutter. This cutter has the shape of the gear's tooth with respect to profile. One pass through the gear blank by the cutter leaves a profile identical to the cutter. Slots in machine tables are milled with the *T*-slot cutter. Dovetails are milled with the dovetail cutter. An angle is milled into a surface by means of the angle cutter. Keyways are machined with a slitting saw or the key cutter, depending on whether the square keyway or the semicircular keyseat is needed. When corners of parts or machine tables require rounding, the corner rounding cutter is used. Then, there are other form cutters such as the double angle for machining a *V*-shaped groove, the reamer fluting cutter, the convex cutter for tap fluting, and hobs for gear cutting.

Regardless of the milling operation, the basic principles are essentially the same. When milling, the workpiece must be securely locked to the machine's holding fixture so that it will not be snatched loose and ruined. Usually, as much time is spent in the setup as in the milling operation. Proper setups should be the aim of all machine operators and technicians as good setups rarely produce scrap. It is often required to use parallels between the workpiece and the holding device for the purpose of raising the workpiece away from the bottom or contact surface of the vise or fixture. *Parallels* are small rectangular shaped hardened steel bars of various dimensions. There must be clearance from the holding device so that the milling cutter will move freely across the surface, therefore, the bars are smaller in at least one dimension than the dimensions of the workpiece. Caution must be used to make certain that the cutter will not touch the holding fixture and that all the surfaces to be milled are accessible. After assuring accessibility and clearances of the cutter, table dogs are set, if needed, in order that the table will be reversed automatically.

The workpiece usually rests on the fixture's base or a parallel. In this respect, care is used to assure proper seating of the workpiece by tapping it with a rawhide hammer, after checking that all nuts, bolts, levers or handles are firmly tight. Sometimes the dial indicator is run across the top and sides of the workpiece to assure proper setup in the holding fixture. Because of high stresses occurring in the workpiece during milling operations, the workpiece must be securely held in position so it cannot slip.

Grinding

Grinding is an operation which removes small particles of metal by the cutting actions of many abrasive particles imbedded in a grinding wheel. The grinding action is a combination of compression loading and shear forces which cause very small particles, or chips, of metal to be removed from the work. Grinding depends on the fast moving abrasive to do its work. Therefore, surface speeds are high. Grinding is performed while both work and grinding wheel are moving. Round parts rotate while the wheel spins at a much higher r/min. Flat work reciprocates across the face of the wheel, or in the case of very large work, the face of the wheel moves across the work. In form grinding, the wheel relatively moves inward to grind the grooved recess in the workpiece. Edges of cutting tools move slowly across the face of the rotating wheel. Grinding operations remove only a very small quantity of metal compared to other operations, consequently, grinding or precision grinding is often thought of as a finishing operation. In this respect, there are many types of operations which are accomplished with several different types of grinding wheels.

ABRASIVES AND GRINDING WHEELS

Grinding wheels are combinations of abrasives and bonding materials which are molded into different shapes compatible with high velocity rotation. The molding or bonding material securely holds each small piece of abrasive in a random position. Because thousands of tiny particles are held closely together by the bonding material, a combination of factors result in the structure of the wheel. First, the bonding operation provides either rigidity, flexibility, or a combination of these two factors in the makeup of the wheel. Secondly, abrasive particles scattered throughout the bonding matrix provide the cutting tools, there being many shapes, sizes, and kinds. Thirdly, the air spaces existing between bonding and abrasives provide clearances for the protruding abrasives to cut as they swing in an arc through the material of the workpiece. Because the abrasive particles are randomly oriented in the wheel's total structure, cutting actions result from a combination of both positive and negative rake angles. When the rake angle in the particles is reduced to ineffectiveness, the compressive forces between wheel and work grab the particles from the bond, leaving new particles to carry on the cutting actions. The total grinding wheel's makeup is composed of five main factors—type of abrasive, the grain size of the abrasive particles, the grade of the wheel, the bonding material in the wheel, and the final structure of the wheel.

Abrasives

Grinding wheels contain several different types of abrasives, but the two most common are aluminum oxide and silicon carbide. The compound of aluminum and oxygen ranks high on the scale of hardness, being just slightly less in hardness than the compound of silicon and carbon. As an example, the diamond ranks as 10 on Moh's scale of hardness, the highest, while SiC ranks better than 9, and $A1_20_3$ ranks just slightly less, still being in the hardest categories by relative values. Moh's scale of hardness is a comparative method of rating the various minerals in a hierarchy from 1 to 10. Talc is the softest and is rated as 1, while the diamond is the hardest and is rated as 10. Other minerals fall between 1 and 10, depending on their hardnesses. Silicon carbide is brittle, while aluminum oxide is less brittle, therefore, the carbide is predominantly used in the grinding of soft materials such as aluminum, low strength nonferrous metals such as copper and its alloys, and brittle materials such as cast iron and ceramics. Aluminum oxide is used to grind the harder and higher strength steels as it does not disintegrate as fast as the silicon carbide. Aluminum oxide lasts longer than silicon carbide under most conditions due to the inherent brittleness of the carbide. When maximum hardness is needed, the diamond is used. It is used in particular situations where the carbide or oxide is not satisfactory. Diamond dust is used in wheels while larger particles of diamonds are used as individual cutters. Most diamonds used in industrial applications are of commercial quality which is reflected by their discoloration.

Abrasive Grain Sizes Any abrasive, like a powder, is measured in screen sizes, that is, the size of the particle that will barely fall through a hole in the screen or just barely be retained, the screen's hole size being measured as a given number of holes per linear inch. The meshes per linear inch vary from eight to 600. Those having up to 24 meshes per inch are considered coarse. Medium sizes range from 30 to 60, and fine sizes range up to 600. The very fine, or flour, sizes range from 240 to 600.

Bonding The abrasive particles are held in place by the bonding material. Several types of materials are used, the vitrified being used in the majority of wheels. Vitrified wheels are hard and strong, as well as porous. Commercial usage such as contact with water, oil, chemicals, and varying temperatures do not adversely affect the vitrified structure. Silicate bonded wheels are much softer than the vitrified and produce smoother finishes. The silicate bond releases worn grains periodically, keeping the wheel's face clean and sharp. When rough grinding operations are required and higher speeds are desired, the resinoid bond is used. A cut-off wheel is frequently bonded with this material. In other words, the wheel will flex somewhat before breaking, therefore, such operations as gouging and snagging of large protrusions of metal are common. Rubber wheels which are impregnated with abrasives have many uses in industry. They are used at high speeds and are flexible for such operations as cut-off, angular cutting, and surface touch up. Shellac wheels are flexible and produce a very smooth finish.

Wheel Grade

The grade of the wheel refers to the holding strength of the wheel's bond. Soft wheels release their grit readily and are, therefore, used in the grinding of hard materials. The sizes of the bonding areas between the abrasive particles refer to the hardness or softness of the wheel. The soft metals are ground with hard wheels, while hard metals are ground with soft wheels. The hardness and softness factor refers to the ease with which abrasive particles release from the bond. Hardness of abrasive grains has little effect on the hardness of the wheel.

The actual density of the wheel pertains to the wheel's structure. Fine finishes are accomplished with fine grains, while coarse grains allow greater clearances for the abrasive particles and greater metal removal capacity. Closely spaced particles have small clearance cavities which result in only very small opportunities to shear a small chip from the workpiece. Soft materials are normally ground with coarse structures, while hard materials are ground with fine structures.

Wheel Identification

After World War II, the Grinding Wheel Manufacturers' Association established a standard working system for grinding wheels. The system includes six parts—abrasive, grain size, grade, structure, bond, and manufacturer's record. Abrasives are coded by letters to indicate type; grain sizes are in numbers to indicate coarse, medium, or fine; hardness is indicated by letter (*A–M* are soft and *N–Z* are hard); structure is denoted by number (1–8 are dense and 9–15 are open); bond is shown by letter which reflects from the specific type; and the manufacturer's record is variable. Wheels are marked on their sides near the hole, the short code indicating all of the above information.

Wheel Sizes and Shapes

Wheel sizes are indicated as diameter in inches, hole size, and face width. Shapes of wheels are standardized and refer to the shape of the entire wheel, with emphasis on the shape of the cutting face. Some common wheels include straight, cylinder, cutoff, straight cup, flaring cup, dish, beveled face, and saucer. As a result of the differently shaped wheels, work is ground either on the outside periphery which is parallel to the shaft or at some angle from the shaft such as 90°. Figure 9–13 illustrates some typical grinding wheels.

FIGURE 9–13 Sizes and shapes of grinding wheels. (Courtesy of Norton Co.)

SPEEDS AND FEEDS

Cutting speed is the speed that the wheel's periphery travels, and it is measured in surface feet per minute. Control of surface speed is determined by size of the wheel diameter and revolutions per minute. Surface speeds recommended by the manufacturer should be adhered to in order to produce best grinding efficiency and for safety purposes. The wheel's speed is shown on the side of the wheel, along with its other identifying data. In this respect, some motors drive a spindle at twice the speed of other motors. This factor must be considered when mounting wheels on the spindle. Wheels are not interchangeable, in other words, as the diameter of the wheel increases, the r/min must be decreased to maintain the sf/min speed. The machine's r/min must then be maintained in order to prevent a wheel explosion. The vitrified wheels range in sf/min speeds of 4000–6500, this range being variable due to the specific type of grinding to be done. General grinding practice will require about 5000 sf/min. Surface speed is determined by multiplying the circumference of the wheel in feet by the r/min of the spindle. If this value is equal to or less than the recommended speed, then it is safe.

During cylindrical grinding operations, both the wheel and workpiece are turning. Machines are set to provide a certain r/min and, in turn, this allows a satisfactory surface speed. The workpiece must also rotate to allow from approximately 50 to slightly more than 100 sf/min. Headstock spindle speeds provide for several r/min settings to allow for the varying hardness of different metals. Very soft materials are ground at the rate of more than 200 sf/min. Work speeds must remain within a reasonable range in order to produce acceptable finishes.

Work travel across the face of the wheel is controlled by a dial on many grinding machines. Generally, rough grinding allows table travel of two-thirds of the wheel's face for each wheel revolution. Finishing operations are reduced to approximately one-half this value. When the work speed is multiplied by the distance the work should travel per revolution, the travel rate is obtained. Because the wheel must contact the workpiece, a depth of cut must be determined. Good practice depends on variables such as capability of the machine, type of finish desired, and the type and condition of the material being ground. Generally, a roughing cut of 0.003–0.004 inch is acceptable, while the finishing cut is from 0.0001–0.0007 inch. Obviously, when 0.010–0.020 inch of material has been left for grinding due to heat treating requirements, several passes across the wheel are required. The last few passes are finishing cuts accompanied with coolant.

SAFETY PRECAUTIONS

On receipt of a grinding wheel, place a screwdriver in the hole and lightly tap the wheel with a small metal tool such as another screwdriver. The ring between a sound wheel and a cracked wheel is distinct. Never use a cracked or damaged wheel. Wheels are mounted snugly on their holders; too much pressure will crack them. Wheels are dressed on their shafts to eliminate vibration. The metal dressing tool, consisting of a group of ridged rollers, is a fine roughing tool. The wheel's surface is finished with the diamond which is mounted in the end of a steel shank. Goggles or a face shield are worn while grinding. Refrain from standing in the wheel's line of rotation, roll up sleeves, and tuck in loose clothing, and never turn a wheel faster than its specified cutting speed.

GRINDING MACHINES

Grinding machines perform operations which are not obtainable by any other method, that is, the smoothness of the surface finish and the close tolerance capability. It is possible to produce glassy smooth surfaces with the grinding wheel. Also, it is possible to remove a quantity of metal so small that it cannot be feasibly measured. Grinding operations requiring tolerances smaller than 0.0001 inch are not common, but one-half of a ten-thousandth of an inch can be produced. Many grinding machines perform external grinding operations while others perform internal. Some external operations are cylindrical while others are flat. The type of operation determines the kind of grinding machine to use.

Universal Grinder

The universal grinding machine is capable of performing many routine types of grinding operations (fig. 9–14). The wheel stand unit houses the wheel for external grinding

FIGURE 9–14 A universal grinding machine complete with internal grinding attachment. (Courtesy of Summit Machine Tool Mfg. Corp.)

and the spindle for internal grinding. The wheel spindle head can be swiveled. A similarity to the lathe exists in that a headstock is mounted on the table's left while a footstock is mounted on the right. When centers are provided for these two assemblies, the workpiece can be mounted between them, whereby it rests in front of the grinding wheel and at the center height of the wheel. The headstock may also be swiveled for taper grinding.

The workpiece is driven by a chuck, collet, or faceplate and dog. Normal grinding operations provide the rotation directions of the wheel and workpiece to be in the same direction in order that the directions are opposite at the point of contact. A coolant assembly is provided so that the liquid will play onto the work's surface at the point of contact. Similarity with the milling machine is structured in the machine's table since the feeds include longitudinal and cross. Trip dogs on the table are adjustable to provide for reciprocating lengthwise grinding. In this respect, at least one-third of the grinding wheel should remain on the workpiece during grinding. Once the setup is complete, the grinding operation becomes semiautomatic insofar as lengthwise travel is concerned, the trip dogs performing the several movements. Being universal, the machine's table can be swiveled to allow taper grinding to be conducted.

Much of the external cylindrical grinding operations are performed with operations combining principles of the lathe and milling machine (fig. 9–15). Calibrated dials on

FIGURE 9–15 Landis 14" x 36" Type 3R universal with 14" swing and 36" between centers. Machine is equipped with 2238 W motor, 14" diameter wheel. Headstock speeds 25 to 500 r/min. Headstock motor 559.5 W. (Courtesy of Landis Tool)

the feeding mechanisms provide precision grinding. When the wheel contacts the workpiece, a small shower of sparks occurs. These sparks are highly heated chips. The greater the spark shower, the more metal being removed. Proper setup is naturally required while grinding can be performed by hand or by semiautomated processes.

Cylindrical Grinding Cylindrical grinding is accomplished by feeding the turning workpiece into the rotating grinding wheel. After properly providing the required r/min and wheel, proper workpiece setup between centers and its r/min, proper center alignment, and proper trip dog setting, then the proper feed and depth of cut are determined. Pushing the switches engages the wheel and workpiece and commences grinding operations, including the spark shower. During grinding, there must be a provision for a proper coolant. When the roughing cuts are completed, after micrometer measurements are made, and the finishing cut is carefully completed, then the wheel is allowed to spark out. Usually, when using a coolant, the spark shower is not observed. Keep in mind that hot metal measures larger than metal at 70°F. Reversal by the trip dogs provides reciprocating actions of the table. Correct rate of travel and depth of cut by the cross feed dial allow the workpiece to be brought to a satisfactory completion. Special types of cylindrical grinding operations are shown in figure 9–16 and figure 9–17.

FIGURE 9–16 Cylindrical grinding a 13-ton forged and hardened steel roll. (Courtesy of Bethlehem Steel Corp.)

Taper Grinding When taper grinding is to be performed, the same operations are repeated as for straight grinding, allowing the pressure of the wheel to be in the direction of the headstock. Swiveling either the table, the headstock, or the wheel spindle head the proper number of degrees for the taper completes the setup. Grinding is accomplished as in normal grinding.

Internal Grinding Internal grinding is accomplished by swinging the internal grinding spindle into proper position, (fig. 9–18). The setup is similar to external grinding, only the wheel rotates inside the hollow workpiece, moving longitudinally

FIGURE 9–17 A cutter grinding operation. (Courtesy of Cleveland Twist Drill Co.)

according to the trip dog settings. Cross feeds are performed to bring the workpiece to its proper dimensions. Care must be exercised in turning the longitudinal and cross feed dials to assure proper wheel contact. Special care is needed during this operation because the operation is blind to the operator at times. In these circumstances, trust is placed in the dials and dog settings.

Cutter Grinding Cutter sharpening is also accomplished on the universal machine and other machines. The cutter is placed on a mandrel and mounted between centers. The tooth rest is adjusted from 4° to 7° below the wheel's center in order to remove metal behind the cutter's edge and provide clearance for the tooth. When the cutter is held against the rest and passed across the face of the wheel, clearance behind the edge occurs and the tooth is sharpened. When each tooth is properly brought across the wheel's face, the cutter is sharpened. Figure 9–17 illustrates the grinding of a cutter.

Thread Grinding The grinding of threads on either soft or hardened steel is a grinding operation that imparts smooth and accurately ground threads on the rod.

FIGURE 9–18 Internal grinding operation on Landis 10" x 20" Type 1R universal grinder. The internal attachment is mounted on the grinding wheelhead for easy swing up or down with locking arrangement for both positions. Spindle speed is 16,500 r/min with a drive motor developing 746 W at 3600 r/min. (Courtesy of Landis Tool)

Several methods are available, however, one of the most common is similar to thread cutting on the lathe. The main difference is the grinding wheel which has been dressed to the shape of the thread. When the grinding wheel moves along the axis of the rod, its depth of cut and lead are fixed by the operator and machine. Parts can be completely heat treated and then ground to specifications. A special wheel having multiple cutting ability is also used in thread grinding.

Surface Grinder

Surface grinding is performed on flat work by means of a surface grinding machine (fig. 9–19). Basically, many of the same grinding priniciples apply when surface grinding as in cylindrical grinding, except that the workpiece is held on a surface chuck. Many of these chucks are of the permanent magnet type whereby the steel is held securely to the surface of the chuck which is mounted on the machine's table. The magnets can be adjusted to give the on and off results. Grinding operations in surface grinding then follow the same general rules as in cylindrical grinding, the workpiece reciprocating under the wheel while the automatic crossfeed allows a series of new cuts until the surface is ground smooth. Coolants are used as in other grinding operations. Another form of surface grinding is accomplished with the use of rotary tables. Many parts can be ground at one time by this type of process. When properly

FIGURE 9–19 A surface grinding machine. (Courtesy of Summit Machine Tool Mfg. Corp.)

connected by appropriate mechanisms, this grinding operation can be automated in both its rough and finish grinding operations.

Centerless Grinder

Round parts can be ground to size with the use of the centerless grinder. This machine consists of the grinding wheel, the regulating wheel, and a workrest for the purpose of grinding cylindrical parts (fig. 9–20). The workpiece is not supported on centers, but is maintained in position by both the workrest and the regulating wheel. The purpose of the regulating wheel is to support the workpiece against the force of the horizontal thrust of the grinding wheel, to provide the feed of the workpiece, and to rotate the work slowly. The grinding wheel rotates at the normal speed while the workpiece turns at slow speeds, often not more than 225 feet per minute.

Pedestal Grinder

The pedestal grinder is one of the most needed machines in the general metal shop or machine shop. Even though tools and parts are ground by hand manipulation, the grinder performs many roughing operations such as burr removal, saw mark removal, weldment edge preparation, and general metal hand grinding operations. Often, the

FIGURE 9–20 High production bearing race grinding on two Landis centerless grinders in tandem, stock removal is 0.015" in three passes. Production 60 per pass per minute. Tolerance: 2 point roundness 0.001"; 3 point roundness 0.00015"; taper across OD 0.001" and 15 RMS finish across OD. (Courtesy of Landis Tool)

machine's shaft provides for a roughing wheel on one end and a finishing wheel on the other. Normally, the grit sizes are much larger than on precision grinders. This type of grinder requires a support on which to rest the workpiece while grinding, the ground surface having slight curvatures at points of contact with the wheel's periphery. Supports must be kept tight and perpendicular to the wheel's periphery. Wide wheels and steel backed wheels have side supports so that flat surfaces can be hand ground. In this respect, heavy pressure must never be exerted against the side of a thin wheel due to danger of wheel explosion. Grinding wheels are brittle, but stiff. Some pedestal grinders may only have one grinding wheel while the other end of the shaft holds a wire buffer or polishing cloth. Tool bits are frequently ground on the pedestal grinder while other tool bits are ground on the cup shaped tool grinder.

COATED ABRASIVES

The use of coated abrasives is steadily increasing due to the ability of the abrasive to remove large quantities of material in a short period of time. The same abrasives used in grinding wheels are used on strips of paper, cloth, and combinations of different materials. Again, silicon carbide and aluminum oxide are primarily used due to their cutting efficiencies. The grit size on these cloths varies from very fine to coarse. Some of the coarse types are used for snagging operations. When resins and glues are used to bond abrasives to tough cloths, a durable abrasive belt is produced. When the bonding material completely covers the abrasive grains, a heavy duty metal remover or wood sander exists. Sheets and belts may be used dry or wet depending on the type of backing material and bond. Abrasive grain size is like those sizes used in the grinding wheels. Abrasive belts are produced with very coarse grains, size 12 and up to 600 which is the finest practical size. One of the finest abrasive cloths is *crocus*, an iron oxide. When finer abrasives are required, the levigated alumina is used, usually mixed with clean water which produces very superior polishing.

Abrasive cloths are used in sanding operations, the degree of surface smoothness depending greatly on the size of abrasive grit used. Again, grits are measured according to mesh sizes per inch. The coarse grits literally snatch patches of metal from a surface. This series of scratches then remains unless finer abrasives are used to gradually wear away the metal on both sides of the scratches. Care then must be taken when using very coarse grits. On the other hand, the finer grits wear away or scratch away smaller pieces of metal, but always leave their marks. Obviously, by using finer and finer grit sizes, scratches may be reduced to only those of the last grit used. Alumina, measured in microns diameter, provides one of the smoothest possible surfaces. A polished surface produced with alumina in water presents a surface comparable to the glass mirror.

SPECIAL GRINDING OPERATIONS

Buffing and Polishing

Buffing and polishing operations utilize cloths such as flannel and muslin which have been impregnated with a polishing compound such as rouge. Polishing speeds vary according to the situation and depend heavily on operator judgment. Very small amounts of metal are removed, such as very shallow scratches.

Honing

The honing operation is a grinding operation which removes only a very small quantity of metal. The honing stone is used by hand operations when very fine finishing is required on parts adaptable to hand honing by their shape. Machine honing is more common, especially when many similar parts are to be honed. Equipment is especially designated to hold the particular stones and, at the same time, hold the work and perform the finishing operation. Stones with the appropriate fine grit are shaped to the workpieces surface. During the reciprocation action of the stone, very little pressure is placed against the workpiece as the stone moves across the work's surface. Fluids are used in most instances as lubricants. The honing speed varies with the setup, but seldom exceeds 300 sf/min. Cylindrical workpieces are honed with a rotary motion as the up and down movement continues. Surfaces can be brought to mirror finishes with very close tolerances with the hone.

All metal surfaces are rough to certain degrees and consist of high and low ridges. Honing tends to equalize the heights of ridges by bringing them closer to the valley depressions. In other words, the smoother the rubbing surfaces, the less friction and wear of the.mating parts. Honing increases wearability of the moving parts and allows closer tolerances for better fits. Lubrication is not a substitute for rough surfaces. When an acceptable surface is provided with a film of lubricant under reasonable pressure, moving parts last longer. The term *superfinishing* refers to rotary and flat surfaces which have been honed to necessary degrees of smoothness between mating parts.

Regardless of how well the surface has been superfinished, a degree of roughness persists. This roughness can be measured, for example, by the height of its ridges, and then expressed as an arithmetical average in microinches. Roughness height is shown as the arithmetical average (*AA*) distance from a center line theoretically drawn through a cross section of the hills and valleys of the rough surface. Conversion of surface roughness indications to a microfinish comparator assists in determining what surface smoothness is needed. The comparator reflects surfaces of different smoothnesses produced by several cutting operations. Also, special instruments are available for determining surface roughness measured in average microinches. A General Electric Surface Roughness Scale indicates surface roughness by sight and feel. A very smooth surface is seen and felt at an average roughness of four microinches. Visible scratches appear at an average of 32 microinches while roughness appears at 125. The scale increases in roughness up to a value of 2,000 microinches whereby the surface appears to be rough ground with a portable grinder.

Lapping

The lapping of a metal includes the abrading of one metal by an abrasive compound which is held in a semifixed position on the surface of a softer metal. A smooth cast iron lap having grooves at equal intervals across its surface for storing both chips and excess compound is mounted on a bench to enable the lapping of flat parts. The softer metal provides a fixture for abrasives. When a harder metal is moved across the lap, the abrasive, being lodged in the lap's surface, removes a very small quantity of metal from the harder metal. Continuous lapping brings the surface to the necessary smoothness which also is measurable in average microinches of roughness. Lapping operations remove less than a thousandth of an inch of metal, therefore, lapping is reserved for surfaces requiring this degree of smoothness. Only fine particles of abrasives are used in a greaselike lubricant. Grits may be as fine as 600.

Miscellaneous Cutting Operations

BROACHING

Numerous tools and parts must have slots such as keyways or splines along a shaft, or lands within a wrench, or the part may require round or square holes. One of the quickest methods in producing internal and external shapes on a part is with the broach. For example, if a rectangular hole is to be placed through the center of a round steel bar six inches long, the broach should be used. A study of figure 9–21 illustrates a common broach. First, a hole is machined through the mass of metal, being careful to maintain the center of the longitudinal axis all the way. The hole's

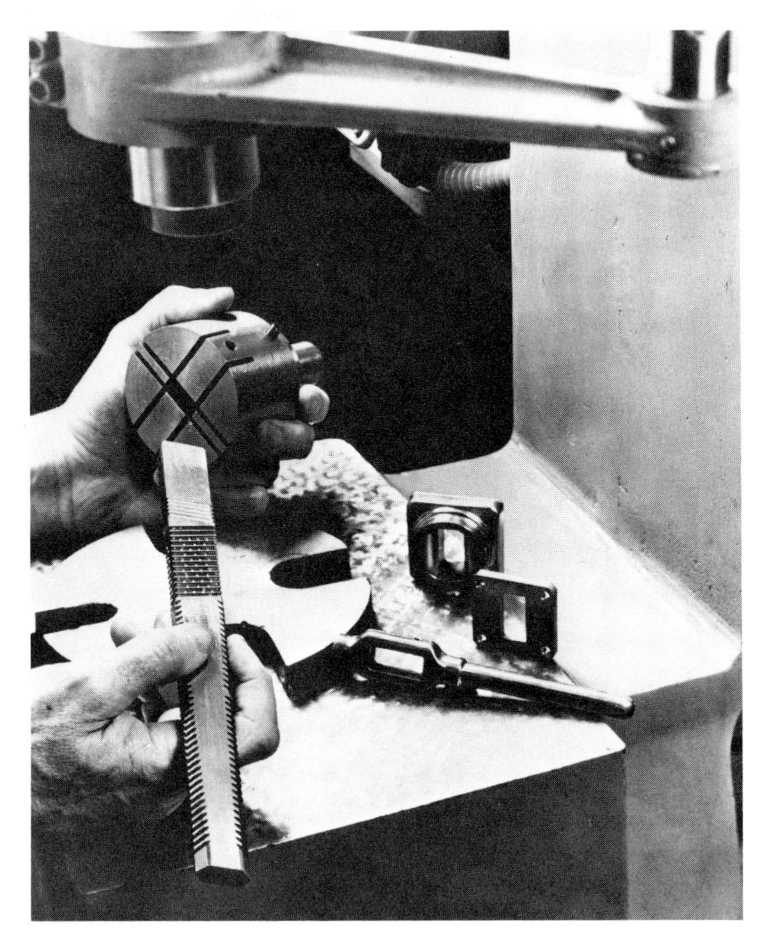

FIGURE 9–21 A ram adapter and rectangular broach required in broaching a section of metal. (Courtesy of The duMont Corp.)

diameter must be large enough to allow the pilot to enter, and the broach's length or shank length must be long enough to allow the pulling, or forward, end to extend through the hole before the roughing teeth make contact with the material to be cut. According to figure 9–21, as the broach is pushed through the hole, each succeeding tooth on the broach extends farther outward around the entire periphery, approaching the proper shape near the end of the broach. The last few teeth are finishing teeth and leave the hole within its needed tolerances. Only a few seconds are required to pull or push the broach through the hole. Figure 9–22 illustrates how the hole in a collet's face is produced by broaching.

A study of the above procedure shows that each tooth must be spaced enough distance apart, the pitch, whereby chips will accumulate and can be held for the duration

FIGURE 9–22 The broaching of a lathe collet with a hexagon shaped broach. (Courtesy of The duMont Corp.)

of the single broach operation. It should be understood that each tooth all around its cutting edge, removes only a small chip, approximately 0.005 inch during roughing and about 0.001 inch or less during finishing cuts. This means that each succeeding tooth extends farther into the material all the way to the final shape of the last tooth. The step of the teeth gradually increases from minimum to maximum. Each tooth is shaped to provide clean shearing action, therefore, rake and clearance angles must be appropriate to allow for the land and back angle. Broaches, then, are multiple-tooth cutting tools. The lathe tool is a single-point tool, whereas the broach performs continuous cutting, the cutting assigned to the teeth which automatically move into the shear planes of the metal to be cut.

Broaches are used in both hand and machine methods. Often, many small gears are purchased without keyways or set screws. With the use of the appropriate broach, the keyway can be quickly pushed through the gear's hole by the ram of a press. Accordingly, the broached hole can be nearly any shape, the last few teeth determining the final shape. Broaches may be pulled, pushed, or pulled and pushed through or across a workpiece. The principles of internal broaching are applicable to external broaching such as the broaching of splines on a shaft. Broaches are made of the same materials as cutters, the high speed or high carbon–high alloy steel being commonly used.

When many parts are to be produced by broaching methods, special machines are constructed. Broaching is accomplished both vertically and horizontally with nearly any shape of broach. Also, the broach may move in a straight line cut or it may rotate as a helix, either internally or externally, cutting a helical grooved pattern.

SHAPING

The shaper is a powerful machine tool that removes a chip of metal by either pushing or pulling a single-point cutting tool across a flat surface, usually across the top of the material being cut. The most common shaper is the horizontal push type that has a variable length of stroke of just a few inches up to 36 inches (fig. 9–23). The cutting

FIGURE 9–23 A shaper. (Courtesy of Western Machine Tool Works)

tool is similar to the lathe cutting tool in that a chip continues to be removed from the surface of the workpiece, but only during the cutting stroke. Shapers are reciprocating cutting machines and have operations similar to the power hacksaw, except that the feed on the shaper is lateral, whereas on the hacksaw it is vertical. In other words, the shaper's tool shears a fairly thick chip from the workpiece, the chip peeling from the material across the tool's cutting edge and face (fig. 9–24). The tool pushes the chip ahead in pusher type shapers. During the return stroke when no cutting occurs, the work table feeds the appropriate width of cut, the depth of cut remaining constant. The cutting cycle is repeated as so many strokes per minute.

After the surface of the workpiece has been reduced in size by the tool's depth of cut, another depth is adjusted by turning the micrometer type feed dial to extend the tool deeper into the work. The length of stroke is maintained or readjusted when necessary. Because the typical shaper does not cut approximately 40% of the time, allowances must be made for this factor. Production type shapers are very powerful, the ram moving steadily ahead, curling the chip in front of the tool. Cutting speeds during shaping operations are reduced to a number of strokes per minute. Because the length of stroke is adjustable, the number of strokes can be quickly calculated in relation to the length of stroke. Low carbon steel can be machined at approximately 100 feet per

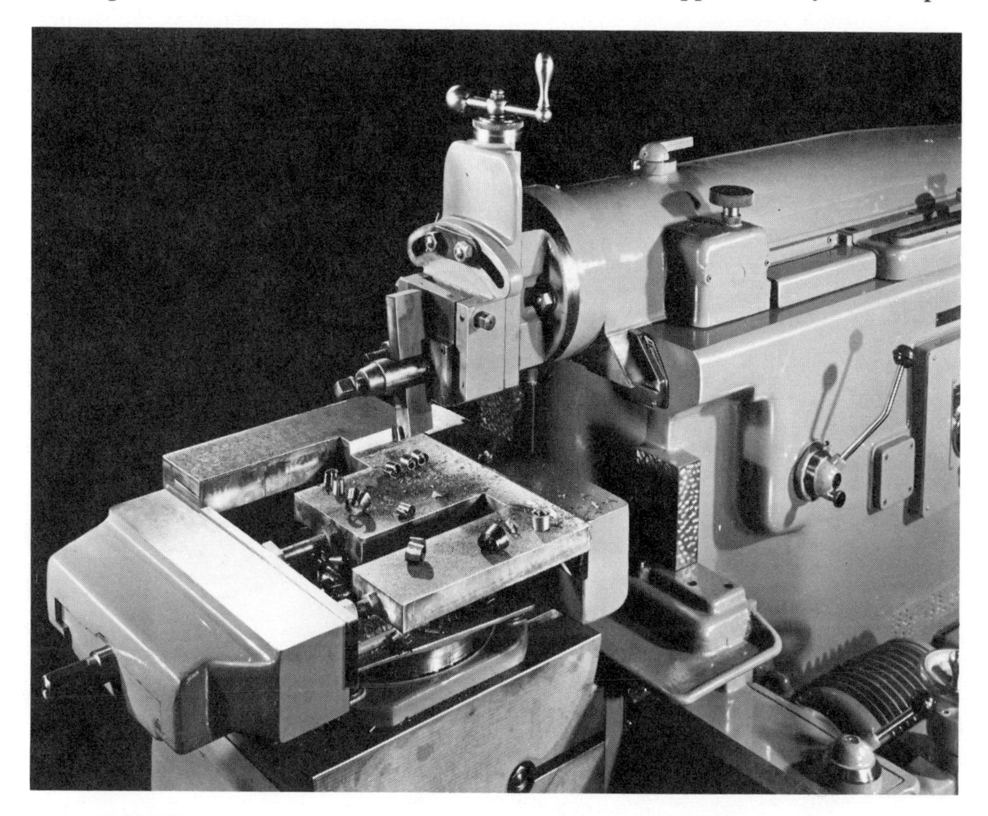

FIGURE 9–24 A ram-type Rockford Hy-draulic Shaper shaping an irregular shaped workpiece. "For clarity, guarding required by the Occupational Safety and Health Act has been removed." (Courtesy of Rockford Machine Tool Co.)

minute. After measuring the necessary length of needed stroke which is about one inch longer than the length of the workpiece, the number of strokes per minute is found by multiplying 100 by 60% of 12 and dividing this value by the length of stroke. If the length of stroke is 10 inches, then the number of strokes per minute is 72. The shaper, like other machines, is not capable of providing any number of strokes, therefore, the stroke setting is adjusted to the nearest setting which it can deliver, and this is 70. A word of caution with respect to shapers and their reciprocation capability is given—do not set a long stroke with a high number of strokes because the machine may be severely damaged as it attempts to leap from its tie downs.

Other types of shapers include the vertical which performs edge work and angular cutting. Also, the shaper is capable of slotting and grooving operations, both internal and external. Careful setups are required, and the workpiece must be tightly secured in its holding device. Many other types of shapers are available for performing various types of machining operations on flat surfaces which may be arranged in any position where cutting is feasible. Even though the horizontal shaper has been an excellent cutting tool for many years, its capabilities are being rapidly absorbed by the milling machine. Large slab mills remove larger quantities of metal than the shaper in a given period of time. Even so, many special operations are being performed by the shaper.

PLANING

The planer is a machine tool that functions in an opposite manner than that of the shaper. Basically, the planer is built to handle large sections of metal where flat areas require machining (fig. 9–25). The workpiece is secured to the table. During opera-

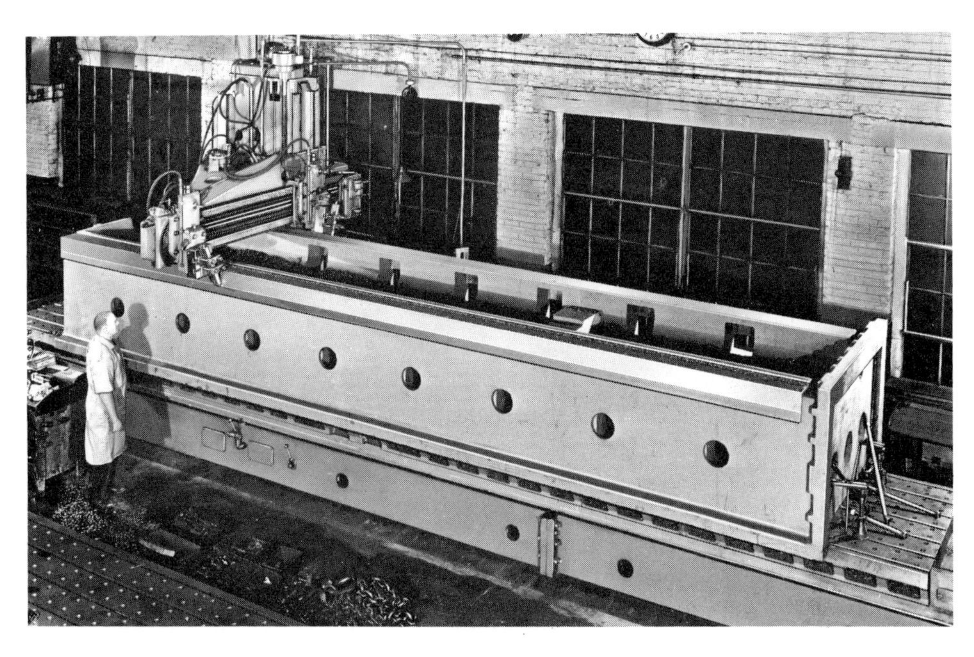

FIGURE 9–25 A Rockford Hy-Draulic Openside Planer planing a slotter bed. "For clarity, guarding required by the Occupational Safety and Health Act has been removed." (Courtesy of Rockford Machine Tool Co.)

tion, the heavy table reciprocates longitudinally beneath the stationary cutting tool, causing the chip to be removed relatively similar to the horizontal shaper. The table can be arranged to travel long distances and then reverse, but the shaper's ram length is limited due to excessive overhang of the ram on extending during the cutting stroke. Otherwise, the planer produces work in a similar manner to the shaper, not cutting on the return stroke of the table. To increase machining effort, some planers are equipped with multiple heads to produce multiple chips during each cutting stroke. Also, some planers are equipped with reversible cutting heads to accomplish cutting on both strokes. A modification of the planer includes the substitution of large mills for the single-point cutting tool (fig. 9–10).

Automation

As the quantity of identical parts increases, the engineer and technician look towards mass production devices with constant repeatability. When an operation is performed in an established sequence, the operation can be automated. When large numbers of parts are required, then the economics of the production capability appear reasonable and encouraging, especially when the market is available. Automated and semiautomated machines are increasing in large numbers because precision parts can be produced with less skilled operators. Automation is the control of a machine's and a mechanism's movements on command from a tape controller or computer, but all the work must be done by the machines, mechanisms, handling devices, and controls.

DESIGN CONSIDERATIONS

Before any manufacturing process can be automated, careful consideration must be given to the feasibility and economics of the processing procedures. The study must view the whole process from start to finish in some kind of a pattern, that is, whether the workpieces move in an intermittent series of processing cycles or move at a constant speed in a long and continuous line. The product must lend itself to being processed by one of these two methods. Further, the number of steps or processing stations must be minimized to impart simplicity in the overall operation. Length of time at each station becomes a factor to balance long operations together and keep the shorter timed processes in a separate grouping of machines. The total cost of the automation package must assume that quantity of products over a long time period will result in profit. Then, too, the kind of material to be processed must be compatible with the available machines. Standardization and maximum tolerances become part of the feasibility study. Where it is shown that a train of parts can be moved through a mechanized series of processing stations economically for profit, then a justification exists for automating the process when the market is ready to receive the products.

TYPES OF AUTOMATION

Basically, there are the constant travel process and the intermittent process types of automation. This means that the workpiece either moves through the production line at a constant movement or the workpiece stops intermittently for processing which requires different lengths of time.

Constant Travel

Operations lending themselves favorably to this process include grinding and milling of the workpiece. These operations may be in a straight line series, or they may be arranged in a circular pattern. The workpiece is normally secured in a fixture one time and is held in place while one or more machining operations are performed singularly or concurrently. When this process is used, consideration must be given to the practicability of removing the workpieces from moving fixtures. Another limitation is the possibility of line breakdown which means stopping the whole line. Of course, the advantage is time saved as a continuous series of workpieces move in sequence for designated processing.

Intermittent, or Indexing

The movement of the workpiece in a circular pattern or in a straight line as it stops and starts along the processing stations provides advantages related to the processing to be done. A circular type of processing machine clamps the work in its specially designed fixtures so that a series of machine operations can be performed. An example is the single- or multiple-spindle types of drilling and reaming operations. The workpiece is drilled, reamed, and countersunk in a series of fast operations. The fixture quickly returns to its starting point and the process is repeated. Intermittent cycling occurs when uneven time periods are required for drilling small and large holes. Such uneven cycling can be balanced by grouping the operations by time consumed and by sequence. Circular methods function in any plane, that is, they may be vertical or horizontal.

Straight-line indexing allows the workpiece to move from station to station in accordance with the processing requirements. Machines can be grouped according to function and sequence, time being alloted according to need. Many machines are self-contained, that is, they process the workpiece to its completion. Such a machine can completely process a small bracket by drilling and tapping its seven holes in predetermined places. On the other hand, a series of grouped machines may be extended in the production line and require automatic feeding mechanisms for the workpieces. Such a continuous feeding array of machines and equipment handling devices requires exactness and completeness in its design. If the entire processing line is machine and mechanism operated, then it is automated. Obviously, some means must be incorporated to process the workpiece through the entire operation. Such a means changes ideas and facts into electromechanical impulses whereby each and every aspect of movement of the workpiece and the tools must be plotted and programmed into a controlling device. Automation, then, uses sophisticated mechanization to function in a predetermined manner in response to impulses of energy. Computers and tapes are commonly used in the control of manufacturing processes.

NUMERICAL CONTROL

The movements of the processing machines are controlled by impulses from another mechanism. Man programs the impulses on tapes or in the storage banks of computers. Then, on command, impulses move by electrical means from the electronic controller to interface at the machine and mechanism which converts electronic energy into mechanical energy. At the several interfaces, motors turn, and switches, clutches, and brakes function to control, guide, and time every movement of the processing tools on the workpiece. Lathes, milling machines, drilling machines, and other machines are capable of being controlled by tape or computer. An automatic boring machine, for example, is shown in figure 9–26.

Numerical control is actually control by a numbering system which uses letters and numbers to indicate an entire machining operation. Two basic systems are in use, the point-to-point, or positioning, system and the continuous path. Operations such as drilling, reaming, tapping, and boring at a single point utilize the positioning system, whereas the continuous path system is used during milling, turning, and continuous metal removal operations.

Positioning System

Positioning control is the simpler of the two operations, and it is usually controlled by the tape controller. Basically, there are three axes used in many machine operations—horizontal, the x; lateral from x in the same plane at 90°, the y; and vertical from x,

FIGURE 9–26 Cast assemblies being machined in an automated production line. (Courtesy of Cincinnati Milacron)

the z. From a common 0–0–0 point of origin along the horizontal plane, measurements and machining operations to the right are $+x$ while operations to the left are $-x$. At 90° from the x axis are the plus and minus y axes, inwards being plus and outwards being minus. At 90° to the x–y axes is the z axis. Operations upwards are $+z$ while operations downwards are $-z$. This is the basic concept of the Cartesian coordinates system of rectangles.

Accordingly, dimensions in the x–y–z axes are known and are consequently recorded in terms of arithmetic and simple mathematics by a programmer. Such a person must be cognizant of manufacturing processes, materials, engineering concepts, drafting and design, and mathematics. Also, he must understand the whole machining operation, step by step, as it is this step-by-step sequence that is translated into the numbering system, or *program*. Because the manual machining operation involves human thought, calculations, and actions, all these things must be translated to logic and then to action on a time basis. This means that the programmer must write the manuscript to include all actions and instructions to the machine operator. The tape, when control is by tape, is typed with a special typewriter and is checked for accuracy, being subsequently placed in the controller. (A numerical control device is shown in figure 9–27.) Then, the machine must be equipped with the necessary tools in its turret, and the workpiece must be inserted. On command, the work machine

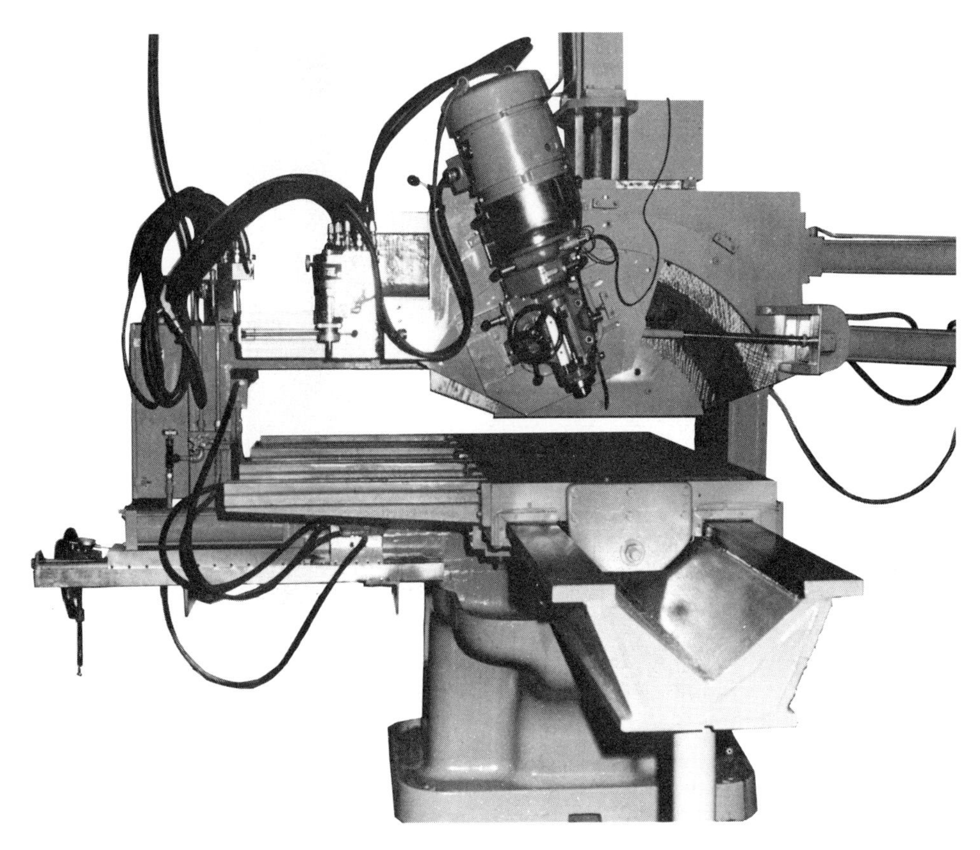

FIGURE 9–27 A typical numerical controlled milling machine having 3-D tracing and five axes milling capabilities. (Courtesy of Putoma Corp.)

proceeds to cut the workpiece in a fast sequence until all operations are completed. The operator, unless full automation exists, replaces the workpiece, and the operation is repeated.

Continuous Path Method

When continuous cutting is to be accomplished, a contour control method or system is used. Much of the principle in the point-to-point system pertains to the continuous. The programmer programs cards for the computer, the data being subsequently fed into the computer. Computer punched cards are then translated to the controlling punched tape. Due to the complexity of this method, highly skilled and trained programmers are used. Such details as curves and axes require precise understanding of the entire operation because the machine will do what the operator programs.

AUTOMATED MACHINERY

A plant that is completely automated has few personnel manning the machines. Frequently, the more sophisticated the machine, the less manual skill is required by

FIGURE 9–28 An automatic die sinking synco-trace machine. (Courtesy of Putoma Corp.)

the operator. Actually, then, the skill of the operation is contained within the total machine, being transferred there by engineers and technicians. Accuracy and repeatability are critical factors governing the movements of a machine tool. When raw materials are mechanically fed into an automatic sensoring mechanism which moves the materials to the processing machines in proper sequence so that complete manufacturing of the part is accomplished, an automated system exists. Tolerances of dimensions are usually close and repeatability is constant. Frequently, however, a technician observes either a part of or the entire operation and lends assistance when it is needed, such as the replacement of a broken drill. As has been pointed out, if the process is repeatable in a given sequence and thousands of parts are needed, the automated process is recommended. Figures 9–28 through 9–36 on pages 311–18 indicate the several types of automatic machine tools, each being electronically controlled. By placing the tools in a sequence and by furnishing conveyor systems and supporting mechanisms from the beginning of the process to the end, the machine and plant can be completely automated.

Questions

1. What is the main difference between a plain and universal milling machine?
2. How is it possible to divide a part into numerous equal divisions?
3. How is a right-hand helix on a cutter determined?

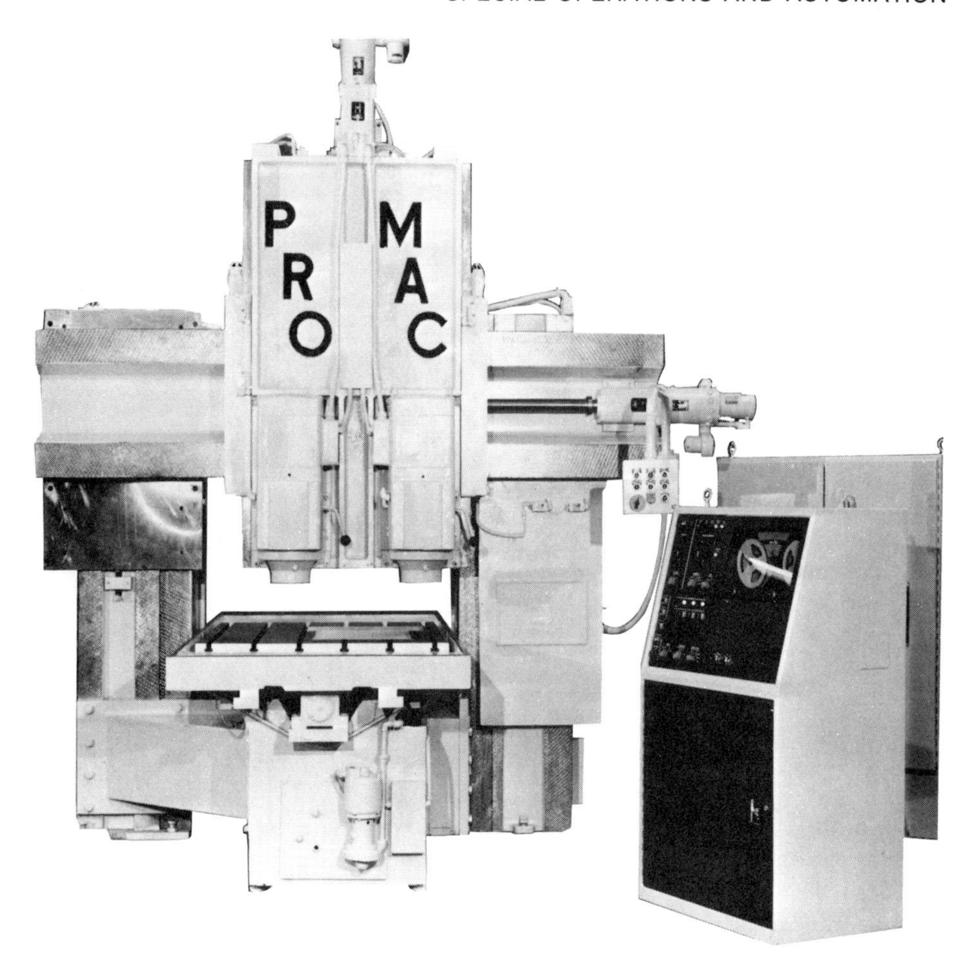

FIGURE 9–29 Two-spindle continuous path numerically controlled milling machine. (Courtesy of Putoma Corp.)

4. Describe the difference between up and down milling.
5. How is a helix machined?
6. Describe two basic types of milling operations.
7. How can a milling machine replace the work of a shaper?
8. Why is the conventional horizontal shaper gradually leaving the machine industry?
9. Differentiate between the operations of a planer and a shaper.
10. What is a planer mill?
11. List some machining operations being performed by the shaper that still make this machine needed.
12. Describe the operation of a broach.
13. List several advantages of the broach.
14. What is automation?
15. During cylindrical grinding operations, how can a taper be ground?

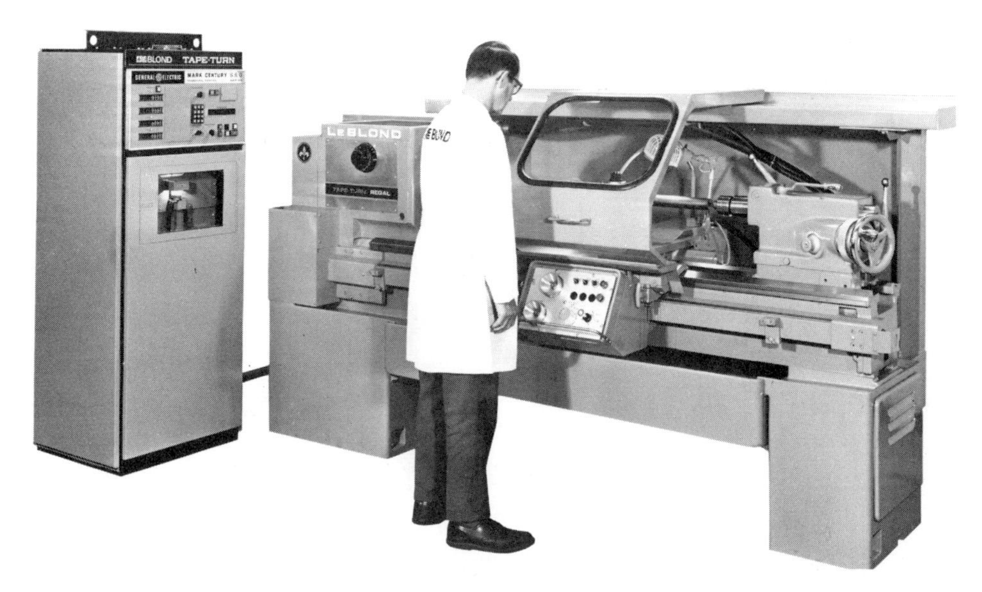

FIGURE 9-30 A numerically controlled lathe. (Courtesy of LeBlond)

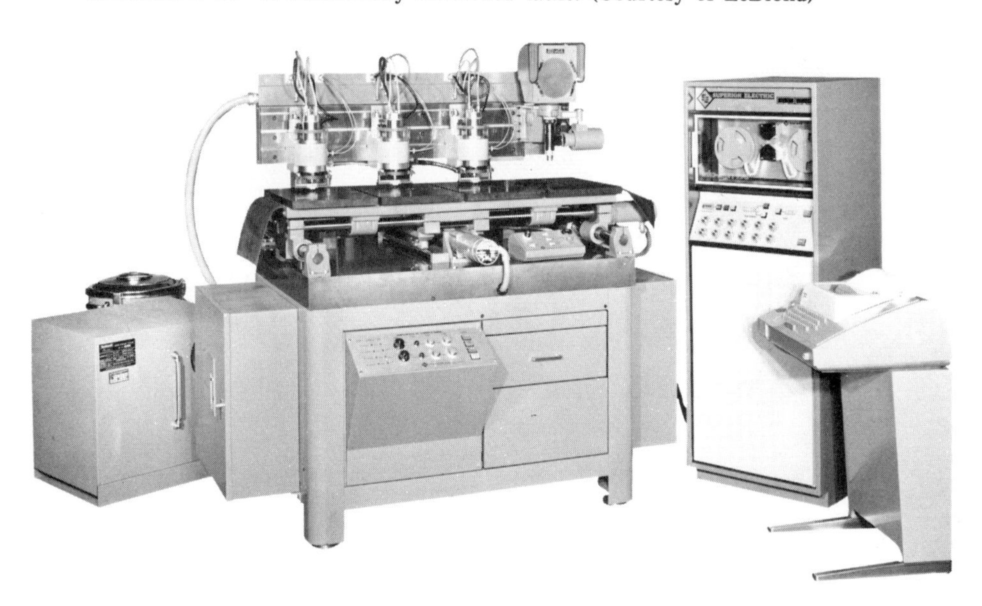

FIGURE 9-31 A numerically controlled drilling machine. (Courtesy of The Superior Electric Co.)

16. Explain how a reamer's teeth are sharpened.
17. Describe the marking on the side of a grinding wheel.
18. Why must the surface speed of a grinding wheel not be exceeded?
19. What is the purpose of coated abrasives?
20. When very small tolerances are required during a grinding operation, why must the coolant keep the workpiece from heating?

FIGURE 9–32 An inline transfer machine for production of compressor shafts in air conditioners. (Courtesy of National Automatic Tool Co.)

FIGURE 9–33 A two-way automatic tool machine with a trunnion type fixture. (Courtesy of National Automatic Tool Co.)

FIGURE 9–34 A hydraulically operated adjustable spindle and rotating table type production machine. (Courtesy of National Automatic Tool Co.)

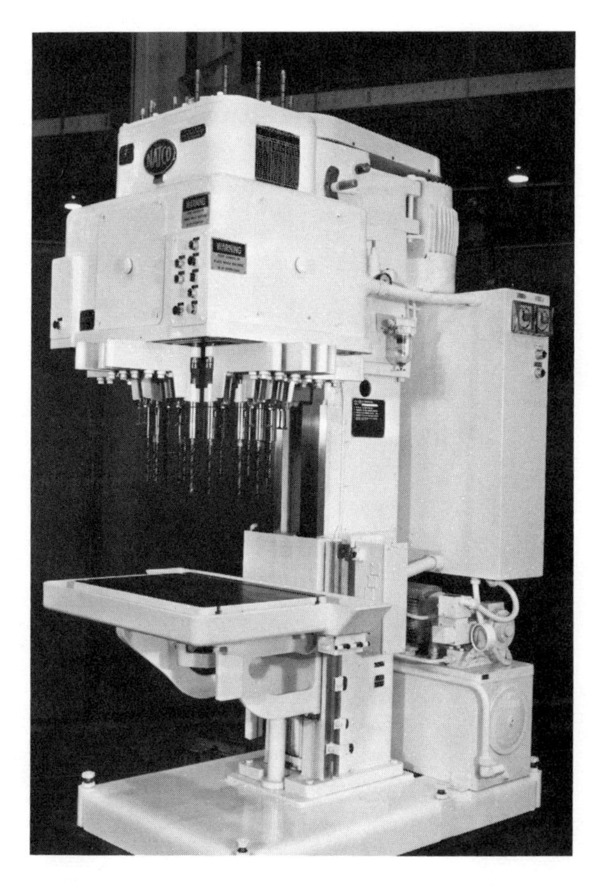

FIGURE 9-35 A vertical multiple spindle drilling machine with a table feed. (Courtesy of National Automatic Tool Co.)

FIGURE 9-36 Automated multiple spindle machine with a slip spindle plate and a slide. (Courtesy of National Automatic Tool Co.)

New Methods and Techniques

Conventional methods of metal shaping have not satisfied the growing needs of manufacturers. In some instances, very small parts requiring extremely small holes have been needed. Due to the hardness of the material, often a metal, the design was either scrapped or production delayed pending the discovery of a machine operation which could produce the holes. At the other extreme, industry was finding a need for very large domes to be used in tank and dome construction. Due to the size of the dome, no known practical method was available to shape such a large hemisphere. According to history, when the problem has been clearly defined, some kind of solution has usually been provided. As an example, the laser is now being used to produce holes as small as 0.005 inch in diameter in a ceramic, while a controlled explosion forms the large hemispheric shaped parts in a single blast. These extremes from very small to very large shaped parts have been brought about as the result of both sophistication and massiveness in part's design.

A Need for New Methods of Manufacturing

As technology grows and new fields of endeavor emerge, additional ways of doing things need to be found. Due to the ever increasing demand for high temperature materials, ceramics have moved into the field as both heat and abrasion resistors. Aluminum oxide, for example, resists extremely high temperatures and is used where unusually high temperatures exist under adverse conditions. Tungsten is also very hard and resists temperatures beyond 5000° F. This element is, therefore, used in rocket engine exhaust systems as well as for electrodes in numerous electronic devices. Machining requirements for small diameter tungsten rods and intricate shapes have introduced the electron beam and plasma arc machining processes. In this respect, some conventional methods of cutting are ineffective in many situations such as in the cutting of certain stainless steels. The plasma arc heats in excess of 20,000° F. nearly instantaneously whereby small craters of material are vaporized and the material is parted.

One of the greatest developers of new methods of shaping materials has been the aerospace industry. Because the aircraft's design has been forced to sustain excessively high stresses while fighting gravity at the same time, intricately shaped tapered parts came into use. Certain lesser stressed areas of the skin, for example, have been removed by chemical milling, resulting in a weight loss. Such a machining process by conventional methods would be prohibitive insofar as cost is concerned, but with a chemical attack at decreasing rates, desired shapes and varying thicknesses have become realities in numerous parts. A modification of the electroplating process is now used to remove large quantities of metal in controlled areas.

The search for new and faster ways of fabrication continues, the search being driven by the pressures of technological needs. All forms of energy have been scrutinized in an effort to utilize more of nature's powers. The uncommon elements are being introduced into the industrial world, such as in the electronics parts industries. Germanium wafers for use in transistors are sliced by ultrasonic machining. Titanium and ceramics are now quickly pierced by masses of energy (electron beams, for example) which are so great in their pinpointed paths that the target material boils and vaporizes away from the many small pits induced by the concentrated energy. Both hard and soft materials succumb to this pulsating point of heat. In essence, any shape can be produced by the cutting heat of new uses of energy, however, depths of cut and sizes of parts are still limited to the very small categories in some instances. In the meantime, the conventional shaping processes continue to process the bulk of industrial parts. However, what is limited today by laser cutting is to be less limited within the coming decades. Casting, forging, bending, welding, and machining will continue to play their great parts in the fabrication industries, but the newer methods are moving in.

Laser Machining

One of the most recently developed cutting methods is the laser. Due to the need to accomplish additional machining on very hard materials such as tungsten, ceramics, and diamonds, a new method involving heat from a light beam has been developed. The beam of the laser is so intense that it will produce a very small hole in the hardest of material nearly instantly. As the beam contacts the material to be penetrated, a tremendously great concentration of energy is released as a pulse onto the target material when using the ruby as the source of extremely high energy. Other materials are also available as laser sources. Such a concentration of heat on a target's surface causes instant melting and vaporization of the material, leaving a small cavity. By controlling the direction of the beam pulses with a lens, systematic cutting can be accomplished. As an example, the diamond is penetrated instantly when brought into the laser beam. Any material can be cut with the laser.

Energy of the laser results from the monochromatic light characteristics of extreme intensity and coherence emitted from a charged ruby. The optical oscillator releases parallel light beams onto an optical lens. In turn, the lens receives the light beams and focuses them onto the target material or workpiece, whereby nearly instantaneous vaporization of the parget results. The beam of energy leaving the ruby is practically parallel, as little divergence occurs even over great distances. The beam is easily directed onto the target material. The ruby is a basic laser material which is used in the more powerful laser devices such as those used in metal cutting. The principle of operation involves the pulsating discharge of a high burst of energy from the ruby configuration which when focused on the target causes immediate penetration into the target material. Because of the extreme limitations of the laser beam process of material cutting, only very small parts are processed, dimensions mostly being in the few thousandths of an inch category. Thin sheets are parted, as well as shallow holes formed, however, some lasers penetrate more than 1/4 inch in a material. Due to the several variables yet to be controlled, the sides of the penetrated material are often irregular and entrances are sometimes bell mouthed. Holes, for example, can be formed at 90° to the surface or at any angle. In this respect, it is very difficult to drill a small hole at an acute angle in work hardenable metals by conventional methods.

The laser beam process is fast because of the many particles of matter being removed from the material or workpiece in time intervals of milliseconds. As an example, a 1/8-inch diameter hole can be formed in a copper alloy shim stock in less than a millisecond. Examples of the laser's use include vent holes in diamonds, ceramics, glass, or any other material. Because of the initial high cost of equipment, the laser should not be considered unless no other method is available for processing the material. A plant specializing in microforming operations of many parts, however, will possibly find the unit cost acceptable. Due to the minute area being cut, a microscope is essential, along with a precision three-axes working stage. Focal length of the beam is variable within a span of a few inches. Figure 10–1 is a schematic of a typical laser machining apparatus.

Plasma Arc Machining

With reference to industrial applications, a plasma is considered a superheated ionized gas. Because temperatures in excess of 10,000° F. exist in the plasma, the cutting of all metals is merely a matter of directional control of the plasma. In order to form a plasma, a combination of some type of gas in the presence of an electric arc is required. Due to the exceptionally high temperatures generated, the plasma nozzle must be maintained in a constant state of cooling, usually by water flow through the electrodes of the torch.

An electric arc can be maintained at the nozzle of the plasma torch in much the same manner as in conventional arc welding by direct current, except that the initial arc is sparked by a high frequency current between the tungsten cathode and nozzle anode. When a gas such as a mixture of hydrogen and nitrogen is caused to flow from the orifice of the nozzle in the presence of a temperature which is more than twice that found in the electric arc, ionization occurs in the gas arc stream within the nozzle of the torch. When the section of metal to be cut is also made an anode, the initial arc moves out to the workpiece to complete the circuit. Control of the torch tip in regard to both distance for maintaining the arc and manipulation of the arc result in an immediate removal of the target metal. An illustration of the plasma arc process is presented in figure 10–2.

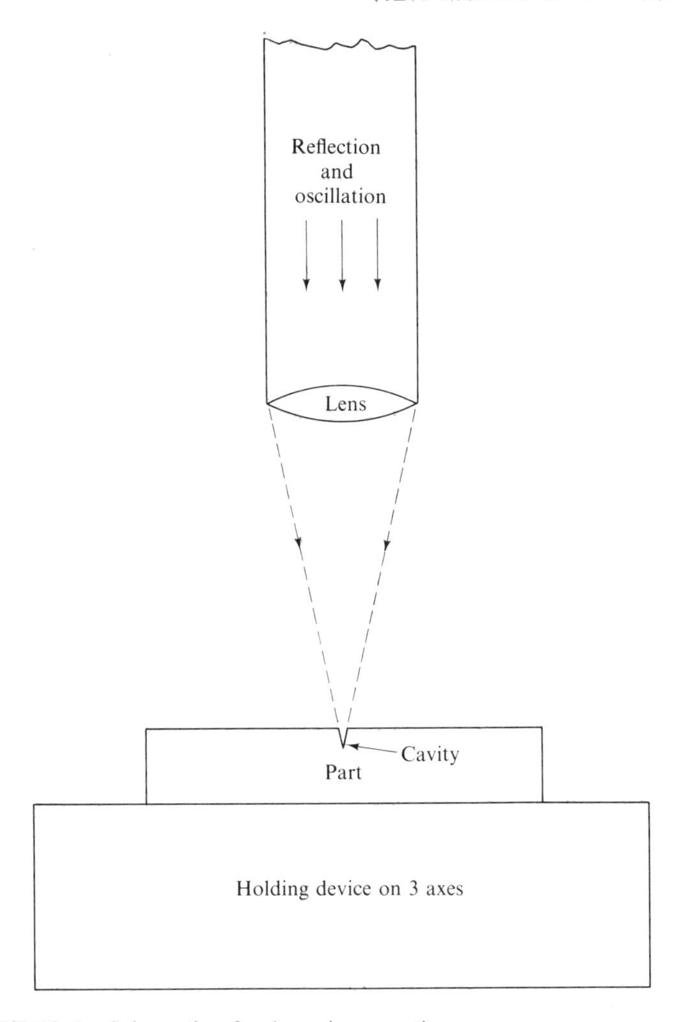

FIGURE 10–1 Schematic of a laser in operation.

When the principle of plasma arc cutting is synchronized with the techniques of tungsten inert gas welding, a metal cutting process becomes available for torch cutting of plates up to six inches thick and for some metals even thicker. Stainless steels, for example, defy severence by regular cutting techniques without an additive such as iron powder to the flame, but when attacked by plasma, cutting becomes routine. Aluminum and its alloys, as well as other metals not subject to cutting by conventional torch cutting methods, are easily severed by the plasma arc method. Depending on the setup, gas flow and amperage are adjusted to the situation. With 100 to 300 cubic feet per hour of gas flow and a current flow up to several hundred amperes, cutting potential is very rapid in terms of inches per minute cutting speed. Four-inch thick stainless steel plate, for instance, is cut at eight inches per minute, while 1/4-inch thick plate is severed at a rate of approximately 200 inches per minute.

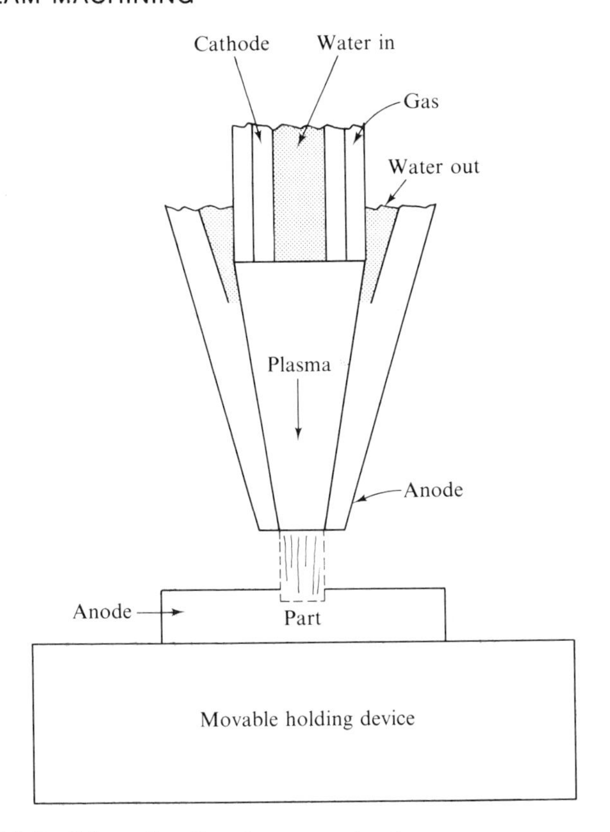

FIGURE 10–2 Schematic of a plasma arc torch.

The potential capability of this process is very great. Hand cutting processes can become automated in the severance of flat and round stock, with either the workpiece or the torch moving. Plates can be moved along the stream of the plasma, while round stock can be parted as it turns. The severance cavity is very narrow, only a few thousandths of an inch wide, while the heat affected zone often averages only less than 0.006 inch. The exceptionally high heat cutting processes—laser beam, electron beam, and plasma arc—have merely been introduced into the industrial field. Their capabilities are tremendous in the areas of materials cutting and melting.

Electron Beam Machining

Another fairly new method of cutting a material is with the use of a fast moving stream of electrons. Kinetic energy of collision between the workpiece and the beam of electrons vaporizes the material of the workpiece nearly instantly. The principle of cutting originates from the speed of a controlled diameter path of electrons. Being in a vacuum, the electrons move from the tungsten electrode toward the gate in the anode at a very high voltage ranging from 60 kV to as high as 150 kV. However, the concentrated beam of energy does not strike the anode, but passes through a hole and continues to accelerate until it collides with the workpiece. At this instant, the heat causes material removal through vaporization. No material will withstand the focused beam of electrons; the target material disintegrates instantly. Because of the tremendous speed of the electron stream at impact, some X-radiation occurs, therefore, proper radiation shielding is required.

Control of the electron beam is accomplished with the use of electromagnetic coils, while cutting is precision controlled, being observed with a microscope. Either holes or slots are formed in the workpiece, some slots being formed which are only 0.001 inch wide. Both thin and thick sections are capable of being severed. Machining is performed on parts greater than ¼-inch thick. For example, ¼-inch thick stainless steel is machined at approximately 20 inches per minute. The cutting action of the beam results from the presence of electrons leaving the cathode in a vacuum. The vacuum assures the absence of air molecules so that no deviation from the focused beam occurs due to collisions of matter. Figure 10–3 illustrates an electron beam cutting device.

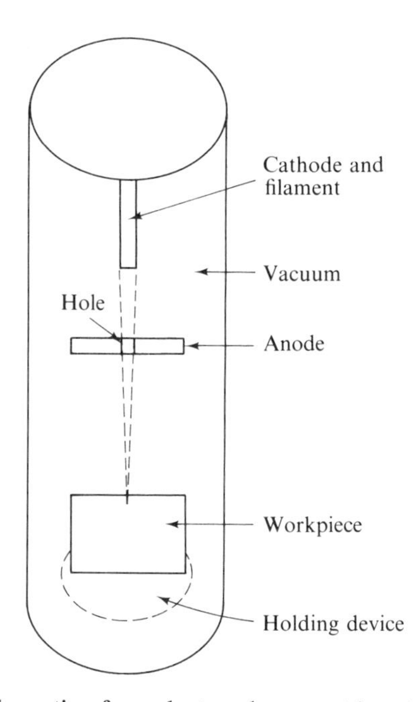

FIGURE 10–3 Schematic of an electron beam cutting device.

Because any material can be penetrated by the electron beam, consideration must be given to the power factor in cutting different materials. As an example, tungsten requires about three times as much as aluminum for metal removal, whereas steel requires approximately twice the power for aluminum. A voltage of 150 kV is sometimes necessary which pulsates at a low frequency. On impact of the beam, the cavity occurs. In this respect, deep holes may sometimes be tapered.

Electron beam equipment is self-contained in a box so that a vacuum environment can be maintained during the cutting operation. Obviously, only small and selected parts are processed, for example, foil is cut by the beam. When no other method of cutting is available, the electron beam method is considered, but its cost is high and the rate of production is low. However, advancing technology has brought forth the need for such sophisticated machining.

Ultrasonic Machining

Ultrasonic machining is the use of high frequency sound waves to cause vibration in a slurry grit which, in turn, causes collisions between grit particles and the workpiece, resulting in surface material being removed from the workpiece. The process is particularly adaptable to the nonmetals and very hard materials, however, softer metals can also be machined by this process. In regard to this process and in terms of hardness and softness, materials ultrasonically machine more effectively when their hardness values are greater than RC 48. As hardness increases, rigidity normally increases, usually accompanied with an increase in brittleness. An RC 65, for example, resists penetration of a load more than an RC 46, therefore, this resistance to penetration increases the ease of surface removal by grit bombardment. Materials in the low Rockwell C values and all those in the B-scale of hardness are subject to plastic deformation and are, consequently, more difficult to machine by this process. When hard materials require holes and other cavities in their surfaces and heat cannot be applied, the ultrasonic cutting process is used. Keep in mind, however, that only small quantities of material are removed.

The workpiece is securely attached in a vise or other fixture which has sufficient mass in damping transferred vibrations. The fixture, which is in a tank, is surrounded by a water-grit slurry, the fixture having the usual three-axes movement capability. When a vibrating tool is brought into very close proximity of the workpiece, the grit particles are bounced against the surface of the workpiece, each grit particle leaving a cavity. Due to the high frequency vibrations of the grit, cavity forming accelerates quite rapidly. When the working tool is properly controlled, a predetermined cavity in a very hard material can be formed. Figure 10–4 is a schematic of an ultrasonic cutting device.

The principle of ultrasonic cutting is based on resonance within an elastic material such as a ferromagnetic metal. The alpha iron body of a magnetostrictive shape in the mechanism is permanently connected to a different kind of metal which provides additional elastic length changes in the total vibrating length of the moving part. High frequency electricity alternating around the mechanism's electric coil causes mechanical vibrations in the shank of the working tool. When this working tool and holder are attached to the magnetostrictive mechanism in the presence of the slurry, an ultrasonic cutting machine exists. Approximately 22,000 Hz are applied to the coil which is wrapped around the ferromagnetic shape in such a manner that lengthwise vibrations occur in the total mechanism. This length change at the surface of the working tool is only about 0.001–0.002 inch, consequently, the slurry gap is very small, also being approximately 0.001–0.002 inch. The slurry gap is the grit impingement area.

An alloy containing nickel makes an excellent body for the mechanism which is fabricated as part of the magnetostrictive mechanism. A toolholder is attached to the mechanism. The low carbon steel working tool is machined to the shape of the desired cavity and temporarily attached to the holder. When the vibrating tool is brought near

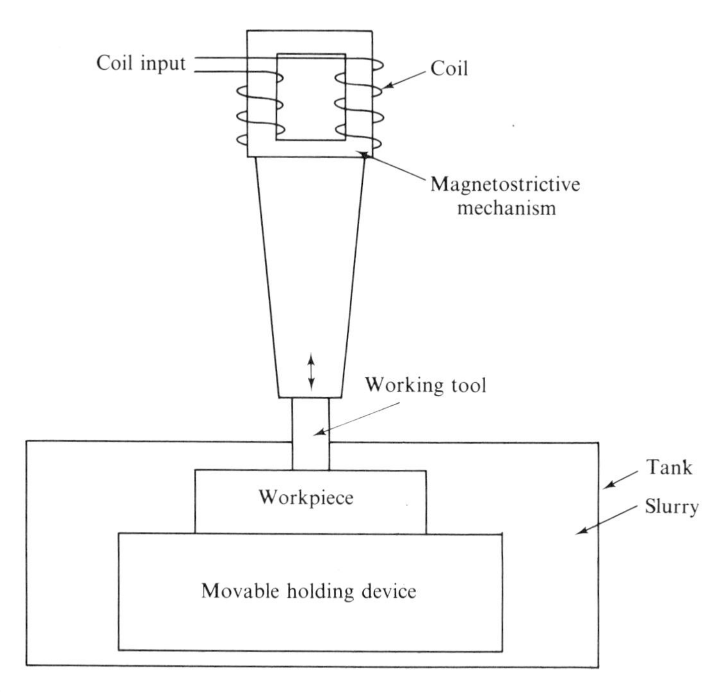

FIGURE 10–4 Schematic of an ultrasonic cutting device.

the workpiece, machining occurs geometrically similar to the shape and size of the face of the tool, plus approximately the size of the grit particle on each surface. Such precision machining on hard and delicate materials is accomplished by the precision of the total ultrasonic mechanism and skill of the operator. The same abrasives are used in ultrasonic machines as are used in grinding wheels. Roughing the cavity is accomplished with approximately a 320 grit size, while finishing requires the very fine particles such as an 800 grit. High frequency vibrations generate heat, therefore, the slurry must be cooled to approximately 40° F.

Brittle materials which are to be machined, as well as thin materials, should be cemented to a backing plate to prevent chipping and fracture of the part. Short length tools are used, the tool moving up and down only 0.001 inch in many situations. Too long a tool stroke accelerates fatigue failure. A depth of cavity in the vicinity of two inches is practical. In this respect, the width of the cavity must accommodate the entry of the tool. Grits must not be in mixed sizes, nor should the moving slurry be over 30% grit. Boron carbide has shown excellent cutting capabilities, therefore, it is used in many ultrasonic cutting operations. Ultrasonic cutting operations can produce finished parts to within a tolerance of plus or minus 0.001 inch.

Electrical Discharge Machining

Many types of heat treated dies are machined by the electrical discharge process. Basically, any material can be machined by this process, whether hard or soft, if it conducts electricity. The process consists of metal removal by means of pulsating electrical sparks. Each spark causes a small cavity to form in the workpiece, and when the sparks occur rapidly, metal is removed at a significant rate. Due to the high initial cost of the machine, other processes should be used unless no other method is available. When metals with hardness values in the RC 60s require machining, or when complex cavities and cuts must be made, the electrical discharge process becomes necessary.

A typical electrical discharge machine consists of a workpiece holding fixture fixed within a suitable tank in which the dielectric fluid flows under a low pressure (fig. 10–5). A vertically operating tool having other directional operating movements is

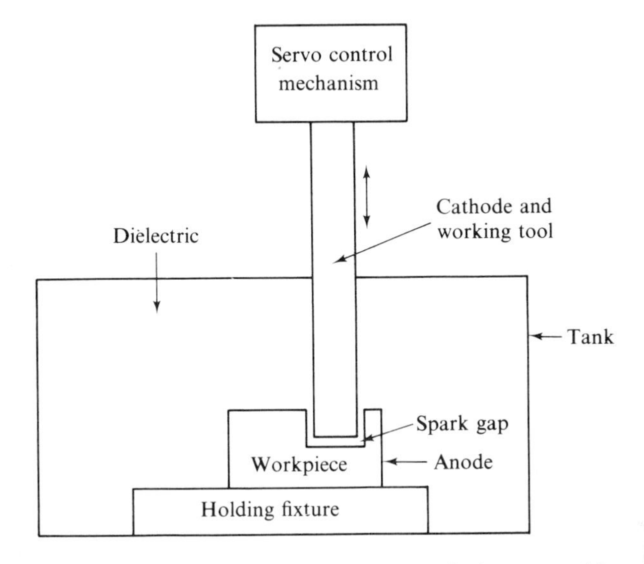

FIGURE 10–5 Schematic of a typical electrical discharge machine.

arranged so that the tool can be brought in near contact with the workpiece. The workpiece, which is firmly secured in the machine's fixture, is electrically positive, while the cutting tool is normally negative. When the electrical system is on and ready to operate, the tool is lowered to its predetermined gap from the workpiece, and a spark completes the circuit. Due to the extremely high temperature created at the spark, the oily dielectric becomes ionized in the vicinity of the spark, enabling the following sparks to vaporize the surface of the metal, reducing its thickness at the sparks' points of contact. Sparks are emitted at predetermined pulses of direct current, the pulses having a large variable frequency to accommodate different machining situations. Voltage varies from a few volts up to several hundred, while the current varies up to less than 500 amperes.

Feeding of the tool occurs as either a quill or ram type advance. Due to closeness of the spark gap, less than 0.020 inch in many situations, an automatic servo-control maintains a constant spark gap for continuous and effective operation. The tool's face helps determine the final cavity shape, but a very small overcut occurs. When an identical part is required, the tool shape is copied from the design of the part. Due to dangerous electrical conditions existing at times, the tool is insulated from the machine.

Because the sparking affects both tool and workpiece, the tool must be replaced at intervals. Tools are made from several materials including graphite, tungsten, steel, and copper alloys. The graphite tool is the longest lasting and is easy to machine into the correct shape. Due to the sparking actions between electrodes, the straight polarity setup is more often used. Reverse polarity, in which the tool becomes the anode, is used when roughing with graphite, when electrodes are steel or aluminum, and when using copper-tungsten electrodes during the cutting of steel.

Because the tool is subject to wear due to sparking actions, the tool's shape at its face will gradually lose its corners and tend toward roundness. This wearing action leads to a wear ratio which is the volume of material removed from the workpiece divided by the volume of material removed from the tool. A side effect of electrical discharge machining is the resulting surface hardness of the workpiece in the zones where machining has occurred. Because the hot surface is oil quenched in the spark gap, a high surface hardness will result if the chemistry of the material has hardening capability. However, the surface hardness is very shallow, a few thousandths of an inch thick, and may be insignificant in many applications. A disadvantage of this process appears to be in the loss of surface resistance to fatigue. In this respect, many stressed parts should be shot peened after the cutting operation to induce compressive stresses in the surface layers. A typical electrical discharge machine is shown in figure 10–6. The working area of an electric discharge machine is illustrated in figure 10–7.

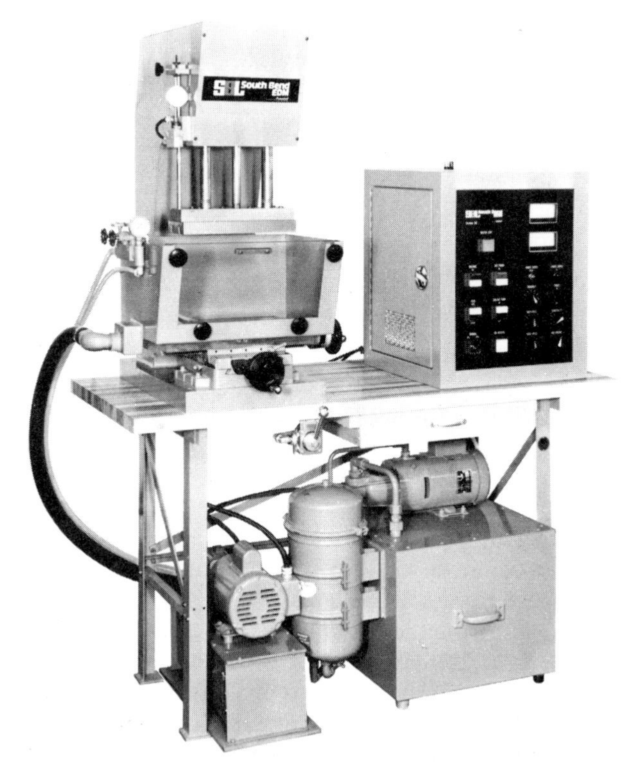

FIGURE 10–6 An electrical discharge machine showing electrode, tank, and working area. (Courtesy of South Bend Lathe)

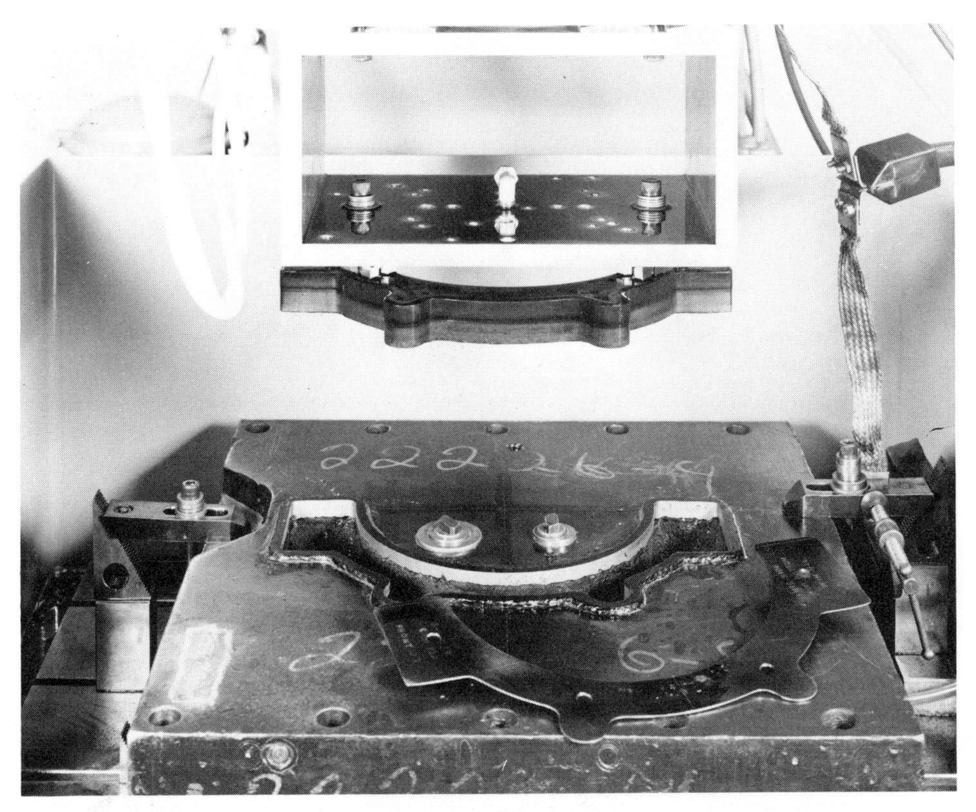

FIGURE 10–7 A close-up of the electrode, workpiece, and template of the electrical discharge machine shown in figure 10–6. (Courtesy of South Bend Lathe)

Electrochemical Machining

The principle of electrochemical machining is based on the dissolving of a portion of the metal workpiece in an electrolyte. When a direct current flows from the anode through an electrolyte to the cathode, metal is removed from the anode into the electrolyte and is deposited at the cathode. Should the flow of material be intercepted and washed away just prior to reaching the cathode, the cathode will remain clean, but the loss of metal will continue at the anode. Based on this process, large amounts of metal can be removed when the workpiece is made the anode. In this respect, the workpiece must be able to conduct electricity. In order to use the process, the cathode becomes the work tool and is brought in very close proximity to the workpiece, or anode. The shape of the tool's face is transferred to the workpiece, but when lateral movement of the tool occurs, a larger hole is made in the surface of the workpiece, depending on the traveling direction of the tool. The process is very efficient, but the initial cost of equipment is high. However, the machine's capability to shape extremely hard metal, along with its capability of forming oddly shaped holes, makes the process economically worthwhile in many situations. Figure 10–8 is a schematic of a typical electrochemical cutting machine.

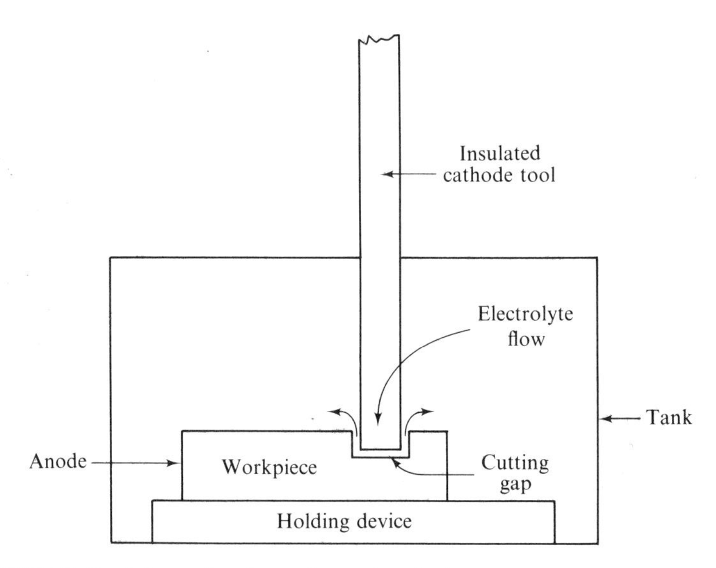

FIGURE 10–8 Schematic of an electrochemical cutting machine.

The electrochemical machine is arranged so that the work holding fixture and the tool with its directional travel are capable of performing the metal removal operation. Voltage and current settings are related to the needed current density per square inch between the working tool and the workpiece. The gap between the tool and workpiece is small, often being in the variable vicinity of 0.015 inch. From 10,000–20,000 A are often used at low voltages when large volumes of metal are to be removed. Because the working gap is small, a high pressure pump circulates the electrolyte through the gap. The returning electrolyte is clean, the metallic sludge having been trapped and drained away.

The electrolyte varies according to the job situation and consists of aqueous solutions such as potassium or sodium chloride or other inorganic salts. The purpose of the flowing electrolyte is to immediately wash away the reaction material from the vicinity of the anode, along with keeping the gap area fairly cool, and to conduct the direct current to the cathode. In regard to the concentration of the electrolyte, care must be used in maintaining a proper solution of salt in water. Often, when using common salt, about two pounds per gallon of water provide good results. The circulation must also provide for temperature control, in addition to removal of the sludge.

The working tool must be shaped to the shape of the desired cavity in the workpiece. This infers no movement of the tool except vertical feed which is constant. Compensation must be made for the slight overcut of the tool; possibly trial and error are best suited in finding the proper size tool. Tool materials include copper and its alloys, titanium, and the stainless steels. When a tool is moved across the surface of the workpiece, a machined dimension occurs in the workpiece. However, the flow of material from the workpiece seeks and follows the surface shape of the tool. Therefore, trenches can be formed in the workpiece, as well as many different types of configurations. The penetration rate of the tool varies with the current, and metal removal rate across the face of the workpiece is several inches per minute at a given depth of penetration. In order that the original tool contour will be maintained as the tool moves into the hole, proper insulation of the tool's sides is essential.

With a fixed voltage and the need to maintain a constant metal removal rate from the workpiece, the feed rate must be steady and uniform. Feed rate depends mainly on the current density. The electrochemical process of metal removal is almost unique in its many unusual applications.

Chemical Machining

The chemical machining of metals is accomplished by a selective chemical attack on the metal's surface. Because different chemicals attack different metals at different rates, careful planning and preparation of workpieces must be done. Due to the nature of chemical machining, there is no competition with conventional machining processes. Therefore, it is used when other methods are not satisfactory. Basically, a section of metal is quickly attacked when in the presence of a chemical that reacts with the metal. Both acid and alkali solutions are used in the chemical machining of metals. The machining process consists of two main procedures, namely, chemical blanking and chemical contouring.

CHEMICAL BLANKING

Chemical blanking has been used for many years in the production of thin sheet metal parts such as are normally blanked from a press. During the process, the metal workpiece or part is carefully coated with a photosensitive or other masking material which provides protection for certain areas of the metal while allowing chemical attack on other areas. Prior to a masking material application, the workpiece is thoroughly cleaned and dried. Then, the maskant is applied, usually by dipping, and dried.

Several methods are available with regard to the use of the master, which is a template or drawing, to cause the desired areas of metal to be etched away during a subsequent operation. The use of ultraviolet light during exposure in conjunction with the master and workpiece is a common procedure to build up the necessary resistance to the developer in certain areas of the photoresistant mask. After exposure, the workpiece is again exposed to a developer solution which causes the unprotected developer to be removed. Then, heating to about 225° F. for a few minutes prepares the unprotected surfaces for etching, ferric chloride being a common etchant. Finally, the blanking operation is completed by etching and rinsing the completed part in water.

The other method, the photographic process, uses the negative of the drawing to expose the desired areas of the mask so that subsequent etching will remove the designated metal from the workpiece. When the mask has been adequately exposed to an ultraviolet light or a medium wattage bulb in the photographic process, the workpiece is developed and subsequently etched, whereby the desired metal is removed from the unprotected areas. Finally, the remaining maskant is removed, and the part is ready for use. An alternate method includes the scribing of the design on the maskant followed by peeling away the areas which are to be etched.

A typical example of the chemical blanking process is the production of an electronic circuit board. First, the basic board is acquired, being made of phenolic, glass-epoxy, or some other insulating material. Copper foil is laminated to the surface. Then a drawing is produced of the circuit, and a photographic negative is made. A photoresistant sensitizer is then thoroughly coated on the copper surface and dried. When the negative rests on the board and the assembly is exposed to the proper light, the nonexposed areas are unable to remain intact and are later washed away during developing. Subsequent etching of the copper occurs when the chemical attacks the bare copper. Ferric chloride is an example of an etchant. The electrical circuit remains after stripping and washing which are required to remove all foreign materials from the board. Figure 10–9 illustrates the remaining circuit after a solid sheet of copper has been etched away.

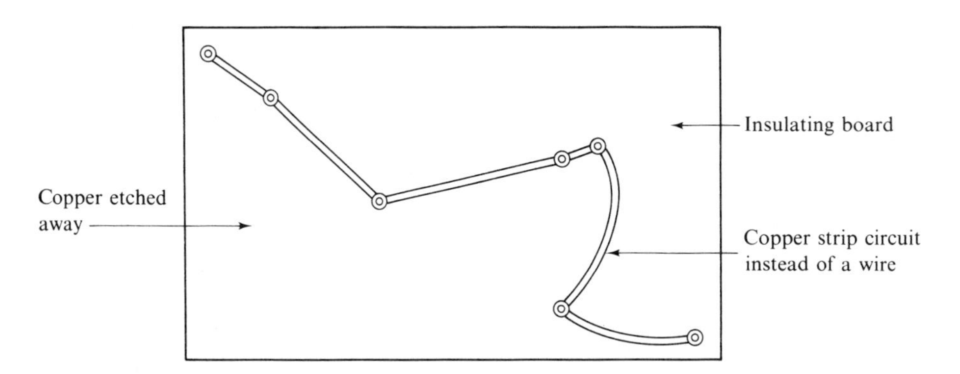

FIGURE 10-9 Part of an electrical printed circuit board.

CHEMICAL CONTOUR MACHINING

In chemical contour machining the etching process usually applies to larger parts, some being more than 50 feet in length. Basically, the chemical milling process was developed for the aerospace industries in order to allow strength in a part at a reduced weight. As an example, if the center portions of the thickness of a structural member are removed, adequate strength will remain around the perimeter to resist bending and other loads while the part will be much lighter in weight. Also, many structural parts, especially the 2024 aluminum alloy, are etched to produce tapers along a dimension. The taper allows adequate strength, but reduces the weight by an appreciable amount. Because all aerospace vehicles contend with gravity, any weight reduction in the structure allows a greater payload.

Fundamentally, the chemical machining, or milling, of large parts is very similar to the blanking process, except that the maskant is usually scribed with the design and the unwanted areas peeled off to allow the chemical to remove the metal. Because no mechanical stresses are induced into the part being etched, the process lends itself favorably to fragile materials such as honeycomb panels. Also, hard and brittle materials can be processed, as well as soft or heat treated parts. The part to be milled is cleaned, dipped into a liquid maskant, and dried. Then, the pattern is placed on the maskant and areas of metal to be etched are carefully peeled away. The part is next placed within the liquid etchant and timed until the desired thickness and shape has been attained. Subsequent washing for neutralization of the etchant is accomplished, and the remaining maskant is peeled away.

Chemical milling produces a uniform and smooth surface, but there is an undercut at the surface whose radius is approximately equal to the depth of etch. This undercut is accounted for during the layout on the maskant. Step etching is frequently performed in order to provide for several thicknesses in the cross section of the part. Because the etchant attacks the metal equally, corners are rounded with smooth radii. When tapering is to be performed, the exposed metal is slowly and uniformly withdrawn from the etchant at a rate commensurate with the needed taper. Because the area of metal which leaves the bath last will be attacked the longest, it will also be the thinnest.

Maskant materials must be capable of adhering to the metal part, even at the edges where scribing occurs. Any lifting of the maskant will allow metal to be removed in these areas. Materials for maskants include the plastics and elastomers such as polyvinyl chloride and neoprene. Qualities of the maskant must also include ability to be easily peeled away before and after the milling operation.

Etchants include numerous chemicals which are applied while hot. Ferric chloride is used on several metals such as aluminum, copper, and steel with a high effectiveness. Sodium hydroxide is commonly used on the aluminum alloys at temperatures around 200° F. Depths of etching include cavities up to ½ inch with tolerances equal to conventional machining. A main advantage of the chemical contour process is its ability to remove large areas of exposed metal at the same time. However, the rate of metal removal is slow, being in the order of 0.0015 inch per minute. A disadvantage is the requirement for metals having homogeneous structures. A welded part, for example, is not usually etched in a uniform manner. Figure 10–10 illustrates a masked part entering the etching tank. The effects of undercutting are shown in figure 10–11*a*, while selected etching is illustrated in *b*. Figure 10–11*c* shows the effects of controlled withdrawal of a part from the hot etching solution, resulting in a taper.

Explosive Forming

The use of TNT and other high energy explosives has demonstrated the capability of this process in forming large parts which are not practical to form in any other way. Two systems are being used, one for a small sheet metal part confined between dies and the other for much larger parts where only one die is used. Economics and practicability are two important considerations when a decision is made to use this process. Because an inherent danger exists when explosives are used, proper safety precautions must be taken such as the elimination of spark- and fire-causing objects and the elimination of electrical mechanisms, except for the detonating device.

The one-die, or unconfined, explosion forming system is used to shape very large parts such as hemispheres which are not formable by other methods. When a very heavy die, on which rests the workpiece or plate of metal, is placed on a substantial foundation within a thick-walled tank containing water, the force of the explosion lays the workpiece against the surface of the die. For symmetrical shapes, the explosive is centered above the workpiece a short distance. The space between the workpiece and die must be evacuated of air so that the workpiece will lie smoothly against the die and not on a cushion of compressed air at the moment of explosion. Vacuum pumps are used after the workpiece has been attached properly to the die's shoulders. Some

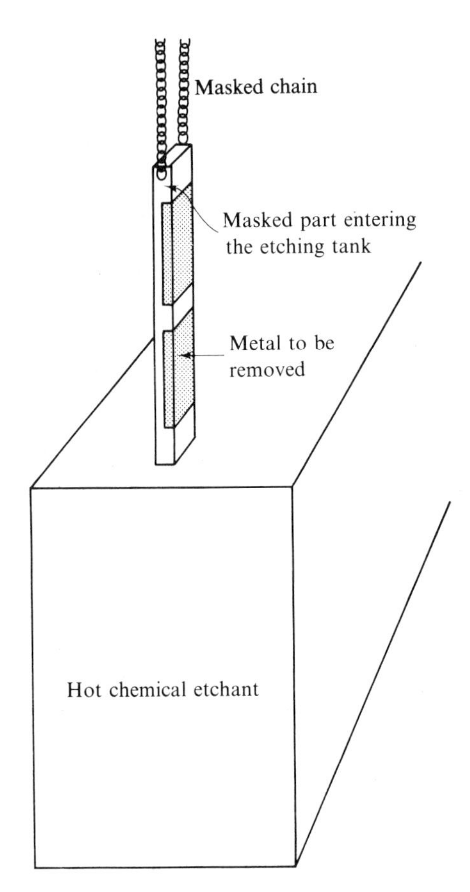

FIGURE 10–10 A masked aluminum alloy entering a hot bath of caustic soda solution.

means must also be provided to allow the flat plate of metal to slide inwards as it moves downwards into the die under the force of the explosion. Figure 10–12 illustrates how a hemisphere is formed from a flat plate by a controlled explosion.

Metallurgically, the explosive process results in no unusual microstructure other than the expected cold-worked structure. Because force is exerted evenly in all directions, the workpiece moves quickly until it is confined. Plastic deformation occurs, therefore, permanence in shape results. Tolerances are reasonable and acceptable, being comparable to other forming operations. However, subsequent edge machining of small deviations may be necessary, depending on the shape of the part. An example of microstructure is the stretching of a welded seam during the shape changing. Cast characteristics are reduced, and strength characteristics are improved. Metals which lend themselves to plastic deformation are formable by explosive means.

Questions

1. List several advantages of the laser machining process.
2. For best results, why should the electron beam machining process be conducted within a vacuum?

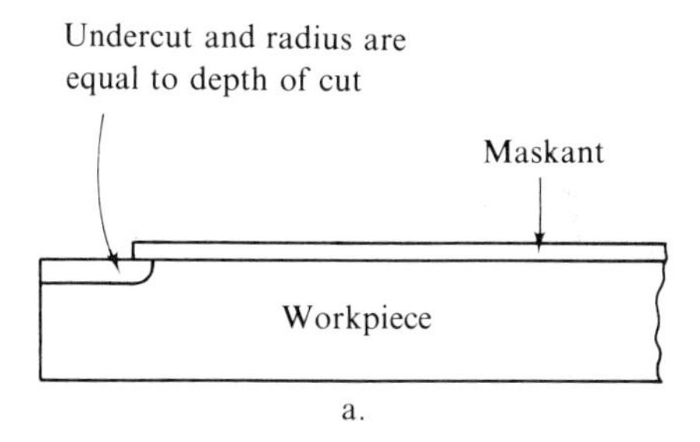

Cross section of a part showing effects of undercut. Radius of undercut is equal to depth of cut.

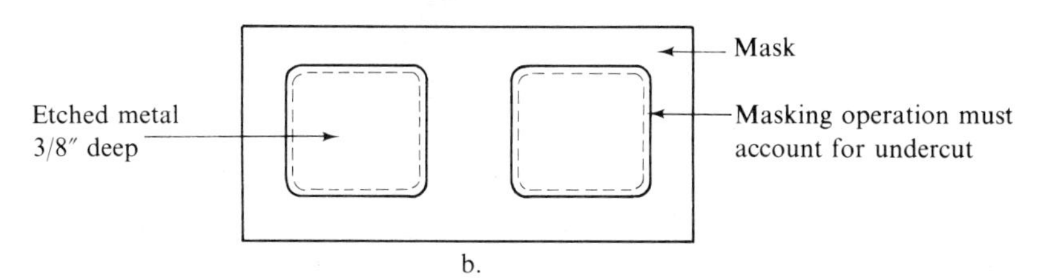

Accounting for undercut during masking.

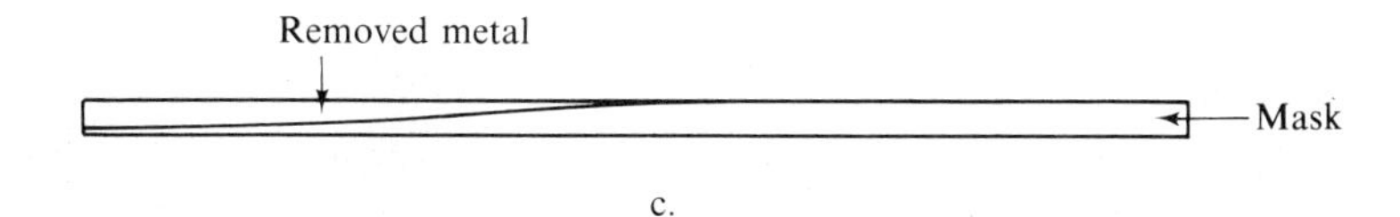

A tapered structural part formed by controlled withdrawal.

FIGURE 10–11 Effects of protective maskants.

3. What process will form a hole in a diamond?
4. What principle gives the laser its powerful beam of light?
5. What is a plasma?
6. Why do hard materials machine more easily than soft materials when the ultrasonic machining process is used?
7. Describe the electrical discharge machining process.
8. How does the electrochemical machining process operate?
9. How is a taper made on a sheet of aluminum alloy when using the chemical machining process?
10. Give an example of a part made by explosion forming and explain how the shape is made.
11. What factor in technology has demanded the availability of new methods of shaping materials?
12. Explain a main limitation of laser machining.
13. Explain a main advantage of the electron beam machining process.
14. Explain a main advantage of plasma machining.
15. What effect does grit size have on the final dimensions of ultrasonically machined parts?

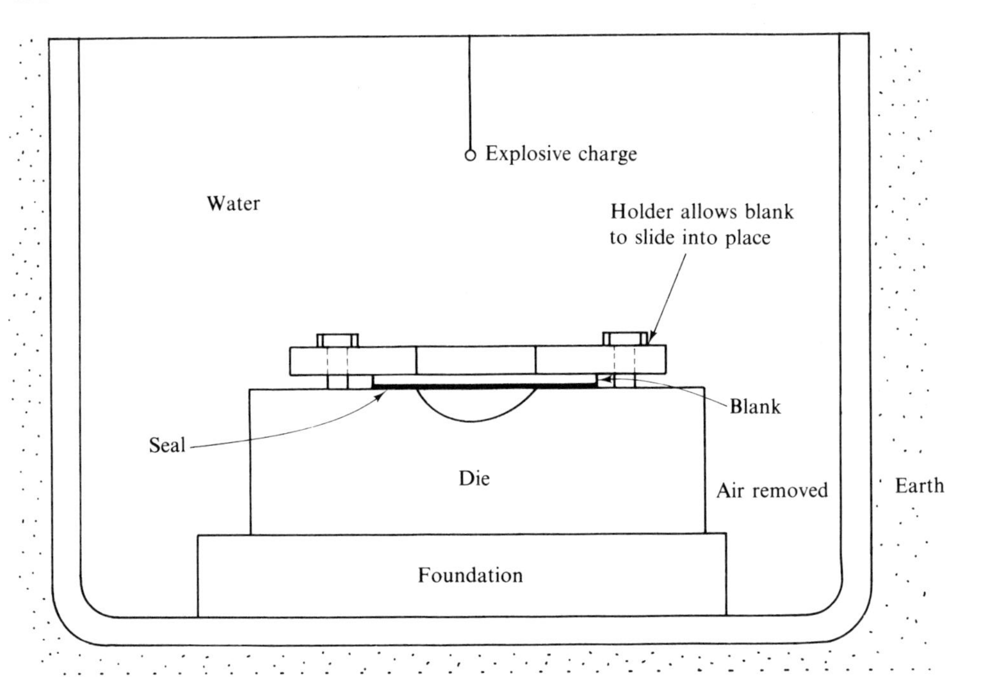

FIGURE 10–12 An unconfined explosion forming system.

16. Explain an important advantage of the electrical discharge machine process.
17. Why must the working tool be shaped to the shape of the cavity during the electrochemical machining process?
18. What is chemical machining?
19. What new method of machining generates X rays?
20. Name two very high temperature use materials.

11

Heat Treatment of Metals

The heat treating of metals is a series of processes which include the heating and cooling of metals for the purposes of imparting specific properties to the metals. The purpose of heat treatment may be to soften a metal for machining operations, or the purpose may be to increase its strength so that it will be capable of holding a given load while in service. In this respect, it must be pointed out that all metals do not respond to heat treating processes in equal ways, that is, copper can be hardened and its strength increased by cold working, but it cannot be hardened or strengthened by heat treatment. When cold-worked copper is heated, it is softened, and will be softened regardless of how fast it is cooled. None of the elements listed in table 11-1

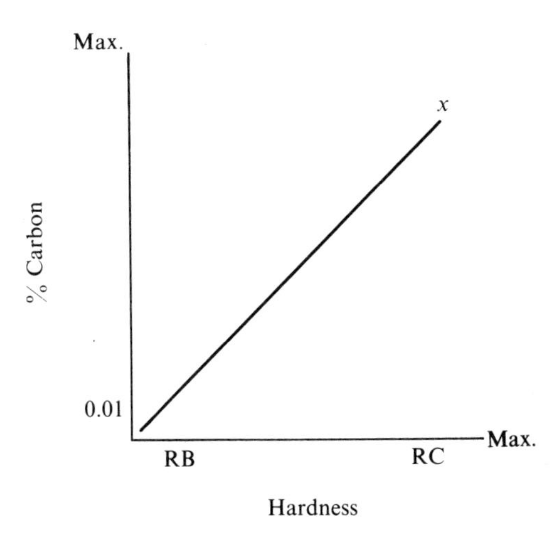

FIGURE 11-1 Carbon effects on potential hardness.

TABLE 11-1 Common Elements Used in Manufacturing

Element	Symbol	Melting Point °F.	Typical Characteristics
Aluminum	Al	1220	Soft, weak, ductile metal
Antimony	Sb	1167	Hard brittle metal
Argon	A	- 309	Inert gas
Beryllium	Be	2332	Hard metal used as alloy
Bismuth	Bi	520	Metal used in low melting alloys
Boron	B	3690	Nonmetallic used in alloys
Cadmium	Cd	609	Ductile metal used in plating
Carbon	C	6740	Nonmetallic used in steel
Cerium	Ce	1479	Malleable metal used as an alloy
Cesium	Cs	84	Soft active metal
Chlorine	Cl	- 150	Gas used in purifying metals
Chromium	Cr	3407	Hard, corrosion resistant metal
Cobalt	Co	2723	Hard metal used in steel
Copper	Cu	1981	Soft ductile metal
Gallium	Ga	86	Metal similar to Al
Gold	Au	1945	Noble metal
Helium	He	- 453	Inert gas
Hydrogen	H	- 434	Explosive gas
Indium	In	313	Malleable metal used in alloys
Iridium	Ir	4449	Hard metal used as alloy
Iron	Fe	2798	Malleable metal in all steels
Lead	Pb	621	Malleable metal
Lithium	Li	357	Lightweight metal
Magnesium	Mg	1202	Lightweight structural metal
Manganese	Mn	2273	Hard metal used as alloy
Mercury	Hg	- 37	Metal used in thermometers
Molybdenum	Mo	4730	Tough metal used as alloy
Nickel	Ni	2647	Malleable metal used as alloy
Niobium	Nb	4474	Metal used as alloy
Nitrogen	N	- 346	Gas used in heat treating
Oxygen	O	- 362	Gas used in manufacturing
Phosphorous	P	111	Nonmetallic found in steel
Platinum	Pt	3217	Malleable metal
Potassium	K	147	Soft metal used in manufacturing
Radium	Ra	1292	Radioactive metal
Rhodium	Rh	3571	Somewhat hard metal used as alloy
Selenium	Se	423	Element used in photoelectric cells
Silicon	Si	2570	Element used in steel making
Silver	Ag	1761	Malleable noble metal
Sodium	Na	208	Malleable metal used in manufacturing
Tantalum	Ta	5425	Hard metal used as alloy
Thallium	Tl	577	Soft metal used as alloy
Thorium	Th	3182	Radioactive heavy metal
Tin	Sn	449	Soft metal used in plating

TABLE 11-1 (continued)

Element	Symbol	Melting Point °F.	Typical Characteristics
Titanium	Ti	3035	Lightweight structural metal
Tungsten	W	6170	Heavy metal used in steel
Uranium	U	2070	Radioactive metal
Vanadium	V	3450	Malleable metal used in steel
Zinc	Zn	787	Malleable metal used in plating
Zirconium	Zr	3366	Metal used in steel

NOTE: Melting temperatures provided through the courtesy of American Society for Metals.

will respond to strength increases by heat treatment. On the other hand, a high carbon steel will be softened when cooled slowly from a bright red temperature, but when it is quenched in water from the bright red temperature, it becomes glassy hard and extremely strong in tensile strength. Carbon, the most critical of the elements in steel, has certain effects on hardness, as illustrated in figure 11-1. As the carbon content increases, the hardenability increases up to a maximum potential. Basically, the chemistry of the metal must be known before any heat treatment can be performed.

Purposes of Heat Treating

Metals include most of the elements, and when several metals or a metal and a nonmetal such as carbon are alloyed with iron, manganese, and silicon, a metal called steel is produced. There are hundreds of alloys, each alloy having its own particular heat treating capability. In view of these many different capabilities among the metals, specific purposes of heat treatment are applicable to specific metals. The hardness results of two different steels are compared in figure 11-2 after a hardening

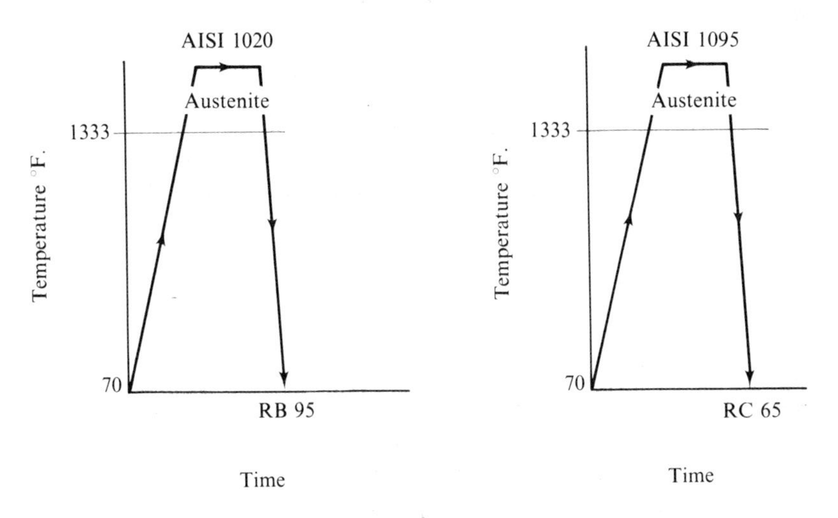

FIGURE 11-2 Low and high carbon steel specimens quenched in water from austenite.

attempt. As shown, the low carbon steel did not harden. In other words, all metals cannot be hardened by heat treatment, nor can all metals be softened. A specimen of copper is cooled at various rates to show the inability to harden this element, as illustrated in figure 11–3.

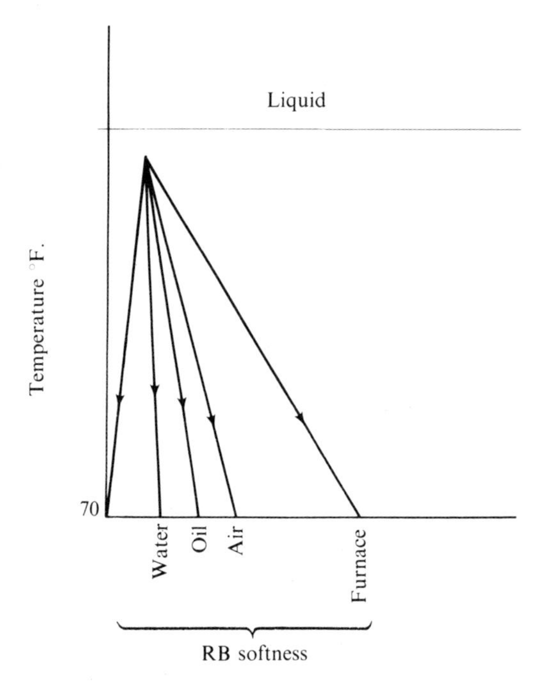

FIGURE 11–3 Attempts to harden copper by varying rates of cool from different temperatures.

The purpose of heat treating, then, applies to a specific metal in which certain strength or other properties need to be obtained. In general, the iron base, or ferrous, group of metals respond to one or more of these heat treating procedures: annealing, normalizing, hardening, tempering, spheroidizing, stress relieving, and case hardening. The nonferrous metals respond to one or more of these heat treatments: annealing, solution treatments, and precipitation treatments. An important point in regard to heat treatment is that the last heat treatment imparts the final properties in the metal, even though a prior treatment is necessary.

Chemistry of the Metal

The chemistry of a metal relates to its type and quantities of constituents. One group of metals includes those containing one or more elements only. Another group includes those metals containing an element and a compound or a combination of elements and compounds. If only one element is present, the metal is a *pure metal.* If two or more constituents are present, the metal is an *alloy.* An alloy may exist in a condition whereby its constituents are in solution, or it may exist as a mixture of its constituents. When the solution type of alloy occurs, it contains many of the characteristics of a pure metal, and its heat treatment will be similar to that of a pure metal. In these respects, the pure metal and the solid solution type of metal can be softened by heat treatment, but not hardened. The alloy occurring as a mixture will normally respond to some degree of hardening by heat treatment and softening. Table 11-2

TABLE 11-2 Typical Alloys

Alloy	Heat Treatment	
	Will harden	Will not Harden
Aluminum-copper	x	
Aluminum-manganese		x
Aluminum-zinc-copper-magnesium	x	
Copper-aluminum-tin (bronze)	x	
Copper-beryllium-tin (bronze)	x	
Copper-chromium	x	
Copper-nickel-silicon	x	
Copper-zinc (brass)		x
Magnesium-aluminum-zinc	x	
Magnesium-aluminum-manganese	x	
Nickel-chromium		x
Nickel-copper		x
Nickel-Cu-Mn-C-Fe-Si-Ti-Al	x	
Steel—low carbon		Not appreciably
Steel—high carbon	x	
Steel-nickel	x	
Steel-nickel-chromium	x	
Steel-nickel-chromium (austenitic)		x
Steel-molybdenum	x	
Steel-chromium	x	
Steel-chromium-vanadium	x	
Steel-tungsten	x	
Steel-nickel-chromium-molybdenum	x	
Steel-silicon-manganese	x	
Titanium-aluminum-molybdenum	x	
Titanium-aluminum-vanadium	x	

lists some typical alloys, the table indicating if the alloy can be hardened. In other words, the chemistry of an alloy includes a designated quantity of elements, the percentage of each element reflecting the total capability of the alloy.

The presence of an element in an alloy does not mean that the alloy is heat treatable for increased strengths. In most instances, a minimum amount of an element must be present in the alloy if the purpose of heat treatment is to harden the alloy. As an example, a small amount of carbon in steel, 0.08%, will not allow the alloy mixture to harden any appreciable amount, but if 0.80% carbon is present, the alloy has the capability to fully harden. Figure 11-4 illustrates the effects of varying quenching rates on a high carbon steel. As illustrated, the water and brine quenches produce the hardest values.

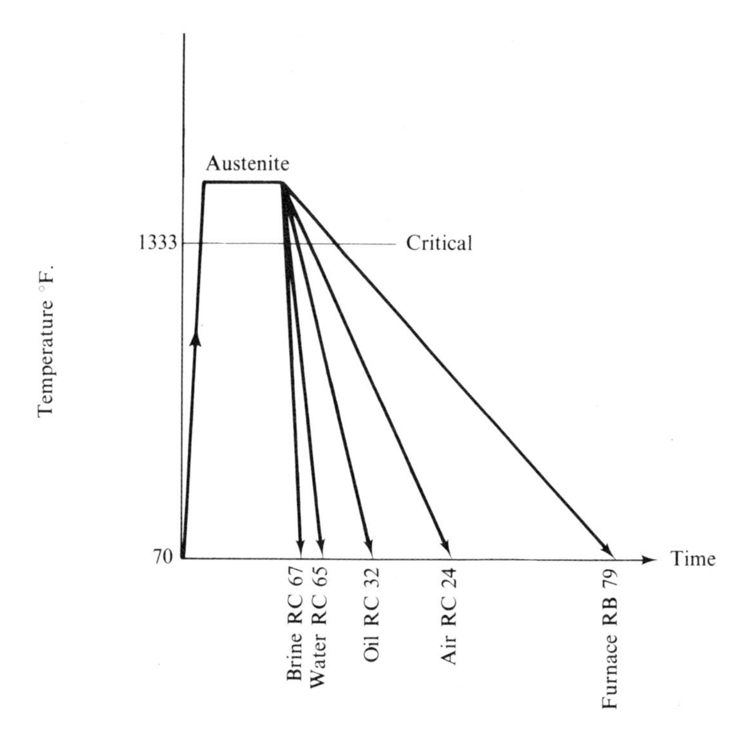

FIGURE 11–4 Effects of varying quenching rates on high carbon steel.

Chemistry and capability go hand-in-hand in stating the purpose of the heat treating operation. It is useless to state that the purpose of heat treating AISI 1020 is to impart a tensile strength property of 150,000 psi. However, if the purpose of heat treating AISI 1095 is to impart 150,000 psi, then this purpose is realistic. A study of the metal's chemistry reflects its heat treatment responses and, therefore, its ultimate capabilities in regard to mechanical properties such as hardness and strength. Figure 11–5 illustrates the potential hardness of a metal according to its chemistry.

Geometry of the Part and Heat Treatment

Once the chemistry of the metal is known, its potentials are established, but with a possible variable existing. In a hardenable steel, for example, when the chemistry infers a maximum hardening capability, the capability exists only with respect to the steel's mass and shape (fig. 11–6). For instance, the presence of 1.0% carbon in a plain carbon steel infers the possibilities of attaining an RC 65 in surface hardness. But when the diameter is determined to be more than one inch, the hardenability drops to a lower value because the large mass of metal will cool too slowly when quenched from the bright red temperature for such a high hardness to exist. On the other hand, a smaller diameter rod such as one inch can produce an RC 65, but only on and near the surface. Lower hardness values will exist in the core. Cooling rates relate to the total mass of metal because both thin and thick sections must be considered. Thin sections cool more quickly than thicker sections and may crack away from the thicker regions on cooling. On the other hand, thick sections must be cooled fast enough for proper structural transformation. Therefore, when heat treating purposes or objec-

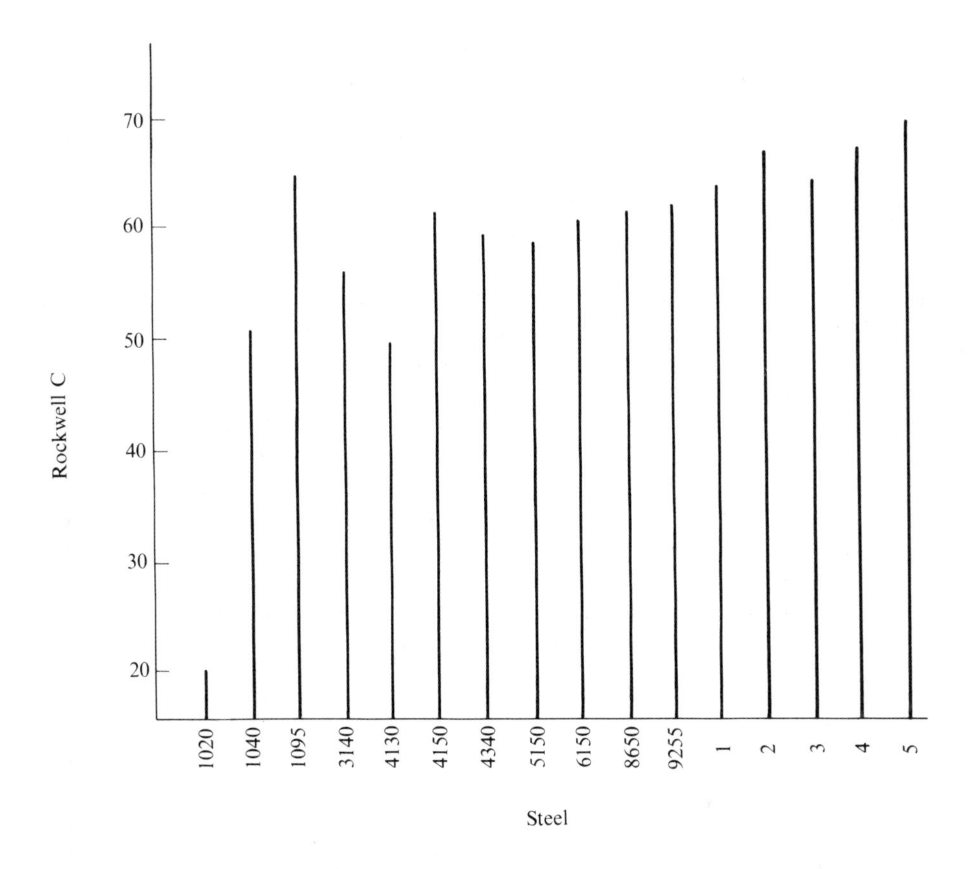

FIGURE 11–5 Potential hardnesses of one-half inch steel rounds.

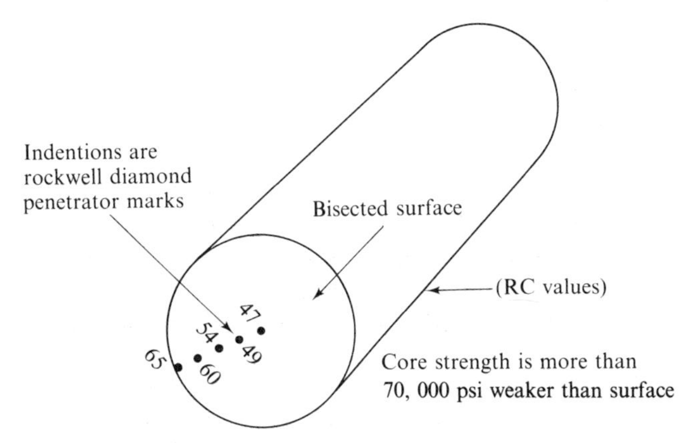

FIGURE 11–6 Mass vs. hardness in a one-inch water quenched AISI 1095 specimen and bisected.

tives are stated, both the chemistry and the mass of metal must be considered. In general, the smaller the cross section of metal, the harder it can be made up to a certain value, if it has the hardening capability. Then, as the cross section increases in size, the hardenability decreases from a given chemistry, the exception being some special alloys. Basically, most alloy additions to steel provide for higher hardenability across the cross section of the metal.

When the geometry of a part shows length dominating over diameter, care must be exercised during cooling from the elevated temperature, especially during quenching in water or oil. Very hot metal is plastic and, when subjected to sudden cooling, may warp due to uneven stresses occurring in the metal. The greater the length-diameter ratio, the greater is the possibility of warping. Long parts are vertically quenched to reduce warpage.

Another heat treating problem relates to cracking of hardened steel due to geometry of the part. Metals which have a high hardenability must be designed so that even cooling occurs during the quench from austenite. That is, right angle sections must contain reasonable fillets to prevent a crack from occurring at the intersection of the legs of the angle. Cracks are illustrated in figure 11–7. Also, a hole which is designed too close to an edge often invites cracking of the metal between the hole and the edge due to the mass differential. Mass and shape of metal, along with its chemistry, must be considered prior to heat treatment of the metal.

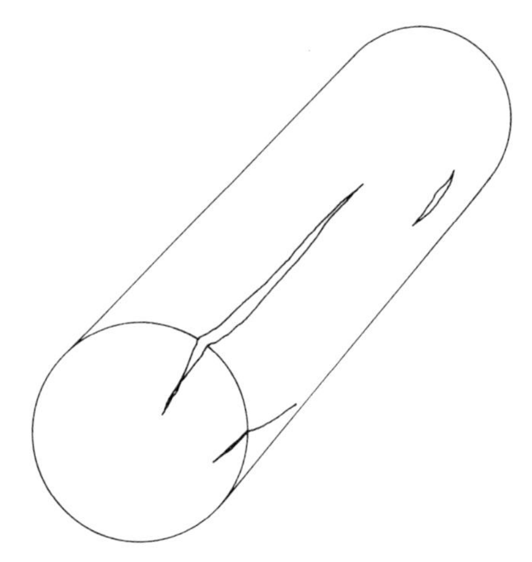

FIGURE 11–7 Cracks in a high carbon steel rod resulting from water quenching.

Heat Treating Furnaces

The heat treatment of metals is partly accomplished with the use of furnaces or related equipment which are especially designed to help obtain the objectives of heat treating. Three primary considerations include the proper size of furnace for the part, the temperature capability, and the temperature control system. Furnaces may be either gas, oil, or electrically heated, each having its advantages. A typical furnace is shown in figure 11–8.

The size and shape of the furnace must be appropriate to the size of the part or parts. Too large a furnace is wasteful, while too small a furnace limits the size and

FIGURE 11–8 A box type electric heat treating furnace. (Courtesy of Lindberg, Division of Solo Basic Ind.)

quantity of parts to be heated. The shape of the furnace depends mainly on the type of work being processed. Most small furnaces have a lifting door for placement of the parts on the hearth and for removal after the soaking period. Basically, the size and shape of the furnace should suit the parts being processed. The box, or batch, type furnace is capable of handling many parts a day, while the continuous furnace is more suitable for increased production quantities.

ATMOSPHERES

Several distinct atmospheres are available for the heat treater in the processing of ferrous and nonferrous parts. The particular atmosphere depends on the kind of material being heat treated and the type of surface finish desired after removal of the part from the furnace. An atmosphere is merely the gaseous environment within the hot furnace or the lack of an atmosphere such as a vacuum. Metal parts in a vacuum have no gaseous components with which to react. Because furnaces are heated mainly by combustion of gases or by electricity, special equipment is necessary to maintain an atmosphere compatible with the surface chemistry of the metal part, ferrous or nonferrous.

In gas-fired furnaces, it is possible to have an excess of oxygen in the furnace atmosphere or an excess of the fuel. When air or oxygen is the excess component, an oxidizing atmosphere results which has detrimental effects on the surface of many metals, especially steels. As an example, oxygen combines with the iron in a steel part and produces a loose scale at red temperatures. This means that the steel's surface chemistry has changed, a loss being the iron in the oxide scale. Also, a chemical reaction causes decarburization by reducing the quantity of carbon in the steel's surface. High carbon steel can be changed to low carbon steel on the surface due to prolonged heating at red temperatures in the presence of an oxidizing atmosphere.

When an excess of fuel is present in the atmosphere, a reducing situation exists in that the surface chemistry is not appreciably altered unless equilibrium does not occur, and carburizing results. When too much carbon-bearing fuel occurs in the atmosphere, a chemical reaction occurs at the surface, and iron carbide is formed. Low carbon steels can be made high carbon at the surface by this method. The neutral atmosphere is desirable in many instances whereby the products of combustion and the metal's surface show no reactions. Oxidizing atmospheres produce a surface scale at all temperatures, but loose scale forms above 1300° F. Neutral and reducing atmospheres produce clean surfaces. Dry molecular nitrogen, argon, and other inert gases produce neutral atmospheres. Also, a proper balance between a fuel gas and air produces a neutral atmosphere. The effects of an atmosphere are illustrated in figure 11–9.

Electric furnaces contain an oxidizing atmosphere, unless an atmosphere generator is part of the furnace system. However, many electric furnaces are used in industry without atmosphere control and produce satisfactory heat treatment when the surface is to be slightly ground. Oversize parts are ground following heat treatment to bring out the desired surface hardness which is just below the surface. Normal soaking periods in austenite do not cause excessive decarburization, but on thin sections such as threads, damage occurs. Many steel parts require grinding after heat treatment in order to bring the dimensions within tolerances, therefore, the slightly decarburized surface is removed. In this respect, many machined parts are left a few thousandths of an inch oversize and, together with any warpage and dimensional change during heat treatment, the decarburized and excess metal is removed.

Induction Heating

Another common method of heating for purposes of heat treating is the induction coil. When a high frequency electric current is passed through a coil of metal such as copper tubing, flux lines coming in contact with the metal part to be heat treated cause rapid heating of the part. For example, a water cooled copper tube is coiled around the periphery of the teeth of a spur gear, but not in contact, and connected to the source of high frequency current. Nearly instantly, after the circuit is closed, the surfaces of the gear's teeth are heated to the proper red temperature, the gear then being dropped into the quench tank and hardened only in the surface areas of the teeth. The surface hardens when the material has enough carbon or alloy present and when quenched from austenite. No response occurs below the hardened case.

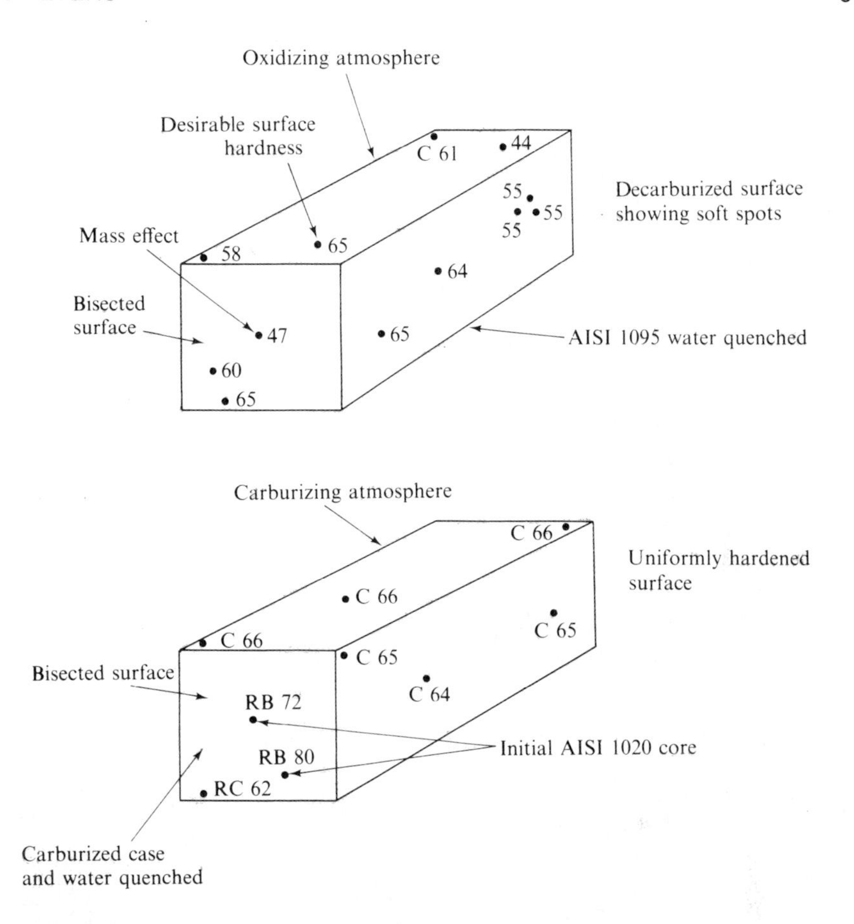

FIGURE 11-9 Surface hardness effects of oxidizing and carburizing atmospheres pertaining to one-inch water quenched AISI 1095 and carburized AISI 1020 bars.

Pyrometers

In order to perform heat treating operations, furnace temperatures must be exact and must be controlled. The pyrometer is used to accomplish both of these functions. Basically, a pyrometer system includes a controlling instrument and temperature indicator, a heat detection device called the *thermocouple*, and lead wires connecting the thermocouple to the pyrometer.

When two dissimilar metal wires are twisted together at one end and fused by welding, a thermocouple exists. Two common wires which are used in thermocouples up to approximately 2,000° F. include the alloys chromel and alumel. Wires or extension leads are connected from the couple to the instrument. Immediately, a temperature indication is shown on the dial of the pyrometer, because a direct electric current flows in the completed curcuit as the result of the junction of the two dissimilar metals. As temperature increases from room temperature at the welded zone of the couple which is inside the furnace, the millivoltage in the system increases and is recorded as temperature. The purpose of the thermocouple is, then, to detect temperature and generate electricity.

Two typical types of pyrometers are the *millivoltmeter* (fig. 11–10) and the *potentiometer* (fig. 11–11). Both instruments use the lead wires and thermocouples. In pyrometer systems, the thermocouple (fig. 11–12) provides the millivoltage direct to a coil in the instrument. As the voltage increases, the pointer and coil deflect, indicating a change in voltage. The instrument's dial is calibrated in degrees temperature, how-

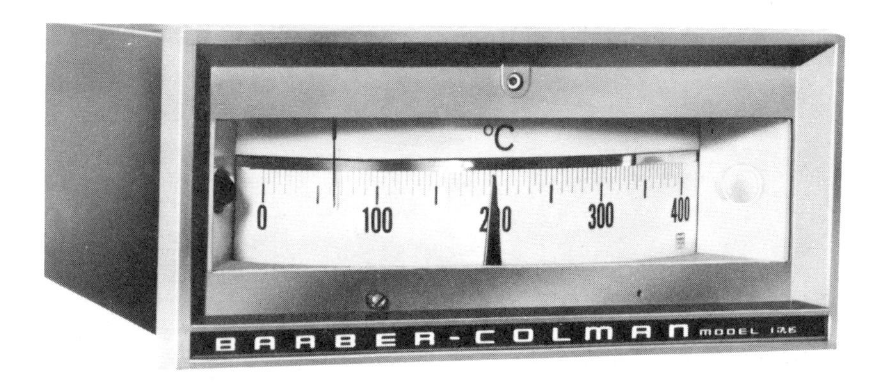

FIGURE 11–10 A millivoltmeter type pyrometer. (Courtesy of Barber-Colman Co.)

FIGURE 11–11 Heat treating furnaces with thermocouples mounted in furnace lids. Extension lead wires connect the potentiometer pyrometers with the furnace temperatures. Pyrometers are wall mounted. (Courtesy of Leeds & Northrup Co.)

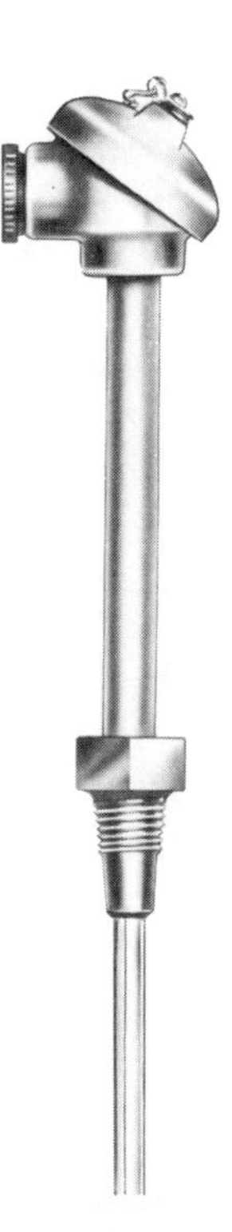

FIGURE 11-12 A typical thermocouple. The couple is protected with a protection tube. (Courtesy of Leeds & Northrup Co.)

ever, instead of millivolts. In the potentiometer, the voltage is connected to a *galvanometer* within the pyrometer which, in turn, balances the incoming voltage with the system's voltage. As balancing occurs, deflection of the instrument's temperature pointer shows furnace temperatures. This type of pyrometer also provides a printed record of the operation. Pyrometer systems hold automatic control over furnace temperatures. A temperature setting of the dial when the system is on causes the furnace to come up to temperature and hold the temperature until manually or automatically turned off. A pyrometer system is illustrated in fig 11-13.

Large and expensive electric furnaces are protected from overheating should the controlling thermocouple system fail as a result of a broken thermocouple. After long periods of time, the couple will deteriorate, and finally separate, resulting in an open electrical circuit and complete loss of temperature control. Because the electricity is on, the temperature will increase until the heating elements melt, causing severe damage to the furnace. In this situation, a safety thermocouple is installed to shut off the electric power automatically when the temperature rises to a predetermined point well within the safe operating temperature range of the furnace.

Heat Treating Equipment

Besides the furnace and pyrometer control system, there needs to be quenching tanks containing different coolants such as brine, water, and oil. Brine provides the fastest

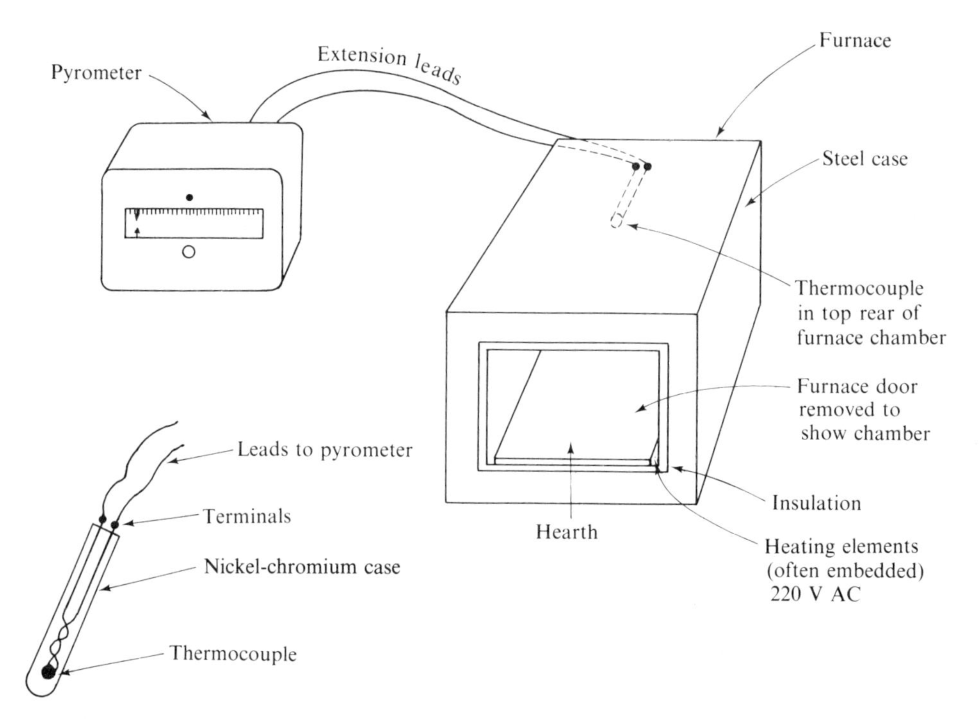

FIGURE 11-13 A typical electric furnace system.

cooling medium, followed by water, oil, air, and furnace cooling, respectively. Also, special baskets to hold small parts are needed, as well as a crane to handle some parts in large installations where very heavy parts are heat treated. Several shapes of tongs are also needed for grasping the hot parts. Because many parts often warp a small amount, some kind of straightening equipment is essential.

Support equipment such as hardness testers and microscopes are needed to verify attainment of desired microstructures. Saws, grinders, cut-off wheels, polishing and etching equipment, and the microscope give support to the heat treater in processing metals to meet specification requirements. Further, a universal tensile tester provides the means for ascertaining ductility and strength factors in a sample of the part.

Safety Precautions

There is a degree of danger in most industrial operations, however, during some heat treating procedures, excessive danger exists. A steel part, for example, can be 1000° F. and yet appear cold to the eyes. When touched in an effort to handle it, a severe burn will result before it can be dropped. Also, should cold or damp metal parts be accidentally placed in a molten salt solution, an explosion frequently follows. A very dangerous operation is cyaniding. Cyanide salt is sometimes used in heat treating shops for case hardening purposes. This material is deadly poison and must not be allowed to enter the body by any means. One of the worst hazards is the relationship between acid and cyanide. If sulfuric acid, for instance, which is used in metal shops contacts cyanide, a deadly gas results.

Testing and Inspection of Parts

After heat treatment, parts must be tested and inspected to assure compliance with specifications. Often, a sample such as a tensile specimen is given the same heat treatment as the part in order that destructive testing can be performed on the sample. Many times, the tensile and yield strengths of the metal must be verified along with the percentage of reduction in area and elongation factors. Hardness values are also nearly always indicated on the job sheet. In order to confirm a required hardness, a sample of the metal is parted with a cut-off wheel and its hardness checked from surface to core. Further, many parts require nondestructive testing to verify soundness of the part. Some of these nondestructive tests include penetrant and magnetic particle, ultrasonic, eddy current, radiographic, and infrared. Figure 11–14 illustrates the hardness test results from bisecting a sample of metal. Notice that in carbon and low alloy steels, maximum hardness is at the surface while hardness decreases in the core. Toughness of the metal sample is measured with impact testers and is measured in foot-pounds of energy. Grain size and other microstructures are verified with small specimen samples of the material which have been heat treated with the part, polished and etched, and then examined under the microscope.

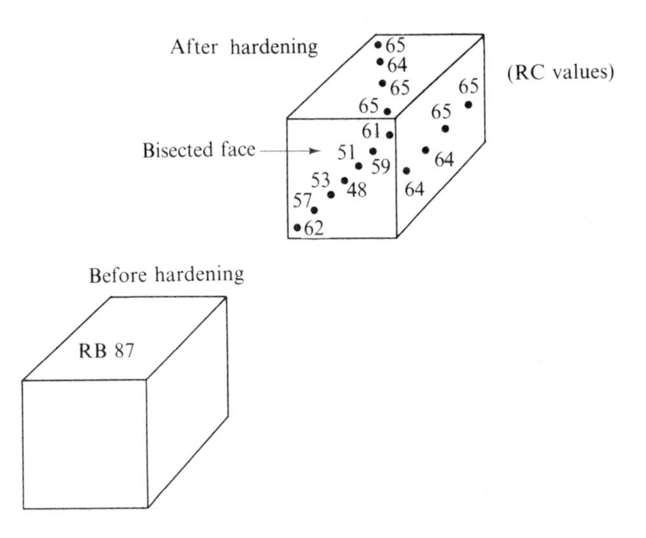

FIGURE 11–14 A bisected one-inch AISI 1095 specimen after quenching showing surface and core hardnesses.

Specifications may originate from the individual producer of the parts, or he may make reference to ASTM, MIL, or others. Many national organizations such as the American Society for Testing and Materials (ASTM) publish specifications. The armed forces also publish specifications. In addition, the American Iron and Steel Institute (AISI), as well as the Society of Automotive Engineers (SAE), publish their particular specifications. Even though many of these documents have similar characteristics such as the definitions of objectives and purposes, their specific coverages are different due to the particular purpose of each publication. A specification indicates, in many instances, how a test is to be made or includes a table showing the exact chemical analysis of a material in order to be acceptable. Therefore, personnel involved in heat treating operations must be cognizant of the many AISI and SAE types of metals and the procedures needed to comply with any of the pertinent specifications.

Heat Treatment of Ferrous Parts

The heat treatment of parts made of ferrous metals includes the heating of the parts to specified temperatures, followed by predetermined cooling rates. A subsequent treatment is usually necessary to impart the exact mechanical properties desired. The ferrous group of metals includes the steels, cast irons, and wrought irons. Wrought iron obtains its mechanical properties during the several processes in its production and not by heat treatment, but the several cast irons and many types of steels frequently receive a special type of heat treatment following their initial production. Ferrous metals include ferrite, carbon, manganese, silicon, sulfur, and phosphorous in their chemical analyses. Alloys of this basic matrix are produced by adding one or more additional elements or increasing the content of one or more of the basic elements. Basically, with reference to the plain carbon group of metals, it is the percentage of carbon which constitutes the most critical factor in subsequent heat treating processes.

CRITICAL POINTS

In the solid state (of steel, for example), exist two types of atomic structures, alpha and gamma. Because the constituent of greatest quantity is iron, or ferrite, the atomic arrangement of its unit cells must be recognized. Iron is cubic in nature and its millions of cubes are arranged as either nine or fourteen atom types, designated as alpha or gamma iron, respectively. The nine or fourteen atom cell pertains to an isolated cell, and when joined by adjacent cells, sharing of the atoms occurs along the corners and surfaces of the cells. The normal dividing point between alpha and gamma iron in carbon steels on heating is called the *critical point*. Steels with temperatures below the critical point are alpha, being body-centered cubic in their cell structure or, basically, the nine atom type. Steels existing at temperatures above the critical point are gamma, being face-centered cubic in their cell structure or, basically, the fourteen atom type.

Chemically, the carbon is converted to iron carbide which was formed at the time of metal solidification and is closely associated with the ferrite, or iron. With respect to cell structure and other elements in the matrix, alpha iron mixes with carbide, while gamma iron absorbs the carbon-iron compound into solid solution. Therefore, a mixture of ferrite and carbide at room temperature becomes a solid solution in the bright red heat range, the critical points varying according to the steel's analysis. Figure 11–15 illustrates a simplified version of the iron-carbon diagram of a series of alloys after slow cooling from bright red temperatures.

Usually, on heating, 1333° F. is considered the lower critical point for steels. The upper critical point increases for steels having less than 0.80% carbon to as high as 1670° F. in the very low carbon steels. Steels with carbon contents ranging from 0.80% up to 1.7% have the upper and lower critical points occuring at the 1333° F. point, respectively. Consequently, in order to heat treat steels, their critical points

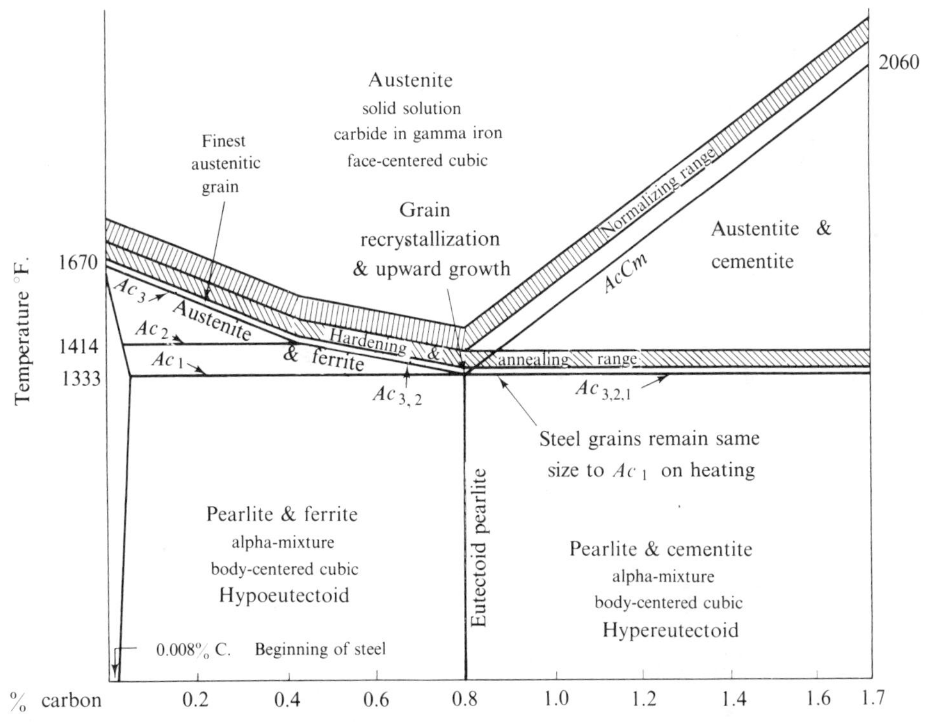

FIGURE 11–15 A simplified version of the iron-carbon diagram for steel.

must be slightly exceeded and the temperature reversed in a controlled manner such as slow or fast cooling.

It is the changing from alpha to gamma and then back to alpha conditions of cells and microstructures that gives steels their variable strengths and hardnesses. In this respect, hardness and tensile strengths are directly related in most steels up to approximately RC 55. Sketches of the microstructure of hard and soft steel are presented in figure 11–16 along with their photomicrographs.

TEMPERATURE AND TIME EFFECTS

Automatic control of temperature uniformity occurs during heating because the furnace environment is uniform as temperature increases to the desired setting of the pyrometer. However, it is good practice to preheat the part, rather than place the cold metal in the red-hot furnace. Uniform heating reduces stresses in the metal which may occur at the critical points when heated too fast, especially when one section of the part is much smaller than another. Preheating at 800° F. allows a better transition from alpha to gamma iron at the critical point. On the other hand, some intricately shaped parts are preheated again at 1250° F. To assure satisfactory structural conditions at the desired elevated temperature in austenite, from 50° to 200° are added to the steel's critical points. This added temperature is essential in causing all the constituents to go into solid solution and to allow the solution to be retained for a few extra seconds during the cooling cycle between the time the hot metal leaves the furnace and enters the quenching medium. In other words, heating only to the designated

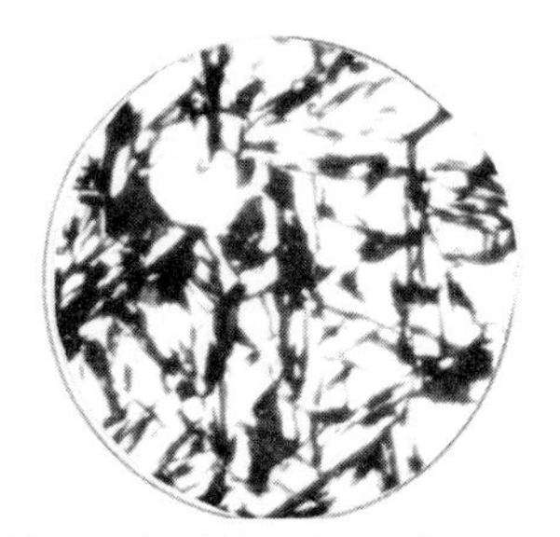
Martensite 400X photomicrograph

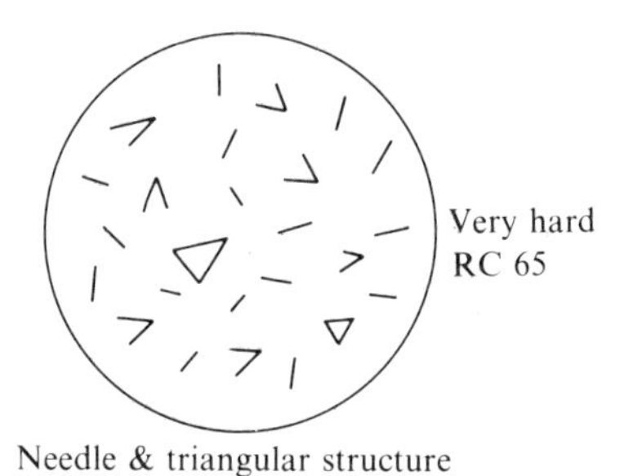

Needle & triangular structure

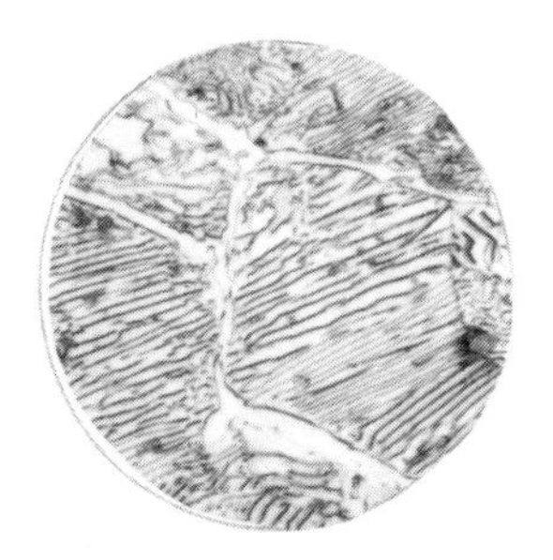
Pearlite 400X photomicrograph

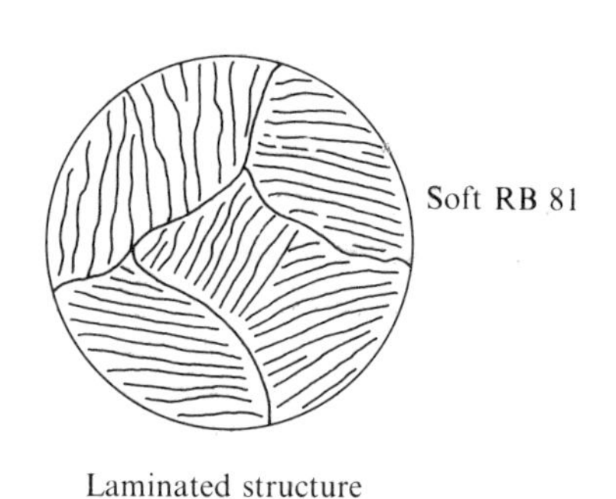

Laminated structure

FIGURE 11–16 Hard and soft steel microstructures.

critical point is not satisfactory to obtain a complete phase change in the metal. The steel must be completely austenitic as it enters the cooling medium, austenite being the structural condition existing above the upper critical point as a solid solution of carbon in gamma iron (fig. 11–15).

As the analysis of the steel changes, the upper critical point also changes for steels having carbon contents below 0.80%. When alloy steels are heat treated, both the lower and upper critical points are changed somewhat, depending on the element or elements added. Table 11–3 lists critical points of several steels. An important point in respect to temperature is that the thermocouple indicates the temperature at the couple, therefore, the part should be close to the couple. A very important factor is the mass of the metal. The greater the thickness of the part, the longer it will have to soak at the desired temperature in order for a complete solution to occur.

The cooling cycle determines the maximum hardness of the part when the part is cooled from austenite. With reference to cooling, the metal's hardening capability

TABLE 11-3 Critical Points of Steels

	On heating 50°F./hr.		On cooling 50°F./hr.	
AISI	Ac_1	Ac_3	Ar_3	Ar_1
1020	1335	1555	1500	1260
1035	1340	1475	1425	1255
1060	1340	1375	1340	1265
1095	1350	1415	1340	1290
1112	1355	1660	1560	1250
1141	1310	1400	1340	1210
1340	1320	1430	1330	1150
2330	1280	1375	1190	1040
3120	1350	1475	1455	1230
3140	1355	1410	1330	1220
3240	1330	1400	1310	1220
3310	1330	1435	1240	1160
4042	1340	1460	1350	1210
4130	1395	1490	1390	1280
4140	1350	1480	1370	1255
4340	1335	1425	1310	1210
4427	1330	1540	1425	1200
4520	1310	1555	1525	1220
4620	1330	1475	1380	1190
5045	1300	1450	1370	1290
5120	1410	1540	1470	1290
5140	1360	1450	1340	1280
5160	1310	1410	1320	1250
52100	1340	1415	1320	1270
6120	1410	1530	1420	1300
6150	1380	1450	1370	1280
7140	1450	1580	1480	1330
8124	1330	1510	1420	1230
8620	1350	1525	1415	1220
8630	1355	1460	1370	1220
8720	1350	1530	1420	1220
8750	1340	1415	1310	1210
9260	1370	1500	1380	1315
9310	1320	1510	1230	1080
9440	1330	1420	1280	1190
9763	1330	1370	1280	1230
9850	1330	1405	1270	1210

SOURCE: Republic Steel Corporation

must be considered, that is, the metal's analysis, shape, and mass predetermine its potential strengths. If the analysis does not have the potential strength capability, then going through the process is useless. For example, when a ½-inch thick high carbon steel in the austenitic condition is rapidly quenched in water, a hardened structural condition results known as *martensite*. The steel, by the way, may be cast or wrought. Before hardening, the soft and weak steel in the pearlitic condition had an RB 84 hardness value. The pearlite condition is a mixture of carbide particles and ferrite in plate form. After quenching, the hardness value increased to RC 65, nearly maximum hardness in metal. Its tensile strength increased from approximately 78,000 psi to over 300,000 psi merely by proper heating and cooling.

On the other hand, when a similar sample of high carbon steel is cooled slowly from austenite such as in furnace cooling, the austenite transforms to pearlite, resulting in a tensile strength of approximately 80,000 psi with a hardness of RB 85. Rockwell B hardness values allow machining and bending of the metal, whereas Rockwell values in the upper C scale indicate high hardness, rigidity, and high tensile and yield strengths. Many metals, then, are variable in their heat treatments, from weak to high strengths. Some metals, however, are not heat treatable for increased strengths because their analyses forbid it, while others respond to smaller degrees of changes in their mechanical properties when only small alloy or carbon additions are made.

Microstructure of a metal includes grain size and shape and the structural condition within the many grains (fig. 11–17). All metals are granular, the tiny crystals

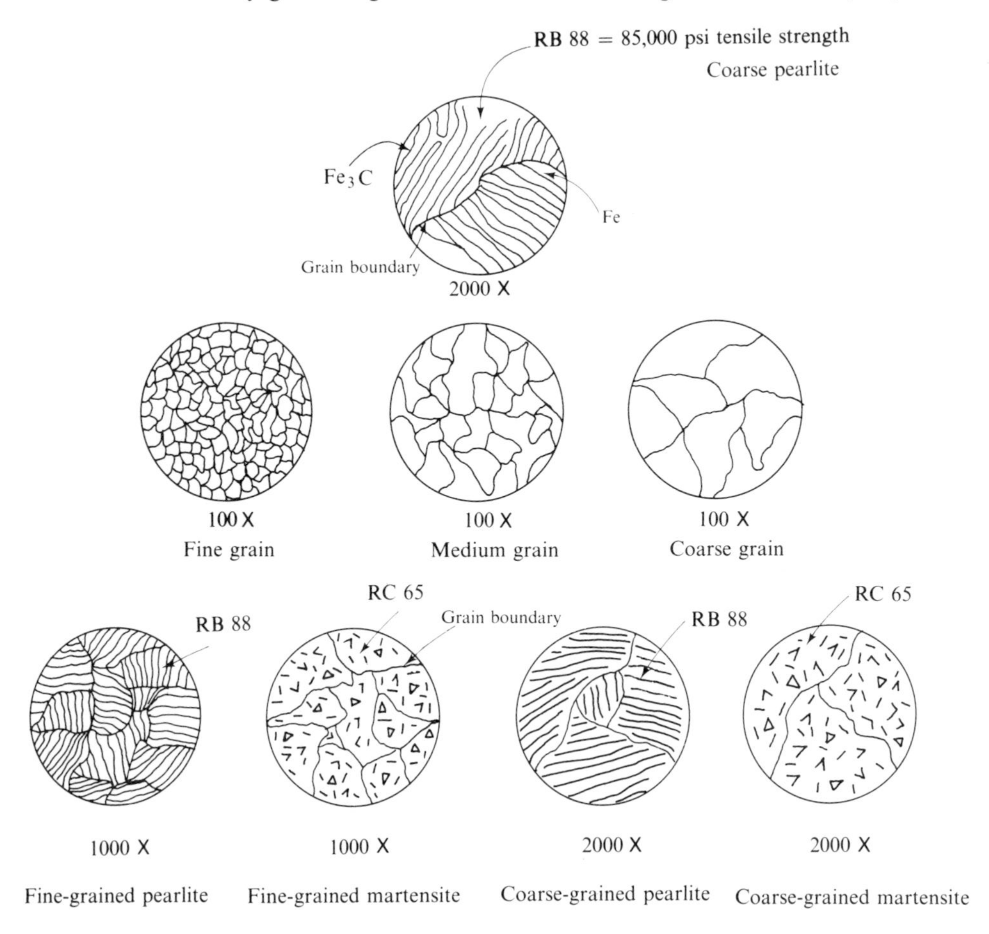

FIGURE 11–17 Microstructures of annealed and hardened eutectoid steels.

being many sided and existing from very fine particles to coarse. Cast grains are equiaxed, but when cold rolled, for example, the grains are elongated in the direction of pressure. Grains may have many shapes. Usually, the fine-grained metal is most desirable because from a given hardness, the finer the grain structure, the tougher the metal. Coarse grains tend toward brittleness. Within each steel grain, whether fine or coarse, a microstructure of austenite, martensite, bainite, or pearlite exists. Two main factors of a metal's microstructure are, then, grain size and condition within the grain. Microstructures of nonferrous metals have similar patterns to the ferrous in many instances. Grain size pertains mainly to the ability of the metal to resist loading shock, while condition in the grain refers to hardness or softness of the metal. Fine-grained steel can be hard or soft, and coarse-grained steel can also be hard or soft (fig. 11–18). Fundamentally, grain size is determined on heating, the austenitic grain size existing at the maximum temperature reached mainly determining the size of the grains at room temperature. The condition existing in the grain will be determined mainly by the rate of cooling from austenite; the faster the cool, the harder the metal will be up to approximately RC 67.

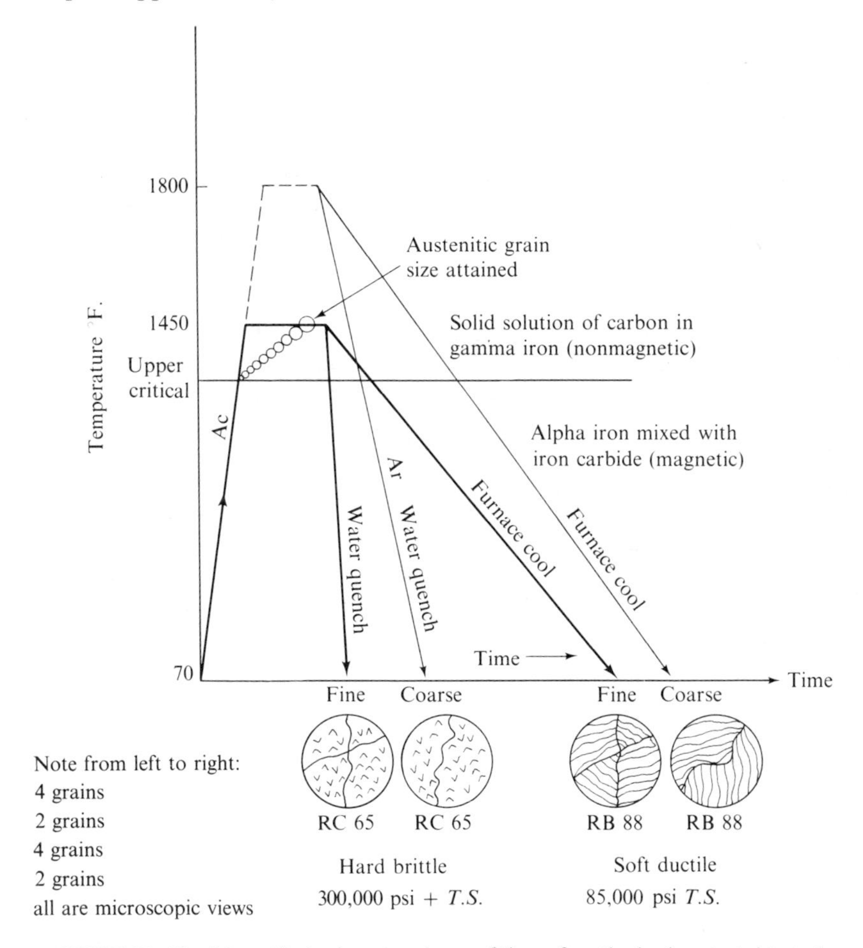

FIGURE 11–18 Magnified microstructures of three-fourths-inch eutectoid steels.

Critical points on cooling occur in reverse order of heating, but at lower temperatures. Actually, three critical points exist on heating and cooling. The lower critical point on heating, Ac_1, initiates the beginning of the change from any structure to austenite, along with the beginning of change in any grain size to its finest condition. The next change occurs at Ac_2, the loss of magnetism occurring as alpha transforms to gamma iron. The third change occurs at Ac_3 when all austenite forms, and it is at this temperature that the finest austenitic grain exists. These changes are illustrated in figure 11–15. Theoretically, the part should be cooled from this point, but practically, at least 50° should be added to assure proper end results. As temperature increases above the upper critical point, austenitic grain size also increases, therefore, parts should not be heated any higher than necessary in order to get the finest possible grain with the desired microstructure of hardness.

On cooling, the faster the quench, the lower will be the critical points' occurrence, Ar_3 to Ar_2 to Ar_1. The more Ar_1 is suppressed toward room temperature, the harder the steel becomes with a given carbon content. Brine will lower Ar_1 closer to room temperature, as gamma iron changes back to alpha, faster than water or oil or air. Therefore, if a high hardness is desired in a high carbon steel, a brine or water quench is used. If a medium hardness is desired, an oil quench is used. Often, this microstructure appears as a feather shape under the microscope and is, in reality, a very fine laminated structure of iron carbide and ferrite known as *bainite*. Bainite is somewhat softer than martensite and transforms from austenite at higher temperatures than martensite. Air cooling leaves the metal in a fairly soft condition, while furnace cooling provides nearly maximum softness. These varying hardnesses are illustrated in figure 11–19. Different alloys have different rates of required cooling, each depending on what structure is desired, the mass of metal, and its analysis. Some steels can be cooled in air and attain a high hardness. Exceptionally fast cooling which produces martensite results in the body-centered tetragonal structure.

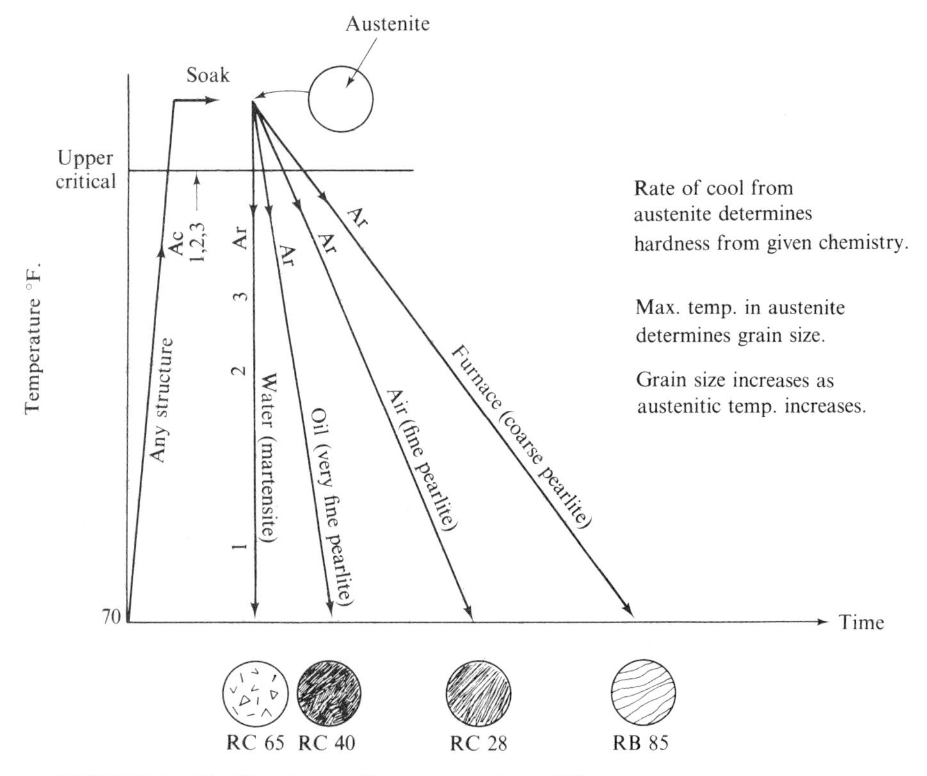

FIGURE 11–19 Varying cooling rates produce different microstructures in eutectoid steels.

THE CAST IRONS

Because the cast irons and steels belong to the ferrous group of metals, many of the general principles pertaining to steel also pertain to the cast irons. However, in cast iron, the carbon content exceeds 1.7% and varies as high as 5% and 6%. An iron with a carbon content of 4.3% is typical of the cast irons. Basically, cast irons are mostly used in the "as cast" condition, being either of the gray or white types. Gray irons are soft and somewhat brittle in nature because of the excess graphite forming during slow cooling in the mold, while white iron is hard and brittle due to its being cooled faster during the casting operation. Some alloys of cast iron are heat treatable, however, being treated in similar ways to the steels. The malleable type of cast iron, however, is produced by holding the metal at a bright red temperature for several hours in order to relieve the ferrite of much of the carbon and transform it into spheres of graphite. A softer but tougher type of cast iron results. Again, the metal's analysis determines its capability and heat treating processes, including critical points, soaking temperatures, and cooling rates.

ANNEALING

The annealing process induces softness in the metal through a process of very slow cooling, such as in furnace cooling from austenite. Annealing is performed in order that fabrication or machining can be accomplished. Because the time required for the furnace to cool to temperatures below the critical points is often a day's length, carbide and alpha ferrite precipitate very slowly from austenite in the pattern of plates of carbide and plates of ferrite. When viewed in the microscope, the resultant pearlite resembles finger prints. The gamma-alpha transformation is very slow, therefore, transformation occurs near the Ac_1, resulting in a soft microstructural condition of the constituents, or pearlite (fig. 11–15 and fig. 11–19). Hypoeutectoid steels, when annealed, include pearlite and ferrite, while hypereutectoid steels include pearlite and cementite. Eutectoid steels contain all pearlite. Rockwell hardness values will usually be in the upper B scale. As an example, AISI 1095 is annealed by soaking at 1450° F. until uniform in temperature, followed by cooling in the furnace overnight or removing the hot metal and cooling in some medium like ashes so that cooling is very slow. Alloy steels are annealed in the same manner, only a higher austenitic temperature is required so that the additional elements can go into solution.

HARDENING

The purpose of hardening is to cause the "as rolled" or annealed steel to become harder and stronger in its three basic strengths—tensile, yield, and shear. As an example, an AISI 4340 steel is hardened by oil quenching from 1500° F. On the other hand, an AISI 1095 is water quenched from only 1450° F. to induce hardness. Both steels reflect their maximum capability in hardness with a given mass, but each is much too hard and stressed to use in the "as quenched" condition. Hardening results in the pearlite changing to austenite and then to martensite during the quenching operation. Figure 11–15 indicates hardening ranges for carbon steels, the alloys requiring a small increase in temperature for the purpose of transforming the more complex carbides. Some of the quenching stress, hardness, and brittleness is subsequently removed by a tempering treatment. With regard to quenching, a general rule states that alloys are usually oil quenched, while plain carbon steels are water quenched from austenite. Quenching alloys in water often causes cracking, while some special air hardening alloys will reach their maximum hardness by air quenching. With reference to austenitic temperatures, the hardening and annealing temperatures are often equal.

TEMPERING

Tempering cannot be performed unless it has been preceded by the hardening operation. Two very important factors pertain to tempering—never heat the metal at a temperature higher than the tempering temperature and never exceed the lowest critical point of the metal. Therefore, tempering is always conducted below the metal's critical point and is preferably conducted for at least one hour at temperature. Thicker parts require longer soaking periods, at the rate of one hour per inch of cross-sectional thickness. After hardening the AISI 1095 in the above example, it is tempered at 430° F. when an RC 61 is desired or at 700° F. when an RC 45 is desired. When the hardened AISI 4340 is tempered at 450° F., approximately 261,000 psi tensile strength is obtained. The tempering temperature varies according to the metal and hardness desired. After tempering, parts are normally air cooled from the tempering furnace, unless otherwise specified, to prevent brittleness in some steels.

Tempering of carbon and most alloy steels reduces hardness and strength while it increases toughness. One of the strongest and toughest steels known is the D-6AC. It is a medium carbon steel with small amounts of nickel, chromium, and molybdenum, in addition to the other five basic elements. When tempered at 300° F., its tensile strength is approximately 340,000 psi with an elongation of 7% and a Rockwell hardness of C 63. This steel, when heat treated to reduced strengths, is used in the central carry-through fitting of large aircraft and has shown tremendous capability in sustaining high flight stresses under fatigue conditions.

Table 11–4 lists some typical mechanical properties after tempering several steels. With respect to the high speed category of steels, tempering at 1050° F. results in an increase in hardness because of some retained austenite from the quench transforming to additional martensite. The resulting Rockwell hardness value increases from about RC 56 as quenched to approximately C 67, while all other steels are reduced in hardness by tempering. Tempering produces a tempered martensite microstructure.

NORMALIZING

Fabricated or heat treated metals are stressed. When reworking of the metal is necessary, or prior to heat treatment of fabricated parts by hardening, the normalizing operation is often performed to remove stresses and place the metal in a natural condition (fig. 11–15). Normalizing usually precedes hardening and is performed by air cooling from the normalizing temperature in austenite. A peculiarity of normalizing is that the austenitic temperature is from 50–200° higher than the hardening temperature for steels having less than 0.80% carbon, or the hypoeutectoid group. This group includes most of the structural steels. For example, an AISI 1070 is normalized at 1550° F., while the AISI 4340 is normalized at 1650° F. The heat treating sequence for a part is, then, annealing when necessary for machining, followed by normalizing, hardening, and tempering. According to the chart in figure 11–15, all hypereutectoid

TABLE 11-4 Mechanical Properties of Steels

AISI or Other	Normalizing °F.	Hardening[a] °F.	Tempering °F.	Tensile Strength psi	Reduction in Area %	Hardness Rockwell
1045	1570	1520[b]	1150	100,000	65	B 95
1095	1520	1450[b]	750	200,000	46	C 44
2330	1510	1470	950	125,000	55	C 27
3140	1620	1510	770	180,000	50	C 40
4130	1625	1610	1045	125,000	55	C 27
4130	1625	1610	700	180,000	50	C 40
4140	1620	1560	1110	125,000	55	C 27
4140	1620	1560	1020	150,000	53	C 34
4140	1620	1560	670	200,000	46	C 44
4340	1610	1520	1210	125,000	55	C 27
4340	1610	1520	1055	150,000	53	C 34
4340	1610	1520	845	200,000	46	C 44
51210	1540	1775	1100	125,000	55	C 27
6135	1640	1610	800	180,000	50	C 40
6150	1620	1600	1000	150,000	53	C 34
8735	1640	1540	775	200,000	46	C 44
D-6AC	1650	1550	300	339,000	28	C 63
D-6AC	1650	1550	400	301,000	30	C 56
D-6AC	1650	1550	500	280,000	38	C 53
D-6AC	1650	1550	600	268,000	42	C 52

[a]Oil quench
[b]Water quench

steels are normalized at temperatures just above the $A_c C_m$ line. However, unless the carbides are not homogeneously mixed, it is best to avoid heating to the extremely high temperatures. In most cases, the excess carbides, or cementite, are properly distributed in the matrix of metal when initially produced, therefore, hardening operations usually produce satisfactory microstructures.

SPHEROIDIZING

Sometimes, high carbide content steels require severe forming operations such as bending or cupping. Tightly wound springs with small radii form very well when their microstructures are spheroidized (fig. 11–20). The operation is conducted by soaking the part at about 1200–1275° F. for 12–18 hours. The microstructure subsequently becomes ferrite embedded with pellets of carbides. Softness is at its maximum, as well as plasticity. A spheroidized structure is several points softer on the Rockwell B-scale than the pearlitic structure formed by annealing. Spheroidized steels bend easier than annealed steels with the same chemistry.

STRESS RELIEVING

When used as *stress relieving*, the process refers to a particular type of operation. Cold worked parts can be stress relieved or partially softened by an application of heat, possibly 600–800° F. Gas welded structures, on the other hand, are often stress

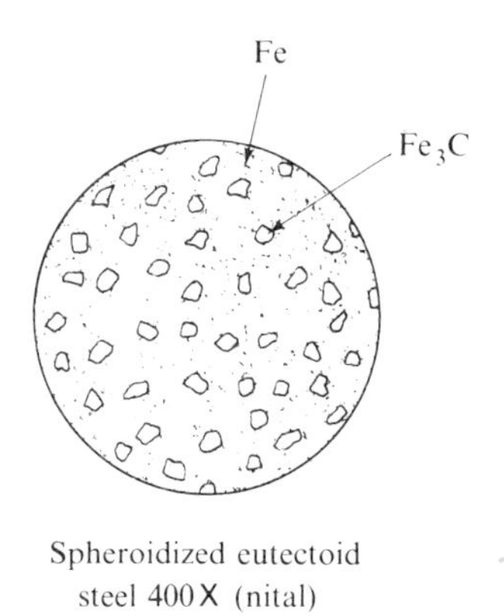

FIGURE 11-20 A spheroidized eutectoid steel.

relieved by heating to 1200° F. and air cooled. A complete stress relief, however, requires normalizing or annealing such as is the situation when fully hardened tools are to be reworked.

CASE HARDENING

The term case hardening means exactly what it states, that is, causing a hard case of metal to be wrapped around a softer or tougher core. A carburized and case hardened part is illustrated in figure 11-21. Such an operation requires the steel to be of a low

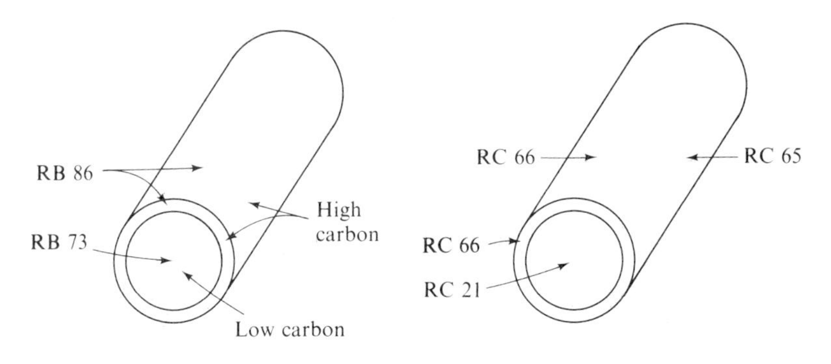

FIGURE 11-21 A carburized low carbon steel illustrating the carburized and case hardened conditions.

carbon content such as AISI 1020. When the low carbon steel is then exposed to a carbonaceous material such as charcoal or a combustible gas containing an excess amount of carbon at a temperature of 1700° F. for eight hours, a high carbon case of approximately 0.060 inch is formed around the low carbon interior. Because low carbon steels, 0.20% carbon, cannot be appreciably hardened, only the high carbon case will respond to the hardening treatment during quenching.

Usually, three distinct carbon zones occur in carburizing, the hypereutectoid, eutectoid, and hypoeutectoid zones. The line of demarcation is 0.80%, which is eutectoid (figure 11–15). Therefore, when the carburized part, RB 88, for example, with an RB 76 core, is quenched in water from 1450° F., the surface becomes an RC 65, while the core may increase to an RC 23. The case will break before bending, while the core will bend if severely overloaded, but will not break except in very thin parts.

Case hardening is performed on many types of parts such as gears, pins, and cams. A cam is a typical example where a case hardened material is essential in the operation of the mechanism, as the hardened surface prohibits the appreciable loss of metal and dimensional change. Higher quality heat treating procedures include a core refinement operation prior to hardening for the purpose of refining the grains in the core. An air cool from 1650° F. accomplishes this task.

INDUCTION HARDENING

The induction hardening process uses a high frequency electric current to quickly heat a shallow surface of a high carbon steel to austenite. When the part is quenched, a high hardness results on the surface, while the core is unaffected because it does not get hot. Many gear teeth are induction hardened, and the process lends itself favorably to complete automation by means of conveyor systems and moving mechanisms.

FLAME HARDENING

In a similar manner to induction hardening, a high carbon steel can be flame heated, usually with a torch, to austenite and then quenched, causing the steel surface to become hard, while the core is unaffected. A slowly rotating part is heated by the torch and quenched with jets of water spray. Such a process differentiates itself from the conventional case hardening process in that high carbon steel is used, while the case hardening process requires the use of low carbon steel. In effect, a case hardened steel results from flame hardening, because only the outer layer of metal, or case, becomes austenite and, subsequently, martensite when quenched. The use of high carbon steel eliminates the carburizing process which is required in case hardening procedures. Because the case is not heated any appreciable amount, it does not respond, and this is what is expected.

Heat Treatment of Nonferrous Parts

The nonferrous group of metals includes those which do not classify as iron base, or ferrous. Many nonferrous metals contain iron, however, but the iron is an alloying element. As in ferrous metals, some nonferrous metals are heat treatable for increased strengths while others are not. Pure metals or elements such as aluminum and copper can only be strengthened by cold working, but most metals can be softened by annealing. Basically, when a compound is present in the alloy which is capable of being finely dispersed throughout the matrix of metal, there is the possibility of increasing the metal's hardness and strength by heat treatment.

Heat treatable nonferrous metals are strengthened by heating them to a temperature which causes a solid solution of their contituents to form. When quenched from the solution temperature, some alloys retain the solution for a period of time until the aging, or precipitation, process begins. *Precipitation* is the forming of small particles of a chemical compound within the softer matrix. Formation of these compounds, or precipitates, promotes stiffness, hardness, and strength. If the alloy is capable of forming a hardening precipitate from a solid solution, then the alloy is considered heat treatable. The aging process, or precipitation treatment, which follows the solution treatment of heating and quenching requires time, usually several hours for artificial aging and up to a day for natural aging.

ALUMINUM ALLOY

An example of a heat treatable nonferrous alloy is the aluminum alloy 2024; copper is the main alloying element which forms copper-aluminide, $CuAl_2$. When the alloy is soaked at 925° F. for a proper time period to assure the formation of the solid solution and then quenched in water, a supersaturated solid solution occurs at room temperature, or 70° F. At this point in the treating process, the alloy is soft and can be easily shaped, but the shaping must be completed within approximately two hours. Being a naturally aging alloy, precipitation occurs at room temperature and continues for approximately 24 hours. Tensile strengths in the vicinity of 72,000 psi are obtainable, along with an increase in hardness with the associated decrease in ductility. An artificially aged aluminum alloy, 6061, for example, is solution treated by quenching in water from 975° F. followed by artificial aging for seven hours at 350° F. and air cooling.

When aging is to be retarded in the 2024 aluminum alloy, freezing temperatures are used. For example, some parts are heat treated by the solution treatment and placed in refrigeration until needed at a later time. Rivets are often solution treated in large batches, kept at near zero temperatures, and used as needed. When driven soon after removal from the cold environment, the rivet is plastic, and the formed end shapes easily. After a time lapse of 24 hours for the 2024 rivet, expected hardness and strength occur due to precipitation.

Annealing of aluminum and its alloys is conducted at approximately 660° F. when stress relief from cold working is desired. However, full annealing requires soaking at approximately 800° F., cooling slowly to approximately 400° F., and then completion by air cooling. Slow cooling causes the precipitates to grow in size and become fewer in number, resulting in more of the matrix being free of the finely divided hard particles.

COPPER ALLOYS

Some of the copper alloys such as aluminum-bronze respond to heat treatment for the purpose of increasing strength. A copper alloy containing about 10% aluminum, in addition to some iron and nickel, is solution treated by quenching in water from around 1800° F. A tempering or aging treatment similar to tempering steel, approximately 800° F., brings out a tensile strength of around 125,000 psi. Beryllium-copper, another alloy, is given a similar solution treatment and aged to bring out specific properties. A copper-nickel-phosphorus alloy, as well as the copper-chromium alloy, is also heat treatable. Annealing of these alloys also follows the general rule of slow cooling for the alloys and any cooling rate for the element, if the element can be softened.

OTHER ALLOYS

The alloys of titanium, magnesium, and nickel include various combinations of elements which produce heat treatable alloys. Solution and precipitation treatments vary with the metals' analyses, but in all cases, there must be a combination of elements or of elements and compounds to produce the required change as the result of heat treating. The solution treatment in the titanium alloys, for instance, is similar to the austenitic treatment of steel in that temperatures of approximately 1450–1800° F. are required to provide the potential microstructure. When water quenched, the titanium alloy contains a supersaturated phase, and when aged, high tensile strengths comparable to hardened and tempered steel result. The titanium-aluminum-molybdenum alloy is aged at approximately 1100° F., after water quenching from 1750° F. Aging in its general meaning requires time, the Ti-Al-Mo alloy requiring approximately six hours.

Magnesium alloys are heat treated at slightly lower temperatures in a manner similar to some of the aluminum alloys, that is, by the solution and precipitation treatments. Different alloys require different temperatures. Because magnesium burns at a steel melting temperature if ignited, a sulfur dioxide or other protective atmosphere is often used during the solution treatment. However, air atmospheres are frequently used, but if ignition occurs, the 3000° F. temperature created must be quickly extinguished by smothering, such as with an application of sulfur dioxide or even sand.

Heat Treatment Problems

The application of heat to metals often changes the surface chemistry, sometimes to a very harmful degree. Heat may also change the metal's interior to unacceptable conditions, for example, by the production of coarse grains or cracks. Again, heat may have no effects on the metal. Most of the heat treating problems are associated with oxidation of the surface, carburization or decarburization of the surface, warpage, cracking, segregation, uneven residual stresses, and coarse grain structures.

Oxidation results from the combination of oxygen and an element to form the oxides which combine with the surface metal, being tightly combined with the surface at temperatures up to around 900° F. But in the red heat ranges and higher, the oxides, or scale, become loose and result in a loss of surface metal. When the surface can be subsequently ground, the oxidation is removed, and virgin metal is again at the surface. Protective atmospheres such as argon prevent harmful surface reactions, but many heat treating operations are conducted whereby oxidation occurs. Later grinding operations bring the metal to acceptable conditions. In steels, carburization (the increase in surface carbon content) and decarburization (the loss of surface carbon) occur due to metal reaction with the furnace atmosphere. By controlling the chemistry of the atmosphere, the surface metal's chemistry will also be controlled.

Warpage and cracking result from internal stress overcoming the strength of the metal as a result of drastic temperature and volume changes. If some plasticity exists in the metal, warpage will occur when uneven stress conditions exist. However, if rigidity is present such as in hardened steels, cracking occurs and stress is reduced at the crack. A combination of good design and heat treating techniques can reduce both warpage and cracking. High mass differentials such as adjacent thick and thin sections should be avoided in the design of the part and, if present, should be connected with generous fillets. Preheating reduces stress at the allotropic change point in some metals, therefore, preheating is very desirable for these types of metals. Vertical quenching of long parts reduces warpage. Many steels crack when allowed to become cold prior to tempering, therefore, the quenched part should be removed from the quenching medium while warm and be immediately placed in the tempering furnace which is at temperature.

Segregation frequently occurs in castings which are cooled too slowly. Reheating to proper temperatures and adequate soaking allows segregated compounds and other undesirable conditions to become homogeneously dispersed in the matrix. Overheating of metals enlarges their grain sizes. This undesirable condition is removed by reheating the metal to its recrystallization temperature and then cooling according to the procedure applicable for that particular metal. Metals should not be heated higher than necessary due to enlargement of the grain structure. Another problem results from hard and soft surface spots forming from steam pockets in the quenching medium or from decarburization. Movement of the parts in the quenching liquid prevents lengthy contact with steam.

One of the most common sources of trouble during heat treatment is failure of the heat treater to either properly soak the metal long enough at the correct temperature or not to quench fast enough during the quenching operation. As an example, a steel must be austenitic as it sinks beneath the surface of the quenching liquid. If austenite transforms in air between the hot furnace and the quenching medium, fine pearlite will result instead of martensite. Another problem relates to temperature. Periodically, furnace temperatures must be checked to assure correct temperature indications. Portable instruments are available for checking furnace temperatures. Adjustments can be made to pyrometers, such as the correct compensation for room temperature.

Classification Systems for Metals

Metals are classified according to their chemistries and are divided according to ferrous and nonferrous with further classifications to include specific coding systems such as those using combinations of letters and numbers to designate specific chemistries. Some of these systems are included in AMS, ASTM, MIL, FED, AISI, SAE, and others. Individual metal producers also have their coding such as USS. Nonferrous metal producers have coding systems, too, to identify particular brands of metals such as AA. Specifications and coding systems are so vast that only a small sample can be illustrated.

The American Iron and Steel Institute's numbering systems (AISI), for example, account for numerous types of metals. The coding system is convertible into other systems such as SAE, while ASTM, MIL, and others are convertible into metal producer numbers. On top of all the complexity is the intermittent amendment which modifies some previously published specification. Basically, in order to determine the chemical, physical, and mechanical properties of metals, the particular specification or code must be consulted. These systems are published by manufacturers and professional organizations and are available upon request. No single system is available to handle all the needed information about metals.

In the AISI system, a grouping of letters and numbers designates a particular steel. Some codes are three digits such as 302, some are four digits such as 4130, and some are five digits such as 52100. Many of the common steels are grouped in the four-digit system and include a fair sampling of carbon and alloy steels. Because this is a common and well known coding system, it will be explained, however, the principles of coding in the following system are not transferable to the numerous other systems. The four-digit system of AISI is as follows:

1—Carbon
2—Nickel
3—Nickel-chromium
4—Molybdenum
5—Chromium
6—Chromium-vanadium
7—Tungsten
8—Nickel-chromium-molybdenum
9—Silicon-manganese

With reference to the above code, the number represents the first digit on the left in the four-digit code and identifies the basic type of steel. For example, a 4130 is a molybdenum steel with an alloy. The second digit from the left in this example (1) indicates the approximate percentage of alloying element and is associated with the first digit. Besides the six basic elements of steel, there is a chromium content of 0.80–1.10% and a molybdenum content of 0.15–0.25%. In very general terms, the alloy chromium constitutes 1%, although the molybdenum is added due to it being a two-alloy steel. The alloy 4032 has a single alloy of 0.20–0.30% molybdenum, therefore, the second digit is zero. The last two digits in our example (30) indicate the percentage of 1.00% carbon content, which means that the 4130 contains from 0.28 to 0.33% carbon. When the left-hand digit is known, the second digit approximates the alloy content, while the last two digits indicate the carbon content of the alloy in this coding system. A last comment in respect to the above system is that there are hundreds of other metals not included in this system.

Questions

1. Define the heat treatment of metals.
2. What are some major purposes of heat treating?
3. Describe the microstructure in a metal.
4. Describe recrystallization in a metal.
5. Why is austenite essential in order to harden a steel?
6. Describe the total heat treating process required for obtaining an RC 61 in a ½-inch thick AISI 1095.
7. How is the spheroidizing process conducted?
8. Describe the case hardening process and explain why a soft core is enveloped by a hard case.
9. Differentiate among austenite, pearlite, and martensite.
10. Describe the main differences between alpha and gamma iron.
11. Differentiate between annealing and normalizing steels.

12. How does the chemistry of a metal relate to its capability?

13. How does the geometry of a part relate to its heat treatment?

14. Why are atmospheres important during the heat treating processes?

15. Describe the induction hardening process.

16. Compare the solution and precipitation treatment of aluminum alloy with the hardening of steel.

17. Differentiate between natural aging and artificial aging.

18. Why must parts be tested and inspected after heat treatment?

19. Why must critical points be recognized during the heat treatment of steel?

20. What factor enables a nonferrous metal to be heat treated?

Joining Operations

The joining of metals includes the soldering, brazing, welding, bonding, and mechanical joining processes. During the joining operation, the purpose may be to permanently connect one metal to another such as in welding or to semipermanently connect two or more pieces by riveting or bolting. The mechanical type of joint is made with any kind of metal, similar or dissimilar, but in welding, the majority of joints are made with similar metals.

Basically, soldering is a process of joining solid metals with a liquid metal whereby the liquid, or solder, closely adheres to the cleaned surfaces of the other metals. On the other hand, brazing is carried out at temperatures approximately twice those used in soldering and part of the brazing metal flows into the surface irregularities or pores of certain metals. For example, two copper sheets are soldered together at temperatures less than 800° F., while two sections of gray cast iron are joined with a liquid copper alloy at a temperature of approximately 1600° F. When one liquid metal flows into another liquid metal, the joint is fused by welding. Maximum temperatures are used in welding, depending on the metal's melting point. In the bonding operation to be discussed in a later chapter, metals are joined together by the adhering strength of some type of resin or epoxy.

Types of Joints

The joining of metals may be for the purpose of repairing broken sections of a part, or the purpose may be to join one metal to another during fabrication of an assembly. Either way, the joint design is limited in that only a few designs exist to make a joint. Broken parts are usually *butt* joined along the break and then soldered, brazed, or welded. Another type of joint is the *lap* whereby one section of metal overlaps the other, the joint being formed between the surfaces or along the edges of the contacting surfaces.

Each type of joint has its advantages, the butt joint resisting stresses in tensile, while the lap joint resists shear stresses. In this respect, consideration must be given to the kind of material being exposed to the stresses and to the types of stresses to occur so that the strength of the joint can be calculated. When tensile or compression loaded, the joint must withstand the accompanying stresses. A butt joint exposed to tensile stress is illustrated in figure 12–1. When a lap joint is exposed to loading, shear

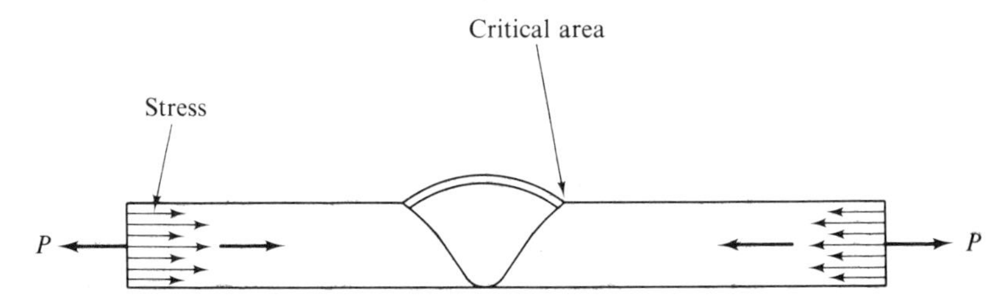

FIGURE 12–1 A butt joint exposed to tensile stress.

stresses develop in the material between the part's contacting surfaces if soldered, for example, or along the material's edges if welded. Shear stresses occur at the joint when plates making the joint are placed in tensile or compression loading. Figure 12–2 illustrates the type of stress induced by shear on a lap joint. Other types of joints such as *T* and corner, along with tubular, react against applied loads in the same manner as the simple butt and shear joints. Stresses may be singular, or a multistress condition may exist.

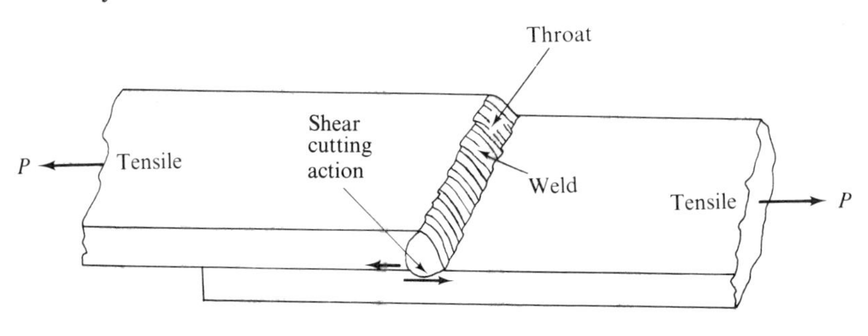

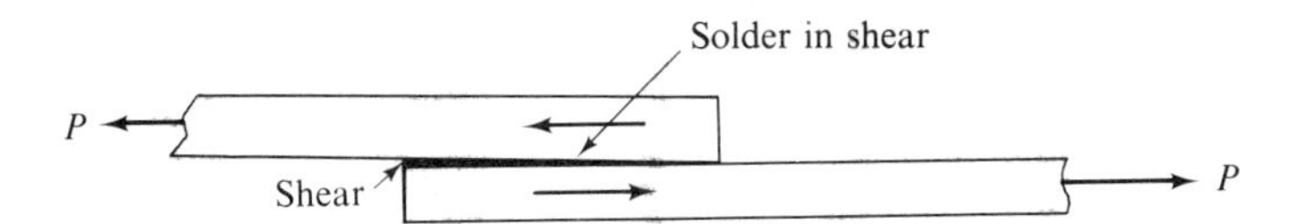

FIGURE 12–2 A lap joint exposed to shear stresses.

METALLURGICAL ASPECTS OF THE JOINT

When metal parts are joined, the metallurgical condition of the metals remains the same if the joint is not exposed to heat, such as in bonding, riveting, and bolting. However, when heat is applied such as in soldering, the metal may only be affected if it has been work hardened or hardened to a high strength by heat treating. This situation depends entirely on the degree of hardness in the metals to be soldered. Soldering temperatures reduce an RC 62 only a few points but do not reduce an RB 80. The reason an RB 80 is not affected by the temperature of 600° F. is that little hardness exists in the metal, therefore, none can be removed. On the other hand, hardened metal may be reduced in hardness if the induced heat is high enough. In this respect, the approximate temperature of 600° F. will reduce the RC 62 about 8–10 points.

Caution must be used when any type of heat is applied to a metal because hardness is also strength, and the heat may dangerously remove a required strength. Silver soldering and brazing, for example, remove most of the hardness in a metal because temperatures in the red heat range are used. High strength steels will be reduced to failure at the joint as a result of brazing when the design calls for a strength factor greater than that of the brazed joint. This condition is illustrated in figure 12–3*a*. Welded joints, on the other hand, completely remove previous metallurgical structures and set a new structure in the metals. Consequently, in welding, wrought structures become cast, and heat treated structures are eliminated in the heat affected zones. This situation is illustrated in figure 12–4. Heat treated parts are not to be welded unless the assembly is heat treatable after welding. As an example, an AISI 4130 heat treated tubular aircraft assembly is repaired by welding. The 180,000 psi tensile strength in the steel of the assembly is reduced after welding to less than 100,000 psi in the joint. If the joint is critical, failure will occur due to about 50% of the steel's strength being removed at the joint. When heat is to be applied to a metal, first determine if the maximum temperature reached will change the metal's microstructure.

Soldering

Metals to be soldered are first fitted together for proper shape and then cleaned in the areas to be soldered. Rust and heavy contamination are removed by emery paper. In order to complete the cleaning, a flux is used just prior to the application of the solder. The purpose of the flux is to remove oxides and prevent them from forming during the soldering or brazing operation and to also assist the liquid solder or brazing material in flowing into the joint. The flux may be a rosin which is noncorrosive such as is used in electrical parts repair or an acid such as muriatic which is used in most industrial soldering other than electrical.

SOLDER

A solder contains various proportions of lead and tin. Commercial solders not requiring a high fluidity are mostly lead, but many solders are 50–50, that is, 50% tin. The solder flows freely at temperatures below 600° F. in which soft soldering operations are conducted. Silver solder, however, contains a large proportion of silver for the purpose of providing a much higher strength joint than the lead-tin types. Silver soldering requires a special flux in the presence of bright red temperatures and is more closely associated with brazing.

PROCEDURE

Parts to be soldered are first cleaned and fluxed. Heat is applied in any manner, but only high enough to preheat the metals and cause the solder to flow freely on the clean

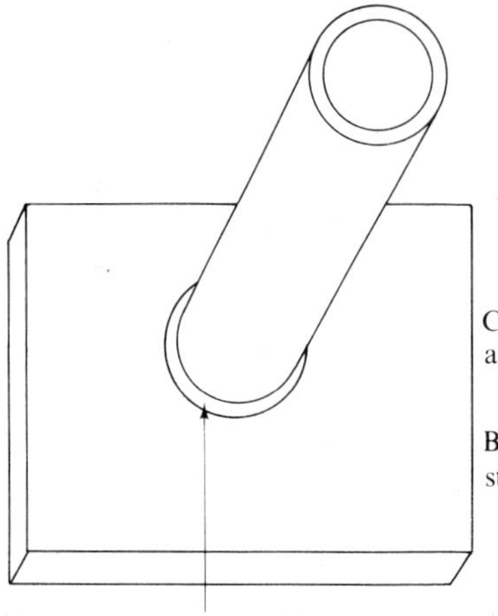

a.

b.

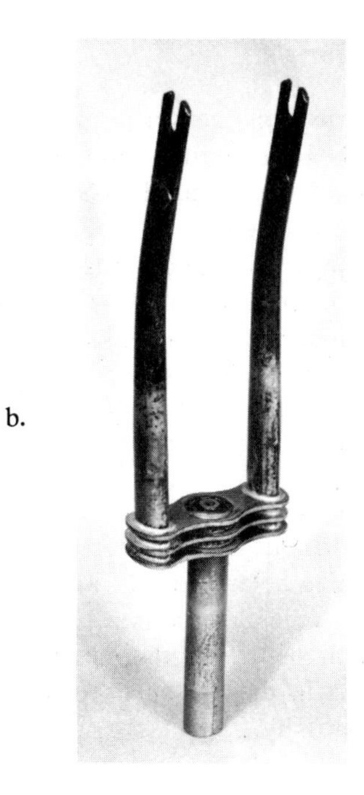

FIGURE 12–3 *a*, A set up for brazing steel tube to steel plate with torch; and *b*, the results of furnace brazing a bicycle fork. (Courtesy of Ajax Electric Co.)

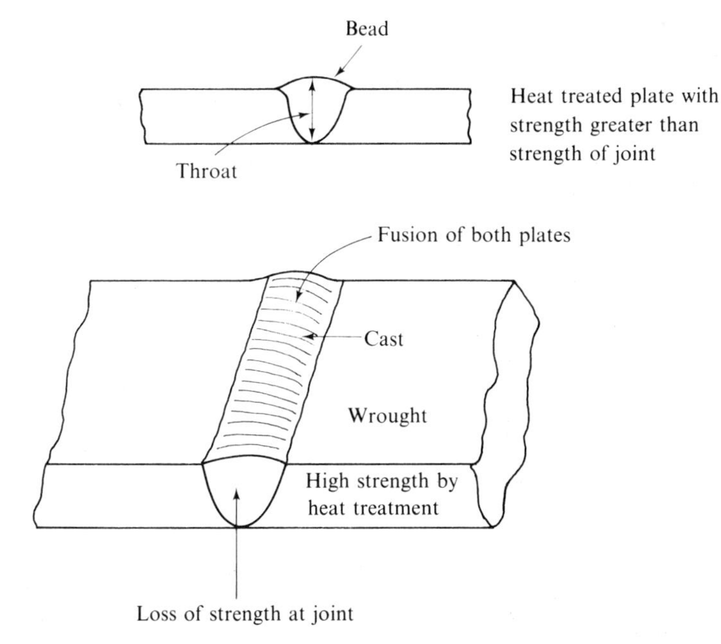

FIGURE 12–4 A welded butt joint illustrating the danger of welding a heat treated part.

surfaces. The simplest procedure involves the application of the solder to each surface to be soldered by heating the part, applying the flux, and then applying the solder. Too high a temperature will not allow the solder to remain, while too low a temperature will not allow the solder to flow. After both parts have been *tinned*, they are mechanically held together and the heat is again applied so that solder on each part flows. The soldering iron or the torch is then removed, the joint not being disturbed until it properly cools in air. Clamps or pressure on the joint are subsequently removed. Because soldered joints are lap joined, the shear strength is low, usually below 275 psi. When higher strength in the joint is required, lock seam and soldered joints are used. Stresses are more evenly conveyed to the parent metal because of the support from the lock seam.

Radiator cores are often soldered by an automated process. A ring of solder is placed between the parts to be soldered and the assembly is run through a furnace. The solder melts and cooling occurs as the assembly moves down the line. Soldered joints must be washed in water or in a solvent, especially the acid flux types, to remove the acid and prevent corrosion. Figure 12–5 is an illustration of soldered joints.

Brazing

The brazing of a metal involves the heating of the metal to a temperature below its melting point so that a lower melting metal will flow into the surfaces and cause coalescence, but not fusion, of the metals. Certain dissimilar metals can be joined by brazing. The nonferrous brazing rod is allowed to flow at a temperature which enables it to penetrate into the numerous surface irregularities of the base metal and to form a union between the base metal and filler rod. Because brazing is not a fusion

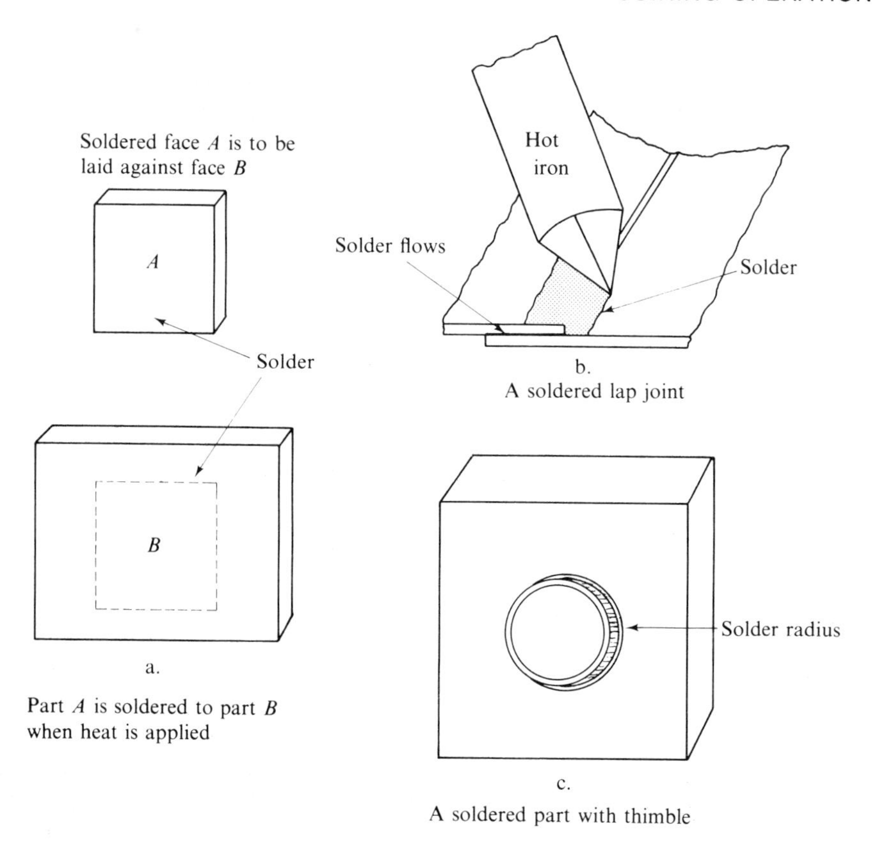

FIGURE 12–5 Soldered joints.

process, but only a surface union process of close adherence, fluxing is required. Borax is a common constituent of many fluxes used in brazing.

MATERIALS

There are several types of brazing rods, for example, aluminum and its alloys are brazed with aluminum alloy rods, while copper is used in brazing steel, but only in a protective atmosphere. Brass and bronze rods are used in brazing the numerous cast irons and steels. Silver alloys join many types of metals such as carbon and stainless steels and the copper alloys.

Basically, the parts to be joined are heated to a temperature which allows the brazing metal to flow freely into and between parts of the affected metals. Fluxing precedes and accompanies the brazing operation in order that the brazing metal will join with the virgin metal. Soldered and brazed joints are as thin as a few thousandths of an inch. As thickness of the metal in the brazed joint increases beyond that quantity needed to enter the joint as a very thin liquid, the strength of the joint decreases.

Brazing is used for cracked parts repair, especially large castings. In this respect, careful preheating of the part is essential so as not to overstress the part and cause additional cracks. Hand brazing procedures are common, especially in the repair business, but when many similar brazing operations are to be performed, automated processes are arranged. As in soldering, the brazing metal with flux can be placed around the joints or between the spaces to be joined and then be carried through a brazing furnace which uniformly brazes the joints and cools them for further processing. A mechanical means must be provided to effectively hold the parts while they are being brazed. When cracked parts are brazed, the repaired joint is again exposed to the same stresses in service which helped cause the crack. Therefore, expertise should be used in making certain that effective coalescence occurs.

PROCEDURE

Heating for brazing is applied through various media. The most common type is the torch which is also used for welding. Some types of parts are best heated in a salt bath, but the parts must be assembled with the brazing metal attached so that slipping of the parts does not occur during the moment of brazing. Similar parts are furnace brazed, often being passed through a continuous furnace. In this situation, the brazing metal and flux are attached, as in soldering, around the joint or between the parts of the joint so that the brazing metal flows during heating. Figure 12–6*a* illustrates an assembly being brazed as it passes through a gas furnace. Also, parts for a pair of pliers are being prepared for brazing (*b*). A typical brazed part is illustrated in figure 12–7*a*, while bicycle parts are prepared for brazing in *b*.

When specialization in part design occurs, parts are often brazed by induction heating. A copper tube is shaped around the joint to be brazed, but does not touch the part. The joint is wrapped with the brazing and fluxing materials. When a high frequency electric current passes through the coil, induced flux at the joint's surfaces causes rapid heating and melting of the brazing metal. Frequency of the coil is adjusted to the melting point of the brazing material only, while water flows through the tube to provide cooling of the coil.

JOINT DESIGN

Joint design provides for effective strength. When parts are lapped at the joint, the lap's length should be approximately three times the thickness of the material being joined. Accordingly, when tensile or compression loaded, the part's material will be exposed to the load in either tension or compression, but the solder or brazing metal will be in shear, the area being measured in square inches of contacting, or lapped, surface. The interface of a soldered joint is illustrated in figure 12–8, while the brazed interface is illustrated in figure 12–9. Notice that the soldered face is nearly a straight line, indicating close surface adherence. In the interface of the brazed joint, coalescence is shown by the very irregular and jagged surface. Some of the brazing material has penetrated beyond a straight line surface and adds strength to the joint.

Welding

Welding is a fusion operation whereby one metal melts into and fuses with another. The result is a joint that is unlike the soldered and brazed joints. Figure 12–10 illustrates the interface between two welded parts. No recognition is displayed to show lines of demarcation as there are in soldering and brazing. Because the welded joint

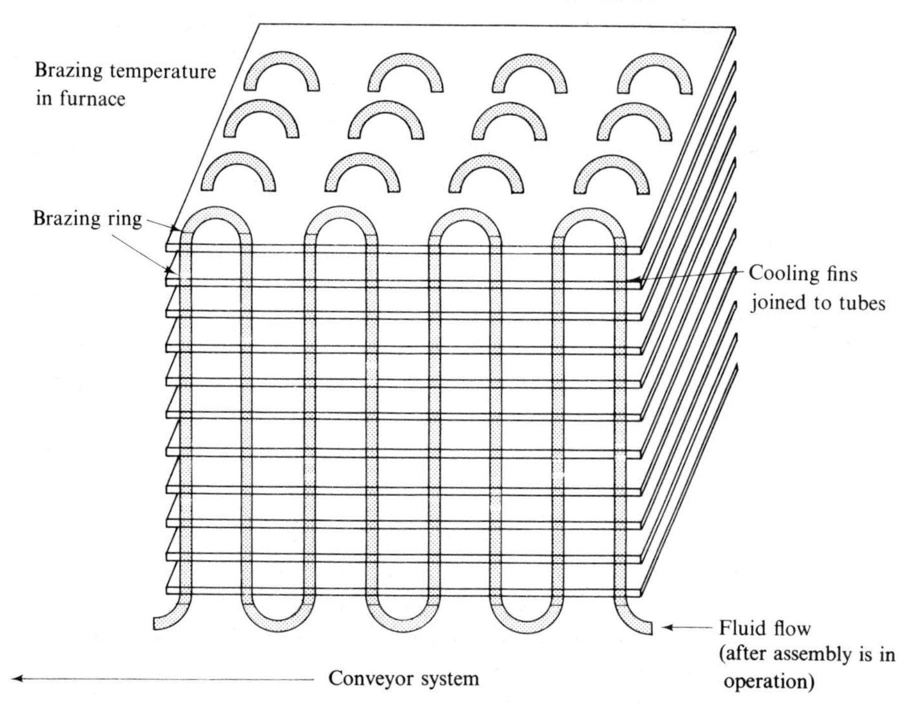

a.

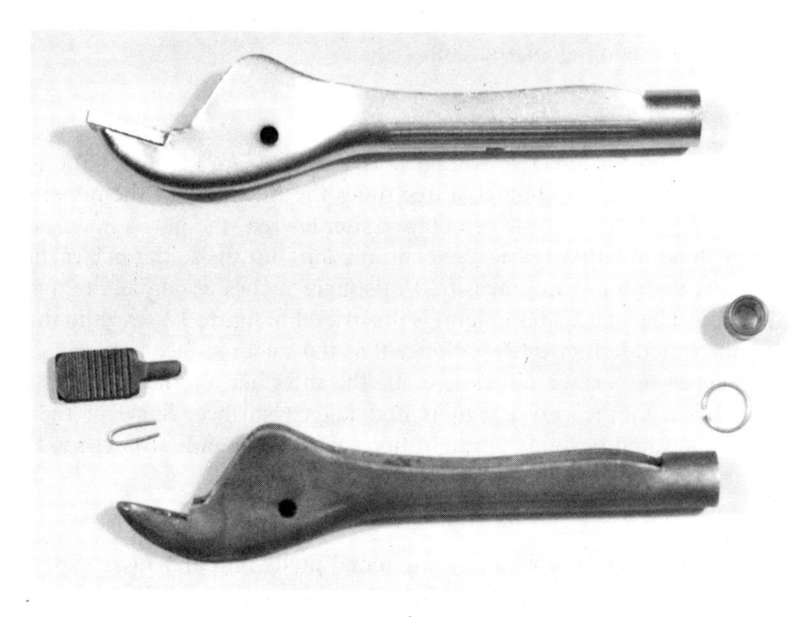

b.

FIGURE 12-6 *a*, Radiator assembly brazed in gas furnace; and *b*, vise-grip pliers parts being readied for salt bath carburizing. The rings of brass melt in the bath and form a strong brazed joint. (Courtesy of Ajax Electric Co.)

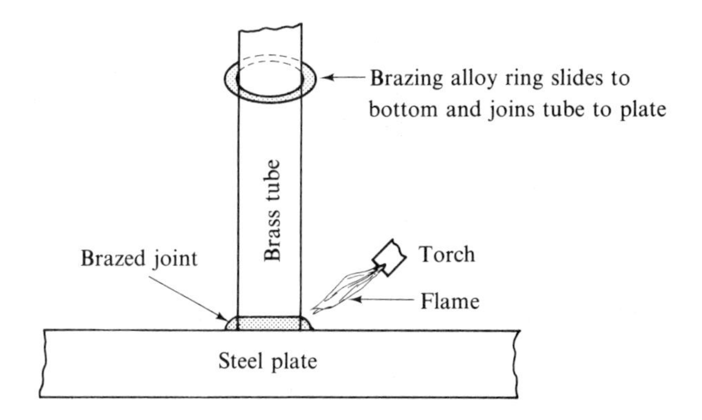

a.

b.

FIGURE 12–7 *a,* A procedure in brazing a part; and *b,* an operator brazes bicycle fork assemblies in a salt bath furnace operating at 2,050° F. (Courtesy of Ajax Electric Co.)

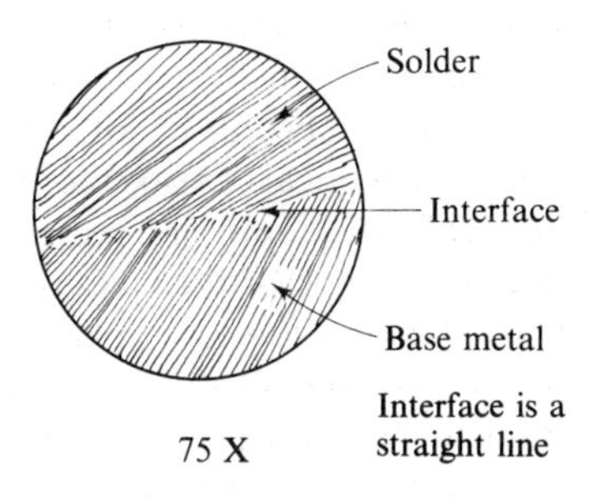

FIGURE 12–8 Interface of a soldered joint cross section.

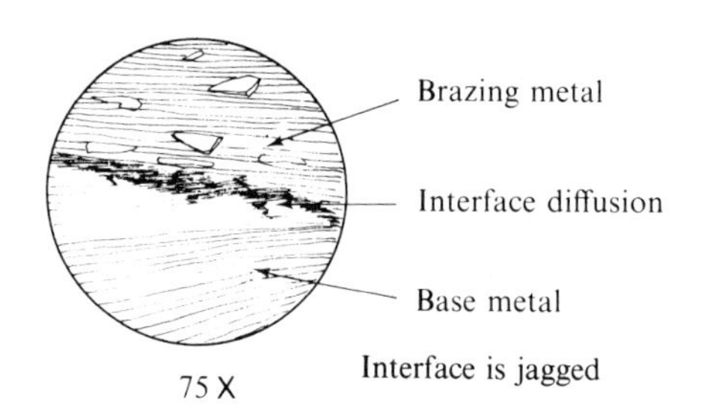

FIGURE 12–9 Interface of a brazed joint.

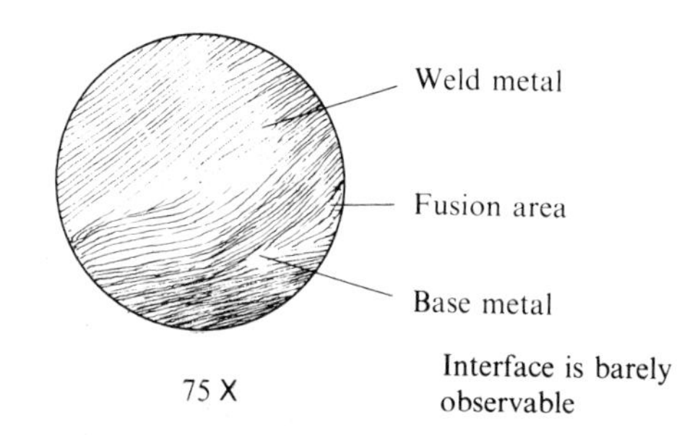

FIGURE 12–10 Interface of a welded joint.

required melting, the liquid solidified as a casting. Being a casting, the welded joint, then, has the characteristics of a casting, usually the air cooled type.

METALLURGY

The metallurgy of the weld must be understood if certified welds are to be made. Both kinds of metal are weldable, the ferrous and nonferrous. Further, two types of each basic metal are weldable, the castings and the wrought products. Because welding causes the metal to become liquid and because the liquid cools at a rate commensurate with its mass and cooling medium, usually air, the microstructure is that of cast metal cooled by various rates of cooling. The grain size of the cast metal and the structure within the grains reflect the high temperature and the rate of cool. The completed weld remains basically a weld, however, some welded structures are subsequently heat treated which refines the grain structure of the weld and imparts more desirable characteristics in the microstructure as a whole. Unless the weldment, the welded structure, is reheated in some manner, the hardness, strength, and fatigue properties of the microstructure remain in the "as welded" condition.

The technique of good welding is reflected in the finished weld. Some basic requirements of an acceptable weldment include the shape of the fusion zone (fig. 12–11). The fusion zone must include enough of all metals involved so that homogeneous fusion results among the chemicals. Also, the fusion zone must sufficiently penetrate the sides of the joint, while the depth of penetration must include the thickness of the part being welded. The shape of the weld reflects a continuous series of cast pools which are interlocked along the length of the weld. Internal structure results from the temperature–mass–cooling rate factors, while external shape depends on mechanical conditions of the entire welding apparatus and the skill of the welder. The external shape must be compatible with acceptable surface stress patterns, that is, stress raisers must not be created at the edges of the fusion zone.

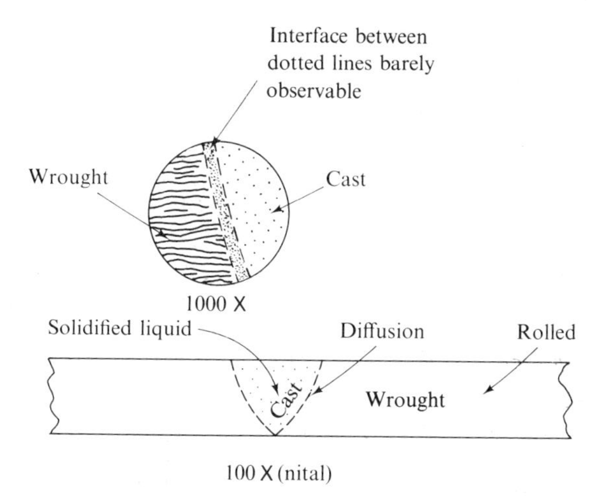

FIGURE 12–11 Fusion zone in a weld after grinding, polishing, and etching.

PROCEDURE

Technique is involved in the chemical and physical distribution of the constituents in the fusion zone. When surface oxides and other unwanted foreign materials enter the liquid, time and technique must allow these harmful materials to float to the surface. If freezing of the liquid occurs and traps these inclusions, the fusion zone is unacceptable when the weld is to be stressed. Lines of stress do not jump through porosity pockets and other cavities or through inclusions such as particles of iron oxide. Instead, stress patterns detour around the discontinuity, resulting in high energy concentration points, as illustrated in figure 12–12. When the stress concentration in a metal is greater than the particular strength of the metal, immediate failure occurs at the point of stress concentration. Welded metal must then be able to carry the stress evenly through the joint, therefore, a sound and homogeneous internal structure must exist.

When the chemistry of the metal to be welded is such that a hardening capability is present, care must be taken to assure slow cooling of the solid metal. In this respect,

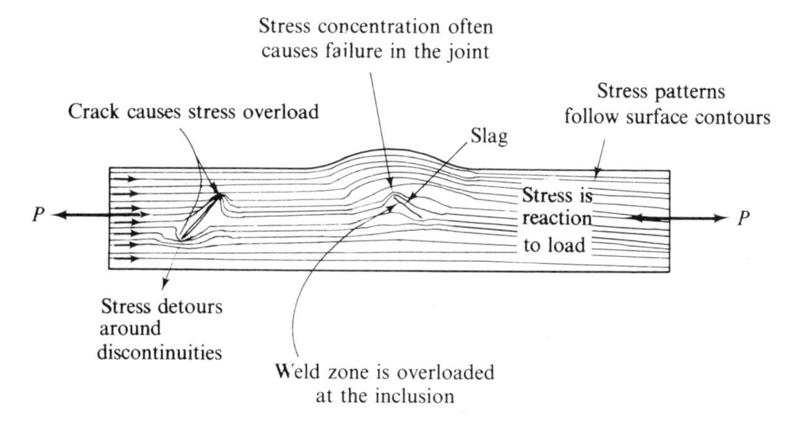

FIGURE 12–12 Stress patterns detour around discontinuities causing stress concentrations.

thin sections cool in air very rapidly, and brittleness can result in the weld. Low carbon steels, for example, are not subject to high hardness by rapid cooling, but medium or mild carbon steels in the presence of certain alloys harden appreciably when air cooled from above the critical point. This cooling rate must be recognized and planned for.

When castings are welded, the welded zone resembles the remainder of the cast type of microstructure. On the other hand, when wrought metals are welded, the microstructure of the welded zone is that of the cast pool linking the wrought sections of metal together. Figure 12–13 illustrates two typical welded microstructures. With respect to the joint, wrought metal is stronger than cast metal due to grain elongation, deformation of the metal's lattice structure, and interlocking of grain boundaries. Consequently, the strength of welded joints must be carefully calculated and then be made capable of carrying designed loads.

TYPES OF WELDING

The welding of metals is accomplished by several methods which include forge and resistance welding, electric arc and shielded arc welding, submerged arc welding, and welding with the use of high energy concentration such as in the electron beam.

Forge Welding

Sometimes, forge welding is accomplished when two large sections of metal need to be fused together. Fusion is performed by hammering while the metals are just below

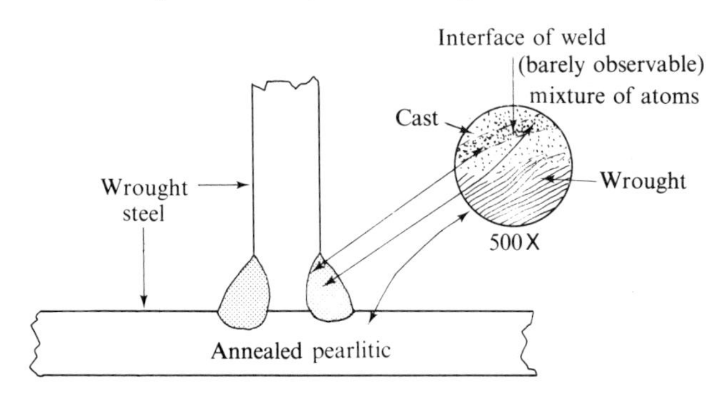

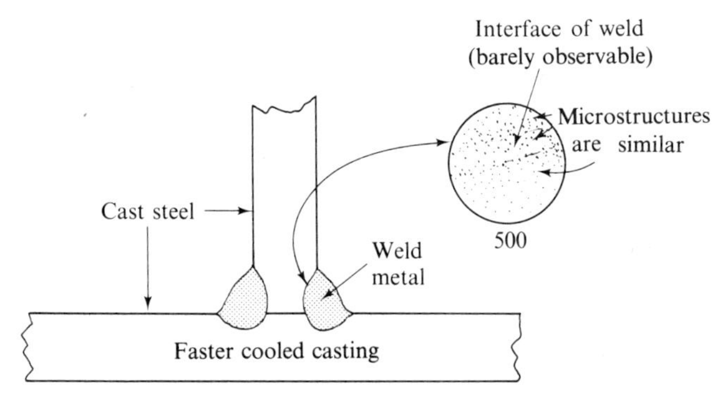

FIGURE 12–13 Wrought and cast steel connected by double *T* welding.

their melting points. While white-hot, steel will fuse, especially when the loose scale is knocked off and a flux is used. Preshaping the mating parts assists the operation. When fused, reheating homogenizes the microstructure to more acceptable standards. This process, however, is subject to contamination by surface oxides and has limited capability, but a refinement of this process is utilized in the pressure welding of pipe.

Resistance Welding

Two metals are joined by coalescence when subjected to a high temperature in the presence of pressure. When one metal is overlapped onto another and placed between two electrodes, joining occurs during current flow and pressure application. Fusion occurs at the interface and inwards towards the surface of the two metals (fig. 12–14).

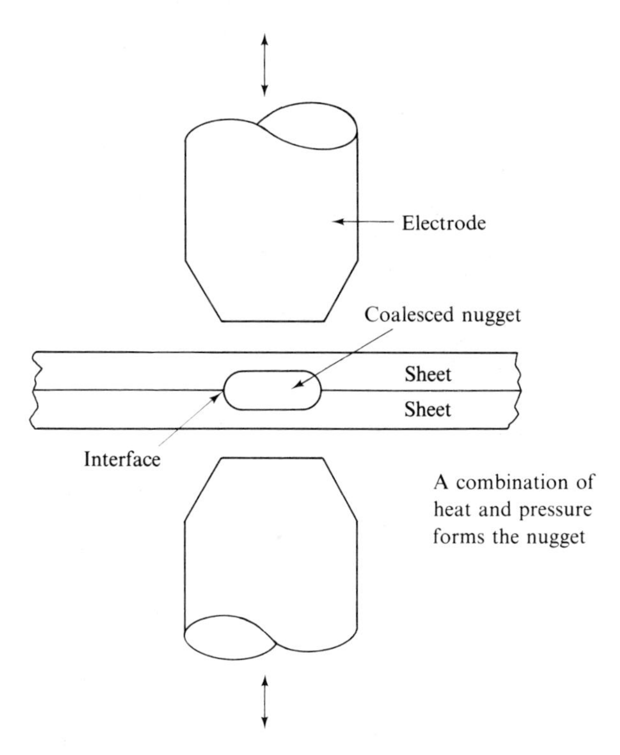

FIGURE 12–14 Two sheets are spot welded between electrodes resulting in the nugget.

Coalescence is not melting, but heat and pressure form a union of the constituents that is equal to or greater than the strength of the surrounding metal. Because metals conduct electricity and because metals resist the conduction of electricity, high temperatures are created in the path of the current flow at the metal's interface.

A controlled alternating voltage with high amperage produces instant heating at the interface of the two metals. Temperature is a function of the square of the current multiplied by the product of the time and resistance of the circuit. Shape of the welded zone is reflected from the faces of the two electrodes plus the amount of pressure exerted immediately following heating. Best results are obtained when the temperature-pressure cycle is electronically timed. When the principles of resistance welding are formed into a machine, a self-contained unit results.

Resistance welding machines are designed to handle and process the specific parts to be welded. The most common type of machine is the spot welder shown in figure 12–15. The clamping tongs are also the water cooled electrodes. The control system provides for timed pulses of current, depending on the pressure signal. Spot welding

FIGURE 12–15 A spot welder. (Courtesy of Otto Konigslow Mfg. Co.)

can be either hand accomplished or automated by means of mechanically controlled devices. Both ferrous and nonferrous metals are spot welded, the thickness of the sheets or plates ranging up to ¼ inch and, in some instances, thicker joints have been welded. As electrical conductivity increases in the metals to be joined, the current also increases. Aluminum and copper, for example, require higher currents than most other metals.

The seam type resistance welder is used in the production of cans and tanks. The principle of operation is similar to spot welding, except that the electrodes are wheels which press onto the lapped surfaces or butt edges as they turn. Timed impulses of current occur along the joint line. The metal's interface results in an overlapping series of welds (fig. 12–16). As in spot welding, pressure, current, and time must be adjusted properly so that no appreciable indention of the surface occurs nor any ejection of liquid metal from between the electrodes.

Resistance welded joints are tear tested by peeling the two welded sheets apart. Properly welded zones will remain intact as the surrounding metal or original metal tears away from the coalescence zone. Such a high strength joint is made possible by the application of pressure during heating, and especially during cooling of the joint. Spots which break through the weld during the test have not been processed properly. In this respect, too low a current factor or an improper current-time factor will produce unsatisfactory and, in some instances, dangerous welded joints. When too high a current flows through the joint or when too great a pressure is exerted, metal loss occurs at the joint, and the remaining metal may not be capable of withstanding future working stresses.

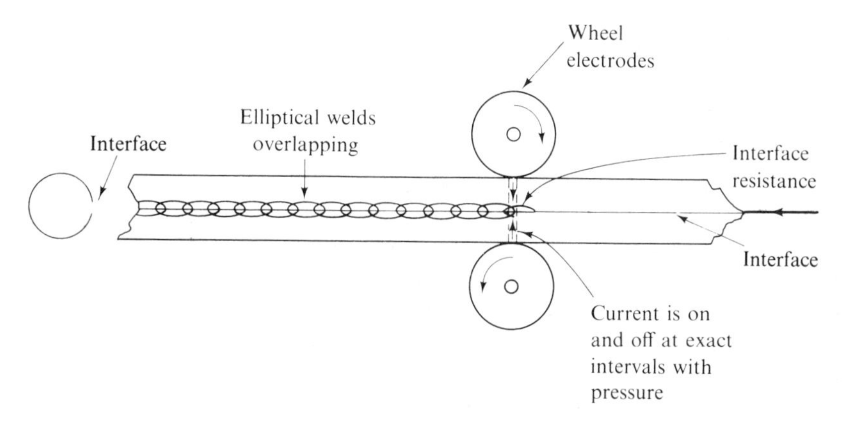

FIGURE 12–16 Two sheets of steel are seam welded.

Flash and Upset Welding

A variation of spot welding is the flash welding of parts such as are used in household and lawn metal furniture. Tubular type sections for chairs, for example, are first fitted for the joint's design and arranged in a special type of fixture. When the two or more mating parts are brought in contact while carrying a current, a flash occurs which induces very high temperatures. Immediately, the heated joint is pressed firmly in compression to fuse the edges, and the current ceases to flow. Precision of the fixture is reflected in the resulting joint.

The upset welding process is used particularly in the pipe and tube industry. Both current and pressure are applied concurrently so that heating results from internal resistance at the joint. A smoother joint occurs than that obtained in flash welding. Shaped rollers, including electrodes and pressure rollers, move the pipe so that welding occurs across the current path between the electrodes.

Gas Welding

The gas welding process uses oxygen and acetylene mixed in a torch and burned at the tip of the torch at 6300° F. This temperature is sufficient to melt all commercial metals except tungsten, therefore, the process is used in many fabrication industries. Gas welding is used in welding thin and thick sections among the numerous ferrous and nonferrous metals, using steel cylinders or bottles to retain the gases.

Gases Two steel bottles are required to store the gases, oxygen and acetylene. The standard oxygen bottle contains approximately 244 cubic feet of oxygen at a pressure of 2200 psi at 70° F. Acetylene is contained in quantities up to approximately 300 cubic feet at a pressure of 250 psi, but the gas is dissolved in acetone to prevent explosion. Acetylene bottles contain a filler material and acetone. During bottle filling operations, incoming gas is dissolved in the acetone, the acetone absorbing many times its volume in gas.

In many situations which require large quantities of gas, several cylinders of oxygen and acetylene are connected to a dual manifold system, oxygen being in one system and acetylene being in another. Necessary additional safety devices must be installed when two or more acetylene cylinders are connected to the manifold. When larger quantities of acetylene are required, the generator is used, the gas resulting from the combination of calcium carbide and water.

Free acetylene under a pressure of more than 15 psi is subject to explosion, therefore, it must never be compressed to pressures greater than 15 psi. When dissolved in acetone, tank pressure can be safely increased, but when the gas leaves the cylinder, line pressures must not be increased beyond the danger point.

Equipment In order to have the gases at working pressures up to 15 psi, the regulator is used. This device reduces cylinder pressures to working pressures in either a one or two stage step and also provides for a uniform flow of gas as the cylinder pressure is reduced. A regulator is required for each cylinder, the two-stage maintaining a uniform flow of gas, while the single-stage requires occasional adjustment of the pressure as the cylinder pressure falls. Regulators indicate cylinder and working pressures. Before releasing cylinder pressure by turning the cylinder valve, the regulator adjusting screw must be turned out to prevent the surge of high pressure from damaging the regulator.

When handling oxygen and acetylene cylinders, care must be used to see that cylinders remain upright and are not bumped.Special care is needed in operation of the valve systems, never allowing oil or grease to contact the system. Cylinders should be protected from the weather. Further, excessive heat must not be projected onto the cylinders. To prevent interchanging of hose lines, oxygen hoses are black or green and have right-hand threads. Acetylene hoses are red and have left-hand threads. Periodic tests for leaks must be made; tests for gas leaks are accomplished with the use of soapsuds. Then, when the welder is ready, the acetylene valve is opened a turn or less, keeping in mind that quick closing may be necessary. The oxygen valve is opened all the way. With respect to cylinder pressures, a built-in safety precaution allows either cylinder to slowly release pressure if exposed to excessive heat.

The oxy-acetylene welding torch provides for three types of flames to be produced. Observation is done with the use of goggles which are especially made for gas welding procedures. The neutral flame is maintained by an equal ratio of oxygen and acetylene, and this flame is used more than any other since it produces no ill effects on the chemistry of the metals being welded. The carburizing flame contains an excess of acetylene and is used, for example, in welding some types of alloy steels, in soldering, silver brazing, and in hard facing operations. When the excess acetylene is reduced to only slightly reducing, or where a small excess of acetylene exists, a typical flame is produced for welding many of the alloy steels. The oxidizing flame, on the other hand, is a decarburizing flame with regard to steels, but it is used in some types of copper alloy welding. For cutting or parting steel, the oxidizing flame melts and oxidizes the metal very rapidly, enabling it to sever thick sections extremely fast. The three types of flames are produced by adjustment of the two torch valves, the maximum volume of gases being controlled by the size of the torch tip. Ignition is then caused by the sparklighter after eye goggles are adjusted.

Procedure When the welder is ready to weld, he has selected the proper size tip and adjusted the regulators for correct working pressures. As an example, a number three tip is used for welding 1/8-inch thick metal parts with working pressures of three pounds for both oxygen and acetylene. Figure 12–17 shows the regulator settings of oxygen and acetylene working pressures. Application of the sparklighter at the tip of the torch causes acetylene ignition when the valve is slightly opened. Opening the oxygen valve then gives proper adjustment of the desired flame. By holding the tip of the

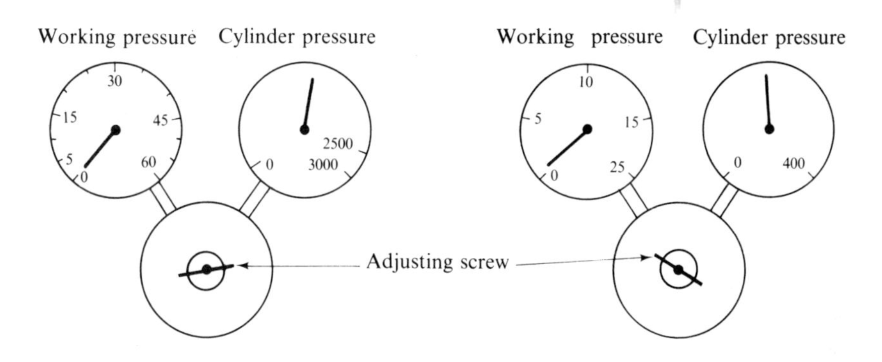

FIGURE 12–17 Cylinder and working pressures for a gas welding operation.

inner cone portion of the flame at the metal's surface, melting of the metal occurs. The welding of two sections of carbon steel is shown in figure 12–18. The actual welding operation is, then, a matter of technique.

When closing the torch valves, the acetylene is closed first, followed by the oxygen. Cylinder valves are then closed, unless additional welding is to be performed. The lines are next drained, followed by closing the needle valves. Adjusting screws on the regulators are subsequently released. With regard to safety, torches should be equipped with flash-back arrestors as well as the acetylene manifold line when a manifold system is used. Also, check valves must be installed in manifold lines for both gases. Further, the acetylene manifold must be equipped with a bursting disc vented to the outside atmosphere. Oxygen equipment must not come in contact with oil.

Arc Welding

The electric arc is another source of welding heat. When an arc forms between an electrode and a metal to be welded, the metal immediately melts due to the very high

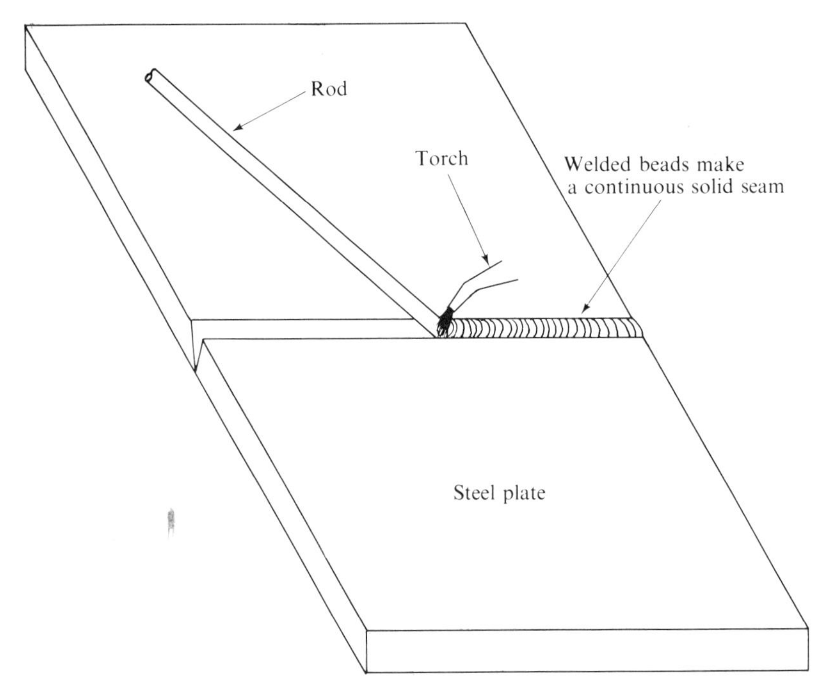

FIGURE 12–18 One position for holding the welding rod.

temperature of the arc. The principle of the arc provides a current path through the air so that a complete electrical circuit exists. Both direct and alternating currents are used, a high amperage being necessary to melt the metal, while the voltage acts as the pressure in the circuit. The switch in the circuit is the welding electrode. By carefully scratching the metal's surface with the electrode and quickly lifting it a small distance, an arc is formed which is then maintained under given factors to support good welding technique. The resistance of the metal to the flow of current causes immediate melting of the metal and the electrode, requiring periodic replacement of the electrode. The weld, then, consists of fused metal from the electrode and metal being welded.

Equipment The source of current is from a welding machine such as the typical arc welder shown in figure 12–19. When direct current is used, straight polarity is maintained when the work is the anode and the electrode is the cathode. On the other hand, when the work is negative and the electrode is positive, reverse polarity is maintained. In this respect, more heat is provided at the work's surface when straight polarity is used. Straight polarity is less penetrating than reverse (fig. 12–20). A switch on the dc welding machine is used to deliver either straight or reverse polarity. Some machines, however, are combination welders, that is, both alternating and direct current are available. The third type is strictly an alternating current welder.

Sizes of electric welding machines are rated in amperes, the capacity being approximately 60% of the duty cycle. During a specified period of time, the welder is rated at

FIGURE 12–19 A typical electric arc welder. Cables are attached to welder at electrodes indicated. (Courtesy of Chemetron Corp.)

a given amperage and voltage output. Ampere values are high, such as 150, 200, 300, and higher. The machine will effectively produce the current indicated for at least six minutes out of ten on many machines, the remainder of the time being required for cooling. A 300 A machine is required for much of the industrial welding applications, but heavy and thick sections of steel require as much as twice the average requirement, or 600 A. Welding machines are energized from 220 V alternating current lines or from independent generator sets. Each welder is self-sufficient in that it contains the necessary transformers, motors, generators, and associated equipment to perform the task.

Procedure Because arc welding occurs within an electric circuit, the welding table must be made from metal, usually a heavy steel plate. One cable from the welder is clamped to the table as a ground to complete the circuit. The other cable from the welder contains the stick electrode in the holder which is used for guiding the line of weld, most electrodes being flux coated for protection of the pool of liquid metal. During welding, the metal and the flux melt, causing excessive smoke and other contamination. Therefore, proper ventilation and fire resistant materials in the welding area are necessary. A special helmet is also necessary to protect the operator's face and eyes; gas welding goggles are not adequate. Long gloves must be worn due to heat radiating from the arc. Also, an apron is worn for clothing protection against the many flying pellets of molten metal. While welding, the arc must never be observed with the naked eye. To protect other workers, the arc should be properly shielded.

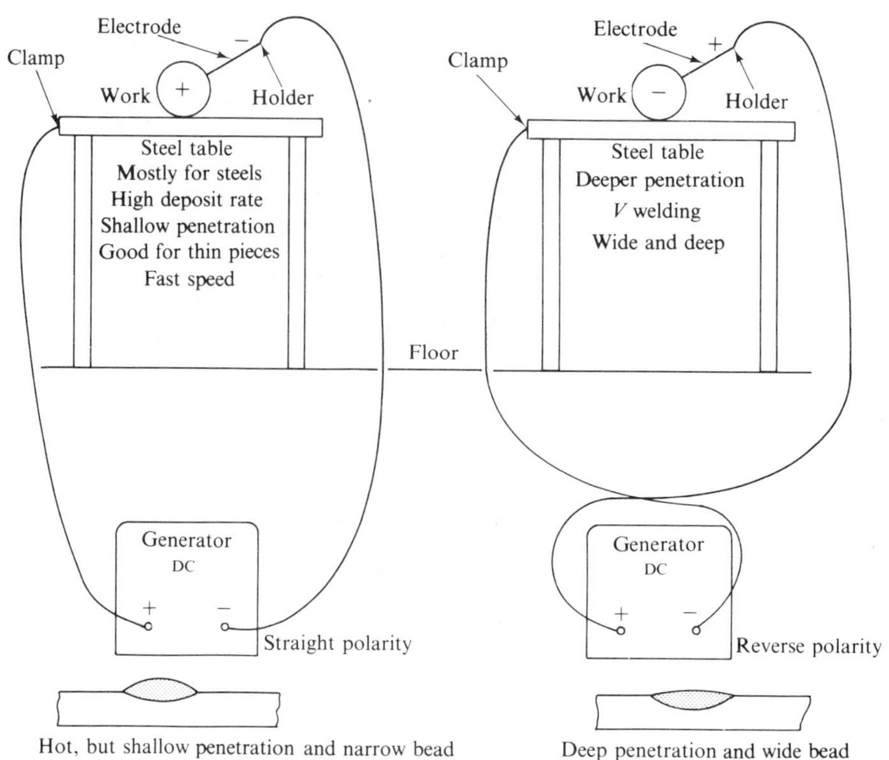

FIGURE 12–20 Comparison between straight and reverse polarity welding procedures using shielded metal arc welding.

Arc welding technique is similar to that in other welding procedures. The work to be welded must be accurately prepared and arranged. Large or intricately shaped parts must be placed in some type of fixture or just clamped into place to prevent warping due to temperature differentials occurring on heating and cooling. As in gas welding, broken parts are repaired, as well as the fabrication of new parts and assemblies. Arc welding is performed on small parts such as trailer hitches to the fusion of larger plates of metal on ships. Figure 12–21 shows the arc welding of a transport vehicle. The construction industry uses a large tonnage of welding electrodes, especially in the beam and column and bridge and building industries.

When welding is to commence, the voltage-amperage settings are fixed according to the thickness of work to be done. As an example, when using a 1/8-inch E-6010 electrode, a machine setting of 25 V and 100 A is typical. The eye shield is flipped into place, and the arc is struck. Proper technique completes the weld. Too high an amperage setting produces too much heat, while incomplete penetration results from too low an amperage setting. Should the electrode stick to the metal, the electrode is quickly twisted in order to break it loose. As in all welding operations, practice increases capability. Upon completion of the weld, the slag is chipped away by hammering the bead. Typical arc welded sections of metal are illustrated in figure 12–22.

Gas Shielded Arc Welding

A modification of the shielded metal arc welding process includes the use of a protective envelope of gas instead of flux around the molten pool of metal. Oxy-acetylene

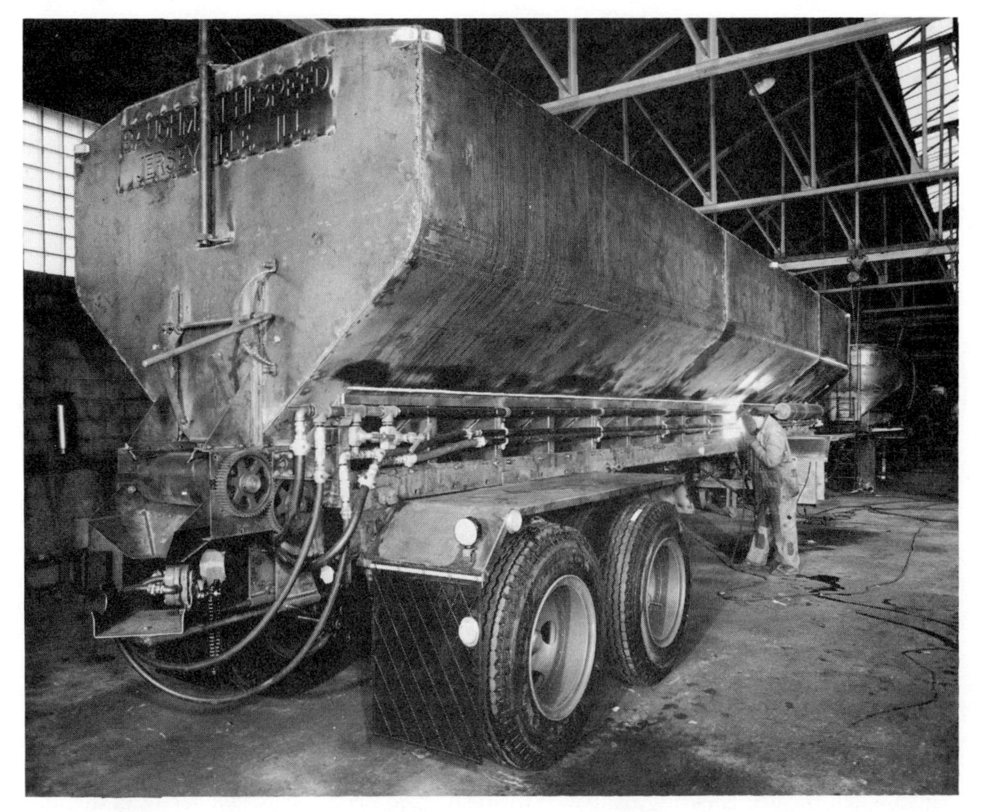

FIGURE 12–21 An arc welding repair to a transport vehicle. (Courtesy of Chemetron Corp.)

and metallic arc welding do not always produce the soundest welds, due mainly to surface contamination in some metals by oxygen and foreign particles entering the pool of metal. By combining the arc and gas principles with good welding technique, sound welds are produced which are not only cleaner internally and externally, but are stronger. Two main types of gas shielded arc processes are in use, the tungsten inert gas (TIG) and the metallic inert gas (MIG) processes. TIG and MIG processes are rapidly absorbing much of the conventional welding procedures. Gas shielded welding is easily automated, as pointed out in figure 12–23.

TIG Welding The tungsten inert gas welding process uses a tungsten electrode to provide the arc heat. It is not comsumable, however, over a long period of time it is gradually consumed and must be replaced. The size of the electrode and its surrounding ceramic cup are determined according to the type of work to be welded. Most metals, as in other welding processes, are weldable by this means. When the arc is struck, a steady flow of inert gas, either argon or helium, flows from the cup and immediately surrounds the arc. Because molten metal exists inside the inert atmosphere, sound and excellent welds are produced, the gas flow being on and off at the signal of the arc. The ac arc will jump the air gap of about 1/8 inch, while the dc must scratch the metal's surface to start the arc. The TIG machine must be capable of producing lower amperage in many welding situations.

FIGURE 12–22 Arc welded brackets to odor control boxes which are used in gas systems. (Peerless Manufacturing Co., Inc.)

TIG machines are similar to arc welding machines, but the gas cylinder and flow meter with its regulator are added (fig. 12–24). As in all electric welding operations, special safety and health protection devices are required. Welding techniques are similar to the other welding processes, but no chipping is required as welds are finished when the arc is extinguished. With respect to argon, it is heavier than air and clings to the base of the weld, giving a better blanket of protection than helium. Helium, however, is used on thicker sections due to its greater penetration ability. A modified TIG welding operation is illustrated in figure 12–25.

Electrode sizes and shapes vary with conditions. Pointed electrodes are used in direct current applications, while electrodes with rounded ends are used in alternating current situations. The size of the electrode depends on the thickness of metal to be welded and the type of polarity when using dc welding. Reverse polarity welding requires a larger electrode than does straight polarity. Some welds are performed by fusion of the edges of the metal, but when filler metal is used, it must be equal to the base metal's mechanical property capabilities. During the welding operation, the rod is placed in the leading edge of the pool of metal, adding the rod to the pool as required. The size of rod is approximately the same diameter as the thickness of metal being welded. TIG welding has good automation possibilities such as in straight-line, template, or circular pattern welds.

MIG Welding The metallic inert gas welding process (fig. 12–26) uses a consumable electrode in the protective atmosphere of an inert gas. The electrode wire is fed through the torch so that no stopping is required as in arc welding. When the arc is struck, the welder controls the movement as in other welding techniques, however, MIG welding lends itself very favorably to automation. MIG welding is faster than conventional welding, resulting in time saved along with improved metallurgical conditions of both the fusion zone and the heat affected zone (fig. 12–27).

FIGURE 12–23 The welding of special alloys such as nickel and copper alloys is partially automated with the use of this integrated and automatic gas-shielded welding machine. (Courtesy of Ransome Co.)

Metal is transferred by a gun from the wire electrode to the pool of metal by spray, globules, and short circuit methods, all depending on the whole condition such as size of electrode, voltage-current factors, and gas. The spray transfer occurs as small drops of liquid metal moving through the plasma to the pool. With argon, a high current density is required, resulting in deep penetration. Globular transfer occurs at current densities lower than those used in spray transfer. Globular transfer causes periodic dropping of a sphere of molten wire onto the work, but only shallow penetration occurs along with metal spatter. In short circuiting transfer, thin and intricately shaped sections are weldable, in addition to the ability of the method to be used in many different welding positions. Short circuiting transfer is an all-around method of metal transfer, especially for jobs requiring less than 200 A and small wire diameter.

A major difference between the MIG method and other methods is that direct current is used. Best efficiency results from reverse polarity in that better penetration occurs. Arc length and voltage must be kept constant to produce good welds. The gun for controlling the arc has a trigger which controls the gas flow, wire feed, and cooling

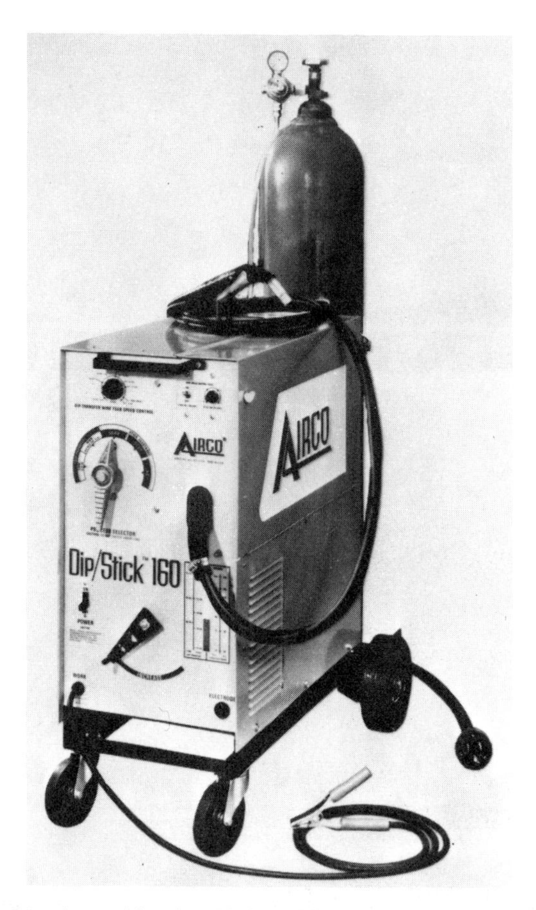

FIGURE 12–24 A combination TIG, MIG, and electric arc welding machine which is capable of performing each of the above welding operations. (Courtesy of Airco Welding Products)

water inside the gun. Different metals require different protective shields, therefore, several gases are used, depending on the metal being welded. As an example, aluminum is welded with argon or an argon-helium mixture. Titanium, magnesium, and copper-nickel alloys use argon, while carbon and stainless steels use a mixture of argon and a small quantity of oxygen to support transfer of metal.

Carbon Arc Welding

When the carbon arc process of welding is desirable, a filler rod of the same metal must be used because the carbon electrode is only the heat source. Carbon electrodes have long tapered points being held in their special holders and require a slightly higher voltage to stabilize the longer arc. Amperage is adjusted to keep the electrode below a red heat at a point not further than approximately 1¼ inches from the tip. When dc is used, the circuit is set for straight polarity whereby the metal to be welded is positive. With respect to this type of welding, some types of nonferrous metals require steel backing plates to assist in a better welding technique.

Carbon arc welding is a puddling process, therefore, the metals to be welded must lie flat. The arc is struck by touching the electrode to the work and lifting to maintain

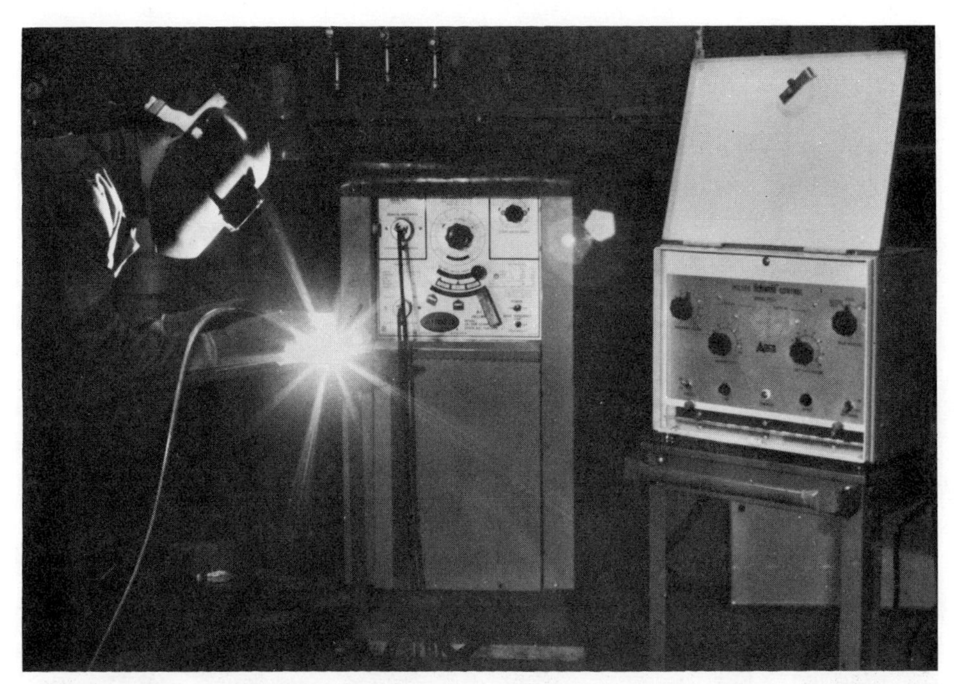

FIGURE 12–25 A pulsed TIG system. (Courtesy of Airco Welding Products)

the desired arc length. The filler rod is added by holding the rod nearly parallel to the bead where the arc touches it.

Thermit Welding

Thermit welding is a limited type of welding which is used mainly on very large ferrous parts, such as cracked castings, or joints of fabricated assemblies. The space between the two parts to be welded must be wide enough for the operation to be effective. A sand mold is built around the joint and dried. Then the joint is heated to around 1500° F. A mixture of one part of finely ground aluminum and three parts of iron oxide is placed within the mold and ignited with a fuse such as magnesium. A temperature of approximately 5000° F. is created, resulting in the fusion of the sides of the joint.

Submerged Arc Welding

A modification of the electric arc process is the use of a bare steel coil of wire as the electrode. The wire's chemistry must be equal to the base metal's chemistry. When either direct or alternating current is used to cause an arc, the wire melts and drops into the joint to be welded. To assure a clean weld, a line of flux is deposited just ahead of the arc which results in a molten pool of flux, electrode metal, and the basic metal. The hot flux is a conductor of electricity, therefore, the arc remains stable even at travel rates of several feet per minute. The setup is often an automated process that produces sound welds. It is also one of the fastest of the welding processes. Submerged arc welding has characteristics of the MIG process, as well as electric arc welding, in regard to slag. Because dangerous electricity is present, necessary precautions must be taken during the welding operation.

FIGURE 12–26 A MIG welding operation. (Courtesy of Airco Welding Products)

Submerged arc welding is being used to weld plates of steel five inches thick for the purpose of holding pressures of thousands of pounds per square inch in large pressure vessels. The steel plates are prepared by machining a *V* at the edges and placing the assembly on a rotating fixture. Some operations use the reverse polarity arc followed by an ac arc, both using 3/16-inch electrode wire. As the assembly turns, the flux is dropped into the *V* opening, while the tandem electrodes fill the opening with a pool of molten steel. The flux protects the pool from the atmosphere. Figure 12–28 illustrates the enormous capability of the manipulator. The manipulator shown in figure 12–29 is equipped to swing into the welding position and adjust its twin arc heads in line with the seam to be submerged arc welded. Flux hoppers precede the twin arcs, allowing the powdered flux to fill the seam ahead of the electrodes. Often, a large oxy-acetylene flame precedes the fluxing operation in order to preheat the seam ahead of the weldment. Rotating fixtures are used in circular welding.

Due to the preciseness required in submerged arc welding, a dry run is frequently conducted ahead of the operation. This means that electrode currents are adjusted (fig. 12–30) to determine that the proper currents are being used. The nozzle end of a pressure vessel is shown in figure 12–31 after welding has been completed. The fixture above the welded seam is part of the automatic process. Submerged arc welded seams begin with the single *V*, consequently, a wide bead results on the outer surface. The

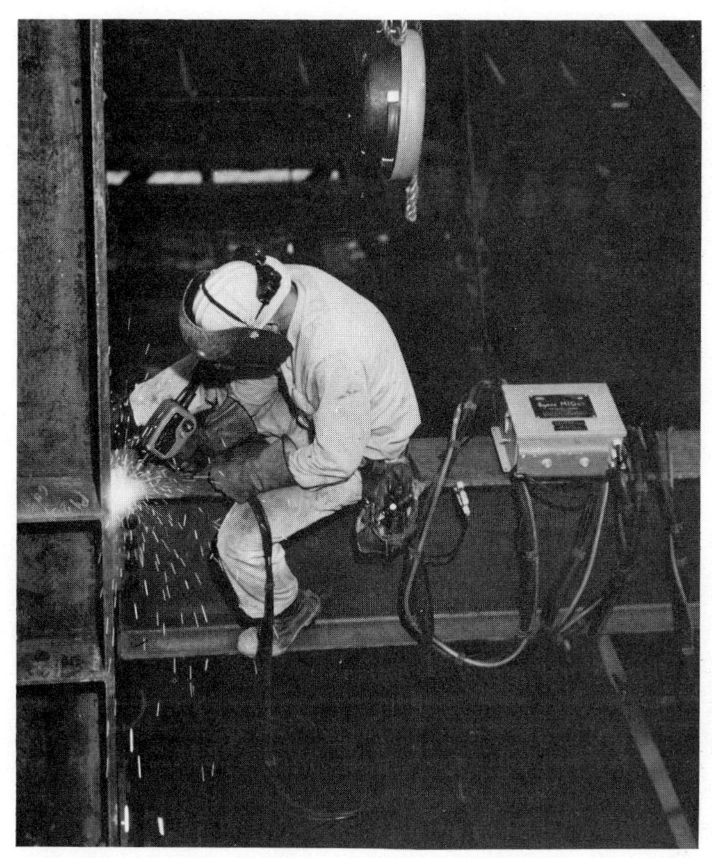

FIGURE 12–27 MIG welding a steel beam to a column with the super midget gun. As the wire is melted into the joint, the gun pulls more wire from the reel. (Courtesy of Airco Welding Products)

With regard to deposit rates, straight polarity provides the highest, but with the least penetration. On the other hand, reverse polarity provides the deepest penetration, making it ideal for welding steel plates greater than seven inches thick. Amperage in both figure 12–31 and 12–32 was 2500A. The leading electrode was consuming 1000 A in a reverse polarity setup, while the trailing electrode was using alternating current at 1500 A. The speed of arc travel was approximately 20 inches per minute. The effectiveness of submerged arc welding is certainly demonstrated in the very large pressure vessel shown in figure 12–33.

Stud Welding

Stud welding is not a new process, but has recently been brought into use due to its unique welding capabilities. Basically, the process provides a special stud welding gun which contains the stud. A ceramic ferrule is placed over the end of the stud and on top of the flux which is contained in the end of the stud. The process is essentially an arc welding process in that a circuit is made among the workpiece, generator, cables, and gun. When the gun is brought into contact with the plate on which the stud is to be mounted and the trigger is squeezed, an arc forms which melts the bottom of

FIGURE 12–28 A fixed boom manipulator mounted on travel car with ac – dc tandem heads for submerged arc welding system. The round vessel is mounted on a turning fixture and is synchronized with the moving arcs. (Courtesy of Ransome Co.)

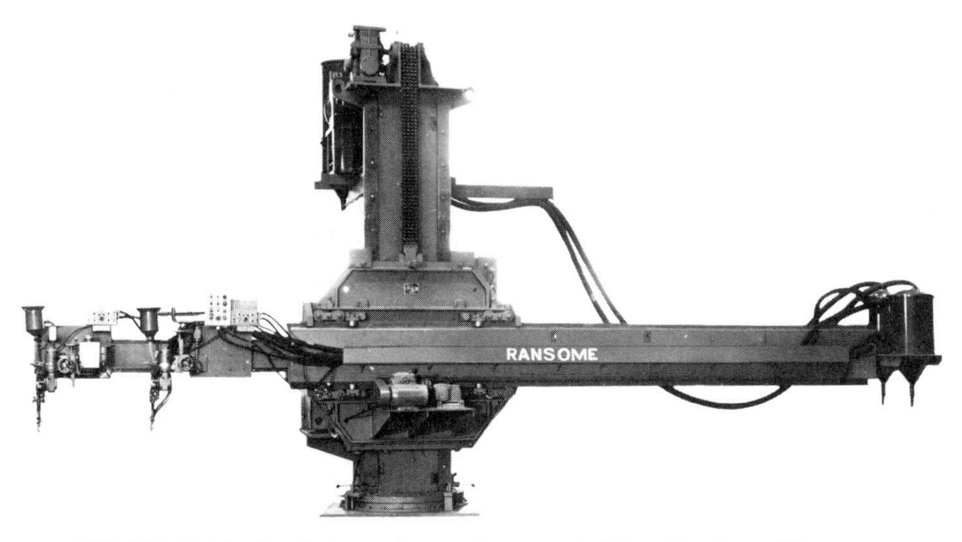

FIGURE 12–29 A twin type submerged arc manipulator. (Courtesy of Ransome Co.)

FIGURE 12–30 Arcs are being adjusted prior to submerged arc welding a pressure vessel. A tube precedes the tandem arcs and lays powdered flux in the *V*-seam to be welded. Movement is to the right. During welding, both arcs are submerged beneath the flux; therefore, they cannot be seen in operation. (Courtesy of Peerless Manufacturing Co., Inc.)

FIGURE 12–31 Nozzle end of pressure vessel welded by submerged arc process. (Courtesy of Peerless Manufacturing Co., Inc.)

FIGURE 12–32 A submerged arc welded seam in a large pressure vessel. (Courtesy of Peerless Manufacturing Co., Inc.)

FIGURE 12–33 A large pressure vessel is being readied for shipment after numerous welding operations. (Courtesy Chematron Corp.)

is slammed into the pool of molten metal, and the weld is completed. Studs can be placed wherever desired when an electrical circuit can be made. Nuts are placed on the studs for holding various parts.

Electron Beam Welding

When a high velocity stream of electrons collides with a metal, kinetic energy is changed to extremely high heat. Electron beam welding is related to cutting, but only with varying techniques pertaining to heat, time, and distance between the sides of the joint. Welding is accomplished either in a vacuum chamber or at normal pressure, however, critical types of welding for which the process was invented are accomplished in a partial or near vacuum. Electrons emitted from a hot tungsten cathode are accelerated toward an anode and onto the target or workpiece by the magnetic influences of deflecting or guiding coils. The electron guns are rated from around 30 kV to more than 160 kV with beam currents of several hundred milliamperes. The electron beam diameter varies up to 0.030 inch and causes rapid vaporization of the target material.

The welding technique originates from the condensation of the metal vapor whereby the liquid fuses and then solidifies. This differentiates it from cutting. By carefully guiding the beam along a weld line, fusion occurs at the joint. Observation of the welding zone is made possible by telescopic and oscilloscopic means. The process is used where other methods are not desirable, consequently, some extra preparations are involved such as metal cleaning and joint preparation. Because the vacuum method limits the size of parts to be welded due to the volume of the welding chamber, remote controls are needed. The work holding fixture allows rotating or linear welding which is controlled by observation and remote devices.

The vacuum is produced in a few seconds to several minutes, depending on the size of the chamber, and the beam is brought to focus at the joint. Depth of penetration is dependent on the total situation, but thicknesses greater than ½ inch are being commonly welded. Metals several inches thick are weldable, such as steel and aluminum plates. Width of fusion is narrow and not *V* shaped, allowing several feet per minute to be welded in many cases. Sometimes, the parts have to be tack welded in place to prevent slipping in the chamber. Figure 12–34 presents a typical electron beam welder. The inside of the vacuum chamber is shown in figure 12–35.

Many kinds of metals are being welded in the vacuum chamber, including the hard-to-weld metals such as tungsten, tantalum, and molybdenum. Intricate and sophisticated parts such as for missiles are welded in a precision manner in the chamber. Hardened steel is also welded, the speed of the welding being exceptionally rapid, while cooling is also fairly rapid. The stainless steels, titanium, and the heat resisting alloys are welded with very good results, and thin sheets of foil are also quickly welded.

Three main disadvantages of electron beam welding include the initial high cost of the installation, the limitations in sizes of parts, and the requirement for radiation shielding against X radiation. Basically, when the normal welding environment is carefully placed in a large box, and the box evacuated of gas, a setup for electron beam welding exists. Obviously, the process is either semiautomated or automated. A very important item in the system is the working fixture which has three movement axes and, in some cases, tilt and rotary axes are added. Once the metal parts are assembled and secured to the fixture in the vacuum, welding becomes automatic as the fixture moves according to the welding needs. A modification of the vacuum

FIGURE 12–34 An electron beam welder. (Courtesy of Union Carbide Corp., Linde Div.)

process is being accomplished outside the vacuum in a normal atmosphere, but collision with air molecules causes some instability in the beam.

Cutting by Flame

The oxy-acetylene cutting torch is used to sever ferrous metals by oxidation. As in welding, the size of the tip is determined by the thickness of metal to be severed. Proper adjustment of the oxygen and acetylene to produce a neutral flame is needed in steel cutting. The type of torch used has a third valve for allowing extra oxygen to leave the tip. It is the excess oxygen that performs the cutting. Flame cutting can be done by hand or by means of automation.

The cutting or parting operation is accomplished by first arranging the layout exactly as desired. Straight-line cutting requires a guide to allow the torch to follow the straight line. Circle cutting requires a radius attachment for cutting various size circles, while cutting by template allows irregular lines to be followed. Chalk lines are easy to follow, and the line should overhang the work table by approximately one inch. When the metal is heated to a bright red heat, the oxygen release valve is pressed and the torch is moved slowly ahead on the guideline. A shower of sparks beneath the workpiece indicates severance. As in welding, good cutting technique is developed by practice.

Both straight and beveled edges are produced by flame cutting. Figure 12–36 shows a steel forging being cut by the cutting torch. Holes in plate can be produced by applying oxygen to the bright red spot where the hole is to be located. When cast iron is to be severed, however, the metal is heated to a molten condition and then the oxygen is

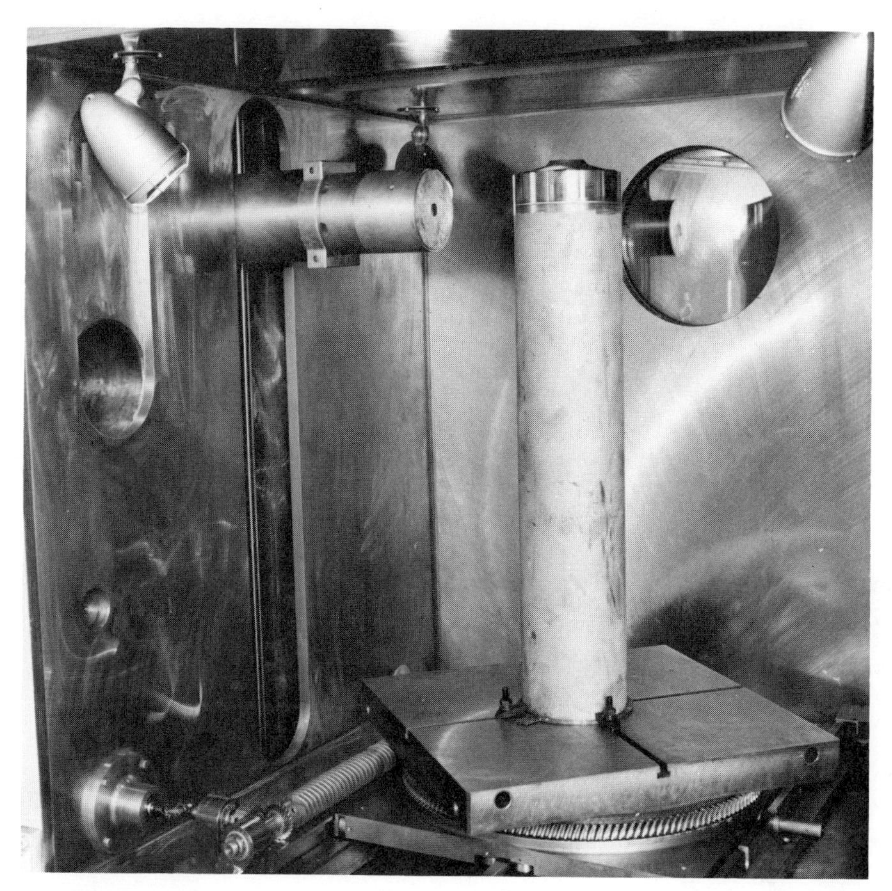

FIGURE 12–35 Inside an electron beam welder showing a part in position for circumferential welding. (Courtesy of Union Carbide Corp., Linde Div.)

applied. A swinging movement of the torch along the line to be cut allows penetration and parting.

A recent use of torch cutting is the production of sprockets from steel plate by automated means. The boom of the torch is lowered to the proper distance from the thick steel plate and cutting commences, being guided with an electric eye. The eye follows the black impression of the drawn sprocket while, at the other end of the boom, the torch's tip responds with high precision and accuracy. In a few minutes, a large sprocket is severed from the plate with such smoothness that no additional machining is required.

Arc Cutting

When smoothness of cut is not important, the electric arc is often used to part steels and cast irons with roughing cuts. The process is used in junk yards, for example, as it quickly severs the metal for further processing. Either AC or DC machines are used with coated electrodes. Mild steel electrodes are usually used, following similar procedures and requirements as in electric arc welding. When the arc is formed, a fast upward motion followed by a downward motion literally pushes the molten metal from the parent metal, and the part is severed.

FIGURE 12-36 A large forging being flame cut into sections. (Courtesy of Peerless Manufacturing Co., Inc.)

Questions

1. Differentiate between brazing and welding.
2. Explain the difference between straight and reverse polarity.
3. Why is submerged arc welding being used for automated welding processes?
4. How does the cutting torch sever the metal?
5. Explain the differences between TIG and MIG welding.
6. What are the differences between the main kinds of gas welding flames?
7. Describe the metallurgical factors concerning the spot welded nugget.
8. Explain a welding operation which could be accomplished by thermit welding.
9. Describe the differences between the welded joints of conventional arc welding and the electron beam welding joint.
10. How can the electric arc cut a steel plate?
11. Describe the basic types of welded joints.
12. Compare the interfaces of soldered, brazed, and welded joints.
13. What precautions are necessary in order to maintain the strength of the joint during welding?
14. What process is especially suitable for welding very thick steel plates? Why?
15. Compare the metallurgy of the solidified cast pool resulting from welding and the region of the original metal before welding.
16. What is meant by a certified weld?
17. During the welding of stressed parts, why must all welds be certified as being acceptable?
18. Compare the procedures and results between an electron beam welded joint and a submerged arc welded joint when five-inch thick plates are welded.
19. How does heat from welding affect the strength of the original metal?
20. What is the relationship between a welded joint and its ability to carry stress?

13

Plastic Forming Operations

The use of plastics in the engineering fields is rapidly increasing due to the great differences between plastics and metals. These differences include the combined chemical, physical, and mechanical properties. Most plastics are organic in origin, while the metals are inorganic. This origin relates directly to the potential uses of plastics which keep their use mostly in the low strength categories when compared to high strength steels. Most plastics are inherently weaker than most metals in their several strengths. Their stiffness factor is only a fraction of the metal's, and they cannot be used at very high temperatures as can some metals. On the other hand, plastics are light in weight, nonconductors of electricity, economical and easy to produce, and have pleasing appearances.

Many of the plastics include additives such as plasticizers, colorants, stabilizers, fillers, and solvents in the basic resin, while the type of resin indicates whether the plastic is thermoplastic or thermosetting. Plasticizers merely improve the shaping qualities and impart an added degree of plasticity. Colorants give different colors to the transparent or opaque plastic, while a stabilizer tends to resist environmental effects on the plastic, such as effects from heat and light. Fillers such as wood and asbestos fibers add strength, and some resist heat and certain chemicals. Solvents cause constituents to mix and become fluid and, subsequently, cause welding by fusion.

Plastics are produced by reorganizing the atomic and molecular structures of the included chemicals. Gases, liquids, and solids are used in the production of plastics and include wood, air, and water. Carbon and hydrogen are only two of the elements present in the numerous combinations available. Because many plastics are manufactured from poisonous gases and other materials, exact control over the production process must be exercised to prevent the escape of these harmful materials into the atmosphere.

Types of Plastics

Two main types of plastics are available, the *thermoplastic* and the *thermosetting.* Each category contains many specific kinds, each having its particular uses and advantages. Plastics are either natural or synthetic resins and are compounds of the large molecule types, either linear linked in chains or cross linked along their chains. The type of molecular linking determines the resultant properties of the plastic. As in the metals, atoms are arranged in patterns, the holding or bonding forces between molecules in the plastics being weaker than the forces within the molecules. Therefore, deformation of the plastic occurs between molecules. It is this deformation factor that relates to the two main categories of plastics.

THERMOPLASTIC

The thermoplastic type of plastic can be formed over and over due to its linear, rather than cross linked, molecular pattern. As temperature increases, the material becomes more plastic because of the weakening of the forces between the molecules. At around 200–350° F., shaping is easily accomplished while the material is in the solid, but plastic, state. As the material cools, it hardens and stiffens, holding its new shape until room temperature is reached. A point of importance, in this respect, relates to the inherent tendency of some thermoplastics to return to their original shape when subjected to an increase in temperature. During forming operations from the plastic raw material, both heat and pressure are used along with controlled cooling.

Some of the important thermoplastic types include the acrylics, acetals, cellulosics, polyamides, styrenes, vinyls, polyolefins, and fluorocarbons. *Acrylics* are tough, transparent, and are not affected in the presence of many chemicals. Automotive and aircraft industries use the acrylics as enclosures and for similar applications due to their toughness and transparent qualities. The clear plastic has architectural applications such as decorations and nonstressed structural parts. Also, it is being used more frequently in the jewelry enterprises.

Acetals have good dimensional stability with high strength and abrasion resistant qualities. The familiar zipper is often made of an acetal plastic, as are many items of common hardware. This plastic is also an excellent electrical insulator, besides being used for low pressure containers.

Cellulosics have fair mechanical properties, but the cellulose nitrate is an exception, being extremely tough. There are several types of cellulosics, including the familiar cellulose acetate which has uses similar to acrylics. The cellulose nitrate is suitable for ball point pens, for example, the plastic providing a high resistance to wear. Other cellulose plastics include the cellulose acetate butyrate which is used for tough parts such as automobile steering wheels and many automotive hardware items.

The *polyamides* include nylon which has many uses due to its higher strength and excellent flexibility factors. The nylons are used for industrial parts such as pins, bushings, gears, and assorted commercial hardware, and are replacing many metal parts within their strength capabilities, even though these nylons are several times weaker in tensile strengths than the common steels.

Styrenes are used in many consumer type products to include covers and housings for portable business machines, dashboards in automobiles, and food containers. The styrenes are good electrical insulators. Also, the toy and picnic supply industries have found many uses for this plastic.

Vinyls vary in mechanical properties, depending on their types, and are used for glues, wall and floor coverings in sheet form, substitutes for furniture upholstery, and for hose and pipe due to their flexibility, toughness, and resistance to many chemicals.

Polyolefins such as polyethylene are flexible and partially transparent. They are not suited to excessive outdoor exposure, however. Much of the packaged consumer goods are sealed in polyethylene bags. The polypropylene is stronger than the polyethylene type and, when produced with fibers, it can compete with some metals in strengths.

The *fluorocarbons* are good insulators from electricity, are heavier than most plastics, resist temperatures up to around 500° F., and are unaffected by most chemicals. The fluorocarbons are used for electrical insulation in wiring and for corrosion resistant linings in tubing. They have low coefficients of friction, requiring no lubrication when used as moving parts.

THERMOSET

The thermosetting plastics are also chainlike in their molecular structure, but when exposed to heat and pressure, the long chains add cross linking patterns to their chainlike structures which sets up permanent rigidity. Prior to molding the thermosetting plastics, a similarity to the thermoplastic group exists in that the long molecular chain pattern is present. However, when heat and pressure are used to shape the thermosetting group, links of molecules form across the chains and, in turn, as curing progresses, permanent setting of the structure results. Therefore, thermosets are processed only one time, whereas the thermoplastics can be reshaped at will. The cross bonding of the very complex network among the molecular chains prohibits plasticity and plastic flow in the thermosets.

Some of the common thermosetting plastics include the silicones, phenolics, urethanes, epoxies, and aminos. The *silicones* are exceptional with regard to heat and electrical resistance. Some of these plastics resist temperatures in the vicinity of 800° F. and, when shaped into electrical parts, make excellent holding devices while acting as insulators. The silicones mix well with the glasses and with rubber to provide unusual and variable mechanical properties unlike the typical thermosets.

The *phenolics* consist of resin and filler material which gives them good electrical insulation properties, as well as rigidity, but with poor heat conductance. Many rigid types of panels and handles for appliances and consumer products are processed from the phenolics, as well as commercial hardware and foamed products.

Urethanes are also good insulators from electricity, making them ideal for use in wire insulation. They have subzero resistance, as well as resistance to heat up to approximately 400° F. Items of clothing, for instance, which require unusual insulation properties from temperature changes are made from the urethanes.

A much different type of plastic is the *epoxy*. When inertness to chemicals is required, the epoxies are used. They also resist temperature changes over a broad range, some up to 600° F. The mechanical properties of this resin are excellent, some having tensile strengths up to 58,000 psi when combined as a laminate. The adhesive property of the epoxy is outstanding, as exhibited by the holding power, weight, and structure of honeycomb paneling which is used in many engineering type structures. The sandwich of aluminum honeycomb is maintained by the adhesive power of epoxy between two sheets of aluminum alloy. Further, very strong lap joints are produced when the epoxy bonds two sections of metal.

With respect to hard type plastics, the *aminos* are produced in several types and are used in such items as plastic dishes and kitchen counter tops. These plastics are fairly good insulators from electricity and heat, and even resist subzero temperatures, but are subject to attack by some chemicals. The laminations in plywood, for example, are often bonded with an amino plastic.

GENERAL USES

Plastics constitute a sizeable portion of the manufacturing processes. These organic base materials have properties compatible with industrial applications such as in the furniture and boat fabrication industries. Many parts are produced from plastics, including automobile and boat assemblies, electrical holders and insulators, mechanisms such as gears, household appliances, toys, hardware, and sporting goods. Even though the mechanical properties are basically less than those of the metals,plastics do have their place in society.

Processing and Shaping

The processing of the numerous plastics first includes the manufacture of the basic constituents from gases, liquids, and solids which are chemically treated in a series of processes whereby the basic resin is produced. Different combinations of hydrogen, oxygen, nitrogen, ethylene, chlorene, or carbon are laboratory arranged by the chemist so that linking and bonding occur as desired. Various treatments of heating, mixing, condensation, and remixing produce reactions which provide for the particular raw material. Combinations of the several raw materials with autoclave and other processing result in the desired plastic. However, the basic constituent produced is the resin. Subsequently, materials previously mentioned such as colorants, stabilizers, plasticizers, fillers, and solvents are added which then result in the pertinent plastic.

Plastics are produced both in the liquid and solid form so that subsequent producers can prepare their particular type of plastic under their trade name. Some plastics are used in the liquid stage as in the epoxies, some are used as solids such as acrylic sheets, while others are produced in the expandable form for subsequent foaming operations. Also, plastics are produced in physical shapes that are identical with the metals, that is, rounds, bars, tubing, extrusions, and sheets, in addition to the foaming types and adhesives.

Plastics are softer than many metals and are easily machinable. They may be drilled, turned, milled, and sawed in much the same manner as wood and metal. Further, some are weldable and, as has been stated, some are reshapable. These nonmentals are available under hundreds of trade names.

Mechanical Properties

Mechanical properties vary considerably among the plastics. One particular polycarbonate resists hammering and deflects many deliberate attempts to shatter it. This very tough thermoplastic is used where superior qualities of toughness are needed such as in the production of window panes, covers for lights, helmets, and consumer products, including kitchen hardware. A typical tensile strength of the polycarbonate, for example, is approximately 8500 psi, while in compression it withstands up to 11,000 psi. In impact resistance, this transparent and acid resistant plastic is one of the toughest. By comparison, one of the cast nylons has a tensile strength of 13,000 psi, but its impact resistance is poor, as well as its resistance to acids. The polyvinyl chloride type of thermoplastic makes excellent water pipes and fittings because of its high strength and flexibility and also its ability to be fused or welded at joints by the application of solvent and cement. These plastics easily withstand water pressures of 70 psi and are not attacked by soils.

Types of Forming Processes

Plastics are formed into their final shape by means of casting, thermoforming, reinforcing, foaming, and molding. Regardless of the method used, the plastic must first be made soft enough to flow plastically, be able to move into a closed container which sets its shape, and then be able to maintain the shape after setting. In order to bring about a final shape, there must be a preferred temperature for causing fluidity or plasticity in both basic types. For thermosets, there must be a means to cause the chemistry to react and cause rigidity while the plastic is under heat and pressure. In this respect, there needs to be a machine to force the plastic material into the desired shape such as a mold, roll, or mechanism. After shaping in the mold, the thermoplastic is usually water cooled, whereas the cooling rate of the thermosets makes no difference. A means must then exist to eject the finished part.

CASTING

The casting process involves the pouring of liquid plastic into a suitable mold and allowing the full mold of plastic to cool, either to a lower temperature or to room temperature. Melting temperatures range from about 250° to 350° F. Molds are made of various materials and include the cavity for shaping the material. The melted liquid is poured into a hole which enters the mold's cavity, the pouring continuing until the mold is full. Upon solidification, the split mold allows the solid plastic to be quickly removed. Sheets and plates, for example, are formed by pouring the liquid plastic into cavities between mold plates, while rods and other shapes are poured into split tubes. A split tube type mold allows the casting to be easily removed.

Cast plastic products are sometimes superior to pressure molded products in strength, whereas most cast metals are weaker than the wrought. Polyesters, epoxies, and acrylics are commonly cast. A type of casting, for instance, includes the covering of electrical components with a cast epoxy. The plastic gives the required strength along with good insulation properties. Another example of liquid pouring is the embedding of objects in a clear plastic, such as items of jewelry. Many geometric shapes are also produced by casting.

Plastisol

A plastisol is a ground mixture of fine particles of plastics such as polyvinyl chloride and plasticizers which fuse into a homogeneous liquid when heated to approximately

350° F. Three main methods of applying the liquid plastic are dipping, slush casting, and rotational casting.

Dip casting involves the dipping of a part into the liquid plastisol and removing. Freezing of the plastic occurs on withdrawal. In this respect, some parts are oven heated at 350° F. after withdrawal from the dip, however, to cause better fusion and then cooled, often by a jet of water. Dipping is a routine process. Many hand tools are dipped in plastisol for purposes of both grip assistance and for electrical insulation. Also, wire mesh screens and kitchen dish drainers are dipped in plastisol, while hollow plastic parts are made by dipping a pattern, which is often the part, in the plastisol and then stripping the flexible part from the pattern.

Slush casting involves the pouring of the plastisol into a hot hollow mold, followed by quickly pouring the remaining liquid from the mold. Thickness of the part's wall is then determined by length of exposure time, usually three minutes giving a useful wall thickness. The entire mold is placed in an oven at 350° F. for a few minutes to cause complete fusion and then water cooled. Slush casting is also a common casting process as many hollow toys and consumer products are produced by this method.

When an enclosed object is to be produced without seams, *rotational casting* is used. A measured quantity of plastisol is poured into a heated split mold. Immediately, the mold is two-dimensionally rotated so that the liquid moves completely around the inside surface of the heated mold. When the mold is separated, the part is removed, being fully enclosed, seamless, and hollow. Hollow balls and numerous kinds of toys are produced by this process.

THERMOFORMING

The thermoforming process involves heating a section of thermoplastic sheet stock to approximately 300° F. whereby it collapses to the shape of the mold. The mechanical shaping process allows the heated sheet to rest on the mold as it plastically deforms to the contour of the mold. When cooled, the newly shaped part is ready for use. Any reasonable shape can be produced by this method.

A modification of the process uses a vacuum. *Vacuum forming* is the use of atmospheric pressure to shape the part. To perform the operation, a heated sheet of thermoplastic stock is placed over the mold and sealed around the periphery. Air is then withdrawn from the space between the mold and sheet, the sheet falling onto the mold and adhering to the fine details of the mold.

A further modification includes the use of pressure without vacuum. When a part is made by a pressure which is higher than atmospheric, the heated sheet is quickly blown onto the mold, taking on the contours and details of the mold. This method is known as *blow forming.* Exhaust holes in the mold allow escape of the trapped air. Pressures vary up to nearly 300 psi in shaping parts by this process. Because the thermoplastic sheet will deform into a shape as if the plastic were nearly liquid, many different shapes are possible. Air cooling after forming normally enables stiffness of the part to quickly occur, but in some situations, water cooled molds are required.

REINFORCING

Reinforcing is the use of some additional material such as a filler or laminate to produce a desired shape. Boat hulls, for example, are often fiber glass structures.

Reinforcing can be accomplished by several methods, the simplest being the application of a layer of fiber glass, or some other material, onto a coating of epoxy and then hand rolling the plastic material to form the bond. A mold or pattern is used which is coated with a release material to prevent sticking of the resin. When the combined resin and fiber material are applied, the plastic begins to cure at room temperature. After curing, the part is ready for further processing according to its use.

Spraying is another method used in applying the mixture of plastic materials. Several coats are sprayed until the proper thickness is obtained. Besides hand brushing and spraying, there is the method of pouring a premixture of fibers and resin into a mold, followed by an application of pressure and heat. Subsequent curing results just as in other similar processes.

Other methods include vacuum bag and pressure bag molding where either vacuum or pressure causes a bag to force the resin mix against the contours of a mold. Hardening follows as in other applications. Another reinforcing process provides a squeezing procedure. When a section of the resin mix is placed between two parts of a mold and squeezed, the match molding results in the desired shape of the part.

Reinforcing the resin provides greater strength properties than the resin alone. Glass cloth or glass fibers mixed with the basic resin provide higher strength materials for molding numerous articles and shapes. Automobile body parts, boats, household hardware and appliances, toys, and sporting goods are some of the products produced by the reinforcing method.

FOAMING

Both thermoplastics and thermosets are foamed, including the urethanes, styrenes, cellulosics, silicones, and phenolics. Any practical shape can be produced that is subject to shaping by pressing or casting. Both physical and chemical foaming processes are used to form the beads, or foam. During the manufacture of the particular plastic, a gas forming material or gas is combined with the plastic's chemistry. Subsequently, during forming operations, and while being heated, gas is liberated and expands, producing the beads, or foam. When the pre-expanded beads of polystyrene, for example, are blown into a mold's cavity and heated to nearly 300° F., fusion occurs among the beads along with further expansion, filling the mold properly. Water cooling of the mold follows. The result is the shaped part.

Chemical foaming results when two materials are mixed which causes a reaction in the form of a foam. A resin expands and forms beads when another constituent causes liberation of gas within the resin. As an example, foamed polyurethane is produced by chemical foaming and is shaped by pouring the gas-forming constituent and resin into a mold, allowing the foam to be made by gas expansion, curing the filled mold by time or with water cooling if the mold was heated, and then removing the finished product. Many shapes of foam are produced by merely pouring the proper mixture of materials into a container having the desired shape, the mixture producing beads and foam until the reaction is complete.

MOLDING

The molding processes are varied and include injection, extrusion, transfer, compression, laminating, cold, and calendering processes. *Injection molding* is used primarily with the thermoplastics. In its simplest form, the pellets of plastic are forced into a heating chamber which causes the material to become liquid. Under high pressure, the liquid is injected into a cold mold having the shape of the part. When cooled, the part is removed. Many molds are water cooled. Two common plastics used in this process are the acrylics and vinyls. Typical products include consumer articles and household wares.

Extrusion is similar to metal extrusion in that a die having the exact shape of the extrusion's cross section is used. Powders or pellets of thermoplastics are moved into the heating chamber at approximately 400° F. and then pushed through the die in the plastic state. A conveyor system moves the length of extruded plastic to another processing station for cutting into the desired dimension. Cooling occurs along the line, and stiffness occurs in the part when cooled to room temperature. Standard types of shapes are producible by this method.

Compression molding is a simple process of pouring a thermoset powder into a heating chamber at approximately 300° F. and then transferring it to a closed hot mold. As the plastic condition turns to liquid, a pressure is placed on the liquid, and curing occurs. Pressures vary according to the size of the mold. A typical one-square-inch mold requires approximately 4000 pounds of pressure while at a temperature of 300° F. to form it. The plastic hardens by curing and is ejected hot. If a set of dies or molds can be made, the part can be produced. This is a fast and economical process because the hot mold is used over and over.

Transfer molding is a modification of compression molding. The thermosetting plastic material is poured into the top portion of a transfer mold where it forms a liquid when heated to its liquifying temperature. This liquid is then pushed through the sprue into the mold's cavity, taking the shape of the part. Being thermosetting, the mold's heat sets the plastic as it hardens, and the part is immediately ejected. Typical parts produced by this process include many of those previously discussed products such as electrical insulators in power transmission lines and in automotive electrical accessories.

Laminating is a common method used in producing higher strength plastics such as the phenolics. Basically, a stack of laminating material such as paper or cloth is arranged so that alternate layers consist of a thermosetting resin. When the stack is placed in a press and heated while being squeezed, bonding and curing occur due to the combination of heat and pressure. Many kinds of higher strength parts are produced by laminating. Structural parts to support electrical transmission are examples of parts made from laminated phenolic. Often, the lamination is a linen cloth and resin. The printed circuit is a typical example of the laminated board.

Cold molding is a simple process which includes the pouring of the thick mixture of plastic into a mold and placing pressure on the mass of cold plastic. No heat is used during the shaping process. After removal from the mold, the part is cured at approximately 425° F. for several hours or even for several days to cause fusion. Products include those which have no special requirements other than the shape. Cold molding produces a wide dimensional tolerance, but a rough surface finish.

Calendering is a sheet producing process that uses several rolls or calenders to form the plastic. Fundamentally, the soft mixture of thermoplastic is poured into a mechanism which feeds the plastic material into heated rolls. As they turn, the distance between the rolls is adjusted to determine the thickness of the thin plastic sheet. A cooling roll subsequently prepares the long length of sheet for the final roll which twists the material into large coils. The vinyls constitute a large percentage of materials for calendering. Products of this process are nearly unlimited and range from flooring to container bags.

New Requirements for Plastics

Research is continually bringing new plastics to the consumer. Some plastics now fulfill the requirement for subzero temperature exposure as well as exposure to slightly elevated temperatures. Such varying temperatures occur in many situations, one being encountered by the several types of space vehicles. These vehicles use various types of plastics in part of their structural makeup, in addition to the several metals. For example, ablative materials such as the polyimide polymers are being used in rocket engines and equivalent environments for timed periods. These plastics have unique heat absorption characteristics that cause only ablation of the material when exposed to elevated temperatures and time. Recent developments show outstanding results with silica in this respect. As an illustration, a phenolic resin makes an excellent ablative material when silica is reinforced with fibers of glass so that the carbon-silica chemical reaction is effective when highly heated.

Questions

1. Differentiate between the molecular arrangements of the thermoplastics and the thermosetting plastics.
2. Explain the cross linking process which occurs in the thermosetting plastics.
3. Why are thermoplastics capable of being deformed when heated while themosets are not?
4. Explain the shaping of a foaming plastic.
5. Explain a method for increasing the strengths of some plastics.
6. Name a plastic that has good ablative qualities.
7. What is a phenolic?
8. Why do thermoplastics require controlled cooling during the setting operation?
9. What is a plastic?
10. What happens when a thermosetting plastic is heated in an attempt to soften it?
11. Which plastic is excellent for water pipe?
12. What is the greatest advantage of the plastics which have the carbon–silica chemical reactions in their makeup?
13. List and describe the several plastic forming operations.
14. Compare the extrusion forming of a plastic material with the extrusion of an aluminum tube.
15. Differentiate between the curing, or setting, processes for thermoplastic and thermosetting plastics.
16. Compare the mechanical properties of the plastics with the metals.
17. The modulus of elasticity factor of plastics is low as compared to metals. What does this mean?

18. What is the purpose in using reinforcement material in a resin?

19. List several limitations of plastics as compared to the metals.

20. List several advantages of the plastics as compared to the metals.

Adhesive Bonding

Adhesive bonding has recently become one of the main means in producing joints of engineering design. This type of bonding is a mechanical process, being somewhat similar to soldering in that adherence is strong at the interface of most materials. No fusion of the metals occurs, but a powerful adhesion results between the bonded parts. The bonding material is an adhesive of some type, being thermoplastic or thermosetting, which secures two or more metals or nonmetals together. Low strength joints not subjected to temperatures beyond room temperature may be bonded with vinyls and polyamides. However, the main purpose of adhesive bonding is to provide a strong and stable joint in engineering structures, therefore, the thermosetting materials are used most frequently. The various epoxy resins are used extensively in several physical shapes, including liquid and tape. Resins are produced in dozens of chemical formulas, therefore, the type of joint to be made helps govern the type of resin to use. Some resins are prepared as adhesive sheets and strips and require cutting to the desired size of the bond. Others are available as a total component in separate containers and must be mixed by weighing the resin with the hardener. Some resins require the careful weighing and mixing of catalysts and accelerators. When both catalysts and accelerators are used, each must be added separately to the resin and mixed before adding the other. When catalysts and accelerators are mixed, an explosion may occur. Inasmuch as strengths of resins are concerned, some of the glass and epoxy sheets after curing have the tensile stengths of low carbon steel in the rolled condition.

Low Strength Joints

When a seal is necessary and the joint is not exposed to temperatures above room temperature, the thermoplastics will suffice as materials for the joint. Because bonding of plastics involves either cohesion or adhesion techniques, care must be used to determine which type of joint is required. The cohesive bond causes the joint to be softened, such as with a solvent, and the joint is made firm by pressing. Actually, this is a type of welding, or fusion, because molecules of both materials intermingle. Solvents and cements join pipes and other shapes made of polyvinyl chloride. An application of solvent to the surfaces of both parts followed by the cement allows a lap joint to be made. Pipes or tubes, for instance, should be quickly twisted after the cement is applied to spread the cement, and then the joint must be held stationary for a few minutes, allowing the joint to set. A silicone adhesive is used to bond and seal various types of oddly shaped joints and connections such as those used in the electronics industries.

Other types of adhesives include those used from animal compounds. The glues, for example, are used in hundreds of applications to include the paper and wood box industries, furniture construction, toy assemblies, and bookbinding. The rubber type, or elastomers, are used extensively in the vehicle industry in such applications as seals around doors and windows. Also, the home appliance industries use the numerous tape and strip type seals which are bonded to any solid material. These types of adhesives are more or less gluing operations, even though they incorporate some principles of bonding. In engineering structural application, however, the stronger type of bonds are essential in transferring stresses between various types of parts.

High Strength Joints

Adhesive bonding causes adhesion at the joint between the two or more parts by a third material which sticks or strongly adheres to the surface of each mating part. Molecules of the materials do not mix, but they strongly attract such as the attraction of some cements and silicones. For strength purposes, the epoxy adhesives are used extensively to bond metal and nonmetals and are used in the metal furniture industries, architectural and construction industries, and especially in the aircraft and space vehicle industries. The majority of the high strength bonded joints are thermoset resins, some requiring 300° to 500° F. for curing, while many of the liquid resins set at room temperature after mixing of the two or more components. In this respect, most all epoxy resins develop cross linking and form a strong bond between two or more parts when proper bonding procedures have been followed. An important physical property of the epoxy resin is its strong resistance to many chemicals and its low shrinkage factor after curing. The epoxies are basically resilient and offer a high degree of toughness so that combination loading in aircraft structural components, as in honeycomb bonding, will be effectively absorbed.

CHARACTERISTICS OF THE JOINT

Parts are bonded together using any type of connection or joint that is feasible. The design of the joint is determined by the types of loading at the joint. For example, the load or stress in the material may tend to cause peeling of the surface, lifting or cleavage of the materials, pulling of the materials as in tensile, or pushing as in compression, or more often, shearing action. If the parts are expected to function in abnormal temperatures such as a range from subzero to approximately 300° F., then the coefficients of expansion of the materials must be calculated in order to determine if the differential will snap the bond. This situation is particularly applicable to the bonding of different types of metals. Also, surface adhesion characteristics must be investigated to assure that the bond will adhere. With regard to the adhesive, it must be thick enough between the mating parts so that the joint is not starved of adhesive, but, also, the adhesive must not be too thick. In this respect, when lap joints are squeezed too tightly during curing, the adhesive material is squeezed out and an inferiòr joint results. On the other hand, when the adhesive is too thick, poor strength factors result. Basically, then, the thinner the adhesive up to a critical value, the stronger the joint will be.

TYPES OF JOINTS

Shear joints which are adhesive bonded produce high strength joints due to the adhesive material being placed in shear when tensile or compression loads are placed on the structure. When high stresses are to be applied, however, joints which produce cleavage and peel tendencies must be avoided. Also, butt joints are not satisfactory, because tensile stresses quickly cause joint failure.

With regard to joint design, figure 14–1 illustrates several types of bonded joints. Notice that the *single lap joint* (*b*) causes off-center loading and eccentricity to occur, therefore, the sheet of material may tend to peel away from the joint. Such a design also causes some bending which tends to promote material cleavage. However, the lap joint is a good joint unless stressed too highly and is used extensively in engineering designs, each design depending on a particular set of factors. The resin is in shear when the members are in tension or compression. A modification of the lap joint can

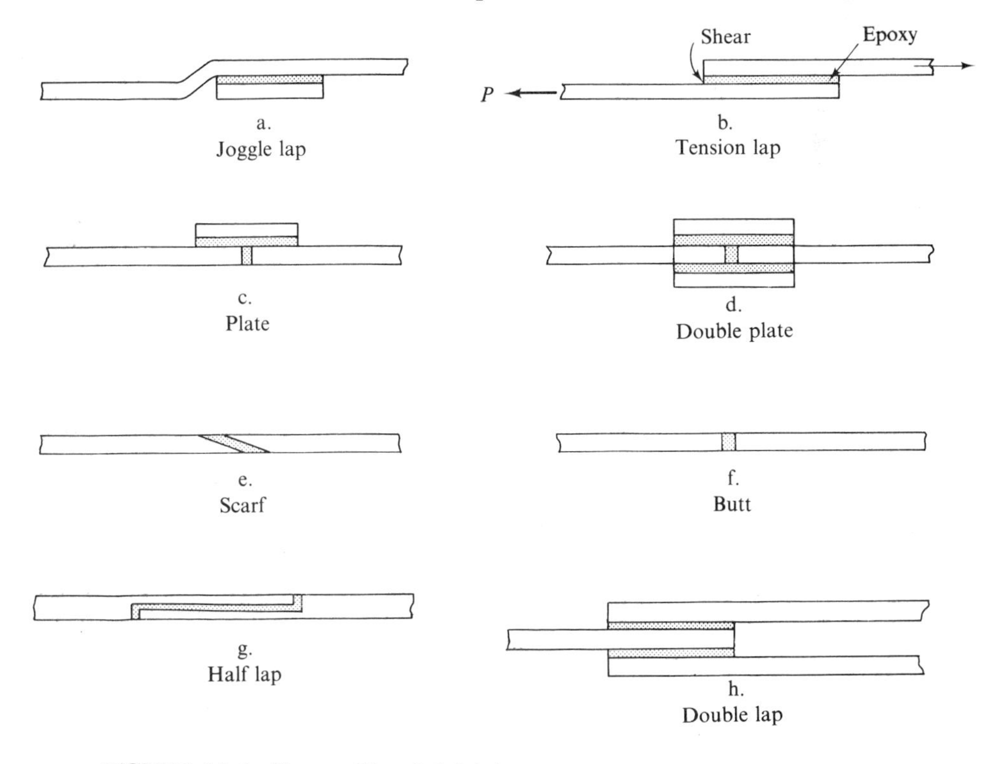

FIGURE 14–1 Types of bonded joints.

be thicker sheets or wider sheets, but, here again, the width of the lap is critically important as wider laps are stronger than the narrow because more material resists the stresses in the wider joint. As the length or depth of lap increases, however, strength also increases up to a critical value and then remains constant as further depth of lap occurs. For example, a four-inch wide lap joint has twice the strength of a two-inch wide joint with equal depths of laps. A *joggle lap* (*a*) or a *scarf lap* (*e*) maintains straight-line, or axial, loading and may be preferred over the simple lap (*b*). In tension loading, there is a tendency for this type of joint to slightly straighten at the offset member, the resin being in shear while the joint assumes the design of the simple lap. The *double lap*, illustrated in *h*, is strong, but, again, loads become eccentric at the joint, each side tending to neutralize the effects of the other. The resin is in shear as in the other joints. When increased strengths are needed, *double plate joints* are used (*d*), the opposite sides being in shear under loading while the butt area is in tension or compression. The *single plate* (*c*) is good, but the tension or compression loading causes a slight bending tendency at the joint. Under load, the resin is placed in shear beneath the plate, but is in tension or compression at the butt. The *half lap joint* shown in *g* is strong and efficient, but shaping of the mating parts increases production costs.

Higher strength joints call for engineering designs which are double lap joints (*h*) and the double plate joints (*d*). When the plates' depths are balanced with widths, a good design exists because the adhesive has a more balanced area, especially at *d*, to resist shear forces. Very often, however, the joint is stronger than the bonded material when placed in stress. In this respect, tests have shown that various metals, the heat treated aluminum alloys, for example, have broken in tension while shear stresses in the joint sustained the load. The butt joint shown in *f* is a poor design because the resin is in tension or compression and cannot compete with applied stresses. Scarf joints are frequently used and are very practical because more surface area for the resin is made possible (*e*). Also, loads are axial. However, this type of joint has its special uses when length dimensions are far larger than width.

THE JOINT AND THE ADHESIVE

Prior to preparation of materials for the joint, several investigations must be made to assure a safe joint. The type of materials to be bonded must be studied to determine their properties and characteristics in given environments. This includes the strengths of the materials, the chemistry of the environment, the temperature of the environment, and the geometry of the joint. Next, loads must be studied at all points in the joint to determine types and magnitudes of stresses. With regard to stresses, alternating loading sometimes induces fatigue failure, so joint materials, condition of materials, safety factors, and design are critical. Economics are important, too, but safety and strength of the joint are more important.

When these factors are calculated, the kind of adhesive must then be chosen that will satisfy these requirements. Stressed configurations such as bonded lap joints in structures and in honeycomb assemblies use the epoxy resins of the thermosetting types. However, some of the modified thermoplastic phenolics also have high strength qualities, but have different uses. With respect to the joint, the adhesive must have high shear and peel strength properties, and be able to resist chemicals, specified temperatures, and fatigue. The epoxies have many of these capabilities, especially the ability to resist subzero temperatures, and some are effective at temperatures above room temperature or within the temperature variations of the four seasons. As an example, the honeycomb assembly is used in aircraft construction. In flight, temperatures as low as 60° below zero occur at altitude, as well as temperatures slightly in excess of 300° F. in some situations. An important point is that the designer and engineer realize that some of the above factors will converge at one time upon the joint, therefore, this situation must be understood and designed for accordingly.

The Honeycomb—An Adhesive Bonded Structure

The most typical example of the adhesive bonded structure is the honeycomb assembly, used extensively in aerospace vehicles. Modern combat aircraft (fig. 4–80) use honeycomb paneling in their structures. The honeycomb sandwich is being discussed in detail because it incorporates the important principles in bonding, but especially includes details concerning the high strength joint. Essentially, the honeycomb is a hexagonally shaped network of openings, or *cells*, within an assembly and offers exceptionally high stiffness and other strength properties along with a low weight factor (fig. 14–2). The honeycomb material may be aluminum, copper alloy, stainless steel, or fiber

FIGURE 14–2 Unexpanded and expanded honeycombs.

glass. Standing alone, the honeycomb will not support its own configuration, but when properly bonded between two outside sheets of materials, the assembly is capable of withstanding normal structural loading for its design, especially being capable of resisting bending loads. In this respect, both metals and nonmetals are capable of being bonded, such as those different materials found in the helicopter rotor blade (fig. 14–3).

FIGURE 14-3 Cross section of a rotor blade showing bonded honeycomb.

THE CORE

On receipt of the honeycomb core, it appears as a solid material. A core of bonded aluminum in the unexpanded condition is illustrated in figure 14-4. The expanded honeycomb core is shown in figure 14-5. The core is fabricated by laying strips of thermosetting adhesive at specified widths along the lengths of extremely thin aluminum sheets or other materials and then alternating the sheets with strips of adhesive. The process is continued until the desired thickness of the expandable assembly is attained. Dimensions between the strips of adhesive reflect the hexagon cell's dimension in the expanded condition, the length, width, and thickness of the core of laminations depending on the requirements in manufacturing the assembly.

Prior to expanding, the aluminum core is sawed to desired dimensions with a very thin six-tooth pitch raker set saw blade at cutting speeds up to 7,400 sf/min and a feed up to approximately 20 inches per minute. The thicker the core, the slower the feed. Nonmetals are sawed differently. A fiber glass core, for example, is sawed with a 10 tooth pitch saw blade approximately ¾-inch wide at a slightly faster feed than for the aluminum. Sanding is then usually performed to bring the edge of the core to the desired tolerance and grade of smoothness for both metals and nonmetals. Also, prior to expanding, the cores may be milled to the contour which, when expanded, will conform to the proper shape of the part.

Following sawing, the core is expanded to take the proper geometry of the resulting hexagonally shaped cell. Figure 14-6 illustrates the expanding process which is used on the "as received" core. Extreme care is used during expanding to avoid overstressing the cells that become stuck; these are hand loosened. When expanded, the section is carefully examined for correct cell shape and is then cleaned and placed in a

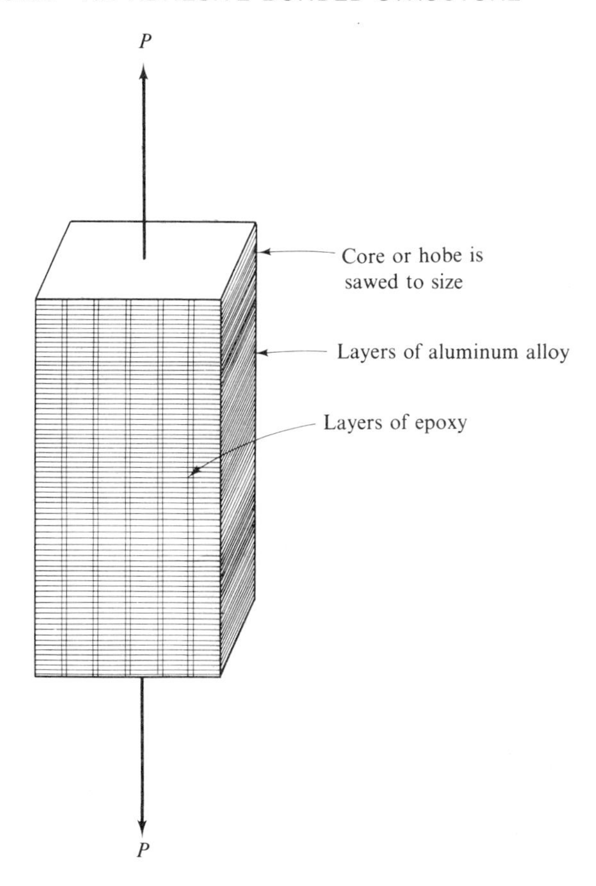

FIGURE 14–4 A core of expandable honeycomb.

special solution which solidifies within the cells and causes a rigid condition throughout the entire section. Such a process enables the machining of the expanded form to the contour requirements.

MACHINING HONEYCOMB

Conventional methods of machining honeycomb are inadequate because the delicate cells will rupture when in contact with conventional cutters. Special milling machines using air foil *T*-shaped cutters reduce the rigid section of filled honeycomb to the correct shape and dimensions within tolerances of approximately 0.010 inch. Figure 14-7 illustrates the positioning relationships of the air foil cutters to the honeycomb. These cutters spin at a very high r/min and literally slit the surface material like a sharp knife with little pressure. The bank of cutters is movable to produce high and low areas in the honeycomb as well as straight lines and curves. When the assembly is completed, with respect to machining, it is prefitted to assure that correct contours will be possible during final assembly. The filler material, often a wax, is then melted and removed from the cells. Cleaning operations follow which prepare the shape for bonding.

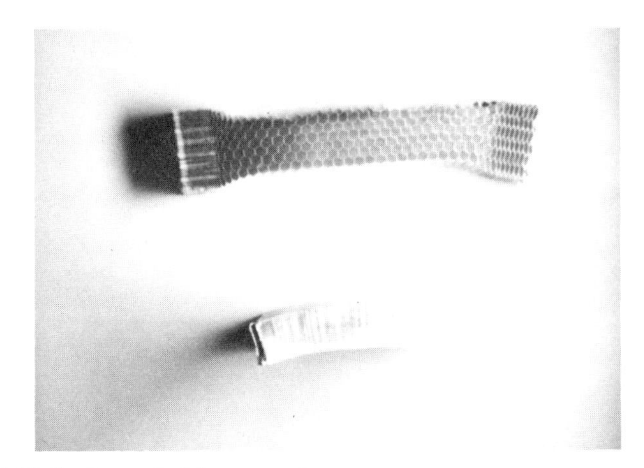

FIGURE 14-5 Expanded honeycomb.

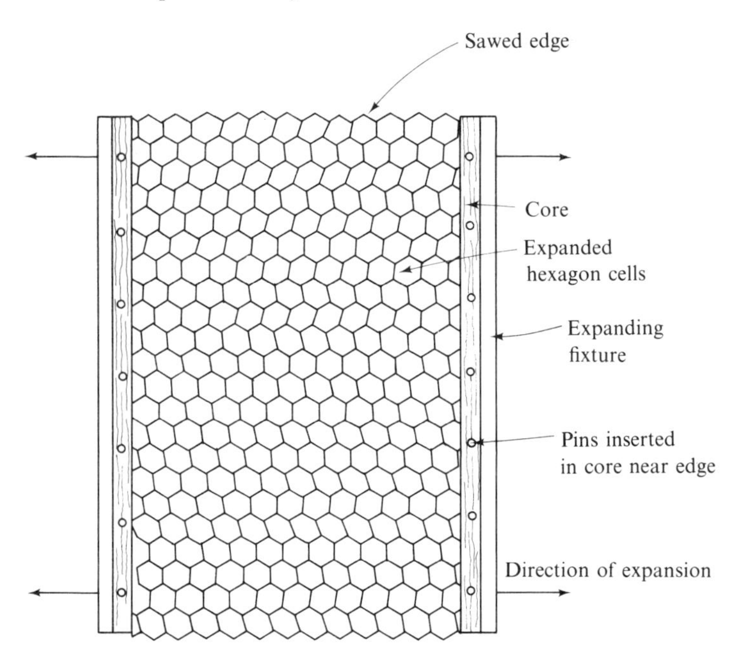

FIGURE 14-6 The expanding process for honeycomb.

PREPARATION FOR BONDING

Cleaning includes the chemical cleaning of all parts of the assembly, that is, the honeycomb and the outer sheets which are to make the sandwich. An example of cleaning aluminum involves a hot vapor bath for degreasing after all foreign materials such as paints have been removed. A solution of trichloroethylene is often used. Following degreasing, the particular surfaces of the metal are processed with an acid brightener and cleaner, each process being followed by rinsing for purposes of neutralization. Hot drying then prepares the metals for bonding. With regard to other materials, the cleaning of fiber glass and titanium have varying procedures. After

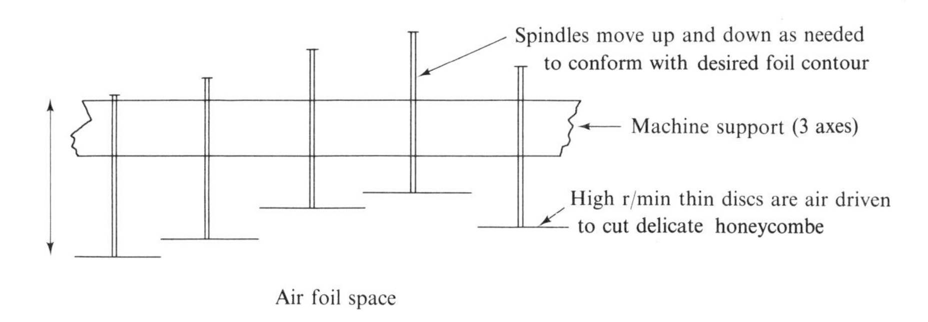

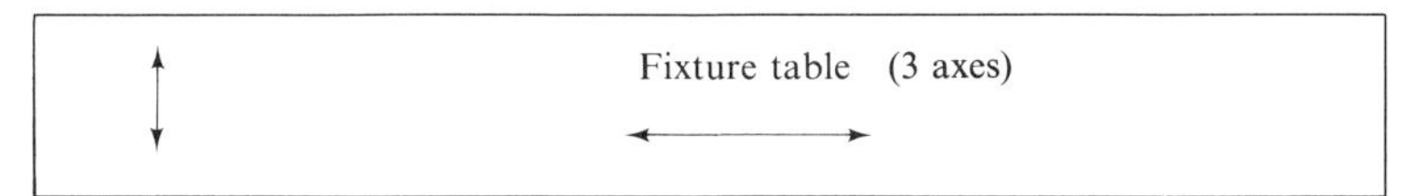

FIGURE 14–7 *T*-shaped air foil cutters.

cleaning, the parts are moved to a climate controlled area where temperature, moisture, and foreign particles in the air are controlled. Personnel responsible for handling parts wear clean cotton gloves whereby no body oil or foreign matter touches the parts. An adhesive primer is then applied when appropriate, care being used to see that the primer is placed only on the areas to be bonded. This procedure depends on the type of adhesive to be used. Areas where cutting of the adhesive tape occurs are cleaned with methl-ethyl-ketone.

BONDING

Adhesives for this type of bonding are kept at zero temperatures until ready for use. When thawed, the pattern is cut with scissors and the separator film is removed. The adhesive is then carefully placed where it is to be bonded and tack bonded to the parts when necessary without a curing heat. The assembly is subsequently placed in the bonding tool or machine where it is bonded by means of controlled heat and pressure. Both the press and autoclave are utilized, depending on the type of adhesive being used. These bonding tools are especially made to the shape and size of the part to be bonded, therefore, the tooling-up phase of production includes the manufacture of these needed tools.

The tools operate on several principles, some by air pressure and others by vacuum in shaped containers or dies. Heat is often applied by steam, both the pressure on the bond and the accompanying heat being closely controlled during a timed period. To protect the assembly, a cork is often used between the surfaces of the part to allow a smooth pressure. A typical sequence of bonding will require approximately 340° F. for up to eight hours at pressures ranging from 30–90 psi.

The other type of bonding tool is the autoclave. These tools are usually more sophisticated than the presses in that rubber blankets are frequently used to hug the assembly at the bonding surface during the pressure–heat–time cycle. Cooling is accomplished in the tool. After removal from the bonding tool, parts are then post cleaned. Post cleaning involves the removal of any flash or other bonding material, the router quickly removes excess adhesive while the MEK dissolves unwanted adhesive.

Bonding of Attachments

Often it is necessary to install inserts in the honeycomb for subsequent attachment of bolts and other holding parts. Holes are drilled into the cured bond and necessary parts inserted. A liquid adhesive is then forced into the hole which envelops the insert while curing occurs at room temperature. Whenever hard spots are required in an assembly, the adhesive is forced into the desired cells and allowed to cure with the other parts of the assembly. Hard spots provide extra stiffness where it is required in the assembly. Some typical honeycomb bonded assemblies are shown in figure 14–8, being cross sectioned to illustrate the bonding principles.

FIGURE 14–8 Cross sections of honeycomb assemblies.

The purpose of the honeycomb assembly is to provide a structural part having necessary strength characteristics, but with a great reduction in weight. When two sheets of metal, for example, are placed in a bending stress, a certain deflection occurs. When the same two pieces are separated and held apart and again placed in a bending stress, the deflection is less, but there must be some means to keep the sheets separated. When the filler or spacing material is very light in weight and constructed as a sandwich, a usable high strength structure is produced. The farther apart the two opposite sides are placed, the stiffer the assembly will be up to a certain limit. When the filler material is structurally bonded honeycomb, a strong part is produced. Much of the helicopter and other aircraft is consumed with bonded paneling. Figure 14–9 illustrates the effects of several forces on the surface of a bonded panel.

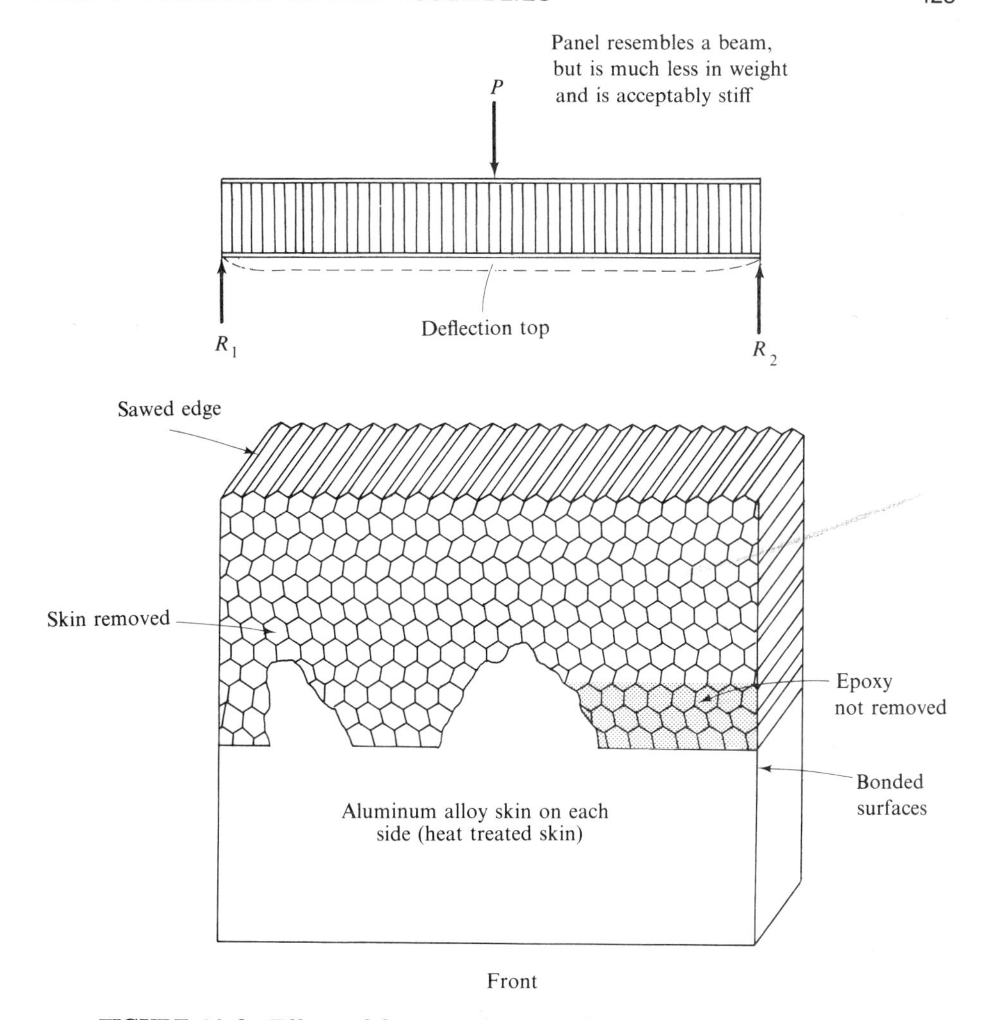

FIGURE 14–9 Effects of forces on honeycomb assembly.

Uses of Adhesive Bonded Assemblies

The use of adhesive bonded parts is so great that limitations are small, being mainly limited to use where temperatures are below 500° F., depending on the type of adhesive and strength properties needed. Even honeycomb paper properly bonded to stiff surfaces makes a good architectural product such as panels in buildings. When foamed plastics are bonded, good soundproofing occurs. Different types of metal parts, other than aerospace, are bonded, such as wrought metal to cast metal and aluminum to copper. Also, lightly stressed structural parts such as light poles and highway marking signs are often adhesive bonded. With respect to corrosion, metal parts having an adhesive at their interface are not subject to galvanic corrosion at the interface or to seasonal temperature problems with the accompanying rains and industrial fumes. Therefore, tubing, wiring, and other solid items can be embedded in another material by adhesive bonding without creating problems. Combinations like wood, plastic, and metal are capable of being bonded, the objective being to obtain desirable physical and mechanical properties of each material along with an increase in stiffness and a decrease in weight of the assembly.

Strengths of bonded joints and assemblies are competitive with those of other joining methods. Because all strengths of materials,including joints, are relative to temperature and environment, the strength of the adhesive compared to its bonded parts is critical. Therefore, industrial systems which involve cryogenic temperatures where chemicals are stored and used are using the adhesive more and more. Some of the polyurethanes, phenolics, and modified epoxies have excellent subzero strength qualities as compared to room temperature strengths. Also, mechanisms are being bonded as an efficient method of joining parts in the automobile industry and in other mechanical type industries. Simple applications like handles, fixtures, and fasteners are being bonded to the parent part. With respect to strength, epoxies containing fibers and laminations have withstood the tensile strengths of low strength metals, some equalling the low strength steels. In the aircraft industry, the helicopter uses a substantial quantity of bonded parts and assemblies. Surface panels and structural parts of the fuselage and rotor blade are adhesive bonded. Because honeycomb bonded parts provide high strength qualities along with low weight factors, aircraft such as the helicopter (fig. 14–10) and the supersonic aircraft *F*-111 (fig. 4–80) are manufactured with greater payload characteristics. Figure 14-11 points out the effectiveness of bonding in that maintenance personnel walk on the center section of the bonded assembly in the helicopter while, at the same time, several structural members are anchored in the bonded assembly.

FIGURE 14–10 The *UH* helicopter utilizes advanced manufacturing processes including honeycomb bonded assemblies in its fuselage and in its rotor blades. (Courtesy of Bell Helicopter Co.)

Questions

1. Differentiate between cohesive and adhesive bonding.
2. Why are thermosetting resins used mostly in adhesive bonded assemblies?
3. Why must clean gloves be worn during the assembly of parts to be bonded?

FIGURE 14–11 A bonded assembly in a military helicopter. (Courtesy of Bell Helicopter Co.)

4. Explain how honeycomb assemblies obtain their strengths.
5. Name an all-around cleaner for bonded assemblies.
6. What types of joints are strongest in bonded assemblies?
7. Why is a bonded butt joint weak in strength?
8. Why is the cleaning operation so very important prior to bonding?
9. How can honeycomb paper produce a fairly stiff part?
10. What is the purpose of using honeycomb assemblies?
11. Name several materials used in honeycomb manufacture.
12. What is the purpose of an autoclave?
13. Which resin is predominately used in the production of honeycomb?
14. Describe a process which is used in machining honeycomb.
15. Why must a surface be chemically clean prior to bonding?
16. How can strong spots or anchor regions be provided in sections of honeycomb?
17. Describe several mechanical properties of honeycomb assemblies.
18. When two dissimilar metals are interfaced with epoxy, what prevents corrosion at the joint?

19. In what main uses are honeycomb assemblies competitive with other structural assemblies?

20. Name several precautions to be used in handling adhesives.

Cleaning and Finishing Operations

Many parts must be cleaned after being processed through one or more of the several manufacturing operations. Further, a large quantity of these parts require surface finishing operations for the purpose of enhancing their appearance, for subsequent surface processing such as plating, for protection against corrosion, or for increased wear resistance. As an example, many of the aluminum parts are anodized, while many steel parts are plated.Cleaning is accomplished prior to plating, anodizing, hard facing, and painting. It may be done in one of several ways, each having its particular effects. Ferrous and nonferrous parts become contaminated on their surfaces during forging, casting, or machining. This surface contamination of new parts consists of pigments from the several processing compounds, grease and oil, and oxides. Many parts which have been in service require periodic or intermittent cleaning whereby grease, oil, dust, and even pigments and varnish must be removed. One type of cleaner will not suffice in cleaning all types of contaminants, therefore, an effective cleaning operation requires knowledge of the material and contaminant.

Because each contaminant is unlike most other contaminants, several cleaning processes are available. One or more of these processes results in the return of the parent metal's surface. In a general way, cleaning is divided into liquid or vapor cleaning and mechanical cleaning. *Liquid cleaning* includes the use of vapors, acid cleaning and pickles, alkaline solutions, emulsions, and solvents. *Mechanical cleaning* requires the use of abrasive blasting, wire scrubbing, and tumbling operations. The choice of process depends on the contamination and the desired surface finish.

Liquid Cleaning

VAPOR CLEANING

Grease, wax, and oil are removed by hot vapors coming in contact with the cooler surfaces of the part. Parts are suspended over a heated tank of solvent so that the hot vapors rise, contact the part, condense, and dissolve the contaminant which drips back into the tank. A special cleaning room is required to prevent atmospheric contamination. Important characteristics of the vapor include a noncorrosive contact with the parts, nonflammability, an adequate boiling point to cause vapors and to subsequently allow contamination separation from the solvent, and a high degree of solvency. Some of the solvents having these characteristics include *perchlorethylene* and *trichloroethylene.*

Several special types of solvents are sometimes used for a particular metal or for a particular circumstance where only part of a contaminated surface needs grease removal. An example is the removal of grease without damage to insulated wiring in an electric motor's parts. Hard-to-dissolve contaminants are processed with the vapors of perchlorethylene, such as the removal of tars. This chemical requires higher temperatures to develop vapors, while it results in a greater attack on some of the higher melting contaminants. The most common vapor degreaser is trichloroethylene, however, a substitute is desirable due to its toxic characteristics. Vapor temperatures are lower and much time is saved in subsequent handling operations.

A Vapor Cleaning System

Essentially, the vapor system includes a large tank which has the space capability to process the largest parts produced by the plant. A heating system is installed in the tank's bottom or a hot solvent is pumped into the tank. The tank's sides are high enough to retain the vapor and cooling zones. Within the vapor zone, holders for the parts are installed. Then, there must be a means of moving parts to the tanks. This is accomplished with mechanically assisted hoists and holders which transport parts from the floor to the tank and return. Figure 15–1 illustrates a typical vapor cleaning tank system.

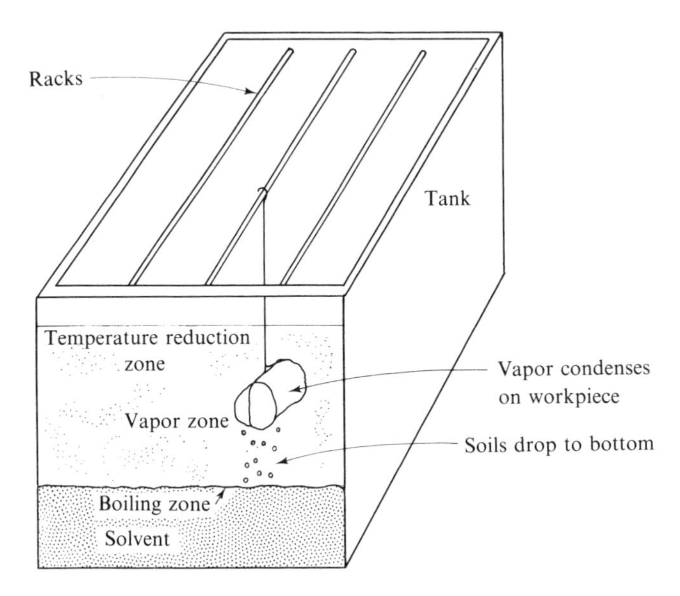

FIGURE 15–1 A vapor cleaning system.

Differently shaped parts are processed in other ways. Parts containing holes and keyways, for instance, are subjected to solvent spray in conjunction with the vapor system. High pressure spray is ejected from nozzles to dislodge internal soils and other contaminants. After surface materials have been removed, parts are dried by the effect of temperature differential between the parts and the vapors.

Possibly the semiautomated or automated system of cleaning is the most common for a plant having a reasonable quantity of parts to degrease. Many of the holders for parts include baskets and hooks which lower into the tank and raise on command from a controlling instrument. Solvents are continuously treated in a reclamation system which distills the solvent for subsequent use, the sludge being drained away. The more difficult soils are removed in the perchlorethylene solvent because of its ability to absorb higher temperatures. Cleaned parts must then be immediately protected from rusting as rust occurs following the drying of the metal.

ACID CLEANING

Prior to plating, painting, and other subsequent processing, many metals are cleaned with a mixture of acids whereby light oxide scale and various industrial soils are removed. The acids from both organic and inorganic sources are prepared either in the mixed condition or as a salt solution in water. Commercial cleaning compounds and solutions include some of these acids used singularly or in various proportions: chromic, nitric, citric, sulfuric, hydrochloric, phosphoric, ammonium persulfate, oxalic, and several others in water or sometimes in alcohol.

The contact of metal with the acid solution causes immediate evolution of gas, therefore, proper ventilation is required due to dangers of explosion and poison. Scale is attacked immediately by sulfuric acid and large quantities of gases are generated. In this respect, the acid cleaning solutions are used particularly on the irons and steels. A statement of caution is given in respect to chemicals, both acids and alkalis. These chemicals burn the skin badly, therefore, handle them carefully. Some procedures require face shielding and special clothing. If spattered by a chemical, neutralize the area of contact immediately with water.

Cleaning Methods

Metals are acid cleaned by immersion, spray, and wiping methods. The *immersion method* is used frequently due to the faster attack by the acid solution. Solutions are usually heated at temperatures ranging from 130–200° F. When heavy scale is to be removed from steel parts, the solution is more concentrated and becomes a *pickle*. Pickling is performed on hot mill products such as bar, strip, wire, and sheet to remove the scale, a product of elevated temperature, time, iron, and oxygen. When increased surface removal is necessary at a fairly uniform speed, the sulfuric acid solution is used. Also, hydrochloric acid and nitric acid are used on special steels such as stainless.

Spray systems are also commonly used because they are less expensive than the more effectively equipped tank systems of liquids. Parts are arranged on hooks and baskets, as in the vapor degreaser system. The line of carriers brings the parts by overhead conveyor systems, finds their space in the spray system, remains according to a predetermined time, and then lifts upwards, moving farther along the line for continued processing. Weaker acids are used during spraying because heavily concentrated sulfuric acid cannot be sprayed into the air.

Older methods such as *wiping* are still being used whereby rubber gloves and protective clothing, including face shielding, are worn by the person doing the cleaning. As in spray cleaning, the cleaner is diluted and is hand applied, followed by rinsing.

ALKALINE CLEANING

Both spray and immersion methods are used in removing contaminants from surfaces of metals with alkaline solutions. The alkaline solutions consist of water plus sodium hydroxide and often other constituents, including sodium bicarbonate and sodium carbonate. Some solutions also contain sodium tripolyphosphate and various wetting agents such as a sodium sulfonate. Generally, the acids are used on ferrous metals, while the alkalines are used on the nonferrous. However, both ferrous and nonferrous metals are cleaned at times with either acid or alkaline cleaners. Solutions are maintained hot, approximately 200° F., as in other cleaning operations, the water rinse following the cleaning operation. Both spray and immersion systems use facilities similar to vapor and solvent cleaning systems.

Electrolytic cleaning merely adds electrodes to the electrolyte system so that current flow between electrodes assists the etching action of the alkaline solution. Metal tanks are insulated from the anodes. During the cleaning process, both oxygen and hydrogen are liberated, the quantity being closely related to the surface area of the electrodes. Good ventilation is, therefore, imperative to exhaust the explosive combinations of oxygen and hydrogen in the presence of a spark. Electrolytic etching, or cleaning, for example, frequently precedes electroplating in order that a receptive surface is available for the plating metal.

EMULSION CLEANING

An emulsion cleaning system provides two liquids, one being highly dispersed in the other, but not dissolved. The amount of dispersion into tiny spheres depends on the chemistry of the mixture. One type of mixture is fairly stable in that both constituents remain in an emulsified condition. However, less dispersed particles require agitation to maintain the dispersion. Yet, another emulsion depends on the two separate layer types whereby the contaminated part enters the tank and plunges through the oil-rich top layer in the immersion system, carrying some of the top layer liquid into the lower water-rich layer.

The solvent is a type of petroleum product having soil dissolving capabilities. To cause oil and water to mix, a special material is added to the mixture to cause these emulsifying conditions to exist and not allow oil to float on water because of their weight differences. Several emulsifiers are available and each must be soluble in the oil constituent to cause independent globules to form in the water constituent. Contact between part and globules allows dissolving of the surface contaminant. Heating of the mixture to approximately 165° F. causes faster removal of contaminants. Both spray and immersion systems are used in the removal of grease and pigment. The cleaned surface is coated with an oily film and this film provides protection against rusting.

SOLVENT CLEANING

The oldest and most common cleaning methods are the spray and immersion processes. Usually, the cleaning is performed at room temperature. The choice of solvent depends on the type of contamination to be removed, many different solvents being available. Because solvents have fire potentials and produce injurious or obnoxious fumes, facilities must be available to prevent fire and provide proper ventilation.

Standard solvents include kerosene, naptha, mineral spirits, acetone, benzol, the several alcohols, and various kinds of chlorinated hydrocarbons, along with other special cleaning solvents. Parts are frequently agitated, or the liquid is vibrated by ultrasonic sound waves. As temperature of the solvent bath increases, cleaning efficiency also increases, but fumes and danger of fire increase. In these respects, only high flash-point solvents should be used. Should the temperature of the solvent such as kerosene become too high, it will suddenly flash, indicating the fire point is near. Also, some cleaning solutions produce poisonous gases when heated, therefore, necessary understanding of the cleaning environment is essential.

As in vapor cleaning, large plants or those plants performing large quantities of cleaning often have automated conveyor systems, some being tape controlled. Parts are brought from various areas of the plant by overhead conveyors to the cleaning tanks where the basket or hook which holds the part is lowered into the solvent. After a given time cycle, the part is automatically lifted and carried to the next processing station.

Mechanical Cleaning

ABRASIVE CLEANING

The abrasive blast operation includes the use of an abrasive which is blasted onto the part's surface by a high pressure stream of air. Sand blasting, for example, is one of the most common cleaning methods and uses an abrasive such as ground garnet or quartz to bombard the surface of the material which causes surface contamination to be dislodged and knocked away. Manufactured abrasives such as aluminum oxide also accomplish good scale removal, while ground nut shells clean with a less impact force and do not affect the surface finish. Harder materials such as ground cast iron and steel shot are sometimes used, each abrasive being chosen by its ability to remove the particular contaminant and not harm the surface finish. Steel shot blasted against many of the nonferrous metals will leave pit marks on the surface and may even fracture small parts. Proper judgment must then be used in choosing the most effective method of cleaning. Abrasive blasting systems are enclosed to prevent abrasives from flying beyond the cleaning area and also for the purpose of reclaiming the abrasive after it has been separated from the contaminants.

WIRE AND FIBER BRUSH CLEANING

A revolving metal wire brush removes loose scale and other contaminants when brought in contact with a part. Type of contaminant, type and construction of the wire brush, speed of brush rotation, and operator technique contribute to good removal of contaminants. Many wire brushes are attached to spindles of pedestal grinders, being partially enclosed for safety purposes. Because parts are differently shaped, the brushes are also differently shaped to conform better with the shape of the parts to be cleaned.

Both dry and wet brushing are used. Dry brush cleaning is used to clean a large portion of industrial parts. Burrs from machining, weld droppings, heat scale, paints and varnishes, and oils are quickly removed by the whirling metal wires. On the other hand, wet brush cleaning utilizes a synthetic fiber in conjunction with a liquid vehicle for carrying cleaning compounds to the parts' surface. Frequently, hot alkaline cleaning solutions are used with the brushes to dislodge many kinds of industrial contaminants. Surface feet per minute of the brush is much faster during dry brushing than during wet brushing, the speed being reduced to less than 2500 sf/min for wet brushing. As in dry brushing, many different shapes of brushes are used, varying from flat discs to cups, to rollers, and to the various shapes used for internal cleaning. With regard to brushing, it must be pointed out that brushing by power means is dangerous, especially in regard to flying wires and clothing contact. A fast whirling wire brush can jerk an individual into the brush should a part of clothing such as a necktie contact the brush.

TUMBLING

The use of a barrel or tub to hold and clean contaminated parts is effective in the removal of burrs, scale, and organic contaminants. Parts to be cleaned and surface finished are placed in a container, the shape depending on the parts, along with a cleaning compound, water, and an abrasive. Many parts are tumbled dry during each operation, depending on the exact situation. The tumbling operation begins with partly filling the barrel, leaving enough air space for movement of the entire contents, and then rotating the barrel slowly. As the barrel turns, the contaminated parts and abrasives collide, causing burnishing, deburring, removal of scale, and dissolving of grease.

Some metal barrels are rubber or plastic lined whereby better surface finishes are produced on the parts. As the barrel turns, parts, abrasives, and water mix by impact, enabling the contaminants to be separated from the parts. Shapes of abrasives vary and include those which are essential to accomplish the job. Several shapes are available, ranging from oval to pin shape. For most effective results, turning must be only fast enough to allow a sliding action of the parts, followed by tumbling and impact. Too fast a turn accomplishes nothing as the contents remain stationary, while too slow a turn only causes sliding. Therefore, the shape and size of container, along with the type of contents, must be carefully studied in order to determine the correct r/min of the barrel.

Electroplating

The electroplating process uses electricity to cause metal transferral in an electrolyte. When a direct current passes through a controlled liquid or electrolyte, metal from an anode can be placed onto a cathode. Then, when the cathode becomes the part to be plated, an electroplating process exists. Accordingly, there must be a tank containing the electrolyte and at least two electrodes, the positive anode and the negative cathode. In order to complete the electric circuit, electrodes must penetrate the electrolyte. The electrolyte consists of a liquid solution of carefully controlled quantities of certain chemicals, depending mainly on the type of plating to be done, while the system's energy depends on predetermined voltage and amperage settings of the equipment. Therefore, these electrical controls are part of the plating system. When a part to be plated is suspended from the cathode bar into the liquid electrolyte, and when anode material such as cadmium is also suspended into the electrolyte, cadmium plating of the part occurs as current flows, causing the cadmium anode to be eventually depleted. Time in the electrolyte is a thickness factor; the longer the current flows, causing depletion of the anode into the electrolyte, the thicker the deposits of cadmium metal become on the workpiece or part.

Electroplating is performed for many purposes which include protection from corrosion, better appearance, and surface buildup. Also, some plating is a prerequisite to a subsequent plating, such as copper on steel which will, in turn, be nickel on copper on steel. Prevention from corrosion, however, is a primary reason for plating, appearance being improved concurrently. The electroplating process uses many metals such as cadmium, lead, tin, zinc, gold, silver, copper, nickel, and chromium. Other metals are also used in the special types of plating requirements. Basically, the metal to be transferred to the part is made into anodes or is added to the electrolyte, as in the chromium plating process which uses chromic acid in the bath. As each metal is different, each electrolyte is also different, the contents being periodically analyzed to maintain required tolerances of the constituents.

When the equipment is prepared and the switch is closed, a current flows through the electrolyte. In order for plating to occur, there must be a current density at the cathode to provide, during a given time interval, a thickness of plate measured in thousandths of an inch. Usually, the plate is only 0.001–0.005 inch thick, but when buildup procedures are used, the surface plate provides thickness to much larger dimensions. In this regard, there is a relationship between current density at the cathode and chemistry of the electrolyte for a given plate thickness. Current density is measured in amperes per square foot of surface area of the part or parts being plated.

A typical electroplating system is illustrated in figure 15–2. In the interests of simplicity, the cadmium system is used to point out the principles governing the electroplating process. Essentially, the process is accomplished with a tank. The plating bath, or electrolyte, is maintained within the tank which is often steel, but sometimes the rubber lined steel tank is used. Rubber insulates the tank and keeps the electric currents within the electrolyte. Connected to, but insulated from, the tank are two anode bars which are usually arranged so that an anode is available to both sides of the parts. The anodes, being a metal to be plated onto another metal, are retained within holders which are suspended from the anode bars into the electrolyte. Parts to be plated are then suspended from the cathode into the electrolyte. When the circuit is closed by a switch, current flows and the plating process begins. Time, then, is an important factor with regard to thickness of the plate.

Plating baths contain the constituents which provide the vehicle and means for metal deposits at the cathode. For example, a typical cadmium bath includes a few ounces per gallon of liquid solution of each of the following: sodium cyanide, sodium hydroxide, sodium carbonate, cadmium oxide, and cadmium. Water consumes most of the liquid. In order for acceptable plating results to occur, a cyanide-cadmium ratio must be maintained within specified limits such as four-to-one, or as high as

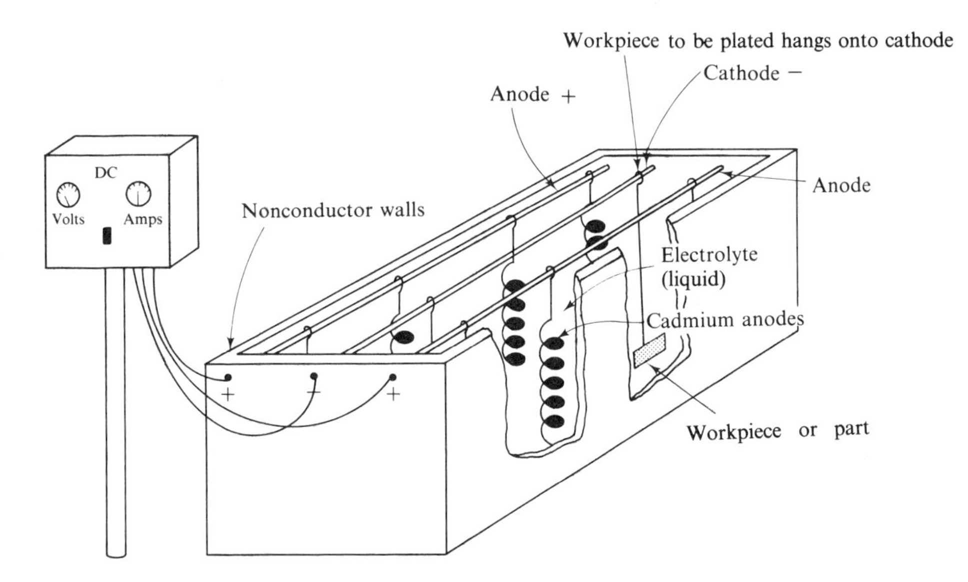

FIGURE 15–2 An electroplating system.

seven-to-one. If the metal cadmium, for example, is 2.5 ounces per gallon of solution, then the cyanide quantity is 10.0 ounces in a four-to-one ratio. Oxide in the solution amounts to only a few ounces, being closely related to the element's weight. Sodium hydroxide forms during the process, while sodium carbonate forms from decomposed cyanide.

With regard to coverage of the part and throwing power, cyanide baths are excellent. During the plating operation, the solution heats to as high as 90° F. Anodes, being consumable, must be periodically added to maintain a proper electrolyte. Cadmium, for example, in anode form is produced as balls, being usually about two inches in diameter. When these anodes are used in the still bath solution and when average conditions exist, approximately 25 A per square foot are needed when the cyanide to metal ratio is four-to-one. When larger tanks are needed, automated processes are easily established.

Other metals such as zinc are also plated onto steel. Zinc, nickel, and copper anodes are used similarly as in the cadmium process. When zinc is used, most of the several bath solutions are cyanide, the acid zinc baths being used mainly in special situations such as in wire and strip plating. Copper solutions use the cyanide-alkaline bath, while the anode used in chromium plating only provides the electrode because chromic anhydride is added to the special type of plating solution. For example, when the decorative chromium plating process is used, the bath solution is basically a chromic acid and sulfate solution. Basic baths must not be mixed.

Obviously, plating baths vary with the process, but all use the same basic principles of electrolysis, the attachment of positive ions at the cathode by direct current. Following the electroplating and electrocleaning processes, parts should be baked at 400° F. for three hours in order to remove hydrogen from the parts' surfaces and prevent hydrogen embrittlement. This procedure is more applicable to thin parts due to the closer ratios between plate thickness and part thickness in thin parts.

The electroplating processes include several of the dangers associated in manufacturing processes. During the plating process, while current is flowing through the electrolyte, harmful gases are generated and must be exhausted from the area. Hydrogen is explosive. Another danger exists in the possible failure to completely wash the plating solution from the plated parts or inadequate washing of the parts following an acid pickle and before plating. When acid and cyanide contact each other, hydrogen cyanide gas forms and is deadly poison. Therefore, never allow cyanide and acid to mix; each should be stored in separate areas. Further, cyanide is extremely poisonous and must not be allowed to enter the mouth or bloodstream. Failure to wash hands before eating can carry small quantities of the deadly salt to the mouth. Personnel engaged in handling cyanide must be proficient in first aid procedures to be used on victims of cyanide poisoning of any type.

Other Surface Treatments

Several other surface treatments are performed on metals, in addition to the electroplating process. A metal's surface is often treated for some reason, plating being done mainly to prevent corrosion. However, plated surfaces are thin, require time to accomplish, and plating is limited by the size of the equipment. Sheets and plates, for example, are more economically coated in a different manner, rather than attempting to build larger plating tanks. Coating of metals is accomplished by several methods.

CLADDING

The cladding of one metal onto another is performed to benefit from both the qualities of the surface cladding and the basic metal. The stiffness and strength of steel, for example, when clad with aluminum provides both strength and corrosion resistance. Cladding of steel is performed by several methods, one being a *diffusion bonding* procedure. The clad metal is placed against the base metal after both surfaces have been cleaned, and the metals are pressed tightly together at a pressure which causes an interface of the two metals to form. Soaking at a temperature below the melting point of either metal while pressure is applied causes a strong bond to result. Another method is performed by *cold rolling* a strip of metal onto the base metal and then subjecting the bond to a high temperature soak.

DIPPING

Dipping is a surface protection process that allows large quantities of metal to be coated in a very short period of time. Tin and zinc are the two most common metals used in the surface protection of steel, however, aluminum dipping is also used in cladding steel. The size of the part or workpiece is limited only to the size of the dipping tanks. Normally, the low carbon steels to be dipped are used in the sheet and structural form, in addition to wire for fencing.

Parts to be dipped are first cleaned by one or more methods, as have been previously pointed out. Solvent cleaning and acid pickling are frequently used to prepare the metal for dipping. After cleaning, the parts are fluxed in order to remove any remaining residue and to allow the molten metal to be coated on the base metal. Flux is applied either on the dry surface of the part or is floated on the molten metal.

Fluxes for tin dipping include the zinc and sodium chlorides, while zinc dipping uses zinc ammonium chloride as the fluxing material. Zinc is applied at a bath temperature of approximately 850° F., while tin is molten at approximately 600° F. When properly applied, the interface allows a good bond, the coating acting as a means of preventing corrosion while, at the same time, the qualities of the base metal are utilized. Galvanized iron, being zinc coated steel, is recognized by the spangles throughout the surface areas, while tinned steel has a smooth and bright surface. Tin and zinc are consumed in large quantities. The canned food industry, for example, uses large quantities of tinned low carbon steel, while the sheet metal industry uses galvanized sheets, zinc coated steel, in duct systems and tanks. Canned drinks are commonly contained in aluminum cans. Many metal cans are currently being sprayed inside with a plastic to prevent food contamination by the metal.

Dipping processes vary from the small parts category whereby individual parts are processed to the automated hook and basket systems which move overhead or on the surface from all parts of the plant. Coils of sheet steel are processed differently. Sheet stock is slowly uncoiled and fed through the dip tanks at a constant rate where time in the molten metal becomes a thickness factor of the resulting coating. Air cooling from the dip provides good solidification. Structural parts are dipped by mechanical assistance from crane and hoist systems, for example.

METAL SPRAYING

Metal spraying is the depositing of molten metal onto a cold metal surface where solidification occurs on the base metal. The resulting metallic structure is essentially an overlay of impacted droplets of solidified metal. Metals are sprayed on other metals for the purpose of corrosion prevention, hard facing for wear resistance purposes, and for buildup of a part's dimensions. Molten or near molten metal is blown from a special type of gun which is powered by compressed air or an inert gas. An oxy-acetylene torch or plasma arc torch vaporizes the wire or powder and blasts it against the part's surface where it flattens, adheres by cohesion, and solidifies.

Questions

1. Why are different liquids used in the cleaning of metals?
2. Explain the procedures used during vapor cleaning.
3. Why must cleaning be accomplished prior to electroplating?
4. How can abrasive blasting damage a metal's surface?
5. Describe how an emulsion cleans a metal's surfaces.
6. Why must cleaning solvents be kept from getting too hot?
7. Describe the several dangers in using cyanide.
8. How does the electroplating process operate?
9. Describe the differences between the anode and cathode in an electrolyte.
10. Why must electroplating and electrocleaning areas be kept ventilated?
11. Explain the principle of surface cleaning.

12. How does vapor cleaning remove surface contaminants?
13. What determines the choice between acid and alkaline cleaning?
14. What is the purpose of an emulsion?
15. Compare liquid cleaning processes with mechanical processes.
16. What is an electrolyte?
17. What is the main function of an anode?
18. Within an electrolyte, describe the flow of ions.
19. What two dangerous gases are produced during electroplating?
20. Compare the dipping process with electroplating.

Corrosion and Corrosion Control

Corrosion is the disintegration of a metal by some kind of chemical attack, transforming the metal into another material. Metals are corroded by either a direct chemical attack or an electrochemical attack. Steel rusts and produces an iron oxide which is a compound of iron and oxygen and is quite unlike iron or steel. The oxide reduces the quantity of metal in the part, thereby reducing its strength and subjecting it to possible complete failure. Different metals corrode at different rates, depending on the metal and the environment. Some metals resist corrosion under varying and adverse corrosive conditions, while others corrode quickly. The causes of corrosion and its control must be well known by the engineer and technician.

Causes of Corrosion

Several types of corrosion result from one or more of the following circumstances: the contact between two dissimilar metals in the presence of moisture, chemical reactions among metals in the presence of industrial atmospheres and liquids, improper heat treatment, stress and uneven stress conditions in a metal, fatigue, and geographical conditions such as contaminating soils and salt. Thus, in order to reduce or eliminate corrosion, measures must be taken to neutralize or eliminate the causes. Such measures require inspections during and after manufacture of a part. For example, a solution of certain alloys like the austenitic steels is produced to resist corrosion, while surface inspections and cleanliness help to prevent corrosion of the metal.

The Corrosion Environment

When a neutral environment changes chemically, the pH value in the environment moves from seven toward zero or from seven toward 14. This means that the corrosive medium becomes either acid or alkaline and has capabilities of metal attack, depending on the metal. Hydrochloric acid will quickly attack steel and corrode it. Sodium hydroxide will quickly attack aluminum alloy and etch away large areas. Acids and alkalis are chemicals and when either is uniformly applied to certain metals, a uniform etch results, the etch being a controlled rapid corrosion. Another example of a direct chemical attack is the oxidation of steel to iron oxide scale when steel is exposed to bright red temperatures. Besides the direct chemical attack, there is the electrochemical corrosive attack which is the couple produced when two dissimilar metals are joined. The more active metal is oxidized and the less active one is protected. The active metal also reacts with the nonmetals, oxygen, for example, while less active metals are more resistant to chemical changes.

THE ELECTROLYTE

Acids, bases, or salts in a water solution are electrolytes and can, therefore, conduct an electric current. Decomposition of materials occurs during current flow, and the resulting ions carry energy to the cathode from the anode. Hydrogen is released at the cathode while oxygen is released at the anode. It is during the electrolysis period that the anode-cathode ion transfer occurs, but the transfer is subject to interchange with other ions, and new compounds are formed at the expense of the original metals. The ionization is the dissociation, or breakup, of the compound in the solution, causing negative ions to flow to the anode while positive ions move to the cathode. Due to the possibilities of ion recombinations in the solution, some reactions move to completion while some do not. During electrolysis, gas escapes and materials are broken up, causing the chemistry to change whereby new combinations of materials are probable in the form of compounds. Basically, the electrolyte can be only a small moist area that has ion flow capability. Such moist areas exist especially at rivet and bolt connections or in blind corners of structures.

DIRECT CHEMICAL ATTACK

Attack on metals by direct chemical processes is common wherever metals exist. Surface oxides form as a direct reaction between the metal and oxygen in the atmosphere or between metal and moisture. Other compounds form from reactions with industrial gases and liquids. Copper produces a blue colored sulfate while copper carbonate is green, aluminum produces a dull gray oxide, steel and iron produce a brownish colored oxide scale, while brass and silver pick up a sulfide tarnish which is darker in color than the base metal. As temperature increases in the corrosion environment, the speed of chemical reaction increases, as in the case of loose scale forming on red-hot steel.

A different type of direct chemical attack occurs when acids are applied to a metal's surface. The products of reaction form rapidly, resulting in discoloration and a residue. Part of the metal is removed or corroded. For example, when hydrochloric acid and zinc come in contact, an immediate reaction occurs whereby zinc displaces hydrogen in the acid to produce a salt, while hydrogen is released from the reaction environment. In time, all the zinc disappears if there is sufficient acid. In such cases of direct chemical attack, a uniform attack, or etch, occurs. That is, uniformity of attack occurs across the metal's surface when the condition of the metal is homogeneous, but when two different types of mixed materials, ferrite and iron carbide in pearlite, for example, are etched, one material is attacked faster than another. Even so, a fairly uniform direct attack occurs across the face of the metal where the chemical is in contact. The longer the corroding contact exists, the thicker or deeper the etch. When scaled steel is pickled, it is corroded and cleaned, the reaction removing the iron oxide all the way to the parent metal over a period of time. The result is a cleaned metal with byproducts of gas and residue. Basically, this chemical environment includes no anode or cathode.

ELECTROCHEMICAL ATTACK

When an electric circuit is generated on or below the surface areas of dissimilar metals, corrosion, or metal deterioration, may occur. In such a situation, there is an anode and a cathode, distance between the electrodes varying. The different materials are the electrodes. When an anode and a cathode are established, corrosion can sometimes result. When an electrolyte exists on the metal's surface which completes the circuit between the anode and cathode, corrosion occurs at the surface. An electrochemical attack includes ion movement between an anode and cathode, the reactions resulting in evolution of gas, the production of a new compound which is the product of corrosion, a transfer of ions, and a loss of the anode or parent metal. On the other hand, surface corrosion which penetrates the metal is very serious, the electrolytic corrodant being localized and resulting in spots of corrosion with unknown depths.

The Electromotive Force Series

The electromotive force series is a heirarchy of metals showing the reducing tendency of elements where the tendency is listed in a decreasing tendency order. Tendencies of the elements are found by measuring the voltages of a special type of electrolytic cell. Single electrode potentials for each element are different, placing the listing of elements in a grouping that indicates whether the element is active, less active, or inactive. When the element is active, it reacts readily with nonmetals such as oxygen. The inactive metals such as gold will not react with the nonmetals, therefore, gold remains pure on its surface, while aluminum reacts with oxygen and corrodes, usually to a passive or protected condition. An example of this phenomenon is illustrated when two different metals are connected in an electrical circuit and inserted in an electrolyte. The more active metal begins to dissociate into the solution, while the less active metal is protected by the solution. An examination of figure 16–1 points out the solution tendencies of the several elements indicated. Those elements listed at the top in the figure have higher electrode potentials and will displace the ions of metals having smaller electrode potentials listed lower. The electrode potential decreases in a descending sequence, as indicated by the list in figure 16–1. Magnesium, for example, becomes sacrificial to iron in the presence of moisture. This electrical potential must then be considered when joining metals.

GALVANIC CORROSION

In accordance with the principles of reducing tendencies in elements, galvanic corrosion of a metal may occur when two dissimilar metals or two dissimilar phases such as the grain interior and grain boundary electrical difference are connected in any fashion that enables an electric current to flow. One of the metals or phases of metal

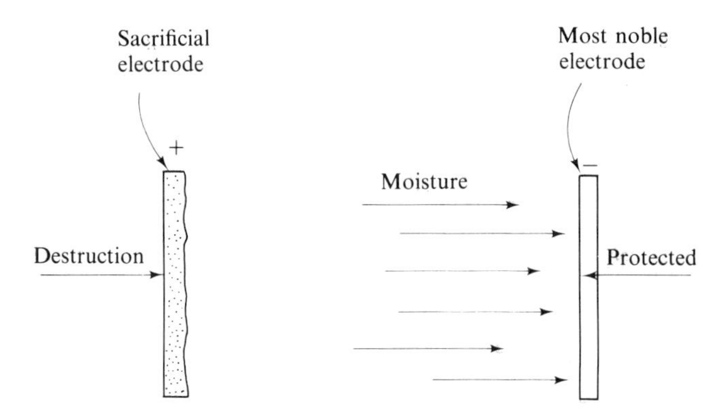

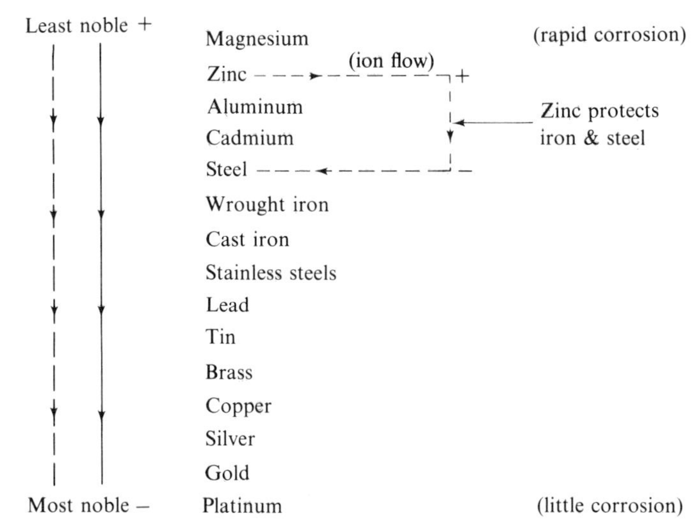

FIGURE 16–1 A galvanic series of metals.

may immediately become anodic to the other through the electrolyte or through the metal itself by means of the large electron cloud surrounding the distance between anode and cathode. The products of corrosion form. For example, when copper sheets are riveted with higher strength iron rivets, a galvanic couple is produced. Iron is higher in the electromotive force series, making it anodic to the copper. The copper then becomes the cathode and provides a very large area for the very small iron anodes. Corrosion of the iron is accelerated two-fold, by the anodic iron and the very large area ratio between the anodes and cathode.

If this situation is reversed whereby iron or steel sheets are riveted with copper rivets for some reason, the very large steel anode will not be seriously affected by the protection given to the small cathodic rivets. When the cathode area is much greater than the anode, the demand from the anode increases proportionately to the areas of anode and cathode, causing a greater current density at the anode when the anode is smaller.

Another example is the joining of two iron or steel pipes with a copper or other less active metal coupling. The pipes are anodic to the coupling and will corrode. A repaired section in a steel pipeline, for instance, can become anodic to the original pipe on each side of the couple. Even though the same metal is used to replace a damaged pipe, the new metal becomes anodic due to a chemical difference on the surface of the original metal. Iron oxide, or rust, is chemically different from steel, therefore, an electromotive force differential exists.

In the case of the electrolyte, the electric current breaks up the anode material which allows the *cations*, or positive ions, to seek the cathode. *Anions* are negative ions and are formed by gaining electrons in the electrolyte, and they seek the anode. Ions are identified by the appropriate chemical symbol and a plus or minus sign to show gain or loss of electrons, for example, Na^+ or CL^-. One of the dissimilar metals at the joint has the larger single electrode potential and immediately undergoes loss of metal as its ions move through the electrolyte to the cathode. In this instance, the metal having the largest single electrode potential is the most active and becomes the anode. During the electrolysis, hydrogen and metal are deposited at the cathode, and the cathodic metal is protected, but the other metal or anode goes through a destruction process. Galvanic attack occurs faster when the anode is small in relation to the area of the cathode. The rate of attack on the anode is directly related to the ratio of cathode to anode areas.

As a result of galvanic corrosion, care must be used when replacing parts or sections of metals and when initially fabricating a joint. Basically, it is best to use the same metal in building an assembly, however, when this is not practical, the connector such as a bolt or rivet should be lower in the electromotive series than the surrounding metal. Then, as a last resort in the connection of different metals, use the metals closest together in the electromotive force series while, at the same time, the connector should be used which is lowest in the series. In other words, the farther apart the metals are in the series of electrical potentials, the more drastic is the corrosion of the more active metal. There are some instances where two dissimilar metals can be separated by a nonmetal such as a plastic. This nonmetallic separation prevents galvanic corrosion.

In the case of electrochemical attack inside a metal, the anode-cathode distance is determined by the chemistry differential in the region of the attack. For example, grain boundaries in a metal may have a higher solution potential than the metal inside the grains; in such a situation, the anode-cathode flow of current occurs through the metal and results in severe destruction inside the metal along the grain boundaries. Figure 16–2 is an illustration of intergranular corrosion as observed through a microscope.

Intergranular corrosion is the most dangerous type of corrosion because assessment of damage is hard to determine as only a small pit may show up at the surface. Severe cases of intergranular corrosion are demonstrated in the *exfoliation* of metals' surfaces. As internal corrosion continues, the products of corrosion force the surface layer of the metal upwards, exhibiting the compounds of corrosion, many being powdery. An extruded part, for example, has closely packed elongated grains, making the potential for exfoliation high if other environmental factors are present.

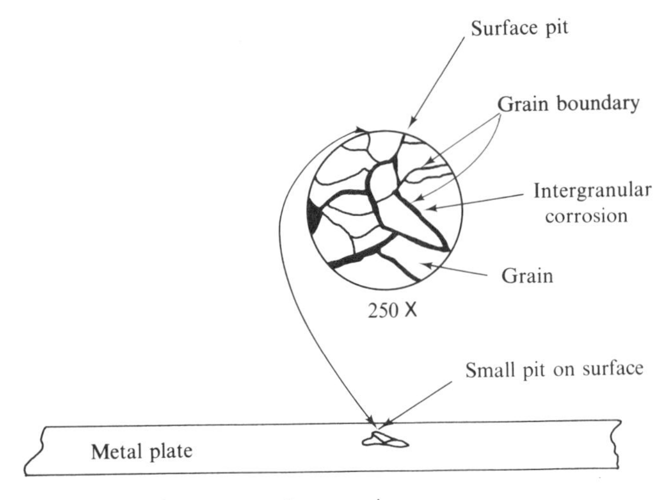

FIGURE 16–2 Intergranular corrosion.

Homogeneous conditions do not always exist inside a metal, especially among the alloys. Brass, for example, when overheated, can loose zinc and, when cooled, a galvanic region can be established which puts the anode-cathode corrosion system in operation. Improper heat treatment can also prepare the way for concentrations of constituents in a metal to be heterogeneous. Selective corrosion may then occur due to factors of time and conditions.

On a metal's surface, the same kind of galvanic attack can occur, resulting in pits. Even pure metals are subject to pitting because the pure metal usually has some kind or kinds of trace elements, that is, a very small quantity of an impurity. The galvanic cell is then established between element and compound or between element and a near insignificant phase of constituent. Pitting of the surface on aluminum and its alloys, for example, moves inward in time, making paths similar to small worm holes in wood. From the surface, only a pit is noticeable, but when a pointed probe is inserted into the pit, white or dull gray powder may be found. Continued probing often shows serious metal separations beneath the surface, being far more corroded than the small surface pit. Structural members showing pitted surfaces must be investigated to determine the extent of corrosion. The strength of the metal is seriously reduced by such corrosion.

An unusual type of corrosion is caused from stray direct currents flowing through a soil from a poorly grounded direct current electrical system. When a steel pipe, for example, is buried near an electrical conductor such as a locomotive rail, stray currents move from the negative rail to the pipe and back to the source through a conducting soil. The corrosion problem begins when the stray current leaves the pipe, causing corrosion on the surface of the pipe.

Surface pitting of a metal also occurs at times as a result of concentration cells and not by galvanic action alone. Again, surface areas of anode and cathode help regulate the amount and speed of attack.

ATTACK BY CONCENTRATION CELLS

Anodic corrosion can also occur when an electrolyte consists of two different concentration regions of solution in which two pieces of the same metal or two areas of the same metal are exposed to the liquid. Two sections of aluminum, for example, can be riveted with an aluminum rivet which provides for the moisture and cell. Also, a sheet of aluminum may contain two or more dents in which the concentration of liquid forming in one cavity is different than that in the other cavity. The region of the electrolyte having the lowest electrolyte concentration causes the contacting metal to become the anode. Therefore, the remaining metal at the higher concentration becomes the cathode, current flows, and anodic corrosion commences. With respect to the above electrolyte, the solution may only contain different degrees of dissolved oxygen and the effect will still be anodic corrosion.

The metal-ion concentration cell establishes electrode voltage in a ratio that is equal to the difference in solution concentrations at the anode and cathode. Mechanically connected joints such as bolted and riveted sheets provide excellent environments for concentration cells. The region of overhanging metal beyond the bolt's surface is compressed less than the two metals' surfaces at the bolt. In other words, the connector tends to lift the outer edges of the sheet of metal at the joint. Consequently, any electrolyte at the edge of the metal is less concentrated in metal ions than that concentration at the bolt's periphery, therefore, the anode forms farthest from the bolt. Figure 16–3 points out the logic of two concentrations in a solution at a joint.

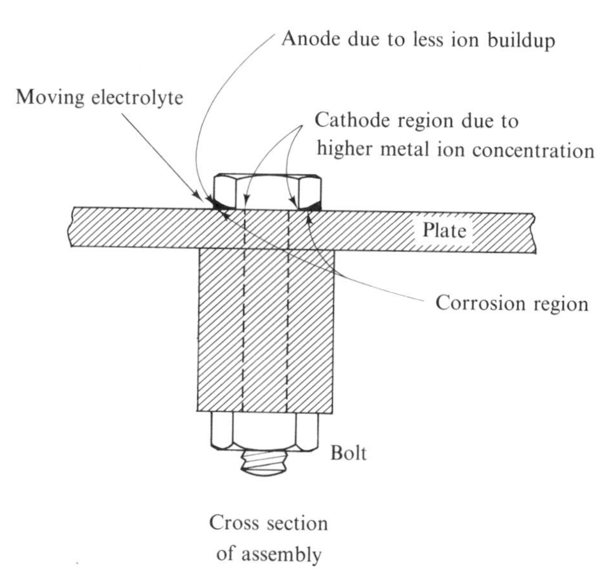

FIGURE 16–3 Corrosion by metal ion concentration cell differential of bolt and plate in a pump mechanism.

The oxygen concentration cell is established in a similar type of joint. The region of the electrolyte closest to the bolt's or rivet's periphery, however, contains less oxygen than the solution at the edge of the joint. With this type of situation, the anode is established at the surface or near the surface of the bolt and between surfaces of the sheets, while the cathode is established at the outer edge of the joint. Concentration cell corrosion occurs deeper into the joint, while ion concentration cell corrosion occurs farthest from the joint's connector, but within the joint. Also, submerged metals are subject to this type of corrosion because oxygen is less concentrated in the deeper regions of the electrolyte and, therefore, the cathode is established at or near the surface, while the anode forms deeper into the electrolyte.

The active-passive cell often begins with an oxygen concentration cell and corrodes through a surface barrier such as a passivated surface of aluminum oxide on aluminum alloy. This closely adhering passive film forms rapidly on some metals as a result of base metal and oxygen contact combining chemically to form the oxide. If the base metal is aluminum, for instance, a dense gray colored coating of oxide forms which helps prevent further corrosion. Such a process is known as *passivation* and produces a passivated metal. However, corrodants such as salt and moisture may frequently break down the oxide film under the active-passive cell attack. In this respect, the oxygen cell often punctures the passivated surface and causes anode-cathode corrosion to begin. The passive area or oxide film becomes cathodic, while the base metal becomes anodic. Again, this is a selective corrosion attack and even happens on the surfaces of stainless steels. Such an attack can become serious due to the very large anode-cathode ratio or the very small anode and large cathode.

STRESS CORROSION

Some metals such as magnesium and its alloys corrode under stress in the presence of some kind of corrosive environment, even though the corrosive medium is weak or mild. Therefore, these metals are often surface treated to resist the corrosive attack when the metal is to be placed in tensile stress by loading. Often, surface areas of several other metals are stressed unevenly during loading and quickly generate anodic and cathodic areas. Cold working operations sometimes produce the exact tensile stress differential in the material, causing corrosion to occur at the higher stressed point or anode when other factors are present. The stress effects at the surface of a metal reduce the metal's tendency to build a passive coating, and selective corrosion occurs. Continued stress, in turn, accelerates further corrosion which, beginning at the surface regions, causes increased depth of material deterioration. In turn, less metal is available, and the metal becomes unable to react properly to further loading. Metal separation develops, being helped by the continued corrosive environment, and subsequent failure of the metal occurs. Cracks developing from stress corrosion run across the stress flow pattern, generally at about 90°. This type of crack is both intragranular and intergranular, the crack separating the metal along the grain boundaries as well as across the grains. This type of crack is differentiated from the intergranular crack because of the orientation to the stress flow.

Proper heat treatment can reduce tendencies of stress corrosion by the application of specific stress relief treatments following fabrication. Copper and its alloys, especially those in the pressure vessel assembly, are sometimes severely subjected to stress corrosion due to the combined environment of the surface tensile stresses which are developed from high pressure fluids in the tubing and pipes. Also, steel piping is similarly affected by stress corrosion because of the high pressure chemicals it transports.

Fatigue Corrosion

Either fatigue or corrosion will cause a metal to fail over a period of time. When both fatigue and corrosion combine, a severe attack on the metal occurs. Fatigue corrosion is a different type of corrosion as it depends mainly on the factors of cyclic loading in a corrosive environment. Because the corrosive environment is constant and the loading factor is cyclic, the time to failure factor is reduced to a value far below either that of fatigue or corrosion failure. Time under changing loads frequently initiates cracks in a metal. Also, corrosion leaves an uneven surface on the metal or stress raisers. As corrosion proceeds, the stress pattern at the surface begins a series of detours around the changing surface contours. When the detours become severely concentrated, the metal separates and invites accelerated stress concentration. Initial cracking is often along the grain boundaries, but final fracture is across the grain or the shortest path to sudden failure. Observance of stress patterns in metals is essential in designs which are to resist both corrosive attacks and concentrations of stress.

Corrosion Control

Each year, millions of dollars are spent in the replacement of parts which have been damaged or destroyed by corrosion. All metals are subject to attack by corrosion to some degree, however, the ferrous group and many of the nonferrous metals such as the copper, magnesium, and aluminum alloys are sometimes damaged heavily, and many parts are destroyed annually. Therefore, metal parts and structures subject to attack by corrosion should be initially prepared to resist corrosion by conditioning and treatment of the surfaces of the metals. Corrosion control includes two main procedures, proper surface treatment and frequent inspections.

SURFACE TREATMENTS

When the corrosion environment is known, a specific material is selected which will, hopefully, resist the corrosion effects and be able to effectively function in its assigned capacity. Corrosion resistance, however, is only one of the properties which is considered in the selection of the material. Along with corrosion resistance, certain mechanical properties must be evaluated such as tensile strength of the material. When the operating environment is known, final selection of material will be based on the ability of all properties to function effectively together. In this regard, some environments are strongly acid and may even include an increase in temperature. Other environments exist at elevated temperatures such as those associated with jet and rocket engine operations, while many environments are anodic and not chemical alone. Consequently, each operating environment should be studied to ascertain all important factors and a decision then be made on the material.

Certain alloys are produced which have the required strength capabilities for specified designs, in addition to their corrosion resistance factors. These include the austenitic stainless steels, for example, which have inherent surface protection against many chemicals, while their strengths are adequate to safely support the particular design. The bronzes resist the corrosive effects of contaminated water and many industrial liquids. Tin and zinc are plated on steel for the food and fencing industries, the plated products to be used at normal temperatures. Also, paint and other coatings, including the plastics, are being used in large quantities to protect metal surfaces against corrosion.

Coatings of paints, asphalts, lacquers, oils, and plastics are used extensively to resist corrosion on metals. Many are organic, while many are mixtures of organic and inorganic material. Powdered lead, zinc, copper, or aluminum dispersed in a liquid base are used to coat different metals. Lead, zinc, or copper powders are used in the painting of ship hulls as a protection against sea water, while aluminum powder forms an excellent coating in the painting of numerous types of structures. Aluminum alloys are frequently coated with primers such as zinc chromate, and some are further coated with lacquers. Steels are coated with zinc or leaded paints such as the familiar red lead used in painting bridges, buildings, and ship hulls. Again, ordinary paint is commonly used to act as the barrier between the metal and the corrodant. Pipes and other metal parts which are sunk in the earth are frequently coated with an asphalt on which a covering of fabric or special paper is placed. Machine parts made of steel such as spindles cannot be coated with paints, but must be periodically coated with a thin film of oil to prevent rusting. One of the finest examples of coated metals is the thermoplastic coatings, being applied often by dipping the metal part in the liquid plastic. As long as the temperatures remain below 200° to 250° F., and a fairly reasonable environment exists, the coated parts protect the metal. Sea water and a sea atmosphere are not normal, therefore, the time interval between inspections is shorter for parts subjected to this environment.

Metallic Coatings

When the strength of steel is required in conjunction with a high degree of corrosion resistance at a moderate cost, zinc and tin coated steels are used. Stainless steels are more expensive, therefore, a coated low carbon steel is often used in many industrial applications. Coatings are applied by plating, dipping, and spraying, the size and shape of the workpiece or part often determining which method is used.

Zinc on steel is anodic to the steel, consequently, zinc is sacrificed for the continued protection of the steel as long as any appreciable quantity of zinc remains. The degree of protection is reduced as the zinc is reduced beyond a minimum value. Galvanized sheet, for example, when scraped on the surface provides base metal protection due to anodic-cathodic action. Because the scraped area of zinc precipitates anodic action, the results are not serious due to the favorable anode-cathode area ratios. As the result of scraping, the area of exposed steel is small compared to the large anodic area of zinc. The scratched surface becomes cathodic as it is protected by the anodic zinc. With respect to anode and cathode areas, the larger the cathode with respect to the anode, the larger the corrosion at the anode. However, in the case of galvanized iron, the scratched area, or cathode, is very small when compared to the remainder of the zinc surface or anode. All types of steel products, including wire, fencing, and piping, are zinc coated for cathodic protection. Hulls of ships often have slabs of zinc or magnesium attached to their lower sides for the protection of the steel plates and other parts. Steel is then cathodic to zinc and, in time, more zinc is required for the anode replacements.

Another cathodic protection method is the use of direct current in an electrical circuit whereby an underground steel structure is made cathodic by connecting the negative terminal to the pipeline or steel columns. The positive terminal of the generator is connected to graphite anodes which are buried in the surrounding soil. Without the use of manufactured electricity, underground steel parts are made cathodic by burying zinc or magnesium anodes at intervals along the structure to be protected. The anodes are electrically connected to the structure. Steel pipelines are protected up to ten years by burying magnesium anodes along the length.

Tin, on the other hand, is cathodic to steel, but is much more resistant to the atmosphere and organic materials than steel. Because of its high resistance to corrosion, tin is coated on steel in large quantities, the canned food industry using a large portion of tin. If the tinned surface is deeply scratched however, corrosion begins and accelerates because of the small steel anode compared to the area of the large tin cathode. In this reverse situation from zinc, care must be exercised not to penetrate the tin cathode; while in the case of zinc, penetration to the base metal is not significant because of the anode-cathode ratio being in favor of the base metal. An examination of the electromotive force series of elements points out the single electrode potential difference between iron and zinc and also between iron and tin. The corrosion problem results when the coating is cathodic and a break in the coating occurs.

Electroplating is used in many situations to prevent or resist corrosion such as the cadmium plating of steel hardware items. Cadmium is anodic to iron. However, when a part is subjected to severe surface treatment such as automobile bumpers, multiple plating is used to combat scratches and resist wear. A triple plate such as copper on steel, and nickel on copper, and chromium on nickel is a very effective means to resist corrosion and wear and, at the same time, present a pleasing appearance.

Metal spraying is a type of plating in that liquid droplets of a metal or a stream of powdered metal is impacted against a surface for the purpose of providing corrosion resistance, wear resistance, or buildup of a dimension. An advantage of spraying metal is its fast surface coverage and the fact that thick layers can be established. For purposes of resisting corrosion, thin layers are applied of whatever metal is considered best. For purposes of resisting wear, only those metals not subject to annealing are used, such as chromium, tungsten, and various alloys of the hard and heat resistant types. Parts which are subject to abrasion are often coated by spraying a metal or metallic powder such as powdered tungsten onto a red-hot surface.

Linings of pipes and vessels are sometimes coated with an oily inhibitor to deter corrosion. Radiator systems use special types of inhibitors, some oily and some of a different mixture which causes passivation of the surface linings. However, a break in the surface film may start an anodic or chemical action and subsequent corrosion.

A common method used in resisting corrosion of the aluminum alloys is an anodizing process which converts the aluminum surface to aluminum oxide. The oxide resists corrosion attacks to a very effective degree, but the process has secondary effects such as causing a dielectric with the ability to better retain subsequent painting or plating. The anodizing process is conducted by making the part the anode in an electrolyte. Prior cleaning is essential, followed by surface etching and rinsing. Anodizing is then conducted by immersing the aluminum part in a warm electrolyte such as sulfuric or chromic acid, the part or workpiece being attached to the anode, while the cathode may be lead plate of area equal to the anode. Approximately 10% chromic acid is used in the solution to maintain a pH value of slightly less than 1.0 when this type of process is used. Current density of approximately 1 A per square foot is used for a given period of time, the voltage increasing slowly to the operating value of less than 50 V. Voltage settings vary according to the particular situation. At the end of approximately one-half hour, the voltage is reduced to zero, and the part is removed from the electrolyte and washed. Sealing of the coating is then performed by soaking the part in a slightly acid hot water for a few minutes followed by drying it in open air. When a colored dye is desired, the part is soaked in the dye before sealing, the resulting color presenting a pleasing appearance which has a metallic luster.

A modified anodizing process uses a cold sulfuric acid electrolyte. A current density of 27 A per square foot in the cold solution produces a much thicker oxide and harder coating than conventional methods which use hot solutions and lower current densities. Conventional methods produce a more porous and softer coating. The hard coating is anodic to the parent metal and is much more effective in wear resistance, the process being known as hard anodizing.

FREQUENT INSPECTIONS

The second procedure in corrosion control is periodic or intermittent inspection of parts and assemblies. More frequent inspections must be performed on those metals which are more susceptible to corrosion than to those which are anodic coated. The inspection consists of a general visual view of accessible surfaces, optical inspection of surfaces and inside tubing, and the nondestructive test for the whole material.

Ferrous parts portray their corroding conditions by the presence of a brown oxide at the surface. This surface presence of oxide does not imply that corrosion is only at the surface, nor does the absence of oxide at the surface imply that corrosion does not exist deeply along the grain boundaries. The presence of pits, organic matter, or moisture presents a suitable environment for promoting corrosion. Therefore, the inspection should be followed by actions to remove small regions of corrosion or to replace the corroded part, followed by cleaning the surroundings. Corrosion should be scraped and sanded from the area or should be removed by special cleaning solutions. Pits and cracks should be investigated to determine depth and character in order to evaluate the affected region in regard to its remaining strength.

Even stainless steels corrode. The austenitic stainless steels, for example, may sometimes be exposed to unusual conditions which cause carbide precipitation from the solid solution which, in turn, promotes corrosion because of the chemical difference between the carbides and the solution. Faulty heat treatment may instigate precipitation of compounds, while proper annealing will help assure an effective corrosion resistance factor.

The nonferrous family of metals varies in corrosion tendencies from near zero for gold in normal environments to rapid deterioration of unprotected magnesium while under stress to the quick destruction of zinc in the presence of hydrochloric acid. Corroded aluminum alloy often produces a small mound of white or gray colored powder in regions of dampness and soil contamination. Also, small pimples may appear on surfaces of structural parts, the rise in the surface being due to the expanding pressure of the hidden corrosion compounds. A pick should be inserted into the pimple or pit to determine severity of attack.

Other metals discolor in appearance, but the discoloration is usually normal or insignificant in the formation of passive films which retard further corrosion. However, unusual deviations in surface appearances such as pits, pimples, contrasting surface colors, and loose oxide are signs of possible corrosion. Again, investigation must be made in order to evaluate the circumstances. Cleaning of the surfaces by either chemical or mechanical means is one of the best means to avoid future corrosion attacks.

Questions

1. Explain galvanic corrosion.
2. Define the difference between an ion concentration cell and an oxygen concentration cell.
3. What is the significance of the electromotive force series of metals?
4. In an electrolyte, what happens at the anode during current flow?
5. Compare the anodic actions caused by the scratching of tin- and zinc-plated steels.
6. When the anode is small, what is the result with respect to the cathode?
7. Explain intergranular corrosion.
8. How does a dielectric prevent galvanic corrosion?
9. What happens during stress corrosion?
10. Name three ways to reduce or prevent corrosion.
11. What is corrosion?
12. What are the requirements for a corrosion environment?
13. In a suitable electrolyte, what happens to a metal's surface when the anode is large and the metal or cathode is small?
14. What is an electrochemical attack?
15. The electromotive force series pertains to materials having potential differences in their voltages. What happens to the voltage when a pure metal and its corrosion residue are tested?
16. Will metals which are very close together in the electromotive force series corrode when in a suitable electrolyte?
17. How does the shape of the anode affect cathodic protection?
18. What is an ion?
19. When the area of the anode is small in relation to the area of the cathode in a suitable electrolyte, what happens to the cathode?
20. How can corrosion be prevented when two dissimilar metals must be jointed?

Safety Procedures and Pollution Control

Many of man's pursuits are accompanied by danger, this dangerous aspect of living having existed since man has been on this planet. The simplest skill such as the making of the fist hatchet by ancient man was accompanied with the danger of flying rock piercing his eye. Many of man's first explosive weapons literally blew up in his face. With the coming of the industrial revolution and the mechanical type inventions, wheels were made to turn rapidly so that their energy would cause motions of tools. The first wheels used in machinery were not guarded by screens, and the long belting whipped and lashed out at the worker. Now and then, a piece of clothing would become entangled in these unsafe mechanisms and the victim was snatched to his death or suddenly lost his arm. No need was apparent to protect the worker because the motive was to produce goods. It was most unfortunate when a woman's hair was caught in the belting, a child's clothing was pulled into the spokes of a flywheel, or a man lost an arm in a winch. This money motive lasted for hundreds of years, whereby no profits were spent for guards around dangerous machinery. During the many centuries of toolmaking, danger lurked in the presence of man's ambitious pursuits.

Need for Personal Safety

Such unsafe practices would not last forever, even though token attempts to practice safety were not successful until recent times. Slowly, small groups of workers in some of the large cities brought these dangerous money-making practices to the attention of the factory owners. After many tries by the working groups, the owners were forced to listen and consider safer ways of producing merchandise. Resistance to unsafe working conditions continued to grow among the working populations of the various nations, until steps were taken by local owners to protect the workers. Guards and screens were placed around some of the whirling wheels and the slapping belts.

But individual attempts to promote safety practices were much too ineffective, therefore, the several levels of government moved in and established legislation to provide safer working conditions. Laws were enacted and fines were provided for those employers who did not comply with the new regulations. Such benefits for the workers made news and, thus, even more benefits. The safe working practices then moved from community to community and from nation to nation.

Many of the safe practice laws were results of dangerous manufacturing practices, that is, people were first hurt or killed before legislation was enacted to prevent or alleviate further suffering. Apparently, little thought was given to the production of a machine that was effective in both its prime purpose and safety. These new safety features were expensive, and expenses reduced profits, but slowly the new concepts in manufacturing began to provide safer operating machines. Unsafe machines had been those which were not bolted to the floor or not properly secured to some structure whereby, at an unfortunate time, the machine fell onto a victim. Other machines had large exposed flywheels or exposed belting turning rapidly around a pulley. Also, many electrically driven machines were not properly grounded and shocks and electrocutions occurred. Some machines even threw their parts into the air because of ignorance of surface speeds, while many machines had no guards covering gear trains. In this respect, a pair of gears whose turning action at the mesh points is away from the observer is a dangerous situation because the tendency of the meshing gears is to pull an object into the grinding teeth.

NEW REQUIREMENTS FOR MACHINE TOOLS

As new laws became effective, machine producers were required to build the essential safeguards into their machines. Such safeguards reduced profits, but also reduced lawsuits, along with bringing an increased satisfaction to the working groups. Today, in many nations it is unlawful to produce and sell machinery that is unsafe. In contrast to earlier times, machines are now being built according to previously prepared specifications which include both the machine's function and its safe operation. Such an accomplishment has taken hundreds of years to attain, but, today, most machine producers and parts manufacturers think in terms of production and safety with a concurrent attitude. It was eventually realized that it pays to work safely, to have safe operating machines, and to practice safety first. Among machine operators today, there is no longer a fear to operate a machine or to venture too close to gear trains because a machine will do only what the operator causes it to do, and whirling gears are enclosed in steel boxes. Also, belting and chains are completely enclosed. However, it is certainly expected that no operator or technician will start a machine until he knows what it will do.

SAFE WORKING CONDITIONS

Along with the requirements for safe machinery came the laws to establish safe working conditions within the vicinity of the machines. In other words, a worker has been injured, for example, as a result of an attempt to move past a long section of bar stock which was being turned in a lathe. Other cases have demonstrated the ability of a shaper to knock a person or an object from its path while the ram was reciprocating. Apparently inexcusable instances have occurred in which the falling hammer of an impact tester has intercepted a person because lack of knowledge of the machine's purpose allowed him to move across the hammer's arc at several feet from the machine. There is also the situation whereby overhead cranes or moving conveyors have hit the unsuspecting workers. Also, poor planning causes many accidents. In attempts to save space, machines have frequently been placed too close together, endangering the operators on both machines while subjecting an observer to severe injury while passing either machine. Consequently, many laws currently require minimum distances between machines, designated areas for conveyor systems, and the marking of floors by color coding so that personnel will be apprised of the existing dangers.

Even though safe machines are being manufactured and danger areas are being marked, other factors in the environment are often harmful to workers. Actually, insufficient safety exists in many working situations where contaminated air is breathed. Many manufacturing plants use steam and chemicals during a processing operation which generate aerosols and contaminate the atmosphere breathed by the worker. Unsuspectingly, a dangerously polluted air is generated which moves from the factory to the atmosphere many miles away. Cleaning tanks, for example, create harmful vapors which sometimes spread throughout the plant, yet these vapors are supposed to be localized and controlled. Consequently, both the health hazard and explosion hazard have sometimes existed for long periods of time before management or legislation took steps to reduce or eliminate these dangerous manufacturing practices. Chemical processes are part of many manufacturing operations, yet the chemical aspects have been considered less often than the mechanical conditions of fabrication and production. Consequently, various kinds of pollutants have been dumped into the living environment.

Recognition of Pollution

It may be difficult to rationalize the compatibility of an advancing technology and improper control of many of its waste products. During the times of early manufacturing the old concept of "money first and safety second" had forced the implementation of better working conditions, but the better environmental conditions were generated mainly in the manufacturing plant. Only in the plant were steps taken to protect the worker from harmful chemicals, just as had been the case with mechanical dangers. On-the-spot corrections were made, and improved environmental working conditions reached the same efficiency levels of the preceding operational conditions. Again, legislation and management had engaged in improving these local conditions pertinent to pollutants in the air. Exhaust systems, screens, and spark proof tools were provided, but the pollutants from the factories were only moved from the plant to the community and even to regions many miles away. As a result, morale in the plants grew because employees were no longer exposed to dangerous mechanical or chemical working conditions. However, such blindness to what was happening outside the plant continued for many years and, in some cases, pollution has existed for centuries. As an example, creosote plants have dumped their residues into the moving waters where sealife once existed. Over a period of years, the sealife was eventually exterminated and this portion of the open waters was sometimes used to coat one's own poles with creosote. Also, during the twentieth century, exhaust emissions from burned fuels and factory smokestacks brought smog to the cities, even death to some innocent peoples, and generated dangerous contamination levels in portions of the atmosphere.

What was happening to many advanced societies around the world was only partially realized, but the wealth from manufacturing was well realized, and most people were not too concerned beyond their places of business and residence. But the total effects of good and bad were operating concurrently in the environment. Suddenly, a new thing of danger presented itself, not to the manufacturer alone, but to a society as a whole. The good atmosphere was becoming poisonous in some areas, while some streams were killing fish by the thousands. Where did the new danger come from? How long had it been around? Answers to both questions exemplify severe cultural lag.

The effects of inventions and mechanization of production have brought beneficial ways of life to those nations having advanced technologies. But advancement in technology has been accompanied with production wastes along with the material advantages. As a result, the end effects of mass production have somewhat reduced the cleanliness of the air and waters near the factories. Recently, however, new legislation has been passed which requires some types of controls during manufacturing to prevent high levels of pollution in the air or in the water. Engineers, managers, and technicians must be cognizant of these new OSHA laws.

Mechanical Dangers in Manufacturing

The story of mechanical dangers in manufacturing is long, but a review of some common dangers will illustrate how legislation cannot remove all dangers. Awareness of these dangers, however, along with reasonable precautions, should guide the production thinking of the specialist, technician, and engineer.

DANGER FROM FLYING PARTICLES

Current mechanical testing practice often requires compression tests on metals and nonmetals. Many of these materials are brittle and, when under destructive stress, failure occurs, often sending small particles of materials flying through the air. During these tests, face shields should be worn to protect the eyes. Failure to wear the shield may cause the loss of an eye, even though it may have been unlawful to conduct the test without a shield. Another typical danger is the grinding of metals without proper face shielding. A small grit particle in the eye can be very serious. Therefore, goggles should be worn while grinding or parting metal with a fast turning wheel. With respect to grinding wheels, caution must be used in selection of the wheel to assure that it does not exceed the safe sf/min speed. Exceeding this factor may cause the wheel to suddenly explode. Also, a cracked or damaged grinding wheel must not be used due to imbalance and subsequent danger of shattering. Another typical example of flying particles is demonstrated when the hardened face of a hammer strikes the face of a hardened tool or the face of an anvil.

Flying particles occur as a result of many manufacturing processes. While steel is being forged, hot scale pops from the surface and will burn clothing and skin, and is especially dangerous to the eyes. Scale also pops from steel during rolling. In these respects, maintaining a reasonable distance from the hot part is good practice, while protective clothing and face shields should be worn. Another operation that can be dangerous is using heat treated hammers. Hammering a hot steel part on an anvil, for example, must be done with caution. Because the hammer's face is hard and the anvil's face is also hard, both being aspects of quality, impact between the two can often send a steel fragment into the air. In other words, the person doing the hammering must hit the hot part and not the anvil.

Another incident which illustrates the danger of flying objects has occurred with the surface grinder. Normally, magnetic chucks safely hold the part being ground, but, occasionally, a part leaves the machine when it is loosened by a grinding wheel. For example, grinding wheels on surface grinders have sometimes been forced into the surface of a part too deeply, and the fast turning wheel has suddenly jerked the part loose, slinging it several feet.

Danger also exists with saws, cutters, and related tools. Cracked circle saws and cutters must never be used due to the probability that a portion of the cracked metal will leave the tool at a high speed. A very common, but dangerous, practice is to fracture a section of hardened steel in a vise with a hammer without a guard. Persons as far as 10 feet away have been severely injured by such practice. In this situation, the flying section of metal should be aimed at a backstop. With regard to plastics, some plastics such as nylon become very brittle during tensile stresses beyond the yield strengths and small particles of hardened plastic fly through the air. Therefore, face shields should be worn during this type of testing.

DANGERS OF BEING CUT

Even though machines are safe to use when used properly, they are nevertheless dangerous when certain principles are ignored. It is common knowledge that a steel chip produced from the lathe is sharp. The continuous chip moving from a part during lathe turning is very hot as well as very sharp. Hot chips have occasionally penetrated the space between a loose ring and the finger. A quick jerk in response to a hot chip can sometimes sever a finger or result in a deep cut. Therefore, good practice states that personnel engaged in certain machine operations such as lathe work will not wear rings or wristwatches. The machine burr which results from many machining operations, especially sawing, can also produce cuts. Failure to recognize the burr on the bottom or back side of a workpiece or specimen can initiate a severe cut when an attempt is made to wipe oil from the part. Even handling the parts must be done cautiously.

Danger of being cut in a machine shop is more probable than in most other shops, other than the wood shop. While a cutter is cutting, the metal operators must never allow their fingers to get close to the cutting tool, whether a single-point lathe tool or drill or saw is being used. The vertical band saw, for instance, can be very dangerous; the eyes must be on the blade at the point of cutting during the entire operation if hand guidance is being done. And, finally, one of the most dangerous things to do is to grab a handful of steel chips to dispose of them. Often, the fingers or the chips slip and a bad cut results. Again, steel chips from the lathe and shaper are sharp; milling cutter chips are often needlelike and quickly penetrate the fingers and hands and then break off in the skin.

DANGERS FROM WHIRLING AND RECIPROCATING MACHINERY

Flywheels, sprockets, gear trains, shafting, and related mechanisms which rotate must be covered with guards when a person is capable of becoming entangled in the mechanism. Likewise, machines which reciprocate or swing in an arc must be covered with guards to prevent collision with a person or capture of a garment. Hacksaws must be placed on the floor so that the stroke is safely operated, and shapers and planers must also be arranged and the floors marked to prevent personnel from intercepting the stroke of parts of the machines. With respect to belting, gears, and chains, they must be encased in a cover to prevent fingers, hair, and clothing from being contacted. Gear trains are very dangerous, especially when the gears are moving away from the observer. In this situation, any pair of moving gears has pulling capability. On the other hand, when sheet metal is being formed by power rolls, great caution must be observed when the operator is on the pulling-in side. Entanglement in the rolls or in a winch can leave a person maimed or can even be fatal. A simple example, but with an element of danger, is the knurling operation. The knurling tool is dangerous when a loose hanging sleeve moves into the knurls.

It has often been implied that these things cannot happen, but they do, nevertheless. Again, long sleeves, loose clothing, rings, wristwatches, and long hair do not fit safely in a machinery or fabrication environment. Even a whirling grinding wheel can be disastrous to a person should a sleeve or long hair be sucked into the periphery. Momentum is high and action is fast, the helpless operator having no time to respond.

DANGERS OF BURNS

Rolling, forging, casting, heat treating, extruding, and welding provide high temperatures which are often in the red heat range and higher, therefore, skin or clothing contact with the hot metal results in immediate burns. During heat treating, for example, steel can be 1,000° F. and yet appear cold. Touching such metal which is near a furnace should be done first with a quick brush of the finger against the metal to determine if it is hot. Heat can also be detected by holding the hand above the metal without touching it. Hot metal leaves a severe and painful burn that is slow to heal. As has been pointed out, hot scale also burns badly, therefore, during any hammering operation, precautions should be taken. The possibility of being burned exists during and after welding. To avoid being burned, attention must be kept upon the pool of metal until it solidifies and the end of the welding electrode due to their high temperatures. Further, the welded bead remains hot for a period of time. With respect to the hot metal which pops from the pool during gas welding, it will penetrate clothing and is also dangerous to the face areas such as the eyes. Consequently, proper eye or face shielding must be worn during the several types of welding, primarily, however, as a shield from the glare. Another hazard is the ring; it should be removed during electric welding to avoid any possible electrical circuit being formed in the ring. Lastly, the pouring of metals must be done with great care. Face shields and protective clothing are worn as a guard against spattering metal. For example, moisture in a cold mold may cause the metal in the mold to spatter, or the mold may even explode.

DANGER FROM CHEMICALS

The most common chemical danger is from fumes, the cleaning solvents producing most of these. Proper ventilation and filtering are, therefore, essential to assure clean air and to avoid disease, explosion, or fire. Most chemicals should be handled with care. Strong acids and alkali solutions burn the skin and are also dangerous if inhaled, therefore, personnel in close proximity to chemicals must be alert to avoid any type of contact. If contact with the skin occurs, water is a good neutralizer and must be used on the burn immediately. When mixing chemicals, face shields are worn when any type of bubbling reaction is to occur. In this respect, acids are added slowly to water or to alcohol.

Different gases are sometimes formed as a result of chemical reactions of different mixtures. Some of these gases are explosive, such as acetylene or hydrogen, while others may be obnoxious or even poisonous. For example, chlorine is injurious, but hydrogen cyanide is deadly. Specifically, a mixture of sulfuric acid and cyanide reacts to form this deadly gas, therefore, a clear understanding must exist as to what is going to happen to the mixture of chemicals after mixing.

When electric etches are used, current sometimes explodes an acid mixture when deviations are too far from required procedures. Then, in heat treating, cyanide is often used for case hardening purposes. A steel part that is red-hot with cyanide on the surface becomes extremely hard when quenched in water, even though the hardness is only a few thousandths of an inch thick. But, cyanide is a deadly poison if taken internally or if absorbed into the bloodstream. One form of chemical etching uses hot caustic soda which is an alkali. This chemical is a powerful alkali and badly burns the skin on contact. Consequently, chemicals must be handled with respect and knowledge, while proper ventilation must exist during their use.

Industrial Pollution

Along with the nearly endless quantity of beneficial products produced by industry are the industrial wastes of numerous types. Man cannot live in these wastes he creates without some type of controls. Manufacturing has indirectly and directly contributed to much of the pollution troubles. Automobiles, produced by manufacturing, create one of the greatest hazards to clean air by the emission of tons of detrimental chemicals. Manufacturing plants have also helped cause air pollution and have contributed in a large way to the pollution of streams, rivers, and bays.

AIR POLLUTION

Air pollution is caused from residues floating in the air, resulting from burned garbage, home fires and trash burning, manufacturing processes, and fuel type transportation systems. From the manufacturing point of view, approximately one-fifth of air pollution has come from residues dumped into the air by smokestacks. An analysis of a polluted quantity of air has shown unhealthy quantities of carbon monoxide, oxides of sulfur and nitrogen, and hydrocarbons. In this respect, automotive engines emit most of the carbon monoxide and irritants such as hydrocarbons and oxides of nitrogen. Damage occurs to living things, both plants and animals, and to metals and many other materials. Manufacturing plants and electrical generation plants have added tremendously to the pollution, both in the air and in water. Possibly, one of the greatest pollutants from all sources is sulfur dioxide, this chemical being very harmful to living things. The burning of fuel and other materials dumps tons of this chemical into the atmosphere annually. Sulfur compounds include the harmful sulfur trioxide which forms an aerosol of sulfuric acid in the presence of moisture.

WATER POLLUTION

Since man's appearance on earth, he has been a polluter, and as the concentration of people has increased, the pollution problems have likewise increased. Because water is the best vehicle for diluting all kinds of wastes and for carrying them away, man has frequently used the earth's streams as the sewer pipes to the bays and oceans.

Domestic sewage accounts for a large portion of water pollution, while industrial organic matter accounts for approximately as much. These wastes are chemically oxygen-demanding and include unused food and animal wastes, degenerated paper and similar products, domestic sewage, viruses, and many types of bacteria. Many tons of agricultural sprays such as the numerous pesticides enter the sewage disposal systems and, eventually, the streams, bays, and lakes.

The inorganic solids and liquid wastes from industry, along with the cyanides, mercury, and the dissolved and mixed by-products of wastes have increased the pollution of some waters to such an extent that living matter such as marine life does not exist. In some instances, radioactive residues have been dumped into containers and facilities which eventually have allowed penetration of the dangerous materials into the water. To unintentionally accelerate water pollution and increase the extermination of marine life, many cities have released their hot waters from generation plants into streams and drainage systems. In effect, such increases in water temperatures reduce the oxygen content and, coupled with the oxygen-demanding wastes, the pollution count in that quantity of water has increased rapidly.

Another pollutant is oil and oil products. Oil spillage, oil dumping, gasoline filling stations, and petroleum processing plants have been responsible for the entry of these contaminants into the numerous drainage systems which eventually provide entry into the waters. On top of it all, the thousands of homes, laboratories, schools, processing plants, and hospitals have poured their share of contaminants into the waste systems. Many of the waste systems have been ditches and small streams.

Again, man cannot survive in the waste he creates. It is not too late. Just as the pollution problem took years to be recognized, the reduction of pollution will also take years. Legislation at all levels of government, along with industrial help, has already provided guidelines for pollution reduction and control. Possibly, a plateau has been reached and, in time, the good earth's fresh air and clean waters will completely return. If need is the mother of invention, then the inventions will be produced to provide clean wastes concurrent with the mass of needed manufactured products.

POLLUTION CONTROLS

Much of the pollution problem is now controlled by one or more of the following pollution control systems: electrostatic precipitators which allow aerosols to enter an electrical discharge chamber and release clean gas, an ultrasonic vibrator to increase vibrations of polluted particles and reduce the solids for further processing, centrifugal separators which remove heavier particles by centrifugal force, screens and filtration traps to trap the pollutants, water scrubbers and high velocity jet blasts of water into the pollutants, and, subsequently, the chemical separators which break up the polluting compounds into harmless compounds and elements. Resulting sludge is protectively contained. One process, for example, converts sulfur dioxide to water and sulfur.

An ultimate goal of manufacturing must be to produce a safe part, assembly, product, or machine at a reasonable profit without creating pollutants and, in turn, the man-made machine must ultimately reduce its wastes, if any, to the basic elements and harmless compounds. Such a goal is justifiable, reasonable, and feasible. The harmless raw materials would be processed into consumer products in such a way that wastes would be converted back to other harmless raw materials. The transition pollutant would be further processed to finish the manufacturing or chemical cycle. Wastes subsequently dumped into the waters and atmosphere would be clean. Ultimately, control and neutralization of pollutants from other than manufacturing sources would keep in step so that the combined pollution control effort between domestic and industrial pollutant producers would be a complete reality.

Questions

1. What is considered to be an industrial pollutant?
2. Name a condition that will generate hydrogen. Why must hydrogen be vented to the atmosphere?
3. List some industrial pollutants that have been dumped into streams and rivers.
4. List some of the main atmospheric pollutants.
5. Describe a possible danger as a result of mixing acids.
6. Name several reasons why goggles must be worn during welding.
7. Describe a machining operation that nearly always leaves a burr on the part at the end of the operation.
8. How can cyanide and acid accidentally come in contact? What is the resulting danger?
9. List several methods which are used to control and reduce industrial pollutants.
10. In industrial operations, when should face shields be worn?
11. How can management and labor promote greater safety in industry?
12. In general, how has safety in the plant been increased and by what group of people?
13. What are safe practice laws?
14. In general, how has pollution been generated to its present conditions?
15. Differentiate between the chemical pollution of the air and the chemical pollution of the water.
16. Describe several mechanical type dangers.
17. Describe several dangers of whirling machinery.
18. Describe the dangers of using cyanide.

19. Water added to acid is often dangerous. Why? What procedure should be used?

20. Besides industrial pollutant control devices, how can pollution be better controlled?

Planning the Manufacturing Process

Before a part is made, there is usually a drawing, and before the drawing is produced, there is the idea or concept of the part or product. This idea often begins as the result of a need for the product. In his mind, the person visualizes what the part is and what it does. Also, some vague configuration eventually appears which is subsequently transformed into a sketch. After further study, several modifications of the sketch often present a three-dimensional view of the general shape of the part to be produced. Knowing what the part is to do and having a reasonable looking sketch, the services of a designer are sought so that an acceptable material will be specified and the principles of engineering will be used in the production of the drawing.

Product Design

It is at this point that the originator of the idea converses with the designer so that the part can be turned into a reality. Once the designer is fully cognizant of the part's function, he begins the process of design whereby all the facts are organized into a systematic approach that will end with a drawing and a set of specifications in support of the drawing. The drawing is an engineering approach to production, which means that the part will function effectively in its assigned task.

The task of the designer is, then, to interpret an idea into a drawing. In order to accomplish this, many questions will require answering, the answers being the facts in support of the drawing. After all the facts are available, the drawing is then completed, having taken only a fraction of the time needed in obtaining the factual data and the transformation of these data into tangible items to support the drawing. In this respect, there is no stereotyped sequence in a design process, but there is a logical and scientific approach to design flowing from the original idea. The following questions must be answered prior to making the drawing.

What is the function of the part?

How hard should the part be?

What weight or force must the part retain?

Is this load a single or multiple load?

Is the load applied statically or dynamically, and is the load tensile, compression, or shear?

What temperature is the part exposed to?

What chemical environment surrounds the part?

Does the part require lubrication?

Is centrifugal force involved?

Answers to the above questions allow the engineer, technician, or designer to determine such things as the kind of material to use; size, shape, and metallurgical condition of the material so that geometry and dimensions can be established, both for function of the part and strength; designation of a safety factor; feasibility of production; the probable market; and costs, both the production and sales costs. Many manufacturing plants have their own designers and, in turn, these designers have access to metallurgical and engineering data and computers.

With regard to the concept of the part and design criteria, no drawing can be produced without first going through some of these criteria because dimensions and strength of materials are directly related. In this respect, one of the toughest questions to answer relates to choice of material because it is the material that consumes the space between the three dimensions, and these dimensions cannot always be established until questions in the criteria have been answered. In other words, will the volume of material be subjected to a loading of 100,000 psi in tension, 200,000 psi in compression, or 40,000 psi in shear? In these respects, the difference between failing loads and safe loads is the safety factor, therefore, filling the part's three-dimensional space with the correct strength of material is mandatory. Then, also, the designated material must effectively resist the total operating environment, not just strength alone, and be able to carry on in its assigned function. After all these things are established, along with dimensional tolerances, clearances, and the part's total association in an assembly, the designer prepares the drawing. The drawing is further processed so that it reflects data in terms of an operation drawing, that is, the manufacturing engineer and technician see the finished part as outlined by the drawing and its supporting specifications. Again, it must be reiterated that the drawing is a follow-up of the data contained in specifications and related criteria.

Planning the Process

On receipt of the drawing and related data, the manufacturing team studies the drawing and interprets essential manufacturing information so that the production process can be outlined and subsequently established. Basically, a sequence of actions is formulated by the team to include the kinds of manufacturing operations involved and the necessary sequence of operations. From the general kinds of production operations, the detailed operations are listed. A study of all the detailed operations allows the team to choose the needed kinds and types of machines, equipment, raw materials, personnel, and support facilities. Quality and sizes of machines are then chosen from many factors, one being the duration of the market for the part. Because a profit must be made, economics of the entire process are again studied to assure that the market provides a justification for mass production and that a reasonable monetary return will be made. Cost of plant facilities and future equipment and facility replacements are part of the initial study and review.

DETAILING THE PROCESS

Manufacturing processes are subsequently designated and defined in detail so that specific equipment can be requisitioned if it is not on hand in the plant. If the plant is to be self-sustaining and accomplish all the processing, then a much larger enterprise is forseen. For example, many consumer parts are machined from rolled stock, forged stock, extruded stock, or cast stock. Some plants requisition the raw materials in these mill stock shapes and save the tremendous cost of a foundry, forge, press, or rolling mill. Forgings and castings in the approximate shape with oversize dimensions can be quickly and easily machined to the exact requirements of the drawing. However, the cost of these raw materials and the quantity required may provoke thought along the line of possibly providing a forge plant or a foundry to produce the initial raw material shapes and save from the initial costs. Large numbers of castings or forgings may be more economical to produce locally. A study of the products produced by the various rolling and extrusion mills, forge plants, and foundries will help in making the decision as to how the raw materials are to be initially produced or procured.

Such a decision regarding the initial procurement of raw materials also entails the procurement of equipment to provide a complete system for the specific processes. In other words, the foundry will require all the equipment and facilities to produce the castings, while the forge shop will need all the equipment and support to produce the forgings. In either case, the basic ingots must be procured first. Then, again, consideration should be given to the possibilities of using rolled, extruded, or other types of stock. Either way, final processing usually requires the services of a machine shop and even a sheet metal or welding shop. The design of the part dictates the kinds and types of equipment needed to completely finish the part. Often, however, secondary operations such as carburizing or other heat treating or plating are required. The whole operation is viewed as a total process and is then analyzed to determine the flow sequence in order that floor assignments for specific equipment and operations can be made.

AUTOMATING THE PROCESS

Mass production should be as machine dependent as is practical in order to take advantage of automated or semiautomated devices and control systems. The automation process begins with the initial raw material entering the production line. A means must be provided to convey the raw material to the first processing unit which may be a machine. Conveyor systems, in this respect, are available in all shapes, sizes, and functions. Because many types of operations are unsuitable for complete automation, technicians and helpers are frequently used to assist the machine in doing its job. For example, it may be necessary to position the raw material in an exact position so that the machine can do its work. This step changes that part of the production process to a semiautomated process. The next station is mechanically arranged to receive the workpiece and accomplish the assigned task. It may be that a secondary operation following the primary machining operations is required such as carburizing prior to the part's re-entering the production line for continued processing.

The layout of the production facility shows exactly where each machine and piece of equipment will be placed. Conveyor system locations to each of the machines are marked to provide the proper sequence of operations. Equipment needed for secondary operations is also indicated, along with the conveyor mechanisms. Finally, each process station with its complete equipment is shown from the beginning to the end of the total process to include all the secondary operations and support equipment and facilities. The production line is then prepared to commence operation after assignment of skilled personnel. In this respect, should trainees be assigned with the technician, training objectives and procedures are pointed out to both the technician and the trainee.

When the planned sequence flow of operations is completed, the written details of what and how are prepared to include detailed instructions and procedures, routing procedures, specifications, guides, work orders, and any other information needed in carrying out the objectives of production. After all equipment, machines, and operation facilities are installed, along with the procurement of raw materials, a practice run is completed to verify capabilities at each station and along the entire line. To assure product quality along the line, various stations are designated as inspection stations. The product is appropriately inspected or tested and returned to the line for further processing. Quality control aims at quality assurance, therefore, several inspection stations are installed at strategic locations along the line. Any product deviation beyond tolerances from the specification justifies investigation into preceding processing with a view toward correction of the deficiency. In other words, the workpiece, product, or part is inspected on several occasions prior to reaching the final inspection. After final inspection, the parts are packaged and placed in storage for subsequent shipment.

Requirements for Specifications and Production Standards

Specifications are prepared to cover information about a material, a process, or some other factor that requires rigid control. Materials' specifications include requirements for chemical and mechanical properties along with the accompanying physical properties. For example, a certain stainless steel's chemistry will be indicated in a specification by one producer to include from 6–8% nickel, 16–18% chromium, and a maximum of 0.15% carbon. Another producer will be more detailed and show the chemistry to be from 0.60–0.80% manganese, 0.33–0.38% carbon, 0.20–0.35% silicon, 0.04% phosphorus, 0.04% sulfur, 1.10–1.40% nickel, and 0.55–0.75% chromium, the balance being ferrite. From a given chemistry, the physical properties are stated. As a result of a certain heat treatment and for a particular analysis, mechanical properties will be listed to include:

Tensile strength	165,000 psi
Yield strength	145,000 psi
Percent elongation in 2 inches	20
Any additional desired properties	

Another type of specification will include a completely detailed procedure for the production of some material. Every aspect of the several operations will be written into the specification so that there will be only one way, or alternates, if indicated, to produce, test, and inspect the material. For example, the American Society for Testing and Materials' standard specifications for molding a plastic material are very detailed in their instructions. When testing specimens of the particular plastic, certain practices "shall be followed." The manufacturer, then, has the various military, industrial, and professional standards on hand to help guide this particular phase of production. Often, the manufacturer refers to a particular standard for compliance. The ASTM produces a complete series of specifications which are listed as standards and are available as parts 1 to 32 plus. Standards do not allow for deviations among manufacturers, therefore, they are universal.

Numerous professional organizations publish their specifications or standards which govern many procedures and materials in the numerous manufacturing operations. Some of these organizations include the American Iron and Steel Institute, the American Society for Nondestructive Testing, the Aluminum Association of America, the Society of Manufacturing Engineers, the American Foundrymen's Association, the American Welding Association, and numerous others. The objective of these associations' specifications is to provide manufacturing guidance in some particular aspect of the material or product so that expected quality assurance will result. Some specifications are very rigid and include the provision for acceptance or rejection of the product by the party issuing or using the specification. Often, before a contract is let, the manufacturer will be inspected to see that he has equipment and facilities on hand to do the job or that the subcontractor has them.

Sometime during the processing of a product, maybe at the midpoint, the product will be tested and inspected for compliance with the controlling specification by the company issuing the contract. In effect, a specification provides the buyer with a high degree of assurance that he will get what he pays for and that the product is in accordance with the drawing. Many specifications include compliance with standards. Basically, there are codes which are laws, standards which have universal control by some authority, and specifications which are more individual in nature.

THE PRODUCTION GUIDE

A specification may refer to a standard procedure for accomplishing some particular process. This process is frequently controlled by a standard such as a *process standard* issued by the specific manufacturer. Such a process standard, for example, may control the hold-down procedures for milling aluminum honeycomb. In this event, the process will be stated in a numerical sequence as to what to do. Each step will be listed simply and clearly and if details are required, reference will be made to a supporting standard. The purpose of process standards is to provide specific and detailed guidance in accomplishing a particular process. During inspection, parts failing to comply with inspection criteria can be traced to the process deviation. Frequently, parts are inspected at several stations along the production line to assure desired quality. In this regard, contractors supply their manufacturers with standards in order to receive acceptable smaller assemblies or parts which, in turn, will be processed into their finished product. Aircraft manufacturers, for example, supply completed airplanes that have been tested and inspected in accordance with governing specifications and standards and other written documents.

ROUTING PROCEDURES

Different manufacturers arrange their production processes and schedules according to the way they desire to produce the product. Simple instructions such as operation sheets, work orders, process sheets, or routing sheets are used to produce a product. A typical process sheet will include information such as identifying titles of the company, department, and worker; the part number being processed; the kind of processing; time in and out to what subsequent station; the specification, standard, or other controlling document; and any other needed information to effectively control the processing of the part. A *bill of material* is frequently used in processing an assembly and includes columns for noting materials and parts by part number or other identifying factor, nomenclature, quantities, and additional needed data.

STANDARDIZATION AND INTERCHANGEABILITY

Standardization is very important in manufacturing. Interchangeability of parts is facilitated and costs are reduced when standard sized parts and standard materials are used. Production is accelerated when controlling manufacturing documents allow standard tolerances, for example. The use of specifications and standards allow different manufacturers to make parts of an assembly that will fit. The specification saves time and misinterpretation because it requires compliance by all personnel engaged in the manufacture of the product. It is the responsibility of inspectors in quality control and assurance departments to assure compliance with the specification, standard, or code.

The Manufacturing Engineer and Technician

The process of manufacturing converts raw materials into finished products for the consumer market, the market being military or civilian. The raw materials which have been studied include the metals and nonmetals such as the plastics and ceramics. In order to take the raw ore and convert it into a tool, part, assembly such as an engine, or a bolt, a vast amount of knowledge is necessary. Depending on the type of manufacturing operation, certain skilled personnel are needed who are capable of arranging a production line and then operating it for a profit. Two of the most important persons in regard to the production operation are the manufacturing engineer and the manufacturing technician. Some manufacturers assign different titles to these persons, but responsibility for getting the work done belongs to the team which plans, organizes, implements, operates, tests, inspects, and packages the consumer product. The team cannot function alone, but is dependent on the total factory for support of the primary mission of production.

The materials and processes used in manufacturing the product point out the kinds of technical personnel required. When metals are the main raw material, then a metallurgist is needed along with the manufacturing engineer or tool engineer and technicians. These person's responsibilities include a review of plans for production which includes sources of raw materials, equipment, and costs. Also, they analyze equipment listings and recommend the procurement of certain machines, systems, special equipment, conveyor systems, control systems, facilities, and fixtures, along with the numerous special tools and common tools pertinent to the operation. Details of the layout and equipment functions are studied by technicians who are specially trained in the functions and operations of equipment and machines.

Because manufacturing includes many concepts and functions of engineering, the engineer and technician must be cognizant of the chemical, physical, and mechanical properties of the materials used in manufacturing. Such a requirement helps assure the needed understanding of the entire operation. Knowledge of machinery and mechanisms, metallurgy, chemistry, physics, and mathematics are essential in one way or another along the production line. Machines must be verified that they are powerful enough to do their work, strengths of materials in the production line's facilities must be adequate, the flow process must be effective and continuous within the specified time periods, labor-saving devices must be used such as automated equipment and facilities when the product output is large, lubrication of equipment must be arranged for, instrumentation must be provided at key points, and power transmission must be understood to include electrical, mechanical, and fluid. Mechanical and manufacturing technicians support the manufacturing operation and assist the engineer in carrying out the objectives of production.

Once the production process commences, responsibilities are assigned for monitoring the process. Technicians are often used as supervisors, foremen, troubleshooters, inspectors, operators, maintenance men, and assistant engineers. Therefore, the technician must be able to understand his part of the process and also be able to keep it operational. Formal education and training in the basic engineering processes provides the beginning technician with an understanding of the language of engineering and technology so that he can effectively function in his assigned responsibilities.

Conservation of Energy

Just as the pollution problem has crept up on society, the shortage of materials and reduction of available energy have also suddenly come to the attention of most of the world. Because manufacturing produces most of the world's products, except those from the farm, the materials and energy problems are of primary concern. Observation points out that extravagance fluorishes in a plentiful situation, and so it has been since the end of World War II. Suddenly, in the seventh decade of the twentieth century, the world recognized the depletion of raw materials, possibly because of the law of supply and demand being way out of balance.

Due primarily to the population explosion, new markets for manufactured goods, new inventions, a rise in the education level, increased health and longevity, shorter working hours, better transportation, and an increase in money turnover by more members of societies, manufacturers have continued to increase their outputs of goods to satisfy the known and apparent demands. Such a steady consumption of raw materials has slowly depleted many strategic raw materials, namely, the fossil family and many of the metals. With a given quantity of available raw materials, it is only a matter of time before the remaining quantity is recognized as being inadequate to allow the same attrition. Consequently, without an input equal to the output of raw materials, simple arithmetic points to the zero factor.

What, then, can be done in technological societies which are suffering from the numerous problems relative to materials shortages? Just as in the case of pollution, the trend must be reversed. Time is the silent factor that accompanies all things; it took time to relegate advanced societies to such cultural lags, and it will take time to bring about stabilization. It is unreasonable to believe that people will be satisfied to do without those needed things which they have previously enjoyed if there appears to be any way to obtain them. Such situations have been demonstrated throughout history, the battles between the haves and the have-nots.

Two things, then, need to be acted upon quickly; neither is new, but the actions must be greatly accelerated. Number one, a conservation of materials must become reality. Number two, new sources of energy must be found and transformed into usable quantities of sufficient magnitudes to sustain the needs of the total society. With the availability of raw materials and adequate sources of power, the present levels of technology can continue and even advance. Without satisfying both of these factors, technology will not sustain its present levels of advancement.

Factor number one, the conservation of materials, must allow for reclamation of used products of industry. A large percentage of usable materials are being lost forever in the garbage dumps. Both burning and burial of these materials are guaranteeing that man will not use these materials again. Systems of disposal use the land and the sea and even the atmosphere as receptacles in which to discard the undesirables. In conjunction with deliberate disposal practices, waste of materials through carelessness and negligence accounts for a sizable quantity of lost usable materials. Conservation of materials through reclamation should be expanded to include refuse and garbage processing plants whereby most of the usable materials would be reclaimed. Such a system of reclamation is not new, but it could be effective in reducing the attrition rate of available materials. As an example, thousands of tons of steel are being buried in the dumps whereby nature changes this usable metal into oxides mixed among the soils. In addition, thousands of tons of tin and zinc are undergoing the same fate. Much of this waste is the end result of the food industry, including the metals, glasses, and fossil products going into the dumps. Metals and glasses are reusable. The fossil products such as plastics (the garbage bag) and the food wastes have tremendous potential uses both in the solid materials industries and the energy production industries. Garbage and associated household unwanted items such as clothing, appliances, light bulbs, tools, hardware, toys, paper, furniture, and the like are convertible to elements and compounds for reuse. Some reclamation is in progress, especially among the metals, but more action is necessary in the garbage disposal industry.

Factor number two, the finding of new sources of energy, is moving in the proper direction, but at an alarmingly slow rate. There is only so much energy available, and man has tapped only a minute quantity. Reliance on fossil energy has been the easy way, but this consumption will continue to be lessened. Consequently, the alternate sources are required to sustain our technology. Feasibly thinking, atomic fusion and fission must be pursued as rapidly as practical, but progress must be accompanied with complete safety. Residues must be converted and neutralized and pollution products converted to harmless elements. Secondly, solar energy must become a technological reality. Conversion must be practical for the home and the factory. Lastly, the winds of the earth must be harnessed as well as the power of the tidal waters as these two forms of energy have tremendous power potentials.

In summary, a challenge prevails for those minds capable of thinking in terms of the practical application of theory. With regard to the conservation of energy, more effective reclamation practices must be implemented in order to increase the resultant supply of raw materials and to increase the sources of power. Lastly, the search for new forms of energy must culminate in the transition of power from the atom, the sun, the winds, and the tides of water. In order for manufacturing to continue in its societal need's satisfactions, and in support to the prevailing technology, satisfactory solutions to the materials and energy problems must be obtained and then placed into operation. The facts have been presented and reasonable and feasible solutions offered. The challenge, then, is directed to those manufacturers and educators who have elevated the technology to its present level.

Questions

1. Describe your concept of engineering design.
2. Why are essential data needed prior to making a drawing?
3. Why must strength of materials be verified prior to fixing the dimensions of a drawing?
4. What is meant by an operating environment?
5. Differentiate between mechanization and automation.
6. What is the difference between automation and semiautomation?
7. Describe the flow process in an automated production line.
8. What is a specification?
9. What is the purpose of a production standard?
10. Explain your definition of a manufacturing technician.
11. Why do some industrial plants use both specifications and standards?
12. What is a drawing and what information supports it?
13. Why must research often be conducted to obtain data prior to making a drawing?
14. All drawings include materials. Why must material capability be verified prior to completing the drawing?
15. What is a manufacturing plan?
16. During manufacturing, what is a secondary operation?
17. What is the purpose of inspection stations during production?
18. List some information that is included in a material's specification.
19. How do professional organizations help the industrial processes?
20. What is the purpose of production guides?

Glossary

Albate To melt away by vaporizing.

Adhesion The sticking together, or adherence, of two or more materials as a result of some treatment such as adhesive bonding whereby there is a strong molecular attraction between the materials.

Aging The changing of a mechanical property in a metal due to temperature and time. A chemical compound precipitates from a solid solution causing an increase in strength of the material.

Agglomerate The combining of small particles of ore into larger pieces for the purpose of increased efficiency in subsequent processing.

Allotropic The reversible atomic cell lattice in a grain of metal which allows the properties of the metal to be changed. The alpha-gamma-alpha transformation is an allotropic change.

Alpha iron The body-centered cubic atomic lattice network in iron which is below the allotropic change temperature, varying from approximately 1420° F. to 1333° F. The alpha cell, if it could exist alone, would consist of nine atoms.

Amorphous A noncrystalline material such as glass. Metals are crystalline, but material at the interface of two grains may be amorphous.

Angle of bite The points on a metal being rolled where maximum force is applied from the rolls.

Anistropy Directional properties in a material.

Anode The positive electrode in an electric circuit. In an electrolysis process, the positive electrode or terminal is the anode.

Atomic structure The lattice network in the granular structure of metals organized as unit cells. In the nonmetals, the atomic patterns group to form molecules as in the plastics.

Austenite The solid solution of carbide in gamma iron and when with nickel, as in austenitic steel, is tough, corrosion resistant, and work hardens.

Autoclave An especially designed pressure vessel that can exert internal pressure and heat on a workpiece assembly for the purpose of bonding a joint or other purpose.

Automation The manufacturing process which produces a part by means of machines, mechanisms, and handling devices without the aid of human hands. The entire operation is controlled by electrical impulses from a computer or tape device.

Axis The imaginary line that runs through an object and about which a turning tendency exists.

Bainite A microstructure in steel having very close platelike or feathery appearances resulting from an intermediate quench such as oil from austenite. Higher magnification identifies the feathery shape as platelike.

Baked The heating of a metal for the purpose of removing hydrogen from the metal's surface such as a low temperature heat of 400° F. for three hours. Also, the drying of metal parts by heating to 250° F.

Beneficiate Special handling and preparation of ores and ore dust for increased efficiency in the blast furnace operations.

Billet The reduced bloom prepares a billet for special fabrication purposes.

Blast furnace The hot air furnace required for the production of pig iron from iron ore, coke, and limestone.

Bloom A smaller section of metal reduced from the ingot and usually larger than the billet.

Body-centered cubic The lattice structure of a metal such as cold iron whereby the central body region of a cell contains an atom in addition to the corner atoms. When heated to approximately 1333° F. to 1420° F., the central atom leaves the body and joins the face-centered atoms.

Bonded The close adherence of a material to another material usually joined by an epoxy as in the bonding of honeycomb paneling.

Bosh The lower portion of a blast furnace where chemical reactions occur and steel making begins.

Boss A raised area of a part, for example, a high place on a casting on which another part is to fit such as by bolting.

Brass An alloy of copper and zinc.

Brinell hardness value The hardness value of a material indicated by the Brinell hardness testing machine. The value is convertible to other testing machines' hardness values such as Rockwell, Shore, Vickers, Knoop, and others.

Briquette The nonsintered shape of compacted powder produced by powder metallurgy.

Bronze An alloy of copper and tin, often alloyed with a third element such as aluminum, silicon, manganese, or phosphorus.

Burnish The smoothing of a part's edges by rubbing actions in an especially made die.

Burr The sharp ridge of metal nearly always remaining at the end of a sawing operation. Also, the ridge remaining at the bottom side of a drilled hole.

Button A small piece of machined metal that acts as a guide during a machining operation. Buttons are used in jigs and are also used in layout work.

Carbide A chemical compound of a nonmetal and some metal such as iron carbide. The new material is completely unlike carbon or iron and is very hard. Its makeup is Fe_3C.

Cast iron An alloy containing the same basic elements as steel, but with a carbon content exceeding 1.7%.

Catalyst A substance or thing that causes a change in a given condition without itself being affected.

Cathode The negative electrode in an electric circuit. In an electrolysis process, the negative electrode is the cathode.

Centrifugal The tensile force produced in a material due to rapid rotation about an axis. With respect to liquids, a centrifugal force causes the liquid to move to the periphery of a hollow spinning mold.

Ceramics A material composed of compounds of metallic and nonmetallic elements existing as phases such as stone, brick, abrasives, glass, and concrete.

Checker system A fire brick network shaped to allow hot air to be moved along passageways whereby the air becomes hotter as it picks up heat from the bricks on the way to the furnace.

Cheek Small parts of a mold which are sometimes required when the cope and drag are not sufficient.

Chemical properties The quantity of all elements constituting a material.

Chill A cold spot or other shape in the cavity wall to allow molten metal to cool more rapidly for the purpose of developing hard areas in the metal or a finer grain structure.

Chip The small ribbon, flake, or fragment of material leaving the workpiece during a machining operation, resulting from failure of the material at the shear plane which is immediately ahead of the cutting tool.

Chuck A holding device for use in holding the workpiece during machining operations.

Clearance The designed space between two moving parts in order that movement will occur as required within a specific temperature tolerance. The space must allow for lubrication as expansion and contraction occur.

Coalescence The welding operation that fuses metals while under heat and pressure, but melting does not occur.

Cohesion The mingling of molecules in a joint such as in cohesive bonding with the use of a solvent. This type of joint is lower in stress carrying capability than the adhesive bonded joint which uses cured epoxy.

Coining The precise pressing operation which permanently deforms a metal into the fine impressions in the die.

Cold heading The cold deformation of a metal bar by impact loads for the purpose of shaping, such as a bolt head.

Commercial anneal The softening of a metal, usually steel, at a temperature below its critical point to avoid scale and also to save time. This temperature is also equal to the maximum tempering temperature.

Commercial diamond These diamonds are discolored and are used for industrial operations such as cutting and in hardness tester penetrators.

Compaction The compressing of metal or nonmetallic powders in a die by the powerful action of a punch.

Compound The chemical combination of two or more elements which results in a new material.

Compression strength The maximum strength of a material while the material is under a pushing load as it collapses in some manner.

Concentric A circular dimension having a constant radius with a central axis.

Coolant A liquid used during machining for the purpose of cooling the metal being cut and the cutter, lubrication of the cutting action, and for washing chips from the cutter zone.

Cope The top portion of a mold, the bottom portion being the drag.

Core The central portion of a material. Also, the inserted object in a mold to allow a hole to result in the casting.

Creep The dimensional change of a material while under load and an elevated temperature. The temperature is mainly responsible for the gradual change in dimension.

Critical point The temperature in a metal that causes a major change to occur (excluding tempering) such as recrystallization or the change from a mixture of constituents to a solid solution.

Cross-sectional area The area of a rectangular shape in square inches is equal to the width of the cross section multiplied by the thickness.

Cryogenic Subzero temperatures usually colder than -200° F.

Crystal The very small grain of a metal that is visible to the eye. Crystals vary in size and have significant effects on a metal's mechanical properties. The word *grain* is preferred in the discussion of metallurgy.

Crystalline A granular material such as metal. All metals are made up of grains, or crystals, very similar to grains of sand.

Cupola The hot air furnace used in the production of cast iron.

Cupping The pressing of a very malleable metal into deep cups whereby the metal flows plastically without tearing.

Cutting speed The recommended surface speed in feet per minute that a material turns during a cutting operation which results in maximum cutting efficiency for both tool and material.

Cycle A rhythmic vibration measured as a given number of direction reversals per second, minute, or any specific time period. Hz now replaces cps with reference to alternating electricity. A cyclic condition exists in a structural beam when loads vary from tensile to compression in a continuous or timed condition.

Deep drawn A cupping operation using the plastic flow ability of the metal to shape the part.

Deformation Under a load, deformation first occurs as elastic, then inelastic as the yield is exceeded.

Demarcation A sharp dividing line between two different conditions in a material such as hard and soft.

Dendrite The coarse grain or "pine tree" shape of crystal which forms in castings as solid nuclei are formed from the liquid. The dendrite has a central axis with differing masses of materials forming at 90° to the central axis.

Density The amount of a material in a given volume. Water is the standard for liquids and solids and hydrogen is standard for gases.

Destructive testing The physical destruction of a specimen or sample in determining the several mechanical properties of like materials and shapes.

Dielectric A nonconductor of electricity such as air. It must be understood that nonconductance exists only in the presence of certain circumstances.

Diffusion The entrance of atoms into the interstices between other atoms such as the carbon atoms placing themselves within the allowable spaces among iron atoms in austenite.

Dog A device for securely attaching the workpiece in a lathe to the drive plate so that turning or other machining operations can be performed.

Dovetail A precision sliding joint arrangement in a machine tool that provides for a movable table or base for other fixtures or control mechanisms and allows movement along a given axis. Many machines have three-way axes movement provided by dovetail joints.

Draft The slight taper on a pattern that enables it to be removed from the mold without damage to the mold.

Drag The bottom portion of a mold, the top half being the cope. Prior to casting, the two parts are locked together to assure the molten metal is confined within the casting system.

Drawing The permanent deformation of a metal by tensile loading beyond the yield. Drawing may be performed cold or hot.

Drift A pointed tool having a sliding handle that is used to remove tapered drills, sleeves, and other tapered tools from the tapered socket of a machine such as a drilling machine.

Drill sleeve The device that fits between a tapered shank tool and the tapered spindle of a machine tool for the purpose of a positive connection between the tool and the spindle.

Ductility The ability of a material to be plastically deformed without rupture as the stress increases beyond the yield strength and below the tensile.

Dynamic A moving load.

Eccentric A circular dimension having a changing radius at some point when measured from the central axis. A lobe on a cam is eccentric to the central axis.

Elastic The ability of a material to be temporarily deformed as the stress increases up to any value below the elastic limit or the yield strength.

Electrolysis The chemical change resulting from the passage of an electric current through an electrolyte such as the changes occurring in the liquid during electroplating. An anode and cathode are submerged in the electrolyte.

Element A pure substance such as oxygen, gold, or mercury.

Elongation The ability of a material to stretch temporarily and/or permanently as the stress increases. The term normally pertains to tensile loading of a material when the stress increases beyond the yield strength, causing an increase in the length dimension.

Emboss To impress a design onto a material by compression force. The quality of the impression is less than the coined impression.

Embrittlement A potential brittleness factor induced in a metal by some factor such as absorption of hydrogen during electrolysis.

Equiaxed Pertaining to the stress free grain in a cast metal that tends to be round in shape but has many facets.

Etch The application of a chemical onto a prepared specimen's surface for the purpose of microscopic examination.

Face-centered cubic The lattice structure of a metal such as aluminum whereby the unit cells have atoms at each face of the cell in addition to the corner atoms. Gamma iron is also face-centered cubic.

Fatigue The combination of a cyclic load and time on a material which ultimately leads to material failure, even though the stresses are far below the tensile.

Ferrite The solid solution of carbon in alpha iron when the carbon content is below 0.008%. Ferrite and the element iron are often used interchangeably.

Ferrous All the iron base metals to include wrought iron, steel, and cast iron.

Ferrule A ceramic cup which shields the electric arc in welding and concentrates the heat in the weld area along with controlling the molten metal in the designated weld area.

Fillet A radius at a change in mass of material.

Fine pearlite The very close layers of carbide and ferrite in the microstructure of steel resulting from an intermediate quench from austenite, such as an air or oil quench. The strength of fine pearlite is much greater than coarse pearlite.

Fixture A device for holding the workpiece during fabrication or machining.

Flash The thin web of solid metal that forms as a result of leakage of molten metal at the parting line of the mold. Also, hot metal fins which are squeezed from a die during forging.

Flaw An irregularity in a material such as a hole, crack, or inclusion.

Flute The machined out areas along the longitudinal axis of a cutting tool which provides the cutting edges of the tool and provides space for chip removal.

Flux A material used in the melting of metal to protect it from the atmosphere and oxygen. Flux floats on the surface of molten metal. Flux is also used in some welding and in brazing and soldering operations to protect the surface metal.

Force The weight or pressure on a material which endeavors to deform it.

Fracture A separation in a material caused by some type of load.

Freezing When used in the casting processes, it denotes solidification from the liquid and may be any temperature.

Frequency The rhythmic cycle in a continuous vibration of energy caused by changing conditions.

Gage blocks Very accurate metal blocks which act as standards of measurements at a given temperature.

Gamma iron The face-centered cubic atomic lattice network in iron which is above the allotropic change temperature. If the cell could stand alone, the network would consist of 14 atoms.

Gangue The discarded quantity of an ore that results after processing.

Gate The region in a mold system that is open to allow molten metal to flow into a cavity for casting purposes.

Grain The very small crystal of a metal that is visible to the eye in fractured metal. Grains vary in size and have significant effects on a metal's mechanical properties. Grains are three-dimensional and have many facets. All metals are granular.

Green sand A raw sand and clay mixture that is used in the damp condition for mold making as opposed to the dry sands which are harder and stronger as a result of heating. Green sands may contain binders.

Hardness The resistance of a material to penetration.

Hearth The portion of the furnace that contains the molten metal during ore refinement or other melting practices. Also, the bottom portion of a furnace which holds solid parts during heat treating operations.

Heat treat The heating and cooling of a metal for the purpose of inducing desirable mechanical properties.

Helix A constant rate curve around the surface of a cylinder which moves along the longitudinal axis identified with a lead. The lead is the distance that the curve advances along the axis in one revolution about the cylinder or rod.

Hexagonal close packed The lattice structure of a metal such as magnesium which is arranged in a hexagon shaped cell. Brittleness is inhered in the material at room temperature by such a lattice arrangement.

High speed steel An alloy containing more than the six basic elements in steel which has the capability to retain its hardness in the presence of 1,000° F. It is used primarily for cutting tools and heat resisting parts.

Impact strength The maximum strength of a material when subjected to a moving load. The load is suddenly intercepted by the object, the material being subjected to a specific stress such as tensile-impact.

Inclusion The nonmetallic such as an oxide trapped within solid metal.

Inelastic The ability of a material to be permanently deformed as the deformation stress increases beyond the yield strength.

Interface The surfaces of two contacting materials.

Involute The shape of the gear tooth profile which provides for efficient power transfer between the driver and driven gears.

Ionize The dividing of a material into positive and negative ions with an electric current during electrolysis. The positive ions seek the cathode due to a loss of electrons. Negative ions seek the anode due to a gain in electrons.

Iron An element, but it is never used as an industrial metal. Iron is often used interchangeably with ferrite which is the base metal of all ferrous metals. Iron and cast iron are completely unalike.

Jig A device for holding the workpiece during machining and then guiding the tool as it machines the material.

Keyway A machined groove in a part, a shaft, for example, that holds the key for connecting the shaft to the driving gear. The keyway may be rectangular in shape or semicircular such as the Woodruff.

Knee The vertical column of a machine tool that supports the table and operating mechanisms. Large milling machines have powerful knees to retain the combined stresses during machining.

Lance A tubular shaped device for injecting oxygen into molten steel. The lance usually is lowered from the roof of the furnace.

Lattice The organized network of atoms in a metallic grain. The lattice builds the unit cells and, in turn, grains form and, in turn, the material is produced.

Liquidus The temperature at which a metal melts on heating.

Load Measured in pounds. Also, a weight or force.

Malleable The ability of a material to be permanently deformed without tearing while under compression loading.

Mandrel A slightly tapered round tool used in manufacturing processes that is inserted in a workpiece for the purpose of securely holding the workpiece during machining operations. A mandrel is also used in the fabrication of seamless tubing.

Martensite The solid solution of carbide in iron having a body-centered tetragonal cell arrangement. Martensite is extremely hard and brittle and has a very high tensile strength.

Mass The quantity of a material, normally the cross-sectional area, with reference to strength or heat treating procedures.

Matrix The base metal in an alloy. Ferrite is the matrix in steel.

Matte A mixture of metallic sulfides produced by heating sulfide copper ores and then melting.

Mechanical properties The variable properties of a material which relate to engineering properties, such as hardness, strength, ductility, and brittleness. The nonvariable property, modulus of elasticity, is also a mechanical property.

Metallurgy The science of metals.

Micron A measurement of one-millionth of a meter.

Microscopic Magnification of a prepared specimen at more than 10X. Often, magnifications at less than 1,000X are sufficient for examination of materials. Magnifications at less than 10X are *macroscopic*.

Microstructure The internal structure of a material when viewed under the microscope after proper preparation of the specimen.

Mil A measurement of one-thousandth of an inch. A measurement of one-millionth of an inch is a *micro-inch*.

Modulus of elasticity The ratio of stress to strain in a material under load when the load is not greater than the elastic limit or does not encroach into the yield property. Formula:

$$E = \frac{S}{\epsilon}$$

Mold The container or flask which holds molten metal as it solidifies. Usually, two parts, the cope and drag, constitute the mold. Molds are made of metal, sand, or other nonmetal.

Moment The product of the force and the perpendicular distance from the force to the axis of tending rotation.

N/m^2 The international symbol for pounds per square inch or psi (now considered a deprecated term). The newton, N, is the kilogram meter per second squared and is the unit of force. N/m^2 is the newton per square meter. The kilogram is equal to 2.204 pounds, while the meter is equal to 39.37 inches. N/m^2 is an extremely small value, being 0.15×10 lb/in^2

Neutral plane The plane of material under a bending load where the tensile stress is reduced to zero as it changes toward compression on the opposite side of the part.

Nondestructive testing The testing and inspection of a material, workpiece, or part for the purpose of determining mechanical properties without harming the material, workpiece, or part.

Nonferrous All the metals which are not ferrous, even though iron may be a constituent in some alloys.

Nuclei A quantity of a material which acts as a central axis, such as the solid masses of metal which form from the cooling liquid.

Oxidation The process which causes oxygen to combine with other elements or substances for the purpose of producing new substances.

Oxide A chemical compound including oxygen and another element such as iron. Iron oxide forms on the surface of hot steel in the presence of an oxidizing atmosphere.

Part To separate a material by cutting action.

Pattern The shape of an object which is placed in a mold for the purpose of making a cavity, as in sand molding prior to casting.

Pearlite The mechanical mixture of carbide and alpha iron existing in layer form which is soft and ductile, as well as low in tensile strength, relative to the family of steels in other microstructural conditions.

Peen To bombard a metal's surface with a material such as small steel shot for the purpose of work hardening the surface and producing residual compression stresses in the surface.

Periphery The outer regions of a turning part or tool.

Permeable The ability of a material to allow penetration of gases to a small degree, such as the sand in a mold allowing the escape of hot gases from the cavity.

pH The value which indicates the degree of basicity or acidity of a solution.

Phase A homogeneous mass of material differing from other masses in the metal system.

Physical properties The nonvariable properties of a material which relate to engineering properties, such as weight, melting point, color, electrical conductivity, atomic lattice, and magnetic properties.

Pickling The chemical cleaning of a metal such as the pickling of steel in a bath of diluted sulfuric acid.

Pig iron The ferrous alloy produced by the blast furnace for the purpose of making steel, cast iron, or wrought iron. Pig iron results from processed iron ore.

Pinion wire Two to six foot lengths of rod having splines of gear teeth shape and dimensions around the periphery for use in making gears by parting the rod at specified thicknesses.

Pitch The distance from a point on one saw (or other tool) tooth to the adjacent tooth. The distance may be linear or circular, as in a gear. Threads also have a pitch.

Pitch circle The point on the gear tooth that is halfway between the top and bottom of the tooth and bears against the mating tooth from another gear as the gears turn. When projected around the periphery, the series of points make the pitch circle which partly occupies space.

Pitch diameter The diameter of the gear at the pitch circle.

Plastic A nonmetallic material consisting of large molecules manufactured from gases, liquids, and solids. Also, the term infers the ability of a metal to deform under pressure without tearing.

Plastic flow The permanent flow or deformation of a material which occurs when loads exceed the material's yield strength, but do not attain the tensile.

Plate When the thickness of sheet stock increases a substantial amount, such as 0.125 inch, it is often designated as plate. The dimensional change from sheet to plate varies with the metals.

Plating The application of one metal onto the surface of another.

Powder metallurgy The science of compacting and sintering powdered metals into specific shapes.

Precipitation The movement of a material from the solution of materials. Copper aluminide precipitates from a supersaturated solid solution of copper and aluminum.

Preheat The heating of a metal at a lower temperature than the processing temperature to eliminate overstressing the metal on heating. Steels are usually preheated at 700° F. prior to annealing.

psi Pounds per square inch. Now being converted to N/m^2 when used with modulus of elasticity, pressure, force, and stress. *See* N/m^2.

Pure metal An element such as aluminum, iron, or chromium.

Quality control The series of inspections performed on a workpiece or part from beginning of manufacturing to the end for the purpose of quality assurance.

Quench Usually meant to rapidly cool a hot metal in water or oil.

Reaction The internal stress occurring in a material as a result of an external load. Also, pertains to chemical changes during specific types of processing.

Recrystallization The creation of new and fine grains in a metal at a specific temperature. Grains may be small or large after various heat treating operations, and they respond to heat with respect to rebirth only at the recrystallization temperature.

Red hardness A mechanical property induced in a steel which resists softening by temperatures as high as 1,050° F. High speed steels have this property.

Reduction The chemical process which changes ores and oxides to metal by means of a non-oxidizing flame and results in a loss of electrons in the oxidized material.

Reduction in area As a section of material increases in tensile stress, its length increases in dimension and its cross section decreases. Permanent deformation occurs beyond the yield to the tensile. The percent of reduction in area occurring as a result of tensile fracture indicates the material's ductility factor and is mathematically determined as follows:

$$\%RA = \frac{(A_o - A_f \text{ x } 100)}{A}$$

where A=area, o=original, and f=final.

Reentrant The angular or undercut portion of a part or workpiece that will not allow metal powders to be compacted except when in direct line with the punch at 90°.

Refractory Heat resistant materials such as fire bricks, some ceramics, and some metals. Tungsten resists elevated temperatures, as does aluminum oxide.

r/min Revolutions per minute of a turning object, formerly symbolized as rpm.

Reverberatory An arrangement inside a furnace that causes heat to be reflected from the roof onto the charge.

Riser The region in a mold system that allows molten metal to rise from the central cavity to signal the filling of the cavity.

Rockwell hardness value The hardness value of a material indicated by the Rockwell hardness testing machine. The value is convertible to other testing machines' hardness values such as Brinell, Shore, Vickers, Knoop, and others.

Runner A connecting opening inside a mold which allows molten metal to move freely from the sprues and gate to the cavity during a casting operation.

Saturate The chemical condition that causes maximum solubility of a substance in another substance.

Section modulus The dimensional properties of material relating to the material's cross section.

Setup The prepared arrangement of the workpiece in its holding device so that machining operations can be performed.

Shear plane The failing plane of material under shear stresses when the shearing stress exceeds the shearing strength of the material.

Shear strength The maximum strength of a material under cutting or shearing stresses. The load is applied across the cross section of the material.

Shim stock Very thin sheets of metal, usually steel or brass, used in bringing two undersize parts to the required dimension. The stock is placed between two metal blocks, for example, whereby the overall dimension is obtained. Shims are available in increments of one-thousandth inch up to a certain thickness.

Sintering The heating of a compacted briquette for the purpose of bonding the individual surfaces of the powdered metal.

Skelp A predetermined width and thickness of steel sheet unwound from a coil in the manufacture of welded pipe.

Slab A large mass of metal rolled into a rectangular shape having a thickness much less than the width.

Slurry A partially fluid substance of a solid material and a liquid which results in a type of fluid mud.

Smelting The heating of ores to produce a liquid metal.

Soak To maintain a metal at a specified temperature for a designated period of time to assure a uniform temperature throughout the mass of metal.

Solidification The freezing of a liquid metal as its temperature decreases to the solidus.

Solidus The temperature at which a metal solidifies on cooling from the liquid or begins to melt on heating.

Solute The material, liquid or solid, that is dissolved in the base material (greatest quantity) or solvent.

Solution treatment The heating of an alloy to a specified temperature which causes a homogeneous mass of the constituents.

Solvent The basic material in a solution, either liquid or solid.

Spheroidize The heating of a metal, usually high carbon steel, at a temperature just below the critical point for the purpose of placing the carbide particles in spheres. Such a process produces maximum malleability in the metal for bending operations.

Springback The elastic factor in a metal that causes it to decrease the angle of deflection after the bending load is removed.

Sprocket A round part having teeth shaped for a chain drive.

Sprue The region in a mold system that allows molten metal to flow through the gate and runner system to the central cavity. Metal enters the sprue and rises in the riser when the cavity is filled.

Static A steady load in equilibrium.

Steel The alloy of iron and carbon. All steels contain at least six basic elements—iron, carbon, silicon, manganese, phosphorus, and sulfur. The carbon content ranges from 0–.008% to 1.7%.

Stock The raw material which is used to manufacture parts.

Strain The change in dimension of a material as a result of a stress. The stress accompanies a load. Usually, unless otherwise indicated, the strain is understood to be linear.

Stress The internal reaction in a material to an external force, weight, pressure, or load measured in psi.

Supersaturated The condition in a metal whereby all or most of the constituents are retained in solid solution following a quench from the material's solution temperature. As time elapses, some constituent precipitates from the solution.

Tang The thin and small end of a drill sleeve that acts as a positive drive should the taper fail. Also, the pointed end of a file on which a handle is placed.

Temper colors The several colors appearing on polished steel when heated below 1,000° F. A correlation exists between color of surface and temperature of metal.

Template A guide used in layout and in machining.

Tensile strength The maximum strength of a material while the material is under a pulling load as it breaks.

Thermal shock A sudden high temperature contact with a material.

Thread A formed helix around a rod with a specified pitch and outside diameter.

Tinned and soldered The soldering process of applying solder to both surfaces of two metals, clamping both sections of metal together, heating the joint to the melting point of the solder, and allowing the joint to cool before removing the clamp.

Tolerance The total deviation of a measurement from the desired measurement such as the distance between the minus and plus factors.

Torque A force which causes rotation tendency and is a moment.

Toughness The mechanical property resulting in a material which resists a combination of stresses acting concurrently and includes impact loading.

T-slots A machine's table usually contains deep and parallel slots with wide bottoms for the purpose of holding *T*-bolts and subsequent locking of a material or device to the table.

Undercut Similar to the reentrant with reference to powder metallurgy.

Vise A holding device having no precision holding capability for the purpose of processing a workpiece.

W Symbol for tungsten. Also, the international symbol for watt. 1 horsepower = 746 watts.

Wide flange Many rolled and extruded shapes have wide flanges, such as columns and beams. The *I*- and *H*-beams and columns are wide flanged.

Work harden The increase in hardness of a metal as the result of working it or forming it beyond its yield strength and below its tensile.

Workpiece The raw material being processed.

Wrought A pressure forming operation that shapes metal, such as rolling, forging, extruding, stamping, drawing, cupping, and coining.

Wrought iron An alloy of iron and slag, the iron or ferrite containing a small quantity of carbon.

Bibliography and References for Further Study

AC Compacting Presses. *Compacting Presses.* New York, undated.

Allegheny Ludlum Steel. *Modern Melting.* Pittsburgh, 1960.

Aluminum Association. *A Guide to Aluminum Extrusions.* New York, 1967.

American Iron and Steel Institute. *The Making of Steel.* New York, 1964.

———. *The Picture Story of Steel.* New York, 1969.

American Society for Metals. *Metals Handbook.* Vol. 1. Metals Park, Ohio, 1961.

———. *Metals Handbook.* Vol. 2. Metals Park, Ohio, 1964.

———. *Metals Handbook.* Vol. 3. Metals Park, Ohio, 1967.

———. *Metals Handbook.* Vol. 4. Metals Park, Ohio, 1969.

———. *Metals Handbook.* Vol. 5. Metals Park, Ohio, 1970.

———. *Metals Handbook.* Vol. 6. Metals Park, Ohio, 1971.

———. *Metals Handbook.* Vol. 7. Metals Park, Ohio, 1972.

———. *Metals Handbook.* Vol. 8. Metals Park, Ohio, 1973.

American Society for Testing and Materials. *Plastics, General Methods of Testing.* Philadelphia, 1971.

———. *Physical and Mechanical Testing of Metals.* Philadelphia, 1973.

———. *Plastics, Specifications.* Philadelphia, 1973.

———. *Structural Sandwich Construction.* Philadelphia, 1972.

American Zinc Institute. *Zinc Die Casting.* New York, undated.

Anderson, Curtis B., Peter C. Ford, and John H. Kennedy. *Chemistry: Principles and Applications.* Lexington, Mass.: Heath, 1973.

Baird, Ronald J. *Industrial Plastics.* South Holland, Ill.: Goodheart Willcox, 1971.

Bethlehem Steel. *Quick Facts about Alloy Steels.* Bethlehem, Pa., undated.

Black, Perry O. *Machinists Library, Basic Machine Shop.* Indianapolis: Theodore Audell, 1973.

Bolz, Roger W. *Production Processes.* New York: Industrial Press, 1963.

Breneman, John W. *Strength of Materials.* New York: McGraw-Hill, 1965.

Budzik, Richard S. *Sheet Metal Technology.* New York: Howard W. Sams, 1971.

Burghardt, Henry D. and Aaron Axelrod. *Machine Tool Operation.* New York: McGraw-Hill, 1953.

Burton, Malcom S. *Applied Metallurgy for Engineers.* New York: McGraw-Hill, 1956.

Campbell, James S. *Principles of Manufacturing Materials and Processes.* New York: McGraw-Hill, 1961.

Chambersburg Engineering. *Forge Shop Modernization.* Chambersburg, Pa., 1970.

Cleveland Twist Drill. *Cutting Tools.* Cleveland, 1973.

Climax Molybdenum. *High-Strength Irons.* New York, undated.

Cole, Charles B. *Tool Design.* Chicago: American Technical Society, 1967.

Davis, Harmer E., George E. Troxell, and Clement T. Wiskocill. *The Testing and Inspection of Engineering Materials.* New York: McGraw-Hill, 1964.

DeGarmo, E. Paul. *Materials and Processes in Manufacturing.* New York: Macmillan, 1974.

Donaldson, Cyril and George H. LeCain. *Tool Design.* New York: McGraw-Hill, 1957.

duMont Corporation. *Broaching.* Greenfield, Mass., 1968.

Eary, Donald F. and Edward A. Reed. *Sheet Metal.* Englewood Cliffs: Prentice-Hall, 1968.

Eclipse Fuel Engineering. *Air Pollution Control.* Rockford, 1971.

Edgar, Carroll. *Fundamentals of Manufacturing Processes and Materials.* New York: Addison Wesley, 1965.

Elco Industries. *Custom Cold Heading.* Rockford, 1973.

Electronicast. *Investment Casting Trends.* Addison, Ill., 1970.

Erie Foundry. *Board Drop Forging Hammers.* Erie, Pa., undated.

General Motors. *Ferrous Casting Design.* Saginaw, Mich., undated.

Gibson, John E. *Engineering Design.* New York: Holt, Rinehart, Winston, 1968.

Grosvenor, A. W. *Basic Metallurgy.* Metals Park, Ohio: American Society for Metals, 1964.

Hine, Charles R. *Machine Tools and Processes for Engineers.* New York: McGraw-Hill, 1971.

Kennedy, Gower A. *Welding Technology.* New York: Howard W. Sams, 1974.

Keyser, Carl A. *Materials Science in Engineering.* 2d ed. Columbus: Charles E. Merrill, 1974.

L. S. Starrett. *Starrett Tools.* Athol, Mass., 1967.

Landis Tool. *Modern Methods of Cylindrical Grinding.* Waynesboro, Pa., 1971.

Lascoe, Orville E., Clyde A. Nelson, and Harold W. Porter. *Machine Shop.* Chicago: American Technical Society, 1973.

Lindberg Hevi-Duty. *Heat for Industry.* Chicago, 1968.

McCarthy, Willard J. and Robert E. Smith. *Machine Tool Technology.* Bloomington, Ill.: McKnight and McKnight, 1968.

McMaster, Ronald C. *Nondestructive Testing Handbook.* Vol. 1 & 2. New York: Ronald Press, 1963.

Munro, Lloyd A. *Chemistry in Engineering.* Englewood Cliffs: Prentice-Hall, 1964.

Norton Company. *Abrasives and Grinding Wheels.* Worcester, Mass., 1973.

Oberg, Eric and Franklin D. Jones. *Machinery's Handbook.* New York: Industrial Press, 1972.

Owatonna Tool Company. *Power Work Holding Systems.* Owatonna, Minn. 1972.

Palin G. R. *Plastics for Engineers.* New York: Pergamon Press, 1967.

Patton, W. J. *Modern Manufacturing.* Englewood Cliffs: Prentice-Hall, 1970.

Patton, William J. *Numerical Control.* Reston, Va.: Reston, 1972.

Phillips, Arthur L. *Welding Aluminum.* New York: American Welding Society, 1966.

Pollack, Herman W. *Materials Science and Metallurgy.* Reston, Va.: Reston, 1973.

Prentice-Hall. *Handbook of Industrial Metrology.* Englewood Cliffs: Prentice-Hall, 1967.

Republic Steel. *Alloy Steels.* Cleveland, 1968.

Rusinoff, S. E. *Manufacturing Processes.* Chicago: American Technical Society, 1962.

———. *Tool Engineering.* Chicago: American Technical Society, 1959.

Schaller, Gilbert S. *Engineering Manufacturing Methods.* New York: McGraw-Hill, 1959.

Small, Louis. *Hardness, Theory and Practice.* Ferndale, Mich.: Service Diamond Tool, 1960.

Smith, Charles O. *The Science of Engineering Materials.* Englewood Cliffs: Prentice-Hall, 1969.

South Bend Lathe. *How to Run a Lathe.* South Bend, 1966.

Summit Machine Tool. *Machine Tools.* Oklahoma City, undated.

Timken Roller Bearing. *Fine Alloy Steel.* Canton, Ohio, 1968.

Turner, Rufus P. *Metrics for the Millions.* New York: Howard W. Sams, 1974.

United States Steel. *Fabrication of Stainless Steels.* Pittsburgh, 1970.

Van Vlack, Lawrence H. *Elements of Materials Science.* Reading, Mass.: Addison Wesley, 1964.

———. *A Textbook of Materials Technology.* Reading, Mass.: Addison Wesley, 1973.

Ward, Darrell. *Power Hacksaw Blades.* St. Paul: Keller-Sales Service, undated.

Weiss, Arthur and Alex Leuchtman. *Properties and Uses of Ferrous and Nonferrous Metals.* Detroit: Royalle, 1967.

Index